EXPERIMENTAL EMBRYOLOGY OF
MARINE AND FRESH-WATER INVERTEBRATES

Experimental embryology of marine and fresh-water invertebrates

Editor

G. REVERBERI

University of Palermo

With an introduction by Professor Brachet

1971

NORTH-HOLLAND PUBLISHING COMPANY
AMSTERDAM · LONDON

AMERICAN ELSEVIER PUBLISHING COMPANY, INC.
NEW YORK

Library of Congress Catalog Card Number 75-126506
ISBN North-Holland 0 7204 4080 7
ISBN American Elsevier 0 444 10065 2

Publishers:

NORTH-HOLLAND PUBLISHING COMPANY – AMSTERDAM
NORTH-HOLLAND PUBLISHING COMPANY, LTD. – LONDON

Sole distributors for the U.S.A. and Canada:

AMERICAN ELSEVIER PUBLISHING COMPANY, INC.
52 Vanderbilt Avenue
New York, N.Y. 10017

PRINTED IN THE NETHERLANDS

Preface

This book is principally addressed to young people who intend to devote their career to embryological research. Embryological research is today at the very centre of biological attention and is the focus of a great deal of work aimed at its central problem, namely, how, from a structure seemingly as simple as an egg, a whole animal with its highly differentiated and diverse organs and tissues is built up. It sometimes seems that molecular biologists feel that for the solution of this problem, descriptive and classical experimental embryology are not required, and consequently they can be ignored. More typically than in other eras, youth of today seeks to break the ties of the past. It is almost as though they do not wish the past to be heard of; as though biology begins with them. However, the validity of this viewpoint has not been proved; science does not grow like mushrooms on a rainy night. It is more comparable to a river which begins as a tiny stream in the mountains and fills with water as it flows along. In other words science is a 'continuum', its present being derived from its past.

The recognition of what has been done before us is an essential condition for being able to carry out what we wish to do in the future. This book aims to be a bridge between the past and future. One of its hopes is to try and convince molecular embryologists that the world does not begin with them, and that their work is not meaningful if it does not follow and carry further the work of the classical embryologist. The chapters contain a large number of references and the reader is continuously advised to go to the 'origins'. There is no doubt that one's scientific background is much enlarged by dipping into the original papers as some of them are the real classics which demonstrate the foundations on which embryology is built.

The reader will notice that some of the chapters are devoted to one animal form in particular (Ilyanassa, Dentalium, Cerebratulus, Mytilus, Amphioxus), others, however, to groups (Coelenterates, Ascidians, Echinoderms).

We have tried to obtain descriptions of a single form from our collaborators each time, but in practice this has proved to be difficult or impossible. Some of the chapters are true monographs and give a very clear and concise view of the development of the animals concerned. Wherever possible experimental details have been included: of the techniques of obtaining material and eggs, fertilization, de-membranation and various manipulations. Up to now invertebrate embryology, and particularly biochemical studies, have been carried out exclusively with sea-urchins. This is a pity as no one type of animal can provide all the answers; each one has its advantages and its drawbacks from an experimental point of view. We hope that this book will stimulate researchers to see what can be done with the animals of the other groups.

The book is the result of collaboration. This gives it a much greater authority. The authors are distinguished experts, all of whom have made well-known contributions to their particular field. Of course collaboration leads to some inconveniences; however they will be easily overcome by a sympathetic reader.

Today it is not possible for any single author to cover all the embryological research being done. On our part we cannot express enough gratitude to the authors, who have given us their cooperation, particularly in a time of university reform, student disturbances and general unrest. That the book is not perfect does not trouble us, we only hope that it will be useful and stimulate interest in these groups.

We would like to add a few words to the young embryologists: do not hurry too much with your research. Remember those who painstakingly worked out the cell lineages: how many hours they used to spend by their microscope!

The optical microscope is almost a thing of the past in a modern embryological laboratory: centrifuges, columns and scintillation counters have taken its place. People are always running after their instruments and there is a danger of a cult of the machine. Remember that the best researchers have never been servants of their machines. In any research it is the brain which is the most useful instrument. In this connection the words of Lazzaro Spallanzani (1729–1799), one of the most famous of all Italian biologists, are very relevant: 'Sperimentare è opera di tutti, sperimentare a dovere è, e sarà sempre, opera di pochi' (Everybody can do experiments, but only a few do, and ever will do, experiments properly).

With these words we conclude this preface. It remains only to thank again most heartily our collaborators, and also the publishers who have carried out the task of producing this book with a skill and care that fully bears up their name and traditions.

G. Reverberi

Introduction

The present book is invaluable, though at first sight one might ask why a treatise on Invertebrate Embryology should be published in 1971, when molecular embryology is entering its Golden Age.

Everybody knows that sea urchin eggs have become a favourite material for those who are interested in the biochemical mechanisms of differentiation. In fact, molecular embryology is becoming almost as elaborate and sophisticated, in the case of sea urchin and amphibian eggs, as molecular biology in that of *Escherichia coli* K 12. A survey of the current literature shows that people are now interested in such problems as: timing and mechanisms of DNA replication, nature and synthesis of the RNA polymerases, localization and identification of ribosomal DNA cistrons, conversion of the ribosomal RNA precursors present in the nucleolus into cytoplasmic ribosomes, origin of the ribosomal proteins, storage of stable messenger RNAs, synthesis of new messenger RNAs, transfer of the latter – perhaps in the form of informosomes – from the nucleus to the cytoplasm, stability of the ribosomes in the unfertilized and fertilized eggs, presence of inhibitors in ribosomes, changes in initiation, elongation and termination factors during development, changes in the proportions of the iso-accepting transfer RNAs and of their synthetases, nature of the proteins synthesized by light and heavy polysomes, etc. Again, why publish today a new treatise on invertebrate embryology, when classical ones already exist?

The answer is simple: most of the literature available at present on sea urchins and amphibians tells us very little about the biochemical mechanisms of *morphogenesis;* furthermore, it disregards the embryological problems raised by the eggs of other species which, as shown in the present book, remain fascinating and still need a solution. For many molecular embryologists, sea urchins and amphibians are the most favourable material because their eggs can be obtained in large amounts, can be fertilized at will, develop in good yield until a selected stage (blastulae, gastrulae, plutei or tadpoles) is reached and can easily be submitted to standard biochemical procedures (homogenization, centrifugation, etc.). Who cares about regulation and induction in these eggs, except a few people, who, like J. Runnström, S. Hörstadius, C. H. Waddington, H. Tiedemann, have a strong background in descriptive and experimental embryology? Still worse, will the meaning of the words regulation and induction still be understood, in their embryological sense, by all molecular embryologists? Nowadays, the word regulation awakes the idea of gene repression and derepression, while the term induction is linked to enzyme formation in the minds of most biologists. But these useful words were coined by experimental embryologists (Driesch and Spemann), a long time ago, in order to describe processes which are of basic importance for cell differentiation.

Embryology is as old as mankind: as soon as man became capable of thinking, he must have asked himself questions about his origin and that of the living organisms which surround him. Artificial incubation was used in the oldest civilization for practical purposes, as pointed out by J. Needham in his 'History of embryology'. Unknown people, very long ago, certainly had the curiosity to break the shell of hen's eggs, before hatching, and to look at the developing embryo. Much later, Aristotle described this embryo and this description remains in written form: he observed, among other things, that the cephalic (anterior) part of the embryo grows faster than the posterior one and, in doing so, he discovered – as pointed out by J. Needham – the morphogenetic cephalocaudal gradient. Despite much work by people like C. M. Child and myself, the origin of these gradients – which are essential for proper development and differentiation – remains unknown. It was a great satisfaction for those who are interested in this problem of gradients to see that such a famous molecular biologist as F. H. C. Crick recently devoted a theoretical paper to this question. But, while his diffusion theory adequately explains the gradient distribution of molecules of medium size (such as steroids or polypeptides), it does not solve the mystery of the establishment of polarity in the yolk laden eggs of the amphibians. This polarity, which corresponds to animal–vegetal gradients in the distribution of ribosomes, glycogen, fats and yolk platelets, is essential for future development: it corresponds to the future cephalocaudal axis of the adult; disturbance of the initial polarity of the fertilized eggs leads to serious abnormalities of development (twinning or microcephaly, after mild centrifugation, as shown by J. Pasteels). The animal–vegetal polarity, in amphibian eggs, probably originates from preferential uptake of the yolk proteins (which are present in the blood and are synthesized in the liver of the females, as a response to hormonal stimulation) by the vegetal pole of the oocyte: but we do not know why or how; all we know is that it is *not* due to the characteristic arrangement of the capillary network or to gravity.

We have jumped fast from Aristotle to the present, because it is the purpose of this Introduction to show that the old and basic problems of embryology (which is the real study of organogenesis and cell differentiation) remain unsolved and that they deserve to find their solution. Let us go back to history.

When man first looked at the developing chick embryo, he became, without knowing it, a 'descriptive embryologist'. When he made the same kind of observation on a duck embryo, he became, still without knowing it, a 'comparative embryologist'. Descriptive and comparative embryology still remain at the root of modern embryology; they grew successfully during the 18th and 19th centuries, using first the naked eye, then better and better compound microscopes and safe histological procedures. Most remarkable has been the work of those embryologists who, very patiently, studied the 'cell lineage' of the invertebrate eggs (Conklin, Wilson, Boveri, etc.): without any other help than the microscope (the local staining method was only introduced much later by Vogt), they followed the fate of each blastomere in the developing embryo and were thus able to establish a correct and precise 'map' of the fertilized or cleaving egg. Descriptive embryology is by no means dead today: thanks to the development of electron microscopy, it goes on at the ultrastructural level and gives very useful information about the distribution of the cell organelles in those blastomeres which play a key role in development.

After man had a good look at the developmental steps which lead to the formation of the adult, he became more curious than ever and he asked himself the question: *how* does this thing develop? The first answers were more philosophical than experimental: granting that, as postulated by the great William Harvey in the 17th century, *omne vivum ex ovo*, the philosophically minded scientists of the 18th century went into long arguments about preformation and epigenesis and became split into 'ovists' and 'animalculists'. Sometimes, philosophical preconceived conceptions blinded men who were exceptional experimenters: Spallanzani, for instance, who beautifully demonstrated the importance of spermatozoa (animalcules), who showed that oxygen is necessary for development, who tried and perhaps succeeded in obtaining hybrids by cross-fertilization, who discovered the digestive enzymes, remained strangely far from reality when he interpreted on the basis of preconceived ideas, the results of his remarkable experiments. But, is it so certain that this never happens nowadays? In discussing our results do we not also tend to express our own preconceived ideas, or even, sometimes, those which are presently so widely accepted as to seem obvious, although in reality they have not necessarily been conclusively demonstrated? Incidentally, the molecular embryologists of today are still met with the old 'preformation *versus* epigenesis' controversy when they talk about 'stable maternal, preformed messengers' and 'newly synthesized populations of messenger RNAs'.

The great period of experimental embryology (which Albert Brachet liked to call '*embryologie causale*', meaning that experimentation will reveal the mechanisms deeply underlying morphogenesis) covered the first 30 years of the present century: W. Roux, E. Schultz, T. H. Morgan, R. Harrison, O. Hertwig and, last but not least, A. Brachet, were those who first unravelled the secrets of the amphibian eggs, discovering the role of the grey crescent and the morphogenetic significance of the first blastomeres in the frog egg. This work culminated in the discovery of the organizer by Spemann. What is the situation today? We know, from the work of A. Dalcq and J. Pasteels, of A. Curtis and others that both the dorsal cortex of the egg and the heavy yolk material play an important role in the future organization of the embryo. But we still ignore how dorsal cortex and yolk might interact biochemically. I have been working on that question during the past 3 years and my unpublished results are still insufficient to draw important conclusions: it is clear that minute punctures of the cortex have far-reaching effects on the mitotic activity of the pricked egg and that they often lead to aneuploidy. There is little doubt that both early and late abnormalities of development are more frequent after dorsal than after ventral pricking of the egg cortex. On the biochemical level, pricking of the dorsal side induces a stronger inhibition of RNA synthesis than similar punctures of the ventral cortex; on the other hand, DNA and protein synthesis seem to be equally affected. But this is obviously still a very crude approach, the significance of the dorsal cortex of the amphibian eggs will not be understood, in molecular terms, unless methods can be developed allowing a physical-chemical study of the proteins and lipids which make up the egg cortex.

No less significant was the progress made by the experimental embryologists who worked with the eggs of the invertebrates: in the sea urchin,

H. Driesch discovered that an isolated blastomere at the 2-cell stage can develop into a whole larva. Part of an egg can thus produce, after isolation, structures that it would not have developed if left in situ; the total potentialities of the isolated blastomere are higher than the real ones, which express themselves when the blastomere is left in situ. The difference between the total and the real potentialities is *regulation*, which widely varies in importance from one egg species to another. On the other hand, as shown first by E. B. Wilson's experiments on *Dentalium*, certain eggs behave like a *mosaic* of morphogenetic territories: if, for instance, the first polar lobe of the egg is taken away, the apical tuft of the larva is missing. In this case, regulation is kept at a minimum. But careful analysis always discloses that the differences between regulatory and mosaic eggs are quantitative rather than qualitative, and that there is no strong reason for opposing sharply the two kinds of eggs.

In sea urchin eggs, the beautiful experimental work of S. Hörstadius demonstrated that morphogenesis is controlled by two opposite, antagonistic, animal and vegetal gradients. Most vegetal cells, the micromeres, act, under certain circumstances, like Spemann's organizer. Despite the intensive biochemical work done on sea urchin eggs, we still know very little about the chemical nature of the animal and vegetal gradients and the specific properties of the micromeres. But a very promising start has recently been made: Runnström, Hörstadius and Josefsson have isolated animalizing and vegetalizing substances from unfertilized sea urchin eggs, while Czihak obtained autoradiographic indications for a transfer of newly synthesized RNA from the micromeres to the adjacent cells. Methods for the isolation, on a large scale, of the various types of blastomeres have been developed in the laboratory of P. Gross and interesting results can be expected in the near future. We still know nothing about the molecular basis of regulation: in fact, Driesch found this phenomenon so mysterious that he ascribed it to the '*entelechy*', a principle which, by definition, has no spatial localization and cannot be experimentally studied. But the molecular embryologists of today cannot satisfy themselves with this kind of philosophical explanation. It might be relevant to the problem that I recently found that, when sea urchin blastomeres are dissociated as was done by Driesch, inhibitory substances are set free in the medium: these substances which almost certainly originate from the morphogenetically important egg cortex, block the development of normal eggs at an early stage. Removal of these inhibitors – their nature is now being studied – might explain why an isolated blastomere develops into a complete larva.

But, as pointed out above, sea urchin eggs are regulatory eggs and those of many invertebrates are of the mosaic type: the reader of this book will learn a great deal about these fascinating eggs and embryos which have been little studied from the biochemical viewpoint. Some of them can be obtained in sufficient quantity to be studied by conventional methods: for instance, those of *Mactra* have been the object of biochemical investigations by A. Monroy and I have been (and still am) working with those of *Chaetopterus* (which display the unusual property of being capable of some differentiation – the formation of cilia, for instance – without undergoing cleavage). Biochemical studies on isolated blastomeres or polar lobes have also been made on the eggs of the molluscs *Mytilus* (Berg) and *Ilyanassa* (Collier; Davidson) and on those of the Ascidians (Holter; De Vincentiis and Ortolani). This is a very encouraging start, which already demonstrates the usefulness of the eggs described in Prof. Reverberi's book: no good biochemical work can be done without a perfect knowledge of the biological material that is used.

This is a lesson I learned from my father when, being twenty years old, I was trying to study the effects of KCN on *Ascaris* eggs. He came to see me in the laboratory and enquired about my work: I told him proudly that I had seen one of the treated eggs dividing into three cells. He smiled and said: 'If you knew the descriptive embryology of *Ascaris* as well as the effects of cyanide on cell respiration, you would know that there is such a thing as a polar lobe in the eggs of the *Spiralia*. If you had looked carefully at your controls (I hope they exist), you would have seen eggs at the trefoil stage. And if your dishes always remain as dirty as they are now, I am not going to believe anything you tell me about your experiments'.

The molecular embryologist of today, after reading the present book will, at any rate, avoid the first one of these three obstacles on the road towards discovery.

University of Brussels J. BRACHET

Taxonomic contents*

Chapter 1. *Cnidaria*

phylum: COELENTERATA

classis	*ordo*
HYDROZOA	Anthomedusae Leptomedusae Trachylina Siphonophora
SCYPHOZOA	Stauromedusae Discomedusae
ANTHOZOA	Alcyonacea Gorgonacea Pennatulacea Madreporaria Actiniaria

Chapter 2. *Ctenophores*

phylum: CTENOPHORA

classis	*ordo*
TENTACULATA	Cydippida (Pleurobranchia)
NUDA	Beroida (Beroe)

* This classification is limited to the groups or forms reported in the chapters.

Chapter 3. *Planarians*

phylum: PLATYHELMINTHES

classis	*sub-classis*	*ordo*
TURBELLARIA		Polycladida Tricladida
TREMATODA	Monogenea Digenea	
CESTODA	Cestodaria Eucestoda	Merozoa Cyclophyllidea

Chapter 4. *Annelids*

phylum: ANNELIDA

classis	*ordo*
CHAETOPODA	Polichaetes Oligochaetes
HIRUDINEA	
GEPHYREA	Echiuroidea Sipunculoidea

Chapter 5. *Nemertinae*

phylum NEMERTINA

classis	*ordo*	*genus*
ANOPLA	Palaeonemertina Heteronemertina	Cerebratulus
ENOPLA	Hoplonemertina Bdellonemertina	

Chapters 6, 7, 8, 9, 10. *Mytilus, Ilyanassa, Fresh water gastropoda, Dentalium, Cephalopoda*

phylum: MOLLUSCA

classis	*ordo*	*genus*
AMPHINEURA		
GASTROPODA	Prosobranchia	Ilyanassa
	Opistobranchia	Bythinia
	Pulmonata	Limnea
LAMELLIBRANCHIA	Protobranchia	Mytilus
	Filibranchia	
	Eulamellibranchia	
SCAPHOPODA		Dentalium
CEPHALOPODA	Dibranchia	
	Tetrabranchia	

Chapter 11 *Crustaceans*

phylum: ARTHROPODA

classis	*subclassis*	*ordo*
CRUSTACEA		
	Branchiopoda	Anostraca
		Notostraca
		Conchostraca
		Cladocera
	Cephalocarida	
	Copepoda	
	Mystacocarida	
	Ostracoda	
	Cirripedia	
	Malacostraca	

Chapter 12. *Echinoderms*

phylum: ECHINODERMATA

classis
CRINOIDEA
ASTEROIDEA
OPHIUROIDEA
ECHINOIDEA
HOLOTHUROIDEA

Chapters 13, 14. *Ascidians*, *Amphioxus*

phylum: CHORDATA

classis	*subclassis*	*genus*
HEMICHORDA		
UROCHORDA	Larvacea	Ciona
	Ascidiacea	Ascidiella
	Thaliacea	Phallusia
CEPHALOCHORDA		Amphioxus

Contents

Chapter 9. *Dentalium, G. Reverberi*

Chapter 10. *Cephalopods, J. M. Arnold*

Chapter 11. *Crustaceans, J. Green*

G. Reverberi (ed.), Experimental embryology of marine and fresh-water invertebrates

CHAPTER 1

Cnidaria

H. MERGNER

Institute of Special Zoology, Ruhr-University. Bochum, Germany

1.1 Normogenesis

The marked tendency of Cnidaria towards asexual reproduction and the stem and colony formation and polymorphism of individuals associated with it are variously pronounced among the different classes. The normogenesis of Hydrozoa, Scyphozoa and Anthozoa must thus each be dealt with separately. In the present state of our knowledge we must regard Anthozoa as the phylogenetically most primitive class of Cnidaria, and Hydrozoa as the most highly evolved. If, therefore, the development of Hydrozoa stands at the head of the discussion below, this is so only for practical reasons. For this class exhibits the majority of the modes of cleavage and germ layer formation occurring in Cnidaria and a great variety of types of vegetal reproduction. Furthermore, the evolution of many species of Hydrozoa perfectly exemplifies these modes.

1.1.1. Hydrozoa

A typical feature of many Hydrozoa is metagenesis, or the alternation between a sessile, asexual polyp generation and a sexual generation of planktonic medusae. In spite of numerous investigations undertaken with this end in view, it has not been possible to demonstrate such alternation of generations in many species of Hydrozoa, partly for technical reasons. As a result, a different nomenclature and systematic classification is often used for the two generations. Moreover, metagenesis is no longer present in most Hydrozoa, a generation being suppressed or missed. Thus, the planktonic medusa generation is often completely suppressed or its development arrested prematurely. Then sessile medusa buds at various stages of reduction, so-called medusoids or sessile gonophores, occur temporarily or permanently. A comprehensive description of the reduction stages and a detailed discussion of the 'regression theory' are to be found in Kühn (1913). In many oceanic forms, such as several Trachylina, on the other hand, the polyp generation is lost and the eggs develop via a young polyp-like stage, the actinula, straight to the medusa.

1.1.1.1. Egg stage

The egg cells of Hydrozoa are of epidermal origin throughout and arise either from interstitial cells or by differentiation of ordinary epidermis cells or from their daughter cells formed by mitosis. For the most part they arise out of the ectoderm, as for example in Eudendrium, Pennaria and Cladocoryne, but occasionally also from the entoderm, as in various species of Coryne. (Details will be found in Weismann 1883; Goette 1907; Kühn 1910, 1913; Mergner 1957; etc.) In several Hydroida the sites of egg differentiation and maturation are far distant from one another in the stem, so that the oocytes will reach their destination only after migrating a considerable distance. Eggs formed in the ectoderm often pass to the entoderm at an early stage, in which, possibly on account of the better nutritional conditions, they wander for some distance to their sites of maturation in the gonophores. The eggs of *Eudendrium armatum* even pass through a veritable stationary phase in the gastrodermis, during which they take up dissolved nutrients straight from the lumen of the hydrocaulus (Wasserthal 1969).

The site of maturation of the egg cells in hydroids with alternation of generations and in Siphonophora is the medusa bud or the sessile gonophore or medusoid. In hydroids showing no alternation of generations and in Trachylina, on the other hand, maturation takes place at the site of formation of the eggs. Sexual dimorphism is predominant in marine Hydrozoa, hermaphroditism being the exception (as in *Eudendrium simplex*), while in colony-producing forms the whole colony is generally male or female. The eggs either remain in the mother animal, where they are fertilised, or are shed into the open water by bursting of the ectoderm and fertilised there, as is the case with most planktonic medusae. If the development of the embryo is completed within the mother animal the sessile gonophores become brood chambers, out of which the embryos migrate only at the planula larva stage. On occasion, however, only one stage corresponding to that of the planula is passed through and still within the embryonic theca, young tentaculate stages like the actinulae of Tubularia develop, becoming free only at this stage (see p. 21 for details).

The egg cells of Hydrozoa are surrounded by a delicate oolemma and after their fertilisation or during cleavage secrete a firm, multilayered embryonic cover when the embryos develop in the shelter of the maternal organism. The yolk content of the eggs varies from one species to another and often influences the course of development: in addition to isolecithal eggs with a distinctly poor yolk content, which develop chiefly by holoblastic cleavage, yolky centrolecithal eggs which develop in a manner similar to superficial cleavage are not uncommon. Ancillary apparatuses to nourish the growing egg cell such as the entodermal spadix tubes of the sessile gonophores are widely distributed, as are the ingestion of ectodermal and entodermal cells as nutritive cells and the development of individual egg cells at the cost of neighbouring ones. Development in the mother animal is however always associated with a reduction in the number of eggs, while planktonic medusae, whose eggs are expulsed into the open water, generally produce eggs in large quantities.

1.1.1.2. Cleavage

Cleavage in Hydrozoa is never determinative: all blastomeres are equivalent to one another (see p. 39 for details). It may be either holoblastic adequal, with radial symmetry in typical cases, or meroblastic; true unequal cleavage is virtually unknown among Hydrozoa, since differences in blastomere size are almost always regulated in later cleavage stages.

In the holoblastic forms of cleavage the first two furrows are always meridional, and the third equatorial. The first furrow begins at the animal pole, whence it can spread in a ring towards the vegetal pole; it is then 'circular', as for instance in *Aequorea forskalea* (Claus 1883) and *Clava squamata* (Harm 1903). Or it descends from the animal pole as a 'cutting' furrow into the depths, thereby dividing the egg, as in the eggs of many planktonic hydromedusae such as *Rathkea fasciculata*, *Clytia flavidula* and *Mitrocoma annae* (Metschnikoff 1886b), or in those of a few sessile gonophores as in *Gonothyraea loveni* (Wulfert 1902), *Hydractinia echinata* (Bunting 1894) and *Cordylophora lacustris* (Morgenstern 1901). The second furrow is only rarely circular and the third can cut in a circular, centripetal or centrifugal fashion.

The first three cleavage divisions generally occur synchronously and give rise to regularly arranged blastomeres of equal size. Only then do appreciable irregularities in the sequence and direction of divisions occur, so that from the 16-cell stage onwards the embryo is often no more than a mass of blastomeres with little pattern of disposition. An exception to this rule is the leptomedusa *Aequorea forskalea* (Haecker 1892), in which the cleavage steps occur completely synchronously until the 64-cell stage and give rise to blastomeres of equal size (fig. 1f).

The typical cleavage of many Cnidaria and thus of Hydrozoa is radial, i.e. seen from the animal or vegetal pole, all blastomeres are arranged in regular radial symmetry around the axis of the embryo between these poles. But other modes of cleavage occur in Hydrozoa which are directly related to radial cleavage or can be derived from it, the modification in the cleavage process increasing as the yolk is more abundant. The most important difference between the various modes of cleavage is whether a blastocoele occurs, and if so at what stage and to what extent. Here too the causes can probably be sought in the different yolk content of the eggs, and in the case of sessile gonophores in the space conditions under which the development of the embryo takes place (Kühn 1913). The different modes of cleavage of Hydrozoa will be discussed below.

(a) *Radial cleavage*. Holoblastic, adequal, radial cleavage is exemplarily developed in the leptomedusa *Aequorea forskalea* (fig. 1). The eggs are shed into the open water and divide with the typical regular alternation between meridional and equatorial cleavage until the 64-cell stage, any small differences in size between the animal and vegetal blastomeres being balanced out again before the blastula stage is reached. At the 64-cell stage the blastomeres form a cylindrical mantle composed of tiers of wreaths of cells one on top of the other, curved in slightly at the top and bottom. With the continued increase in the number of cells the hitherto spherical shape of the blastomeres is gradually lost, they press together more densely, and while cleavage from now on remains centripetal, a spacious hollow cavity is finally formed, on the surface of which furrows are hardly any longer to be seen. This is the coeloblastula, each cell of which bears a flagellum directed outwards, and which as in most metagenetic hydromedusae becomes more and more enlarged to a pelagic hollow vesicle (fig. 1g).

Radial cleavage follows a similar course in the anthomedusa Tiara (Hamann 1883; Metschnikoff 1886b), the leptomedusa Obelia (Metschnikoff 1886b), the limnomedusa *Gonionemus murbachi* (Perkins 1903) and numerous Trachylina. In these cases, however, irregularities in the sequence and direction of the divisions and in the size of blastomeres occur earlier, either after initially normal radial segmentations or from the start of cleavage, which later becomes regular. Such frequently convergent variations in the typical radial cleavage, which must be regarded as the phylogenetically original type, are modes of cleavage in their own right, discussion of which follows.

(b) *Pseudospiral cleavage.* In the first of these modified forms of cleavage a pattern of blastomeres emerges, generally as early as the 4- or 8-cell stage, which is reminiscent of certain stages of spiral cleavage: starting from normal radial cleavage, in which the spindle axes are perpendicular to the polar axis during the transition to the 4-cell stage, and parallel to it at that to 8-cell stage, the divided daughter cells then move into the intercellular spaces of the parental cells. Thus at stage 4 there are two and at stage 8 four cells displaced by 45 degrees with respect to the others. This process is termed pseudospiral cleavage, but has nothing to do with true spiral cleavage, in which the axes of the spindles lie diagonally to the polar axis and the daughter cells immediately move to the intercellular spaces. Furthermore, neither alternation of the direction of division nor formation of crosses or rosettes is observed in pseudospiral cleavage. According to Siewing (1969) it may not therefore be regarded as a phylogenetic precursor of spiral cleavage or as the model for its phylogenetic genesis.

Such pseudospiral cleavage, which may be more or less marked, occurs in the hydroids *Cordylophora lacustris* (Morgenstern 1901), *Stomotoca apicata* (Rittenhouse 1910), *Rathkea fasciculata* (Metschnikoff 1886b), and *Gonionemus murbachi* (Perkins 1903). Even the syncytial cleavage of *Eudendrium racemosum* (Mergner 1957), greatly modified as the result of the high yolk content, being reminiscent of pseudospiral cleavage in the 4- and 8-nucleus stage (see p. 9).

(c) *Total irregular cleavage.* A number of Anthomedusae such as *Turritopsis nutricola* (Brooks and Rittenhouse 1907), *Stomotoca apicata* (Rittenhouse 1910) and *Pennaria tiarella* (Hargitt 1904a), show no regular cleavage pattern from the 4- or 8-cell stage onwards. In Turritopsis (fig. 2) the cells, after initial regular radial cleavage, first take the form of a single-layered, downward curved plate consisting of 8 cells and then display a completely irregular pattern in which racemose clusters are arbitrarily formed, their individual cells hanging only loosely against one another (fig. 2d, e). This condition has been termed 'blastomere anarchy' and the mode of cleavage 'total irregular cleavage'. Once the number of cells has become very large, however, the blastomere mass rounds off once again, while the individual cells press closely together and lose their cell boundaries within the embryo, so that for a time a syncytium develops from which the two germ layers become differentiated (fig. 15a). Thus in spite of the irregular course of cleavage the developed organism is normal, testifying to the marked regulative ability of this hydroid embryo. For the subsequent course of development see pp. 16 and 17.

(d) *Delayed blastomere separation.* With increasing yolk content of the eggs nuclear divisions occur during early cleavage of a number of athecate hydropolyps such as *Tubularia mesembryanthemum* (Brauer 1891b) and species of Millepora, without division of the cytoplasm at this stage. Only from the 16-cell stage onwards do cell boundaries form, with subsequent normal total cleavage, while the entoderm begins to develop as early as from the 30-cell stage (fig. 3). (See pp. 13 and 14 for the further course of development.) In the anthomedusa *Myriothela cocki* cytoplasm division is delayed as late as the 32-cell stage. In the thecate hydropolyps *Aglaophenia pluma* and *A. helleri* (Müller-Calé and Krüger 1913a), on the other hand, the cell boundaries disappear in the middle stages after initial total cleavage, to recur only towards the end of the cleavage period.

This temporary disappearance of the cell boundaries in the course of particular cleavage stages has been called 'delayed blastomere separation' by Siewing (1969). But at first this term denotes only the ontogenetic development. More important for an assessment of the phylogenetic course of development, however, is the fact that in the process syncytia are formed at various stages, as was also the case with *Turritopsis nutricola* (Brooks and Rittenhouse 1907) (see p. 5 and fig. 15a). Here, in relatively yolky eggs, tendencies towards meroblastic modes of cleavage such as syncytial cleavage are discernible, not that it is possible or admissible, of course, to use this to prove the phylogenetic derivation of one from the other. Nevertheless, the developmental processes of Tubularia, Myriothela, Aglaophenia and Turritopsis, also in this sequence, serve to some extent as models in hypothetical speculation about the genesis of meroblastic types of cleavage in Hydrozoa. For this reason they ought more accurately to be regarded as partially syncytial modes of cleavage. Furthermore, on the basis of recent findings by Mergner (1957) and Wasserthal (1969) it is probable, although not definite, that at the origin cell boundaries were once formed in syncytial cleavage, so that the term 'syncytial' is used correctly.

(e) *Shortened development.* Before discussion of syncytial cleavage is started, however, it is necessary to refer to a number of Hydroida whose development is shortened in the sense that germ layer formation begins before the cleavage process has terminated so that these two otherwise distinct phases of development overlap, generally resulting in the omission of a typical blastula stage.

What is more, the onset of germ layer formation can be advanced to increasingly earlier stages of cleavage, as will be described in detail (see pp. 13, 14). The shortening of development, in a similar fashion to the temporary appearance of stages of syncytial cleavage, is frequently bound up with the nutrient content within the eggs. Since it occurs primarily in hydroids with sessile gonophores, that is to say with more or less reduced medusa generations, it may be considered a phylogenetically derived process (Siewing 1969). It may thus be stated that forms of which the yolk content makes an early change to autonomous nutrition on the part of the offspring unnecessary show a trend towards reduction of the medusa generation, shortening of development and the occurrence of syncytial stages.

Shortened development of this type is found in the thecate hydropolyps *Gonothyraea (Laomedea) loveni* (Wulfert 1902) and *Laomedea flexuosa* (Müller-Calé 1913) and the athecate *Clava squamata* (Harm 1903). In *Gonothyraea loveni* the regular radial cleavage gives rise to a blastula composed of few cells, the small blastocoele of which is filled by the daughter cells of a few radial cleavages, while the outer daughter cells remain in the blastoderm (fig. 5a, b). These cells, which have entered the blastocoele at an early stage, are already entodermal cells. Siewing (1969) has therefore rightly criticised the use of the term 'morula delamination' for this process, since strictly speaking no true morula whatsoever is formed. This problem will be discussed in detail in section 1.1.1.3.

The entoderm is laid down in the interior of the embryo even earlier in *Clava squamata* (fig. 4) and *Laomedea flexuosa*, where despite initially radial cleavage no blastocoele at all is formed (see also p. 16). In this connection it is worthy of note that in the early development of *Clava squamata* the first blastomeres to appear exhibit marked differences in size, so that here an – albeit altered – holoblastic unequal cleavage process constitutes the beginning of individual development, an exception among Hydrozoa. The product of cleavage in the examples given can be defined as a 'cleavage morula' ('Furchungsmorula'), which is however already a component of germ layer formation and progressively obliterates the coeloblastula in the course of embryonic development within sessile gonophores.

(f) *Syncytial cleavage*. Further increases in the yolk content of centrolecithal eggs and the tendency towards shortened development finally give rise to a process resembling superficial cleavage, in which no cell boundaries whatsoever occur during cleavage. Syncytial cleavage (Brauer 1891b; Hargitt 1904b; Kühn 1913; Mergner 1957) may with Siewing (1969) be regarded as the culmination of a comparative embryological series.

Siewing replaces the long-established term 'syncytial' cleavage by 'partially superficial' cleavage, on the grounds that we are not dealing here with a syncytium, which is of course defined by the secondary loss of originally present cell boundaries. A number of considerations speak against such rechristening, however: the cleavage processes of Tubularia, Myriothela, Aglaophenia and Turritopsis described above, clearly exhibit the tendency, observed in a great variety of Hydroida families, to intercalate a shorter or longer syncytial cleavage stage during early, middle or late cleavage. This temporary loss of cell boundaries can with increasing yolk content be advanced to such a degree that boundaries no longer occur morphologically:

For example in the genus Eudendrium, too, there are cases in which superficial cleavages of the embryo correspond with the number and position of cleavage nuclei in the interior (2- to 8-nucleus stage of *E. racemosum*, Mergner 1957) or with the delimitation of the individual regions of yolk each surrounding one nucleus by cell-boundary-like, plasmatic septa (4- to 8-nucleus stage of *E. ramosum*, Hargitt 1904b, and 8-nucleus stage of *E. armatum*, Wasserthal 1969). Here too then, an increasing yolk content is associated with an ever more marked derivation in the developmental process from *Eudendrium armatum* via *E. ramosum* to *E. racemosum*. This lends justification to the hypothesis that the syncytial mode of cleavage of extremely yolky eggs developed out of a holoblastic one by loss of cell boundaries, and to the use of the term 'syncytial cleavage'.

Furthermore, in spite of signs of convergence in respect of early superficial cleavage and its origin in the richness of the nutrient store of the eggs, we should not speak of 'superficial' cleavage. In this case, in fact, both germ layers are formed from the cleavage nuclei after migration to the periphery, but in syncytial cleavage only the ectoderm. The qualification 'partially' superficial cleavage must also be rejected, since it does not so illuminate the crucial difference that it may justifiably supersede a term which is firmly rooted in the relevant literature. On the other hand, there are good grounds for proposing the term 'partially syncytial cleavage' for the modes of cleavage mentioned with temporary occurrence of syncytial stages (see pp. 6, 7, 16 and 17).

Siewing rightly points out that *Clava squamata*, *Corydendrium parasiticum* (family Clavidae) and *Tubularia larynx* (Tubulariidae) also display, in addition to holoblastic cleavage throughout, initially holoblastic and then 'superficial' cleavage. This again underlines the possibility that syncytial cleavage is derived from holoblastic and syncytial delamination from morula delamination (see also p. 17). The very early separation of the prospective ectoderm from the prospective entoderm in syncytial cleavage, in *Eudendrium racemosum* (Mergner 1957) as early as the 2-nucleus stage, in fact brings about an even greater shortening of development, as has already been described for *Gonothyraea loveni, Clava squamata* and *Laomedea flexuosa.* Strikingly, an increasing reduction in the sessile gonophores, the seat of the developmental stages, can be observed at the same time: while in *Tubularia larynx* (Schneider 1902) these are still a eumedusoid with rudiments of tentacles and in *Clava squamata* and *Gonothyraea loveni* (Kühn 1910) still cryptomedusoids, *Laomedea flexuosa* (Kühn 1910) displays highly reduced heteromedusoids and Eudendrium and Corydendrium (Weismann 1883; Goette 1907) only simple styloids, even. Increasing nutrient content and greater reduction of the gonophores are thus bound up on the one hand with an increasing tendency towards shortened development and on the other hand with the occurrence of syncytial stages, although of course we cannot clarify the ultimate relationships, let alone draw a definite conclusion as to the phylogenetic ancestry.

Typical syncytial cleavage has been observed to date only in the stylasterine *Distichopora violacea* and in species of Eudendrium such as *E. racemosum* (Mergner 1957), *E. ramosum* (Hargitt 1904b) and *E. armatum* (Wasserthal 1969); it probably also occurs in other Eudendriidae, however. The high yolk content of the centrolecithal eggs of *E. racemosum* impedes its total cleavage (fig. 6). For this reason only the nuclei with their surrounding plasma divide synchronously until the 16-nucleus stage and until the 8-nucleus stage in a strictly orderly cleavage pattern modified from total cleavage: the spindle axes at each mitosis step lie perpendicular to those at the preceding step. This nuclear cleavage pattern is reflected in a chequering of the surface, so-called pseudoblastomeres, but not in the distribution of cytoplasm and yolk. After division of the zygote nucleus a daughter nucleus wanders to the interior of a skull cap structure, which consists of liquid phase yolk; its descendants become vitellophages and prospective entoderm, while those of the outer daughter nucleus form the definitive ectoderm via a preectoderm. The result of syncytial cleavage of *E. racemosum* is the pregastrula, which is however already a component of germ layer formation. Thus a drastic shortening of development with overlapping of cleavage and germ layer formation can be also observed in Eudendrium, bound up with an extremely rich yolk supply in the eggs and a sheltered site for the developmental stages in a highly reduced styloid gonophore.

1.1.1.3. Germ layer formation

In the previous section it was shown that all types of cleavage of Hydrozoa can be ultimately related to the phenomenon of total adequal radial cleavage, of which they represent variations. Normally cleavage of eggs that are not too yolky culminates in the formation of a coeloblastula, which is thus proper to the metagenetic forms above all. Only as the yolk content increases is the blastocoele constricted. But the early putting forth of prospective entoderm cells, so that a cleavage morula is formed, can also suppress the blastocoele; or one might never be formed, as in syncytial cleavage, when the prospective entoderm nuclei remain within the embryo from the start. The result of cleavage is then a pregastrula which, like the cleavage morula, is itself a stage of germ layer formation.

Thus while the cleavage of Hydrozoa demonstrates a wide diversity of forms, this must be all the more true of germ layer formation, for which depending on the viewpoint six or seven possibilities exist, the largest number among all classes of animals: unipolar and multipolar immigration with gradual transition forms between the two; mixed delamination combining multipolar ingression and blastula delamination; typical blastula delamination; morula delamination; syncytial delamination after blastomere anarchy and syncytial delamination after syncytial cleavage. If below we often speak of germ layer and entoderm formation the term 'formation' according to Siewing (1969) does not mean that new material is actually formed, but that already available material is rendered visible.

Entoderm formation very probably originally took place at the vegetal pole, as is still the case with invagination (emboly), epiboly and unipolar ingression, and only then spread to the whole embryo on a multipolar basis. Here too the separating cells at first had an epithelial coherence as in blastula delamination, which then gradually got lost in such forms as mixed delamination, typical multipolar ingression and morula delamination. Entoderm formation probably proceeded at first with a large number of cells and later with only a few or even only a single cell. All these possible types of germ layer formation and transitions between them are still to be observed today. Finally for the reasons already stated a shortening of development must have occurred as the final step, as is achieved in the extreme case in syncytial delamination. Shortened development was however always realised by bringing germ layer differentiation forward to the cleavage period, thus blurring the boundary between the two. Examples of the individual modes of germ layer formation are given below.

(a) *Unipolar ingression.* Unipolar ingression may be considered closely related to invagination, which never occurs in Hydrozoa, and is always bound up with the presence of a coeloblastula. In the typical case the blastula when freeing itself is elongated to an oval shape and develops cilia. It is then termed the 'blastula larva'. The blastoderm cells of the somewhat flattened vegetal pole become highly cylindrical, crowd into the cleavage cavity, where they spread out and fill it more or less completely (fig. 7, 8). Paratangential divisions are very rare (fig. 8a); as a rule only whole cells immigrate, first pushing themselves aborad like a compact plug and gradually obliterating the blastocoele. Occasionally the whole blastoderm of the vegetal pole is transitorily transformed into prospective entoderm cells, which are readily distinguishable from the other blastoderm cells by virtue of their irregular disposition, as for example in the leptomedusa *Aequorea forskalea* (Claus 1883) (fig. 7c). Once the blastocoele has been completely filled by the prospective entoderm cells the resulting two-layered embryo, possessing a simple ciliated outer epithelium – the ectoderm – and a compact, still disordered entoderm within, is known as the 'sterrogastrula' (Goette 1907) or 'parenchymella' (Metschnikoff 1886 b). This is the planula larva, so characteristic for Cnidaria, in the solid entoderm of which the gastrovascular cavity is later formed by the dissolution or dissociation of the innermost cells (fig. 7e). It is only at this stage that the entoderm takes on an epithelial organisation as an archenteron around the gastric cavity. In the process individual cells may separate off and serve as food. The archenteron remains closed at first, since the mouth breaks through only later at the pole of ingression, the oral pole. Thus the result of germ layer formation by polar ingression is a planula larva the young stages of which we know as the sterrogastrula or parenchymella.

The formation of entoderm by unipolar ingression is very widespread in the swarming blastulae of numerous hydromedusae, e.g. the anthomedusae *Tiara pileata* (Hamann 1883), *Tiara leucostyla*, *Rathkea fasciculata* (Metschnikoff 1886b) and *Stomotoca apicata* (Rittenhouse 1910); and the leptomedusae *Aequorea forskalea* (fig. 7, Claus 1883), *Clytia flavidula* and *Clytia viridicans* (fig. 8a, b), *Octorchis gegenbauri*, *Mitrocoma annae* (fig. 8c), *Laodicea cruciata*, Obelia and Tima (Metschnikoff 1886b).

(b) *Multipolar ingression.* Starting from unipolar ingression, a comparative embryological series can be drawn up within Hydrozoa (Siewing 1969), the next link of which is doubtless multipolar ingression. In its pure form, i.e. as only an immigration of whole blastoderm cells, without daughter cells of paratangential divisions, it is not found among Hydroida, but it does occur in Trachylina.

In the narcomedusa *Aeginopsis mediterranea* (fig. 9, Metschnikoff 1886b) a 32-cell blastula develops early by holoblastic, equal, radial cleavage, the blastocoele of which becomes completely filled by immigrating whole blastoderm cells. All cells reaching to the surface of the embryo become ectoderm, while those remaining within become entoderm. The two-layered, free-swimming embryo can again be called a sterrogastrula or parenchymella. Only later, once it has expanded to a planula and gone on to actinula formation, does the entoderm become organised epithelially around the resultant gastric cavity (fig. 30d).

In most cases, however, multipolar ingression does not occur in pure form, but as an intermediate stage tending partly towards unipolar ingression and partly towards blastula delamination. While transitions tending towards blastula delamination are quite common and will be dealt with in the next section as 'mixed delamination', those towards unipolar ingression are an only occasional phenomenon. They nevertheless are of especial interest, since they bridge the fictitiously wide gulf between the two modes of germ layer formation (Siewing 1969): thus in Obelia initial polar ingression can spread in the further course of development to the whole vegetal third of the embryo. In *Melicertidium octocostatum* (Gemmill 1922) polar ingression from the vegetal pole even encroaches on the whole inner surface of the blastula and thus becomes multipolar ingression, and in *Tubularia larynx* unipolar ingression is occasionally seen in addition to multipolar ingression (as mixed delamination).

(c) *Mixed delamination.* Far more frequently, however, multipolar ingression tends towards blastula delamination, with which it is related via numerous intermediate stages. In point of fact, most instances of the occurrence of multipolar ingression are in reality instances of mixed delamination. But although the two are closely related, they must be carefully distinguished from one another. Thus in mixed delamination, in addition to whole blastoderm cells, the inner daughter cells of paratangential mitoses from radial spindle axes enter the blastocoele, which is thus filled by whole immigrated and delaminated prospective entoderm cells. It is for this reason that this mode of germ layer formation combining multipolar ingression and delamination is termed 'mixed delamination'.

Mixed delamination is a widespread phenomenon and is found in several species of Hydra (fig. 11, Brauer 1891a), the athecate hydropolyps *Cordy-*

lophora lacustris (Morgenstern 1901) and *Tubularia mesembryanthemum* (fig. 12, Brauer 1891b), the anthomedusa *Bougainvillia superciliaris* (Gerd 1892), *Hydractinia echinata* (Bunting 1894), the thecate hydropolyp *Halecium tenellum* (Hamann 1882), and the trachymedusa *Liriope mucronata* (fig. 10, Metschnikoff 1886b; Brooks 1886).

In *Liriope mucronata* (fig. 10), at first whole blastoderm cells migrate into the blastocoele of the few-celled blastula, which is derived from radial cleavage, but then, too, daughter cells of paratangential mitoses, until the cleavage cavity is loosely filled by prospective entoderm cells. Only then do the outer cells differentiate as ectoderm from the inner cells, which take the form of an entoderm vesicle floating in the excreted jelly. The gastrovascular cavity forms in its interior. Thus while in the hydroids mentioned above such an entoderm vesicle is never formed, *Liriope mucronata* provides a perfect example of typical mixed delamination (Metschnikoff 1886b) out of multipolar ingression and blastula delamination. Important differences in the process of mixed delamination are also seen, in that the moment of its inception varies from species to species. Entoderm proliferation can in effect be advanced to ever earlier cleavage stages, as will be illustrated by the following examples:

In Hydra (fig. 11, Brauer 1891a) a blastocoele occurs as early as the 8-cell stage, growing constantly larger until at the 128-cell stage whole blastoderm cells and daughter cells from radially placed mitotic spindles converge into the cleavage cavity from all sides simultaneously and obliterate it completely. The two germ layers of the resulting sterrogastrula then separate from one another and in the entoderm the gastrovascular cavity is formed by dissolution of the innermost cells, the entoderm cells arranging themselves around it as epithelium. Then, the embryo breaks free and attaches fast and with the formation of the mouth and tentacles develops directly to the young polyp.

The embryos of *Cordylophora lacustris* (Morgenstern 1901) develop in sessile gonophores and already at the 4-cell stage display a slit-like cleavage cavity, which rapidly enlarges. With 64 cells the coeloblastula is formed, in which mixed delamination now begins. It is pursued as in Hydra and also gives rise to a sterrogastrula, in which ectoderm and entoderm differentiate.

Entoderm ingression begins even earlier in *Tubularia mesembryanthemum* (fig. 12, Brauer 1891b), i.e. already at the stage of the 30-cell blastula. A solid embryo, the 'pseudomorula' (Gerd 1892), results, accumulating a great number of cells as the result of continued divisions in all directions, especially at the periphery, where the ectoderm is laid down in several layers from the outset. The embryo flattens out and produces tentacles very early. The gastrovascular cavity, on the other hand, is formed only in the young actinula by dissolution of the innermost entoderm cells. The earliest onset of endoderm formation by mixed delamination is displayed by *Hydractinia echinata* (Bunting 1894), viz., at the 16-cell blastula stage, at which cleavage already ceases (fig. 56d, Teissier 1931).

Thus, the whole mixed delamination is seen to be associated with a distinct trend towards shortened development as a result of advancing entoderm formation to increasingly earlier cleavage stages. But in all cases the solid pseudomorula or sterrogastrula stage does not signify the end of cleavage, but the result and termination of topographical differentiation of the cells for the formation of entoderm. Only then does specialisation of ectoderm and entoderm cells begin after formation of the mesolamella.

(d) *Blastula delamination.* The previous section described various intermediate stages between germ layer formation by multipolar ingression and that by blastula delamination, modes which are known as mixed delamination. From here to pure blastula delamination ('primary delamination': Metschnikoff 1886b) – albeit a rare phenomenon – it is but a small step. Already in *Liriope mucronata* (fig. 10) an entoderm vesicle, swimming free in the jelly, is ultimately formed from whole blastoderm cells and the daughter cells of paratangential divisions.

If now all blastoderm cells divide at about the same time, with the cleavage spindles lying radially, so that their inner daughter cells enter the blastocoele together, a cohesive inner layer of blastoderm becomes separated, as it were. This layer again constitutes a hollow, very regular entoderm vesicle, enclosed by the temporarily reappearing cleavage cavity (fig. 13). A gelatinous substance is given off into the interstice between the two germ layers, increasing rapidly in quantity and exerting an appreciable influence on the further development of the embryo (see p. 21). The result of this blastula delamination, which is typified in the trachymedusa *Geryonia fungiformis* is an embryo composed of two epithelial vesicles lying one within the other.

(e) *Morula delamination.* We have seen how entoderm is formed in free-living hydromedusae and Trachylina by unipolar and multipolar ingression and by mixed delamination and blastula delamination, that is to say by modes closely related to one another and always proceeding from a coeloblastula formed by radial cleavage. In the process the blastocoele is gradually obliterated by the prospective entoderm cells, which occupy it in a solid mass: the result is a pseudomorula or sterrogastrula, and only then do the two germ layers differentiate. Mixed delamination is also found in hydropolyps with sessile gonophores, such as *Cordylophora lacustris*, *Tubularia mesembryanthemum* and *Hydractinia echinata* (see pp. 13 and 14). At the same time it is possible to observe a trend towards shortened development by a bringing forward of entoderm proliferation to increasingly earlier stages of the cleavage process.

This relatively self-contained group of modes of germ layer formation may be contrasted with another which is almost always associated with development in sessile gonophores and a shortening of development (see also section 1.1.1.2(e)). In the most widespread of these modes, 'secondary delamination' (Metschnikoff 1886b), cleavage from the very first gives rise to a solid multilayered embryo without a blastocoele, the 'cleavage morula' (Metschnikoff 1886b). Release of prospective entoderm cells into the interior is effected chiefly by paratangential divisions, that is to say by delamination. This process extends over a longer period, does not take place concurrently and cannot therefore ever give rise to an epithelial arrangement of the entoderm. Since delamination gives rise to an embryo which is solid from the outset, and which has been somewhat incorrectly described as a 'morula', we also speak of 'morula delamination'. Only when all prospective entoderm cells have been given off does the surface cell layer become differentiated from the entoderm as ectoderm. Strictly speaking only this final step can be regarded as constituting germ layer differentiation, the actual starting point of which is thus the cleavage morula.

Morula delamination indubitably shows a connection with multipolar ingression and with mixed delamination. An important transition here is represented by germ layer formation in *Gonothyraea loveni* (Wulfert 1902). Radial cleavage in this thecate hydroid normally leads to a spacious coeloblastula with 64 cells, in which entoderm proliferation begins (p. 7, fig. 5b). On occasion, however, when a large number of embryos are densely packed in the cavity of the gonophore, no blastocoele or only a very small one is formed. Paratangential divisions then set in very early, so that embryos of only 24 cells already have a solid form (fig. 5c, d), with inner cells which are prospective entoderm. Thus Gonothyraea illustrates the possibility of the formation of a coeloblastula with subsequent multipolar entoderm proliferation and as product of cleavage a pseudomorula which is already an element of germ layer formation. With the earlier onset of paratangential mitoses we then have a mode of development in which cleavage and entoderm formation are no longer successive, but are concurrent, overlapping processes giving rise to a cleavage morula. Thus in both cases the end result is a solid embryo composed of histologically similar cells, differing only in their position relative to the surface, and whose further fate is the same in the case of both pseudomorula and cleavage morula: the superficial cells form the ectoderm, the inner the entoderm. This shows that it is not possible to draw a hard and fast distinction between multipolar entoderm ingression and morula delamination or between the pseudomorula and cleavage morula (Kühn 1913).

Morula delamination is also found in thecate hydropolyps with sessile gonophores such as *Laomedea flexuosa*, *Sertularella polyzonias*, *Dynamena pumila* and *Plumularia echinulata* (Müller-Calé 1913), the trachymedusae Aglaura and Rhopalonema, and Siphonophora. In all these species cells pass into the interior at a very early stage and in general no or only a very small blastocoele is formed. Morula delamination is to be seen particularly clearly in the athecate *Clava squamata* (fig. 4b, 14, Harm 1903). As the result of untypical unequal cleavage a solid embryo of 16 closely packed cells is formed, which already begins to release daughter cells of paratangential mitoses and which at the 32-cell stage comprises several layers. Whole cells may be crowded into the interior in the process. Only after vigorous multiplication of cleavage cells does further differentiation into ectoderm and entoderm in the 'morula' begin. The outer cells become prismatic and separate off the mesolamella. Finally through dissolution of the innermost entoderm cells the emerging planula develops a slit-like aperture, the gastrovascular cavity. The product of germ layer formation in *Clava squamata* is thus a 'differentiation gastrula'.

A typical differentiation gastrula, in which no connection with entoderm formation by premature multipolar ingression is any longer discernible, is found in the anthomedusa *Turritopsis nutricola* (fig. 15, Brooks and Rittenhouse 1907). Irregular cleavage (see p. 5) at first gives rise to an aggregation of blastomeres hanging loosely together without forming a blastocoele, i.e. to 'blastomere anarchy', in which the cells continue to multiply. The blastomere mass then roundens off again, the cells press together densely and for a time lose their cell boundaries. In the resulting syncytium the nuclei are particularly abundant at the periphery. These then become surrounded once again by cellular areas, which represent the ectoderm and separate off a mesolamella against the mass inside. Epithelialisation of the entoderm, formation of the gastrovascular cavity, starting with a slit-shaped cavity, and breaking through of the mouth occur only once the planula has become free.

Differentiation into ectoderm and entoderm thus sets in at a very late stage in the development of Turritopsis originating from a mass of undifferentiated cleavage cells. Although there are no longer any signs of entoderm formation by ingression, the mode found in Turritopsis is probably ultimately derived from morula delamination. A more interesting feature, however, is the appearance of a syncytial stage at the end of the cleavage period, from which only then the two germ layers differentiate, a process reminiscent of 'syncytial delamination', to which we shall now turn.

(f) *Syncytial delamination.* Comparative embryological studies show syncytial delamination (Kühn 1913) to border on morula delamination. It is probably the furthest derived mode of germ layer formation. As with syncytial cleavage, which it follows, certain relationships can be found here too with holoblastic cleavage and morula delamination, although the phylogeny of these lineages cannot be regarded as confirmed (see also pp. 8 and 9). The disappearance of cell boundaries does not appear to be such a momentous step in this case: on the one hand various Hydrozoa had been shown to intercalate syncytial stages during cleavage (pp. 5–7), and on the other Turritopsis shows a mode of producing the differentiation gastrula which resembles syncytial delamination, and also has a syncytium at its origin (p. 16). Typical syncytial delamination goes one step further still: the cleavage process which proceeds it is from the first directed towards creating the syncytium. The sole instance of syncytial cleavage and syncytial delamination is the genus of athecate hydropolyps Eudendrium (Tichomiroff 1887; Hargitt 1904b; Congdon 1906; Kühn 1913; Mergner 1957; Wasserthal 1969).

In *Eudendrium racemosum* (fig. 16, Mergner 1957) the final stage of syncytial cleavage proceeds as follows: from the peripheral cleavage nuclei of the oral half accumulations of cytoplasm with a single nucleus form, spreading in the early pregastrula stage over the entire surface of the embryo to constitute the first, still syncytial, preectoderm. Syncytial delamination has already begun. These peripheral nucleus-plasma zones next separate off against the embryo interior, and the second preectoderm is formed. Only then do the mesogloea differentiate and cell boundaries develop in the outer zone. Unlike the outcome of superficial cleavage, however, this surface layer does not represent the whole blastoderm, but only the ectoderm, which thus arises by differentiation from a syncytium composed of nuclei and cytoplasm at the periphery, so that the outcome of syncytial delamination is a differentiation gastrula. The entoderm is built up only later from the nucleus-cytoplasm masses remaining in the interior of the embryo, while the gastrovascular cavity develops in the freed planula by the formation of a cleft (fig. 16b, 17a).

These events clearly show that syncytial cleavage and delamination must be firmly distinguished from the superficial mode of development, in spite of certain common features initially, and constitute a mode of development in their own right. In Eudendrium, too, cleavage and germ layer formation are in part concurrent, in which respect they also agree with the events of morula delamination: the overlapping of the two phases produces shortening of embryonic development. Finally, it must be pointed out that germ layer formation occurring by differentiation from a solid, undifferentiated mass of blastomeres (morula delamination) or from a syncytium (syncytial delamination) demonstrates the markedly indeterminative character of these modes of development and the considerable regulative powers of the embryos: whether a single cell or a nucleus-cytoplasm zone subsequently becomes an ectoderm or an entoderm cell can be determined only by conditions obtaining later such as their spatial relationship within the embryo.

(g) *Review of germ layer formation in Hydrozoa.* Unipolar ingression was found primarily in the free-swarming blastulae of most medusae, multipolar ingression and the forms of delamination derived from it, on the other hand, primarily in Hydrozoa with sessile gonophores and in few medusae which carry their embryos with them until the fully formed planula. Unipolar ingression must probably be regarded as the phylogenetically original form, while the other modes would appear to be derived and to have appeared only as the sexual individuals became sessile.

The capacity to form entoderm gradually spread from the vegetal pole over the whole blastoderm as primary delamination (multipolar ingression, mixed delamination, blastula delamination), whence all transitional forms lead to secondary delamination (morula delamination, syncytial delamination) as the most highly derived mode of germ layer formation. As thus formulated by Kühn (1913) the concepts of primary and secondary delamination assume a superordinate significance which raises them far above the merely descriptive designations of certain modes of entoderm formation.

1.1.1.4. Postembryonic development

With the exception of a number of Trachylina such as Liriope and Geryonia entoderm formation gives rise in almost all Hydrozoa to the creation within the embryo of a solid mass of cells separated from the ectoderm by a mesolamella secreted by the latter. The slit which is to become the gastrovascular cavity appears within it only very much later, and the entoderm cells then assume an epithelial arrangement around it. At this stage, however, the embryo already swarms about freely as a mouthless bipolar organism, as a planula larva, which is thus nothing other than a free-living gastrula. The mouth opening breaks through only during metamorphosis after the planula has attached.

In Hydrozoa, especially, the free-living stage is generally advanced to the ciliated blastula, which however exhibits no apparatus such as gland and sense cells which would distinguish it as a larva. The blastula appears merely as a means of prolonging the free-living stage, in order more readily to come to terms with the difference in environment of the polyp and medusa generations.

In some species there is only an embryonic planula stage and the swarming period of this larval form is suppressed, as is the case in Hydra (Kleinenberg 1872; Brauer 1891a), or embryonic development is prolonged by the direct passage of the embryo to the tentacle-bearing young form, the actinula, without a ciliated planula stage (see p. 21).

(a) *Typical postembryonic development through the planula.* The typical larval form of the Hydrozoa is the ciliated planula, which is able to swim and creep about, but the free life of which in species with sessile sexual forms is spent completely in the littoral zone, and therefore lasts for only a few hours or days. It swims on in front with its aboral pole, at which most gland and sense cells are located (fig. 17a), and generally attaches at this pole to undergo metamorphosis. Thus the oral pole, at which the definitive mouth is later to form, stands out freely in the environment. Among forms in which the entoderm develops by multipolar ingression and derived modes and by differentiation from a solid embryo, there is of course no longer any relationship between the polarity of germ layer formation and that of the planula.

Attachment and metamorphosis of the planula to the vegetative polyp stage may vary considerably: the simplest form is found in the athecate hydroids, in which the whole planula changes to the young polyp, accompanied by considerable growth, as in *Clava squamata* (fig. 18, Harm 1903) or *Cordylophora lacustris* (Morgenstern 1901). In Eudendriidae and Pennariidae the young polyp rapidly develops a distinct stem, the hydrocaulus, and polyp head, the hydranth, as in *Pennaria tiarella* (fig. 19, Hargitt 1900) and *Eudendrium racemosum* (fig. 17b–d, Mergner 1957, 1970 unpublished). Here the hydrocaulaus becomes the starting point of branching from the stem. The planulae of other athecates, on the other hand, first attach along their whole length; the oral pole then rears up and forms the primordium of the polyp, while the rest of the planula becomes the hydrorhiza, as for example in *Tiara pileata* (Metschnikoff 1886b). In *Turritopsis nutricola* (fig. 20, Brooks and Rittenhouse 1907) the whole planula, attached along its length, is first transformed into the hydrorhiza, and only then does the polyp bud arise from its centre. In *Oceania armata* (Metschnikoff 1886b), the hydrorhiza formed from the planula very soon branches out, and then often sprouts several polyp buds. Here, then, the planula appears to have completely lost its polarity on attaching: one may even speak of budding in the larval state.

Even more multiform is metamorphosis in the thecate hydroids, in which the larvae often pass through a series of regular changes of form. Starting from the aboral pole at which they are attached, they first flatten to a disc which becomes surrounded by periderm. Out of their centre a bud then arises, the apical entoderm of which contains a rich supply of yolk substances. During this time the lobed edges of the pedal disk grow out to rhizostolons. The bud may develop to a hydrocaulus with a primary terminal polyp, as in *Gonothyraea loveni* (fig. 21, Wulfert 1902) or to an axocaulus with a terminal growing point as in *Plumularia echinulata* (fig. 22, Kühn 1909). Larvae undergoing such metamorphosis are always large and rich in yolk, possibly because the sprouting of hydranths which can take care of their nutrition occurs only at a late stage.

Thus planulae can here be separated into those that represent the young stage of a primary polyp, which then produces the animal colony by vegetative propagation, and those which are the origin of an organism of more complex composition (Kühn 1913). These latter are then not really 'polyp larvae' any more and their oral pole no longer bears any relation to that of the polyp. The polarity of the planulae of some thecates such as *Antennularia antennina* (Strohl 1907), is probably even lost when they attach: for they occasionally sprout two axocauli from a single pedal disk. In others such as *Mitrocoma annae* (fig. 23, Metschnikoff 1886b), again, the planula often attaches along its whole side. Then the posterior end rears up to form the primordium of the polyp, or the polyp forms at the side of the highly elongated larva, which becomes the hydrorhiza. Occasionally several polyps grow simultaneously from a single planula (fig. 23a).

The further development of athecate and thecate polyp colonies, which belongs in the realm of asexual reproduction and thus does not concern us here, is extraordinarily variable and often very complicated (Kühn 1913).

(b) *Postembryonic development through the actinula.* Not infrequently the planula occurs only as an embryonic stage and the first free stage to come about is the tentaculate actinula (Allman 1875), which thus does not represent a phyletic stage as does the planula, but is rather a swollen, medusa-like young stage of the polyp. It is thus not a true larval stage; the planula is merely passed over and embryonic development thereby prolonged.

The classic example of postembryonic development by way of an actinula is the athecate hydropolyp *Tubularia mesembryanthemum* (fig. 24a, Van Beneden 1849; Ciamician 1879; Brauer 1891b): after radial cleavage and a delay in blastomere differentiation through the intercalation of a syncytial stage during early cleavage (see p. 6), entoderm is formed as early as in the 30-cell blastula by mixed delamination (see p. 13). Then after the two germ layers have separated by the proliferation forward of first the ectoderm and then the entoderm, the first two definitive tentacles are formed, and after development of the gastrovascular cavity further tentacles. The embryo expands along its main animal-vegetal axis and the mouth opening breaks through. The finished actinula now leaves the sessile gonophore of the maternal animal and settles down.

In *Myriothela phrygia* (fig. 24b, Allman 1875), on the other hand, larval tentacles are first formed on the actinula, and these are resorbed after fixation, while the definitive tentacles already bud at the swarming period of the actinula. On the manubrium of the anthomedusa *Margelopsis haeckeli* (fig. 25, Hartlaub 1899; Werner 1955), too, the embryo develops from the fertilised egg to the actinula, which becomes free and grows to a pelagic polyp, on which the medusa generation is formed by budding. In contrast to this the actinula-like polyps of *Corymorpha palma* and *C. nutans* (fig. 26, Torrey 1907; Hartlaub 1907) develop from eggs which are shed and undergo their development on the bottom. The young polyps attach by means of the rhizostolons peculiar to them and at an early stage form the characteristic peripheral hydrocaulus canals around the central parenchyma.

The planula stage is lacking in a number of Trachylina, too, and the double-layered embryo is transformed directly into the actinula, as in the trachymedusa *Liriope mucronata* (fig. 27, Brooks 1886; Metschnikoff 1886b): here the entoderm vesicle formed by typical mixed delamination (see p. 13) is separated from the ectoderm by a layer of jelly. As the jelly increases in quantity the ectoderm expands considerably and the entoderm vesicle is pressed against the ectoderm at the oral side. The first tentacles, the gastric cavity and the mouth later form where they meet (fig. 10d, 27). Thus an actinula-like young stage is formed which by flattening, deepening of the subumbrella and formation of the canal system and definitive tentacles becomes the young medusa.

(c) *Postembryonic development through the planula and actinula.* In other Trachylina, on the other hand, the planula stage is retained; this first gives rise to an actinula, or at least a state closely resembling it, which then changes directly into the young medusa, without an asexual reproductive stage having occurred at any time. Cases in point are the trachymedusae Aglaura and Rhopalonema (fig. 28, Metschnikoff 1886b), of which the double-layered embryos are formed by morula delamination. Four tentacles at first sprout in the region of the oral pole of the planula. Thus the actinula is formed, which then by sprouting of further tentacles changes directly into the young medusa.

In the narcomedusa *Aeginopsis mediterranea* (fig. 30, Metschnikoff 1886b; Salensky 1911), the two-layered embryo formed by typical multipolar ingression (see p. 12) expands greatly along its length to a solid planula. Its two ends bend downwards and thus become the first two tentacles of the actinula, while its middle piece forms the body, from a split in which the gastric cavity is formed, while the mouth breaks through at the oral end and two more tentacles sprout to form a cross with the first two. Again the actinula passes directly into the young medusa by the addition of more tentacles, the velum, sense organs, and a jelly layer, and by the infolding of the subumbrella.

Finally, actinula-like polypoid stages must be mentioned which become the starting point of larval budding processes and thus give rise to a rapid succession of generations. They play an important part above all in the reproductive cycle of the parasitic narcomedusa family Cuninidae, such as *Cunoctantha octonaria* (Maas 1892), *Cunina parasitica* and *Pegantha smaragdina* (Bigelow 1909). These parasitic actinulae produce a stolon prolifer from which medusae already bud off, even before the maternal actinula undergoes transformation to a medusa. In *Cunina parasitica* the actinula even forgoes its own further development and becomes a 'spike' bearing numerous medusa buds, and in *Pegantha smaragdina* (fig. 29) it buds nothing but actinulae as long as it remains in the gastric cavity of the maternal medusa (for details see the authors quoted).

(d) *Postembryonic development of the Siphonophora.* To a far greater extent than in the hydroid colony the individuals of the Siphonophora colony undergo transformation and regression by division of labour and polymorphism. This is also to be clearly seen in their development, in that they arise on the young planula almost like the rudiments of organs. While in Trachylina the potency of asexual reproduction is present at the actinula stage of development (see p. 22), here already the planula has become the starting point of an exclusively asexual postembryonic development. We thus have to deal only with its beginnings.

Ectodermal thickenings in the young planula already indicate the budding zones of the individual persons. In Calycophoridae such as Galeolaria and Muggiaea (fig. 31, Metschnikoff 1874; Chun 1886) the planula develops the first swimming bells and tentacles from a lateral budding zone, while its posterior end becomes the primary polyp. Thus the siphonula is formed as young stage, from the budding zone – which has remained embryonic – of which new groups of individuals, or cormidia, are continually put forth, the totality of which, together with the stem, form the siphonophore colony. The planula of Physophoridae, on the other hand, has two budding zones, one at the aboral end for the development of the pneumatophore and a second on the ventral side for the formation of the stem, which starts with the first larval tentacle, while the oral end becomes the first nutritive polyp. Thus in *Halistemma picta* (fig. 32, Chun 1886), while in *Agalma sarsii* (fig. 33, Woltereck 1905) metamorphosis begins with the formation of the primary larval bract, which persists until the pneumatophore has become functional.

1.1.2. Scyphozoa

After the detailed description of the embryonic and postembryonic development of Hydrozoa it will be sufficient to indicate the chief lines of development of Scyphozoa and Anthozoa and to deal above all with their divergent modes of cleavage and germ layer formation.

Scyphozoa, which occur primarily as medusae and more rarely as polyps (scyphostoma), generally exhibit alternation of generations in a manner similar to Hydrozoa, but with a very different organisation. For asexual reproduction – when it gives rise to the formation of medusa – normally occurs by terminal budding and transverse separation at the oral pole of the scyphostoma in the form of strobilation. The sexual phase of normogenesis with which the chapter deals is distinguished by the developmental processes from the fertilised egg cell through embryonic and postembryonic development until metamorphosis of the planula into the scyphostoma. The succeeding events will be dealt with only summarily. The life-history of Scyphozoa is exemplified by the discomedusa *Aurelia aurita*. In addition, however, exceptions and deviations from this developmental process will be discussed.

1.1.2.1. Egg stage

The eggs of *Aurelia aurita*, which are poor to moderately rich in yolk, pass out of the gonad, a fold of entoderm of the gastric cavity, into the genital sinus situated below it, and thence into the gastric cavity itself, where they are fertilised. They then wander on through the mouth to the oral arms, in the folds of which they stick fast and undergo their embryonic development until the planula larva stage.

1.1.2.2. Cleavage

In the typical case cleavage of Scyphozoa is holoblastic, adequal and radial, culminating in a spherical coeloblastula with a narrow blastocoele (fig. 34a, Hein 1900). In the semaeostomes *Pelagia perla* (Delap 1905) and *Chrysaora hyoscella* (fig. 35d, Teissier 1929, on the other hand, a large coeloblastula is formed, and in the lucernariide *Haliclystus octoradius* no blastocoele is formed at any time, but a solid spherical embryo (fig. 38a, Wietrzykowski 1912). In various species of Aurelia, in *Thaumatoscyphus distinctus* (fig. 37b, Hanaoka 1934) and in Lucernaria pseudospiral cleavage (p. 5) has also been seen in addition to the typical radial cleavage.

Often however, the original radial mode of cleavage is modified in the sense that blastoderm cells already immigrate into the blastocoele during the cleavage stage, as occurs in various species of Aurelia (fig. 34a), the semaeostomes *Chrysaora hyoscella* (fig. 35c), *Cyanea capillata* and the rhizostomes Cotylorhiza and *Mastigias papua*. This produces the picture of morula delamination, in which the inwandered cells represent prospective entoderm. In the species named above, however, cells generally degenerate before entoderm formation (Smith 1891; Hyde 1894; Hein 1900; Okada 1927b), which always occurs by invagination in the species in question. It is open to conjecture whether this phenomenon represents the first step in multiphasic entoderm formation or the provision of abortive nutritive material which cannot be assigned to a given germ layer (Siewing 1969). Similar phenomena are also seen in Anthozoa (see p. 29).

1.1.2.3. Germ layer formation

Like their cleavage, germ layer formation in Scyphozoa shows a certain uniformity in contrast to the multiple modes seen in Hydrozoa. The only variations result from slight differences in the yolk content of the embryos. Essentially, entoderm formation comes about by invagination (emboly).

Unlike in Hydrozoa, with their various modes of ingression and delamination, in which a primitive mouth is never formed before the planula metamorphoses, one is formed from the beginning at the time of invagination and becomes temporarily closed again only as a secondary phenomenon.

Invagination of the entoderm almost always begins in a coeloblastula, the typical case being that of *Aurelia aurita* (fig. 34a, b, Hein 1900). The vegetal pole of the blastula gradually folds inwards as a coherent blastoderm layer and in the extreme case comes to lie against the parts of the blastoderm which have not been invaginated. In this case we also speak of emboly and of its result as an embolic gastrula. Thus a double-layered embryo is produced the outer layer of which represents the ectoderm and the inner layer the entoderm. The blastocoele is gradually narrowed in the process or, in the case of emboly, completely obliterated. The invaginated cavity is termed the archenteron cavity, its epithelium archenteron and its apertue connecting it with the exterior the blastopore. The two-layered embryo is a gastrula in the narrow sense (fig. 34b); the embryonic phase which gives rise to it is gastrulation. Its mode must be regarded as the phylogenetically most primitive, from which others such as polar ingression have been derived.

We have seen (p. 24) that during cleavage individual blastoderm cells immigrate at an early stage into the blastocoele but always disappear before the formation of the archenteron and therefore probably have nothing to do with entoderm formation. As a result the process of invagination described above is typical for the majority of Scyphozoa, as for instance the semaeostomes *Cyanea capillata*, *Aurelia flavidula*, and *Pelagia noctiluca*, the rhizostomes Cotylorhiza and *Mastigias papua* and the coronate Nausithoë (Goette 1893; Smith 1891; Hein 1903; Uchida 1926).

In *Chrysaora hyoscella* (fig. 35b–e, Hadži 1907; Teissier 1929; Mergner 1970 unpublished) very many more cells enter the blastocoele and may even obliterate it for a time. But after their dissolution a typical coeloblastula is nevertheless still formed, in which invagination takes place. In the coronate *Linuche unguicula* (fig. 36, Conklin 1909) the otherwise extremely typical invagination is accompanied concurrently by polar ingression, cells sloughing off from the roof of the archenteron and wandering into the already narrowed blastocoele. Other Scyphozoa, again, show a more or less marked degree of invagination, but at least the formation of a plug of cells penetrating inwards from the oral pole and hollowing out again to become the archenteron. The primitive mouth may narrow from the outset or only subsequently as in *Pelagia noctiluca*.

In the stauromedusae *Thaumatoscyphus distinctus* (fig. 37, Hanaoka 1934) and *Haliclystus octoradius* (fig. 38, Wietrzykowski 1912) germ layer formation deviates markedly from the modes hitherto described. At no time during the cleavage period does a blastocoele appear and at the end of its primitive development a solid spherical embryo has arisen through the polar ingression of a few but larger cells, filling the interior of the embryo with prospective entoderm cells. In *Aurelia marginalis* typical multipolar ingression is said to occur as an exception, and in the cubomedusa *Charbydaea rastonii* (Okada 1927a) even as the rule.

To summarise, therefore, Scyphozoa almost invariably exhibit radial cleavage and undergo invagination of the coeloblastula. All other modes of cleavage and germ layer formation are ultimately but variations on this normal theme. Thus embryonic development in Scyphozoa stands on a much more primitive level than in Hydrozoa, in which no case of invagination is known.

1.1.2.4. Postembryonic development

The two-layered embryo of *Aurelia aurita* (fig. 34c–g, Hein 1900; Friedemann 1902) expands to an oval shape while the primitive mouth is still closing, becomes ciliated and changes to a free-swimming planula. After attachment at the aboral pole the free oral end broadens out and while entoderm cells are actively multiplied, the mouth cone, on which the definitive mouth breaks through, is formed. The first four perradial primary tentacles bud off at the edge of the peristome and the four perradial gastric pouches bend out from the entodermal gastric wall, their adjacent lateral walls forming the four interradial septa, which continue as taenioles until the pedal disk. The number of gastric pouches later increases to eight and that of the tentacles to sixteen. The septal funnel and septal musculature, the septal ostia and gastric filaments appear and the outer form of the young scyphostoma is completed by development of the stalk, expansion of the gastric region and deepening of the peristome. The scyphostoma (fig. 34h, i) now closely resembles a young scyphomedusa and is ready to produce medusae by strobilation. The postembryonic development of the cubomedusae *Charybdaea rastonii* and Tripedalia (fig. 40, Okada 1927a) follows a similar course.

With strobilation, i.e. the constricting off of ephyra larvae, as they are called, by one or several transverse constrictions of the scyphostoma (monodisk or polydisk strobila, fig. 34k, 35f), the asexual phase of metagenesis begins; it can only be dealt with briefly here. The ephyra (fig. 34l) serves the first propagation of the species; through a number of transformation processes it gradually changes to a young scyphomedusa, which on reaching sexual maturity once again initiates the alternation of generations. We thus have an alternation between a sexual and an asexual generation each with a metamorphosis.

In addition to this typical postembryonic development there are however also aberrant forms, in which one of the generations, either that of the medusa or the polyp, is omitted. Thus the permanently sessile Stauromedusae exhibit no free medusae and undergo only one metamorphosis, i.e. that of the planula to the scyphopolyp. In *Thaumatoscyphus distinctus* (fig. 37, Hanaoka 1934) and *Haliclystus octoradius* (fig. 38, Wietrzykowski 1912) the solid embryo formed by unipolar ingression of a few but larger cells (pp. 25 and 26) expands to a planula with cuboid ectoderm and a solid column composed of a single row of large vacuolated entoderm cells. This unciliated planula is capable only of creeping about, can bud off four more tentacle-like larvae, which behave like sexually-produced planulae, and attaches at its aboral pole. A normal gastric cavity is formed only during metamorphosis; the goblet shape, stalk and pedal disk develop and while the sixteen tentacles grow into tentacle clusters, the peristome hollows out into a kind of subumbrella. The whole process possesses definite features of scyphomedusa formation, comparable, say, to the formation of a monodisk strobila, in which the separation process is suppressed and the medusa arising out of the scyphostoma remains sessile, although of course this cannot be demonstrated phylogenetically.

In *Pelagia perla* (fig. 39, Delap 1905), on the other hand, it is the polyp generation which is lacking and the planula metamorphoses into the medusa. The first step is the formation of a mouth cone at the oral pole and around it a flat peristome, the edge of which notches inwards and develops eight lobes. The main animal-vegetal axis then shortens and the larva becomes disk-shaped. Thus an ephyra is formed, which in a similar manner to Aurelia changes into a medusa, which thus develops from the planula without a true polyp state and without strobilation, i.e. without alternation of generations. Such behaviour occurs abnormally in the scyphostoma of *Aurelia aurita* in unfavourable living conditions. Strobilation is then suppressed, and the whole scyphostoma becomes a sedentary or free-swimming medusa, an 'ephyra pedunculata' (Haeckel 1881). Here there is thus neither asexual reproduction by strobilation nor alternation of generations.

1.1.3. Anthozoa

The limitation of the number of modes of cleavage and germ layer formation to only a few, already discernible in Scyphozoa, is also seen in the embryonic development of Anthozoa. Since the latter also lack the medusa form, their postembryonic development should also proceed more simply. It is nevertheless complicated again by the special modes of septa, skeleton and colony formation, which as asexual processes cannot of course be discussed here.

1.1.3.1. Egg stage

Early development in Anthozoa still takes place in part within the maternal gastric cavity, where, unless they are shed beforehand, the eggs are fertilized. This gives rise to the difference in the structure of the egg shells: eggs which are shed at an early stage, such as those of Urticina, Anemonia and Actinia, generally possess a thick, often prickly integument, while eggs remaining within the shelter of the mother animal have only a delicate oolemma. They generally develop within the parent until the ciliated planula stage, often even until they possess a complete set of tentacles. Many larvae immediately reattach to the outside wall of the maternal body and continue their development there. Sometimes veritable brood chambers sink in around them (Carlgren 1901).

The eggs of many Octocorallia such as Alcyonium, for example, and of several Hexacorallia such as Bolocera and Urticina are rich in yolk content and of large size, while others such as those of the Hexacorallia Metridium and Edwardsia are poor in yolk and small. Cleavage and germ layer formation proceed accordingly.

1.1.3.2. Cleavage

The mode of cleavage of eggs poor in yolk such as those of the actinians *Metridium marginatum* (McMurrich 1891) and Edwardsia is generally total, equal, and radial, while in the actinian *Sagartia troglodytes* (fig. 42a, b; Nyholm 1943) and the cerianthid *Pachycerianthus multiplicatus* (fig. 41a–h, Nyholm 1943) radial cleavage is modified as pseudospiral cleavage (p. 5). The yolk content of the latter was already greater, and as it further increases phenomena occur which must again be interpreted as a delay in blastomere differentiation: at the onset of cleavage only the nuclei divide, but not, at this stage, the cytoplasm. Only later, at the 16-cell stage, do cell boundaries suddenly reappear, with subsequent normal holoblastic cleavage. Examples are the alcyonarian Alcyonium (Kowalevsky and Marion 1883) and the pennatularian *Renilla reniformis* (Wilson 1884), the yolky eggs of which show a certain tendency towards superficial cleavage, but in which cleavage is generally holoblastic. The actinians *Urticina crassicornis* (Appellöf 1900) and *Bolocera turdiae* (Gemmill 1922) also show such a delay in blastomere differentiation until the 16- or 24-nucleus stage, and *Sagartia troglodytes* up to the 4-nucleus stage only.

Yolky eggs thus show a mode of cleavage in which syncytial stages can alternate with periods of total cleavage. This tendency towards superficial cleavage, as already mentioned, is especially marked in *Alcyonium digitatum*, which, in addition to total cleavage, shows 'partially syncytial' cleavage (see p. 8), either first total and then superficial or vice versa. That these modes of cleavage have nothing to do with the 'syncytial cleavage' of the hydroid genus Eudendrium (p. 9, fig. 6; and Mergner 1957) is illustrated by the markedly typical superficial cleavage in the ceriantid *Cerianthus lloydii* (fig. 45a, b; Nyholm 1943). Here too, cell boundaries are no longer found during cleavage, but unlike in Eudendrium all cleavage nuclei wander to the periphery of the embryo and cleave it, while the abundant yolk within remains undivided. The moderately yolky eggs of *Cerianthus lloydii* thus undergo true superficial cleavage, which as in many arthropods is derived directly from holoblastic or even from radial cleavage.

1.1.3.3. Germ layer formation

Yolk content and mode of cleavage naturally have a significant influence on the course of germ layer formation. Especially the release of yolk into the blastocoele, which is an extremely marked feature of a number of Anthozoa, beginning during the cleavage period itself, must be seen as related to germ layer formation (Siewing 1969). Rather as in Scyphozoa this often blurs the borderline between cleavage and entoderm formation. But while in Scyphozoa the blastocoele is invaded by whole cells, the precise significance of which is still not clear (see p. 24), the situation in Anthozoa is unequivocal: here the blastoderm cells give off only fragments without a nucleus into the generally small blastocoele, components, that is, which no longer participate in the cleavage process. This release of nutritive material is evidently intended to facilitate the subsequent course of development. It is certainly striking that the generally fairly yolky eggs of Anthozoa as a rule present holoblastic cleavage and that even after superficial cleavage entoderm formation is still by invagination. The nutritive substances passed into the blastocoele are not lost, but are transferred into the archenteron in the course of germ layer formation.

The release of nutrients and the 'delaminations' of cell fragments into the blastocoele, as seen, for example, in *Metridium marginatum* (fig. 44, McMurrich 1891), often obscures the boundary line between the blastoderm and the blastocoele. This may also have given rise to the reports of unipolar and multipolar ingression, which however still require confirmation. While all these processes are indeed related to the further development of the embryo, they should nevertheless not be interpreted as entoderm formation as such, but only as preparation for it. With few exceptions, in fact, germ layer formation comes about only by invagination.

These exceptions are found above all in Alcyonaria, the yolky eggs of which generally show holoblastic cleavage (p. 28) in spite of the occurrence of syncytial stages of limited duration. The product of cleavage, as for example in *Sympodium coralloides* (fig. 47a–d; Kowalevsky and Marion 1883), is then a solid embryo, a sterroblastula, the peripheral cells of which can be distinguished as ectoderm from the prospective entoderm in the interior of the embryo. The process of entoderm formation must thus be described as morula delamination. Soon after, by active cell divisions, rearrangement of cells and dissolution of the innermost cells, a slit-shaped gastric cavity begins to form, around which the entoderm becomes arranged epithelially. The embryo finally expands and becomes a ciliated planula, still receiving disintegrated cell material in its gastric cavity as nutrient.

In most Anthozoa, however, the product of cleavage is a typical coeloblastula with a blastocoele more or less narrow according to the amount of yolk present. Entoderm formation basically occurs by invagination independently of the nutrient content. It has in fact even been observed in numerous actinians and cerianthids whose eggs vary considerably in their yolk content and undergo a great variety of modes of cleavage, e.g. in Adamsia (Gemmill 1920), Sagartia (fig. 42c; Faurot 1903, 1907), *Epiactis prolifera* (fig. 49a, Uchida and Iwata 1954), Actinia (Cary 1910), Metridium (fig. 44c; McMurrich 1891), Edwardsia (Wietrzykowski 1914), Urticina (fig. 43; Appellöf 1900), Bolocera (fig.48a; Gemmill 1922), Halcampa (fig. 46b), Cerianthus (fig. 45b, c) and Pachycerianthus (fig. 41i, k; Nyholm 1943, 1949). Of course in such different circumstances invagination cannot occur in a completely identical manner, or always give rise to an identical invagination gastrula. Certainly, a typical feature is the turning in of the vegetal blastoderm to a well-developed archenteron with a wide lumen, generally filled with nutrient substance. Occasionally, however, the marked filling of the blastocoele with yolk substances appears to produce particular malformations of the invaginating archenteron, to cause it to loosen its epithelial coherence, or even to bring about a transition to polar ingression. The product of these deviant modes is nevertheless always an invagination gastrula, the entoderm of which is closely appressed against the outer ectoderm and which may thus be described as an embolic gastrula. Discussion of these deviant forms of entoderm formation follows.

Cleavage of the isolecithal eggs of *Pachycerianthus multiplicatus* (fig. 41a–h and p. 28; Nyholm 1943) is total, adequal and radial until the 256-cell stage, occurring as pseudospiral cleavage at the start. The blastomeres give off yolk into the blastocoele even before the blastula stage is reached; after this they are considerably smaller and their boundaries against the central yolk mass are unsharp. Only the cells of the flattened vegetal pole are still markedly larger. At first a circular furrow cuts in at the edge of this large-celled vegetal plate, while invagination is delayed at its centre by the masses of yolk (fig. 41i, k). Then however the whole yolk gradually passes from the blastocoele, which by then is already greatly constricted, into the invaginating archenteron through its intercellular interstices, so that the entoderm also comes to lie against the ectoderm at the centre. Thus while intercellular spaces allow the yolk to pass, the invaginating cells temporarily lose their epithelial connection (Siewing 1969). Invagination also occurs in a similar fashion in the actinians *Urticina crassicornis*, *Bolocera turdiae*, *Actinia equina*, *A. bermudensis*, *Metridium marginatum* and *Sagartia troglodytes*.

Cerianthus lloydii undergoes typical superficial cleavage (fig. 45a, b; p. 29; Nyholm 1943). At its termination the embryo shows superficial cleavage of the periplasm and a central uncleaved yolk mass within. It then flattens somewhat in the direction of the vegetal pole, where in spite of the large amount of yolk it gradually begins to bend in (fig. 45b, c). Invagination continues until a gastrula is produced, the inner cell layer of which separates off as entoderm from the outer, the ectoderm. Here too the yolk passes into the archenteron.

The greatest modification in the mode of invagination is seen in the actinian *Halcampa duodecimcirrata* (fig. 46; Nyholm 1949). After superficial cleavage, whereby the embryo is divided into a layer of peripheral blastoderm cells and a central uncleaved yolk mass, entoderm formation begins at the vegetal pole, initially with invagination. Then however, cells wander from the blastoderm into the blastocoele, become arranged as epithelium and come into contact with the invaginated entoderm epithelium, i.e. with the incipient archenteron, which in this mode also provides the greater part of the entoderm. The initial invagination is soon transformed into ingression. Entoderm formation in Halcampa thus proceeds in two morphogenetic phases, first as invagination and then as ingression.

In conclusion it may be stated that apart from a few exceptions, entoderm formation occurs in Anthozoa by invagination (emboly), which is retained, albeit in altered form, even in yolky eggs and following superficial cleavage.

1.1.3.4. Postembryonic development

The two-layered embryo of Anthozoa becomes covered with cilia, expands along the polar axis and becomes a free-swimming planula larva (see p. 28 for brooding arrangements). After a swarming period which varies from species to species the larva attaches at its aboral pole and becomes shortened once again. In the metamorphosis thereby initiated the primitive mouth can persist as definitive mouth and soon, with the invagination of the ectodermal pharyngeal tube, is deepened into a pharyngeal orifice. In some Hexacorallia and Octocorallia such as the alcyonarian *Sympodium coralloides* (fig. 47c–e; Kowalevsky and Marion, 1883) and the pennatulid *Renilla reniformis* (fig. 50; Wilson, 1884) the primitive mouth first closes. Only then does the pharyngeal tube bend in, at the inner end of which the definitive mouth breaks through.

Very often, and in Hexacorallia virtually always, the formation of mouth and pharyngeal tube, as well as that of tentacles and septa, which begins at about this time, takes place before the larva attaches. Secondary larvae, as they are called, i.e. larvae which are highly advanced in their development, and which are very characteristic for Hexacorallia, are then produced. Various forms are distinguished by the level of their tentacle and septa development as well as by special motile organells such as cilia, e.g. the cerianthid larvae Cerianthula (fig. 53a; Carlgren 1924) Calpanthula (fig. 53b; Van Beneden 1898; Carlgren 1924), Anactinia (fig. 53c; Carlgren 1924) and Arachnactis (Sars 1846; Agassiz 1862; fig. 45d, Nyholm 1943; fig. 45e and 53d; Van Beneden 1891; fig. 53e; Carlgren 1924) and the zoanthid larvae Zoanthella and Zoanthina (fig. 53g–i, Conklin 1909) and others. They often have a long pelagic life, the duration of which is generally characterised by the number of tentacles formed by the time metamorphosis supervenes and which may occasionally last until sexual maturity, as in *Arachnactis albida*. With such advanced development of these larval forms metamorphosis into sedentary polyps is generally relatively simple.

In the Octocorallia eight entodermal mesenteric septa are always formed as folds of the gastric wall (fig. 50a, 51) and in Hexacorallia first one pair of septa is formed and then three further pairs (fig. 45e, 48c, d). The fact that Hexacorallia always develop only eight septa at first led to their being derived from the Octocorallia. This 8-septa stage is termed the Edwardsia stage (fig. 49b, 53f); with the addition of two further pairs it becomes the Halcampula stage (fig. 49c). After these twelve primary septa of the first cycle the next highest cycles are formed, but always in a multiple of six (fig. 49d). The tentacles appear in a similar order over the fans formed by the septa, the primary tentacles developing more rapidly than the others (fig. 52, 53e). After settling of the larva and further differentiation processes the primary polyp passes to asexual reproduction in stem-producing and colonial forms, but this cannot be discussed here.

1.2. Experimental analysis

As the preceding section has shown, we now possess considerable knowledge of the whole morphologically discernible ontogenesis of Cnidaria. On the other hand, results of causal analysis of the ontogenesis of this animal phylum are very rare. The chief reason for this may be seen in the fact that the problems posed by the 'Entwicklungsmechanik' (developmental physiology) are so different from those of morphological embryology that a coherent approach is not generally possible. Further, experimental analysis of the vegetative processes in the development of Cnidaria was early recognised to be a field of study which could more readily be mastered.

Consequently, a great many investigations have been devoted to budding processes and bud growth, colony and pattern formation, the development of sexual individuals and asexual reproductive organs such as stolons, frustules, etc. Further, numerous reports have appeared on regeneration and reparation in Cnidaria, complementing one another systematically and providing a good insight into the great ability of this group of animals to ensure its preservation by asexual means. A further reason for the scarcity of modern studies on the developmental physiology of Cnidaria is to be found in the shifting of interest to physiological mechanisms. Thus we are already familiar with a large number of the processes of feeding biology and metabolic physiology, the sensory and motor physiology and the various modes of behaviour.

Lastly, one is struck by the fact that in all these fields of study species of Hydra are much preferred to marine Cnidaria on account of their ready accessibility and ease of cultivation. At a conservative estimate, at least 80% of all the experimental work on Cnidaria has been carried out with Hydra, which as a fresh-water inhabitant has no place in our considerations.

Only a small number of publications are thus available on which a review of the findings of experimental analytical research on the sexual development phase of marine Cnidaria can be based. To make matters worse, they generally show no inner coherence in tackling the problems or even a systematic form, such as we find, for example, in the analysis of the development of sea urchins or ascidians.

1.2.1. The egg of Cnidaria

Nevertheless, some aspects at least of the early development of marine Cnidaria have been the object of causal analysis, above all the polarity of the embryo from the egg state to germ layer formation, determination and regulative ability in embryonic development, and the process of cleavage. The discussion below will thus primarily be devoted to these aspects.

1.2.1.1. Origin and migration of germ cells

With few exceptions (e.g. Coryne, Weismann 1883; Goette 1907), the youngest egg stages of Hydrozoa are found in the ectoderm, and those of Scyphozoa and Anthozoa always in the entoderm. But while the egg cells of hydromedusae, Scyphozoa and Anthozoa generally arise close to their site of maturation, in hydroids with sessile gonophores the sites of germination and maturation of the eggs are situated at some distance from one another. The germinal zones are almost always situated close below the growth zones of the main axis and the lateral branches and move upwards constantly as these grow. During the reproductive period they provide a continuous supply of egg cells, which soon begin their migration to the maturation zones in the blastostyles. The laws governing this migration have been analysed and confirmed statistically in *Eudendrium racemosum* by Mergner (1957):

From these studies it emerges that in general there is an overproduction of egg cells in the ratio of 6–7 for each egg which actually undergoes development. A minority of the excess eggs remain in the deeper regions of the stalk and no longer develop, but some wander into blastostyles which are already filled with egg cells and there degenerate, possibly serving as food for ripening eggs. This also applies when a second wreath of gonophores, the sac-like outgrowths of which are virtually never filled up by the excess egg cells, is laid down above the first. The reasons for this are unknown, but they may lie in reduction of the blastocyle head, already beginning at this time.

Apart from these eggs which subsequently wander into the blastostyle and degenerate there, true misdirections are a very rare occurrence. The egg cells wandering towards the periphery hardly ever penetrate to the growth zone located under the hydranth head; it seems that the zone exerts a physiological inhibitory effect. This is also supported by the fact that the eggs wander into the young lateral hydranths only once the hydranth stalk is well-developed, so that the growth zone has already shifted upwards. This inhibition need not be complete, as Mergner's observation (1957) that of 5,000 egg cells studied one large migrant egg was found in the basal entoderm of a hydranth goes to show. Weismann (1883) saw even two gonophores in a hydranth head with egg cells which had immigrated by way of the entoderm, a highly exceptional finding. In the ectoderm of the hydranth, on the other hand, they were regularly degenerated. No inhibitory effect can be present during blastostyle growth since here the young buds are immediately filled with egg cells.

1.2.1.2. Growth and nutrition of egg cells

In *Eudendrium racemosum* the nutrition of the young oocytes in the germinal zones presumably proceeds from the terminal hydranth of the branch in question. According to Mergner (1957) this is demonstrated by the fact that the germinal zone of a branch of which the head is cut off develops only very poorly. The secondary germinal zones in the stalks of lateral hydranths, on the other hand, can undergo premature development.

During their migration to the maturation centres the young oocytes grow to several times their original size. At first they are still nourished from the entoderm, even when in the ectoderm. The mesolamella between the two body layers has in fact fine perforations, visible by light microscope, which apparently permit an unhindered exchange of substances and thus the transport of nutrients to the ectoderm (Mergner 1957). Using the electron microscope Hanisch (1970) has even demonstrated an immediate connection between the two layers by way of fine plasma projections penetrating these perforations. In the further course of their development the egg cells of *Eudendrium racemosum* nevertheless pass into the entoderm, where they are nourished by its gastrodermic cells. The eggs of *Eudendrium armatum* (Wasserthal 1969) possibly even pass through a veritable stationary phase here, in which they can take up dissolved nutrient materials straight from the hydrocaulus lumen. They can also take up ordinary adjacent cells from the surrounding tissue during their journey, as Congdon (1906) already reported for the large migrant eggs of *Eudendrium hargitti*.

Nutrition of the Eudendrium egg lying in the gonophore, i.e. its site of maturation, on the other hand, is always by way of the spadix entoderm, although material from the excess eggs which have degenerated in the blastostyle may be incorporated. The spadix, which may surround the egg in a variety of ways, degenerates as soon as the egg has attained its greatest volume at the start of cleavage, has completed storing yolk and has developed the embryonic shell, and is thus no longer fed from the hydrocaulus. This proves that the spadix is in fact a nutritive tube.

In *Eudendrium armatum* Wasserthal (1969) distinguishes three modes of nutrition: (1) transferal of nutrients from adjacent cells or through perforations in the mesolamella from the gastrodermis, as occurs in the hydrocaulus ectoderm or in the gonophore; (2) incorporation of adjacent cells, as during migration in the ectoderm and the blastostyle gastrodermis; and (3) uptake of predigested nutrients from the hydrocaulus lumen during the stationary phase in the gastrodermis. In *Aglaophenia helleri* (Müller-Calé and Krüger 1913b) the young migrant eggs are endowed with symbiontic zooxanthellae from the entoderm, while in *Halecium ophiodes* (Hadži 1911) only the eggs lying in the gonophore are so endowed. Another form of nutrition of the eggs, fusion with adjacent alimentary egg cells, is widely distributed among Hydrozoa. It has been observed, inter alia, in *Pennaria tiarella* (Smallwood 1899 and others), in the gonophores of which the eggs are formed and reach maturity.

Cowden (1964) analysed the cytochemical events in the course of the growth of eggs of *Pennaria tiarella* in respect of the distribution of nucleic acids, proteins and mucopolysaccharides. During early growth the ribonucleic acid and protein content of the oocytes is at first high. The RNA content then gradually diminishes, and carbohydrates appear in the cytoplasm. In the further course of development a few oocytes spurt in their growth by taking up adjacent primary oocytes and undergo the first meiotic division. The number of oocytes is thereby reduced from several thousand to less than ten. In accordance with this process the few eggs reaching maturity contain a large amount of proteins and carbohydrates but little RNA. Basophilic granules are scattered only over the future animal half, which is directed inwards.

According to Mergner (1957) and Hanisch (1970), in *Eudendrium racemosum*, too, the cytoplasm of very young oocytes and spermatocytes in the germinal beds contains various inclusions, which disappear again after the germ cells begin to migrate, and which possibly represent concentrations of RNA. Once the egg reaches its site of maturation, the gonophore, and is fed via the spadix, however, granules appear from the vegetal pole at the same time as the storage and structuring of the yolk and gradually spread over the whole yolk. When the latter is subsequently used up they resist being dissolved and are still detectable at the planula stage. They appear completely unaffected even after six days of treatment with RNA solvents such as ribonuclease or perchloric acid, $HClO_4$. They may be breakdown products of metabolism which are deposited prematurely, while new yolk substance is still continually being formed.

1.2.2. Polarity and regulation in embryonic development

The question of the polarity of the embryo is closely associated on the one hand with the genesis, migration, nutrition and growth of the egg, and on the other with the course of embryogenesis and postembryogenesis. It is the only aspect of cnidarian development which has been at all thoroughly studied by experimental analysis, since numerous details were already known from morphological embryology.

Thus at almost all stages of the ontogeny the two poles at the extremities of the main body axis are distinguished as the animal and vegetal, the apical and basal, or the aboral and oral poles. This holds true not only in respect of cleavage and germ layer formation, but also for the planula and its metamorphosis to the primary polyp. More precisely a polarity is already manifest in the lie of the egg within the sessile gonophore, in the sites of meiosis and fertilisation, and even in the ordering of cytological differentiations in the egg such as the location of the yolk and many cytochemical processes (see pp. 34–36). This polarity is particularly evident in the arrangement of the blastomeres and mitotic spindles during cleavage, such as of the radial type, but also in the close relationship obtaining between the onset of germ layer formation and the vegetal pole, as in invagination and unipolar ingression. Further, as early as the gastrula stage the locations and directions of early cell differentiation in readiness for the succeeding larval stage are determined by polarity, as Mergner (1957) has shown in *Eudendrium racemosum*. Finally, we may recall the strict distinction between the aboral pole as that of sensory function and attachment of the free larva, and the oral pole as the site of formation of the definitive mouth, or of stem growth following metamorphosis (for details see section 1.1).

Thus the course of development of Cnidaria often exhibits a marked degree of polarity, which becomes temporarily hidden only in a few derived modes of development such as irregular cleavage, multipolar ingression and the various modes of delamination, or appears in the modified form of radial polarity (see pp. 5 and 11–16). It regularly reappears before or after the stages in question however. Nevertheless, according to Kühn (1965) polarity ultimately appears to be the first determining force in the development of Metazoa, whereby the direction of certain body axes, especially at their extremities, is determined. The poles appear, so to speak, as early oppositions of the extremities of the primary axes in the various forms and stages of development. For is not the first cleavage step generally directed according to a definite main axis? The establishment of this main axis may however occur in the ripening egg or only after its fertilisation. It thus stands to reason that such polarities should give rise to the question of their origin and cause, their localisation and their nature.

1.2.2.1. Directional polarity in relation to cytoplasmic structures

Polarity is often already discernible during the growth of the oocyte in the distribution of cytoplasm inclusions along the main animal-vegetal axis. It is thus a structural, directional polarity and as such can be traced back to the arrangement of the egg within the tissues of the mother animal. Teissier (1931) studied these questions in the eggs of the thecate hydroid *Sertularia (Amphisbetia) operculata*, which develops by holoblastic, equal, radial cleavage and polar ingression to the planula.

The eggs first lie, arranged epithelially, in the ectoderm of the manubrium of the cryptomedusoids. Here they already show a clear polarity in the accumulation of orange-coloured pigments in the outer half of the cell (fig. 54a). To shed the eggs the medusoids free themselves from the blastostyle and leave the gonangium (Motz-Kossowska 1905). Swimming around freely, they shed their eggs one by one into the sea, and there the latter become rounded off, ripen and are fertilised. At this stage and in their further development the eggs retain their pigment deposits (fig. 54b–k), so that the distribution of the original egg plasma can still be readily seen during cleavage in the various directions, in the spherical coeloblastula, during unipolar ingression and in the planula.

It is then found that the pole containing the pigment becomes the vegetal pole of cleavage, the coeloblastula and the planula. The blastula now expands in the direction of the main axis and the prospective entoderm wanders from the vegetal pole and fills the blastocoele, at first as a still solid, disorderly mass of cells (fig. 54h–k). Thus is formed the young planula in which a central slit now appears, around which the entoderm becomes arranged epithelially. Since all the pigmented cells become entoderm and the ectoderm contains no pigment, the prospective assignment of the egg cells and blastomeres can be discerned from their pigmentation.

This is nevertheless not determination, as isolation experiments (fig. 55) have shown: for if individual cells from the 2- or 4-cell stage are isolated, such half- or quarter-blastomeres develop to normal planulae (they are simply correspondingly smaller), whose polarity remains intact (fig. 55a, b). Separation of the four animal from the four vegetal blastomeres in the 8-cell stage produces the same result (fig. 55c). Here small planulae arise from each of the parts allotted to the presumptive ectoderm and the presumptive entoderm. Even single isolated cells from the 8-cell stage become well-proportioned, if tiny, planulae, irrespective of whether the original material was composed of animal or vegetal blastomeres. An extreme instance is a 16-cell embryo which produced 15 larvae, of which 12 were quite normal and capable of further development. Similar isolation experiments up to the 16-cell stage had previously been performed by Zoja (1895), Maas (1905) and Hargitt (1911) in the hydroids Clytia, Laodicea, and Mitrocoma with basically the same result, from which it was concluded that cleavage of the hydroid egg could not be determinative. The eggs and embryos of Anthozoa also appear not to develop in a determinative fashion and to possess considerable regulative ability.

Even in the coeloblastula of *Sertularia (Amphisbetia) operculata* (Teissier 1931), determination of the animal and vegetal halves in view of germ layer formation is not yet firmly established. For if they are separated by an equatorial section the cut edge of both first closes (fig. 55d_1, d_2) and both then undergo entoderm formation by polar ingression at their respective posterior ends and become small normal planulae (fig. 55d_3, d_4). In the posterior part, half of the cells which under normogenetic conditions would have ingressed provided the ectoderm of the completely pigmented larva derived from the vegetal half. The cells of the equatorial region, which normally inwander last, and which should in fact lie in the entoderm of the subsequent budding tip of the hydrocaulus, now constitute the pole at which the planula attaches (fig. 55d_3, d_4). The front portion of the divided blastula, on the other hand, lets half of its presumptive ectoderm cells ingress as entoderm and thus becomes a completely unpigmented planula (fig. 55d_3, d_4). Even quarters cut out of the blastula still form larvae. In all these experiments the original polarity passed down from the egg cell is preserved and the cells closest to the vegetal pole in each case inwander as entoderm.

At the same time the blastomeres and blastula are found to have considerable regulative capacity, as can also be demonstrated experimentally (fig. 55e). Thus, if a blastula of *Sertularia (Amphisbetia) operculata* is sectioned obliquely (Teissier 1931) and the cut edges dyed with vital stains, the stained cells will lie posterolaterally in the one planula and anterolaterally in the other (fig. 55e). This signifies that the original polarity yielded a normal arrangement of the individual portions of tissue at ingression of the entoderm in spite of the experimental distortion of the tissue. The polarity thus remained intact throughout the experiment.

On the other hand the maintenance of polarity limits the regulative ability of the embryo to become a uniform whole: for if the blastomeres deriving from early cleavage are so displaced in relation to one another that their main axes, recognizable by their pigmentation, point in different directions (fig. $55f_1$), they no longer unite into a uniform coeloblastula. Instead abnormal embryos or groups of cells, each with a similar blastomere arrangement, are formed which for their part can again give rise to normal planulae (fig. $55f_2$, f_3). Here we are thus clearly dealing with multiplying groups of cells, the individual cells of which have retained the polarity taken over from the oocyte.

The following results thus emerge from the isolation and fragmentation experiments of Teissier (1931):

(1) The animal-vegetal polarity of the embryo of *Sertularia (Amphisbetia) operculata* is already determined by the position of the oocyte in the ectoderm tissue of the maternal manubrium and is manifested in the pigmentation of the vegetal half. It may thus be described as structural, directional polarity with suggestions of polar field polarity (Kühn 1965) and is strictly retained through all stages of development and experimental procedures.

(2) The pigmentation of the cleavage cells also indicates their prospective assignment. This is not the same as determination, however. Thus all cells of the embryo right up to the young planula are still capable of any subsequent specialisation. The embryogenesis of *Sertularia operculata* is thus not determinative.

(3) While the embryo is a typical regulative embryo, the retention of polarity through all the embryonic stages nevertheless limits its regulative ability to produce a uniform whole. Blastomeres displaced in different directions result only in abnormal embryos or each blastomere develops into a normal planula.

1.2.2.2. Directional polarity unrelated to existing plasma stratification

We have just seen how a polar orientation of the cytoplasmic structures in the embryo of *Sertularia (Amphisbetia) operculata* is an early sign of a definite directional polarity. On the other hand visible stratification of the cytoplasm substances may be present in the egg or embryo without signifying polarity. Centrifugation is always employed to determine whether stratification of the substances of the plasma is bound up with polarity, for by reinforcing their stratification or altering its alignment such experiments provide information as to the nature of the polarity of the eggs (Kühn 1965). Such experiments were carried out by Beckwith (1914) on eggs of the athecate hydroid polyp *Hydractinia echinata*.

According to Bunting (1894) and Teissier (1931) embryogenesis begins with normal radial cleavage (fig. 56a–d) which as early as the 16-cell stage yields a coeloblastula in which germ layer formation takes place by mixed delamination comprising multipolar ingression and blastula delamination (fig. 56d, e; see also p. 12–14). Thus whole blastoderm cells as well as the daughter cells of paratangential mitoses crowd into the blastocoele, largely filling it as they continue to multiply. The embryo then expands lengthwise, the inwandered entoderm becoming arranged around the central slit, and becomes a planula.

Thus in its entoderm formation *Hydractinia echinata* shows a clear radial polarity of the egg cell, which is manifested morphologically in the multipolar ingression of whole blastoderm cells and in the radial position of the spindles of daughter cells given off by delamination. Yet according to Beckwith (1914) and Teissier (1931) if the animal and vegetal poles are stained with vital dyes very early in cleavage the further development of the embryo shows that the primary axis of the egg also determines the main axis of the planula, in spite of the radial formation of entoderm: the animal pole of the egg always becomes the anterior or animal end of the planula.

Low-speed centrifugation of Hydractinia eggs before cleavage causes their contents to become stratified as follows (fig. 57a): the centripetal pole contains oil drops and the centrifugal pole yolk and mitochondria, with clear plasma between the two. Before centrifugation this plasma constituted a coherent layer at the periphery and a reticulum between the granular contents of the egg interior. After restratification the nucleus may be located in any of the layers, and the first furrow in the centrifuged egg can cut at any angle in respect of the plane of stratification. The size of the resulting daughter cells will thus vary considerably, according to whether they receive yolk, clear plasma, or only oil (fig. 57a–c).

If now the first two daughter cells, which are unequal in size, are separated from one another, each of the two half-blastomeres will develop into a larger or a smaller, but otherwise normal planula (fig. 57a). This is irrespective of whether it has received plasma and one of the two polar layers or a part of every layer. Polarisation and cell differentiation are also normal, although ectoderm and entoderm cells still evidence the differences in content (fig. 57b, c). This result demonstrates that the distribution of the stratifiable – or stratified – contents of the egg influences neither the direction of cleavage nor the radial release of cells for entoderm formation nor the direction of expansion or polarity of the planula. It is thus established that the animal-vegetal axis of polarity is retained in spite of the different planes of cleavage.

Why should this be? Its cause can only reside in a polar, morphogenetic pattern which cannot be destroyed by centrifugation. Consequently, such a pattern is unlikely to be present within the interior of the egg, for this submicroscopic structure, which we must visualise as a 'basic stroma' (Kühn 1965), would doubtless be shattered on centrifugation by the movement of the relatively large, visible corpuscles within the egg. Such a morphogenetic pattern can thus only be located in the cortex of the egg, which is not affected by centrifugation, or in a rapidly reparable polar organisation of whose underlying principles we are ignorant. Through centrifugation experiments with embryos of sea urchins *(Paracentrotus lividus)* and molluscs (Ilyanassa, Dentalium) the ectoplasmatic egg cortex has been identified with some probability as the carrier of a determinant cytoplasm pattern.

It is fitting to recall here the principal results of centrifugation experiments with Hydractinia eggs:

(1) The different axes of polarity in embryogenesis are already laid down in the egg. During germ layer formation, admittedly, radial polarity is predominant, but on the whole the primary axis of the egg determines the main animal-vegetal axis of the embryo. It is retained even after restratification of the egg contents by centrifugation and cutting in of the first furrow in different directions. For this reason fragments of the embryo varying in size and content always give rise to a normally polarised and differentiated planula of corresponding dimensions.

(2) Since neither the original apical-basal polarity nor the temporary radial polarity is disturbed by restratification of the egg contents, its seat is probably in a polar morphogenetic pattern of the ectoplasmatic cortex. Thus while the Hydractinia egg shows typical directional polarity, it is not related to stratification of the plasma, but is oriented by the egg cortex.

(3) Hydractinia eggs and their embryonic stages also show great regulative ability; their development is not determinative.

1.2.3. Analysis of cleavage and germ layer formation

In the foregoing section 'Polarity and regulation in embryonic development' we have continually referred to the process of radial cleavage, and we shall now deal only with questions related to the process of cleavage itself. According to Kühn (1965) cleavage is the first morphogenetic process in the embryonic development of Metazoa, terminating on attainment of the blastula stage as a characteristic phase of embryogenesis.

When this cleavage is analysed a number of questions come to the fore, an attempt to answer which shall now be made. Results of experimental analysis are not always available, so that the possible behaviour of the Cnidaria embryo during cleavage must often be conjectured on the basis of comparable experiments in other groups of animals. For this reason it is to be hoped that future experimental analysis will also be devoted to the cleavage of Cnidaria.

1.2.3.1. Cutting of the first furrow

The first question to occupy us should be: why does the first furrow cut into the egg at a particular location? As we have already seen (see p. 3), the first furrow always occurs at the animal pole, whence it either spreads in 'circular' fashion to the vegetal pole, e.g. as in *Clava squamata* (Harm 1903), or it cuts deep into the egg as in *Hydractinia echinata* (Bunting 1894).

On the strength of this knowledge we can now word our original question more specifically and enquire how it comes about that the first furrow always begins at the animal pole and that, once begun, it regularly advances towards the vegetal pole.

From the results of past investigations (Rappaport 1961; and others) it appears probable that cleavage is initiated by the participation of certain cortical regions in physical processes caused by the mitotic apparatus. On this basis Rappaport and Conrad (1963) have studied the significance of the spatial relationships between the mitotic apparatus and the furrow in eggs whose mitotic apparatus is shifted eccentrically towards the animal pole. The eggs of the athecate hydropolyp *Hydractinia echinata*, inter alia, were selected as suitable material for this purpose.

In these eggs the fertilised nucleus divides immediately beneath the animal pole and the first cleavage furrow cuts directly above and between the diverging telophase nuclei (fig. 58a). If now the basal anuclear region and a portion of cytoplasm from each side of the cleaving egg are removed by means of a fine glass needle (fig. 58b), a uniform geometrical relationship between the mitotic nuclei and the surface of the cell can be obtained: the nuclei lie at an approximately equal distance from the periphery in all directions. Following these operations, division of the nucleated fragments is now always symmetrical and circular (fig. 58c), although the unoperated blastomere would have produced a unilateral furrow. Blastomeres of early cleavage thus divide symmetrically if the distance between the mitotic apparatus and the cell periphery is equalised by amputating cytoplasm, and thus undergo a mode of cleavage which does not normally occur in Hydractinia.

This proves that the location and direction of the unilateral course of the first furrow are determined by the eccentric position of the mitotic apparatus. But the beginning of cleavage is not irredeemably restricted to a narrowly defined region of the surface: for if the mitotic apparatus is located in the centre, the furrow spreads in a ring around the egg and divides it regularly from all sides.

Rappaport and Conrad (1963) further established that there is considerable tension at the surface of the egg and blastomeres at the beginning of each cleavage. In this they see the cause of the further advance of cleavage once it has been set into motion under the influence of the mitotic apparatus. That is to say, that cleavage continues even when it is no longer in the sphere of influence of the latter, and is thus independent of it. The furrow front shows a surging flow of cytoplasmic particles, clearly a manifestation of the mechanical forces reigning there, such as accompany all cell divisions.

We are now able to answer the two questions posed at the outset as follows: the location of the mitotic apparatus within the egg determines the mode of furrowing and the point of its inception, but not the further progression of the furrow. This appears to be governed by other forces which become visible as surface tension and flow of cytoplasm.

1.2.3.2. Simultaneous mitotic divisions

As a rule the first three cleavage mitoses are synchronous and generally give rise to regularly arranged blastomeres of the same size. But just in radial cleavage such as occurs in *Aequorea forskalea* (Haecker 1892) we saw completely synchronous cleavage steps right up to the 64-cell stage (see p. 4). One is led to speculate how these mitotic divisions occur simultaneously to produce the cleavage embryo.

Basically, the cause must reside in a particular temporal rhythm of spindle formation and cytoplasm division, while the embryo follows a certain sequence in its divisions. In sea urchin embryos, which also undergo radial cleavage, the individual mitotic steps can be uniformly retarded by certain extraneous conditions such as isolating individual blastomeres or suppressing nuclear divisions in all cells. This yields partial embryos in which cleavage divisions also proceed regularly and the temporal course of which corresponds precisely to the normogenesis of control embryos. Thus the temporal rhythm is independent of the situation of the embryo as a whole. Similar trials must also be carried out in Cnidaria exhibiting radial cleavage.

1.2.3.3. Determination of the cleavage pattern

The mechanisms determining the radial mode of cleavage or more precisely the position of the spindles have so far been studied only in sea urchin embryos (Hörstadius 1928), but should also come to be investigated in Cnidaria.

From the results of fragmentation experiments in *Sertularia (Amphisbetia) operculata* and centrifugation experiments in *Hydractinia echinata* (see pp. 40 and 42) we have seen that the direction of cleavage cannot be modified by either experimental displacement of tissues or artificial restratification of the egg contents, but that it proceeds in accordance with the polarity of the cell, which thus doubtless plays a significant part in producing a given cleavage pattern.

The spindle positions themselves are determined by an autonomic rhythmic process which takes place in the whole plasma of the divided or undivided embryo in the manner of an autodifferentiation process (Kühn 1965). Thus a variety of conditions are created in the different regions of the cytoplasm, conditions which for their part determine the different spatial positions of the spindles with respect to the main axis of the embryo and thus also determine division of the cytoplasm. Polarity obviously serves here as an aligning factor. It would of course be interesting to study such questions in species of Hydrozoa in which entoderm formation comes about by the radial release of cells into the blastocoele and which simultaneously or successively exhibit radial and paratangential spindle directions.

1.2.3.4. Prospective assignment and determination of cleavage cells

The question of the prospective assignment and determination of cleavage cells has already been presented in the chapter on 'Polarity and regulation in embryonic development' although it really belongs in the discussion of the cleavage process. Nevertheless it is closely related to the problem of polarity and regulative ability and should therefore not be removed from its context. From here on only its significance for the process of cleavage or the preparation of germ layer formation will be discussed.

The prospective assignment of cleavage cells, that is to say their future fate, is studied basically by vital staining of individual blastomeres or wreaths of blastomeres. Occasionally however differential pigmentation of the individual regions of the egg itself provides a clue to their future localisation. The isolation and fragmentation experiments of Teissier (1931) with *Sertularia (Amphisbetia) operculata* offered an instructive example of this (see pp. 38–40). The vegetal half of the egg of this thecate hydroid is pigmented orange and the vegetal half of the cleavage embryo and coeloblastula retains this pigmentation while the animal half remains unpigmented in each case (fig. 54a–f). During germ layer formation by unipolar ingression all the pigmented cells wander into the blastocoele at the vegetal pole as prospective entoderm and can still be recognised as such in advanced postembryonic development, often even after metamorphosis (fig. 54g–k). Their prospective assignment can thus be followed throughout embryogenesis and post-embryogenesis.

The same experiments with Sertularia have shown that the prospective assignment of the cleavage cells is by no means equivalent to determination (see pp. 38 and 39). This is already clear from the fact that by devising the experiment in an appropriate manner unpigmented prospective ectoderm cells can be made to become entoderm cells and pigmented entoderm cells to become ectoderm cells (fig. $55d_{1-4}$). This demonstrates that no cell determination occurs up to the planula stage, i.e. that cleavage is not determinative. Similar experiments in other hydroids have yielded the same result (Zoja 1895; Maas 1905; Hargitt 1911).

All these experiments clearly demonstrated the great regulative ability of the hydroid embryo (see p. 39 and fig. 55e, f). It is kept through all the developmental stages down to the planula and is limited in the individual cells only by the retention of the polarity from the oocyte. Thus only abnormal embryos from early cleavage stages, or correspondingly smaller normal embryos from each single cell (fig. $55f_{1-3}$), will result when the blastomeres have been displaced experimentally in different directions. Regulation to a uniform whole embryo is impeded by the rigid retention of the original polarity. Polarisation of the cleavage cells is thus superordinate to their regulative ability, which, in turn, is superordinate to their prospective assignment.

1.2.3.5. Termination of cleavage

Once the blastula stage is reached the cleavage divisions stop and the embryo goes on to germ layer formation. In the case of radial cleavage this typical stage of embryogenesis is generally represented by a hollow sphere, the coeloblastula (Kühn 1965). Its formation thus signals the end of the sequence of nuclear and cell divisions which began with stimulation of the egg at fertilisation and which then – varying from species to species – always follows a definite rhythm. What mechanism triggers the termination of this regular sequence of divisions?

To clarify this question we must first turn to the relationship between the nucleus and cytoplasm in embryogenesis, or more specifically, between the volumes of their respective masses. In normal somatic cells this relationship is generally constant and varies only briefly with each mitotic division. By the time the oocyte ceases to grow, however, the ratio is greatly in favour of the cytoplasm. During cleavage this disproportion is gradually offset at each division and finally reverts to the normal state, that is to say the ratio of nucleus to cytoplasm in the blastula cells reaches its definitive value. Two principal processes are involved here: the cytoplasm mass is distributed *in toto* among the blastula cells and thymonucleic acid is synthesised from the virtually constant supply of nucleic acids available during the nuclear division growth that precedes each cleavage mitosis. Chromatin substance is thus synthesised at the cost of the cytoplasmic ribonucleic acid. The individual cells thus attain a particular functional state which is manifested by a qualitative change of the cytoplasm as the result of removal of nucleus constituents. With the attainment at the blastula stage an equilibrium is finally established which halts the cleavage divisions and thus constitutes the dividing line between cleavage as an independent process of embryogenesis and the subsequent germ layer formation.

According to Teissier (1929) cleavage of the scyphomedusa *Chrysaora hyoscella* shows this striving towards the state of equilibrium in the nucleo-cytoplasmic ratio. The entire embryonic development of the eggs takes place in the body of the maternal animal and radial cleavage in the gonads, even. In the process the blastomere nuclei constantly diminish in size until the 120-cell stage, so that after about seven mitosis steps the final cell size and nucleo-cytoplasmic relationship of the coeloblastula are attained (fig. 59a–c). The blastulae are still fed by the maternal organism and by dint of continual multiplication of cells often attain an enormous size, while that of the cells and nuclei remains constant (fig. 59d, $60a_{1-3}$, $60b_{1-3}$). After each division, in fact, the cell first grows to its original size before dividing again. Invagination of the entoderm may occur in the blastula while still small or when its growth is well advanced (fig. 60c, d). The resulting gastrulae thus have very different diameters. Accordingly the planulae vary considerably in size, not infrequently by a factor of 300 (fig. 60e, f_{1-3}). The size of the larva is thus a function of the starting size attained by blastula growth. Only when ectoderm and entoderm differentiate in the planula do their cell bodies and nuclei attain the dimensions specific to them (fig. 60g–i).

Analysing this peculiar development we may first note that during the immense growth of the blastula nuclear and cytoplasmic substances are synthesised *de novo* in a quantity fifty to one hundred times that of the original content of the egg. According to Kühn (1965) the blastula cells are in a state of equilibrium in respect to their developmental-physiological conditions rather like cells in a tissue culture. From this it may be concluded that the embryo need not necessarily proceed with gastrulation as the succeeding morphogenetic process after having attained a certain number of cells or a certain nucleo-cytoplasm ratio. But rather, when appropriately nourished from the exterior, as the example of *Chrysaora hyoscella* demonstrates, it can further increase the number of its cells by the continued synthesis of nuclear and cytoplasmic substances, and will thus attain a considerable size. Once the blastula stage has been attained, on the other hand, the nucleo-cytoplasmic ratio always remains the same even if mitosis continues.

We may thus state that the attainment of the final ratio of nucleus to cytoplasm, or recovery of its normal value, marks the end of the cleavage period and inception of the blastula stage. Whether or not mechanisms other than the state of equilibrium of the blastula cells, which this ratio expresses, cause the termination of cleavage, is unknown at the present time.

1.2.3.6. Initiation of germ layer formation

The end of the cleavage period is thus marked by the attainment of the blastula stage. Its blastoderm represents, so to speak, a primitive blastema from which by morphogenetic movements and differentiation processes the systematically arranged basic body form, in which the spatial arrangement of the organ primordia is completed, now gradually arises (Kühn 1965). The first step in this process is the formation of the germ layers, ectoderm and entoderm. As a result of germ layer formation different parts of the embryo are brought into a topogenetically new relationship; they now come into contact with one another. Differentiation of the organ primordia then begins. In this process material that was hitherto equivalent becomes increasingly inequivalent, while its determination is established by the interactions between the different parts of the embryo. This brief summary of germ layer formation is already sufficient to show its significance for the whole development of Metazoa. But at the same time it raises further questions, the answer to which will close this discussion on the experimental analysis of embryonic development: what are the mechanisms underlying the initiation of germ layer formation and how is its location determined?

Once again the example of the scyphomedusa *Chrysaora hyoscella* (Teissier 1929) will serve to answer the first question. As long as the blastula is nourished by the maternal organism, its cells remain in a state of equilibrium with regard to their developmental physiology. Doubtless, they multiply continuously by virtue of the food supply, whereby the size of the blastula is often enormously increased but their nucleus cytoplasm ratio remains unchanged. The embryo starts to undergo gastrulation only once the supply of food to the growing blastula declines or ceases, so that the state of equilibrium of the cells would be disturbed, if they continued to multiply, by the removal of nucleus-constituents. Thus in *Chrysaora hyoscella* the cessation of the food supply whence nucleus and cytoplasm substances are formed is clearly the inducing process which elicits gastrulation. This would signify that in eggs not fed by the mother animal and depending on their own nutrient capital, internal conditions would arise as the result of the continuous changes in the nature and quantity of these substances which would cause germ layer formation to follow immediately upon cleavage (Kühn 1965).

The result obtained by Teissier (1931) in experiments with *Sertularia (Amphisbetia) operculata* (see pp. 38–40 and fig. 55a–f), have shown that in some hydroids the whole blastoderm has the nature of a self-differentiating field: all parts of the blastula isolated undergo a species-typical proportional differentiation into ectoderm and entoderm by polar ingression, in accordance with the polarity already determined in the egg. The regulative capacity is thus almost complete and is limited only by the retention of polarity. But this polarity inherited from the oocyte determines the site of polar ingression of the entoderm: it always takes place at the vegetal pole, irrespective of whether the cells there are prospective entoderm, as is normally the case (fig. 54h, i), or prospective ectoderm, as can be brought about experimentally (fig. 55d). The site of entoderm ingression, like the proportional distribution of the blastoderm cells into ectoderm and entoderm, is thus rigidly determined by the original polarity along the main axis of the embryo.

On the other hand, entoderm formation from all sides, such as is seen in multipolar ingression and blastula and morula delamination, is determined by radial polarity. Such was the case, for example, in *Hydractinia echinata* (pp. 40–42 and fig. 56d, e). But here too the primary animal-vegetal polarity of the egg determines the further development of the two-layered embryo: the original animal pole of the egg and of the cleavage stages always becomes the front, animal end of the planula. Even fragments of embryos of different sizes and varying in their composition give rise to normally differentiated planulae of corresponding dimensions (fig. 57a–c). Thus, at least in the examples cited, polarity determines the site and mode of germ layer formation.

Fragmentation experiments in Sertularia blastulae (fig. $55d_{1-4}$) show the vegetal pole also to be the seat of autonomic morphogenetic movements: polar ingression is initiated there even in very small fragments, no matter whether the inwandering material consists of prospective entoderm or ectoderm cells.

The lack of suitable experimental studies prevents discussion of further questions about germ layer formation in Cnidaria.

1.2.4. Results of experimental analysis

The discussion of the experimental analysis of Cnidaria has shown that causality studies in embryogenesis have been confined almost exclusively to hydroids. The chief results of these studies will be given briefly below:

(1) The entire course of development is influenced by polarity. This is often already determined as animal-vegetal polarity along the main axis of the embryo by the position of the oocyte in the germinal epithelium. It probably has its seat in a polar morphogenetic pattern of the egg cortex and is retained throughout embryogenesis and postembryogenesis. In radial forms of entoderm formation it is temporarily masked by radial polarity, but it reappears when germ layer formation has ceased.

(2) Polar relationships also decide the prospective assignment of the cleavage cells. This is not equivalent to determination however; cleavage is thus not determinative: all cells can differentiate to any specialisation right up to the larval stage. The embryo is thus also a regulative embryo. Its great regulative capacity is bounded here only by the strict retention of polarity. The polarity of the cleavage cells is thus superordinate to their regulative power, which for its part is superordinate to their prospective assignment.

(3) Cleavage, as the first morphogenetic stage of embryogenesis, always begins at the animal pole of the egg. The mode of cleavage and the site of its inception are determined by the position of the mitotic apparatus: if this is eccentric, the first furrow cuts unilaterally, and if central, circularly. The continued progress of the furrow is dependent on forces which are visible as surface tension and movement of cytoplasm.

(4) The sequence of divisions of the cleavage cells terminates at the blastula stage, when the nucleo-cytoplasmic ratio of the cells has recovered its normal value or attained its final one, i.e. when continued multiplication of cells would disturb this state of equilibrium by removing nucleotrophic substances. Cell multiplication in the blastula stage is possible only when food is supplied extraneously and adequately and halts as soon as this supply declines or ceases: the embryo then enters into germ layer formation. For this reason the latter follows immediately upon the blastula stage in all

embryos which are dependent solely on nutrient resources received from the egg.

(5) The site and mode of germ layer formation are again established by polarity: thus animal-vegetal polarity determines the vegetal pole as the site of entoderm formation in polar ingression as well as the proportional distribution of blastoderm cells into ectoderm and entoderm. Entoderm formation from all sides of the blastoderm, on the other hand, is determined by radial polarity. In the first case the vegetal pole is found to be the site of autonomic morphogenetic movements: polar ingression occurs even in small blastula fragments.

(6) Definitive determination of the fate of the cells does not take place before the planula stage. The actual moment of its occurrence is however unknown.

Notes to illustrations

The illustrations have been taken from the following works:

HYMAN L. H., 1940. The invertebrates: Protozoa through Ctenophora (Vol. I): Figs. 28, 38e–g, 42d, 45e, 50a, b, 52b, 53d, i.
KORSCHELT and HEIDER, 1936. Vergleichende Entwicklungsgeschichte der Tiere, Vol. 1: Figs. 1, 7a–d, 9, 10a–c, 11a, b, 13, 27b, c, 29, 30, 32, 33, 34a–g, i, 38c, d, h, 43, 44a, 47, 48, 50c, 51, 52a, 53a–c, e–h.
KÜHN A., 1913. Entwicklungsgeschichte und Verwandschaftsbeziehungen der Hydrozoen. I. Die Hydroiden: Figs. 2, 3, 4, 5, 8, 11c, d, 12, 14, 15, 18, 19, 20, 21, 22, 23, 24b, 25, 26.
KÜHN A., 1932. Coelenterata, in: Handwörterbuch der Naturwissenschaften, 2nd edition: Figs. 10d, 27a, 31, 35f, 39.
KÜHN A., 1965. Vorlesungen über Entwicklungsphysiologie, 2nd edition: Figs. 35a, b, d, e, 54, 55, 56, 57, 59, 60.
KUMÉ M. and K. DAN, 1968. Invertebrate embryology: Figs. 34h, k, 37, 40, 41a–h, l, m, 42a–c, e, 45d.
SIEWING R., 1969. Lehrbuch der vergleichenden Entwicklungsgeschichte der Tiere: Figs. 36, 38a, b, 41i, k, 44b, c, 45a–c, 46.

The remaining illustrations are taken from the original works acknowledged in each case or drawn after original preparations.

All illustrations have been redrawn, modified in a greater or a lesser degree for the present contribution and generally somewhat schematised. The markings on drawings of sections in plates I–X are used as follows:

Sparsely stippled cells and tissue areas	Undifferentiated cells and tissue parts before the start of germ layer formation
Sparsely stippled areas	Jelly
Withe cells and tissue areas	Ectoderm and prospective ectoderm
Thickly stippled cells and tissue areas	Entoderm and prospective entoderm
White areas with dotted boundaries	Yolk substances and yolk portions of cells
White nuclei with or without black nucleoli	Ectoderm nuclei
Black nuclei with or without white nucleoli	Entoderm nuclei
Thick black lines	Periderm membranes, mesolamellae

Further explanations to signs used are given in the captions.

PLATE I

Cleavage of Hydrozoa

Fig. 1a–g. Total, equal, radially symmetric cleavage in *Aequorea forskalea.* 1a: 2-cell stage, side view; 1b: 4-cell stage, polar view; 1c: 4-cell stage, side view; 1d: 8-cell stage, side view; 1e: 16-cell stage, side view; 1f: 64-cell stage, side view; 1g: blastula, section. (Based on Claus 1883.)

Fig. 2a–e. Total irregular cleavage in *Turritopsis nutricola.* 2a: 4-cell stage, polar view; 2b: 8-cell stage, side view; 2c: 8-cell stage seen from above; 2d: 16-cell stage; 2e: late cleavage stage, section. (After Brooks and Rittenhouse 1907.)

Fig. 3a, b. Delayed blastomere separation in *Tubularia mesembryanthemum.* 3a: young blastula, section; 3b: 30-cell blastula with incipient entoderm ingression by mixed delamination, section. (After Brauer 1891b.)

Fig. 4a, b. Atypical total, unequal, radial cleavage with shortened development in *Clava squamata.* 4a: 12-cell stage, side view; 4b: 32-cell stage with incipient entoderm ingression by morula delamination, section. (After Harm 1903.)

Fig. 5a–d. Modified radial cleavage with shortened development in *Gonothyraea (Laomedea) loveni.* Sections. 5a, b: normal development. 5a: 8-cell stage with wide blastocoele; 5b: 64-cell stage with incipient entoderm ingression; 5c, d: development with embryos crowded in the gonophore; 5c: 8-cell stage with highly reduced blastocoele; 5d: 24-cell stage with entoderm formation by morula delamination. (After Wulfert 1902.)

Fig. 6a–d. Syncytial cleavage with extreme shortening of development in *Eudendrium racemosum.* Sections. 6a: 2-nucleus stage; a daughter nucleus of the first cleavage division has wandered into the interior of the skull cap structure consisting of liquid phase yolk; all entoderm nuclei and vitellophages descend from this nucleus, while the descendants of the outer daughter nucleus later give rise to the ectoderm; 6b: 4-nucleus stage; the skull cap structure has closed to become a hollow sphere, the spindle axes of the second cleavage division are perpendicular to that of the first; 6c: 32-nucleus stage; the yolk forms a ball around each cleavage nucleus; 6d: multinuclear stage of late cleavage with incipient cleavage of yolk; in the interior amitotic divisions appear for the formation of vitellophages; in the pheripheral region accumulations of peripheral nucleus-plasma zones as a sign of beginning pregastrula formation. (After Mergner 1957.)

Plate I.

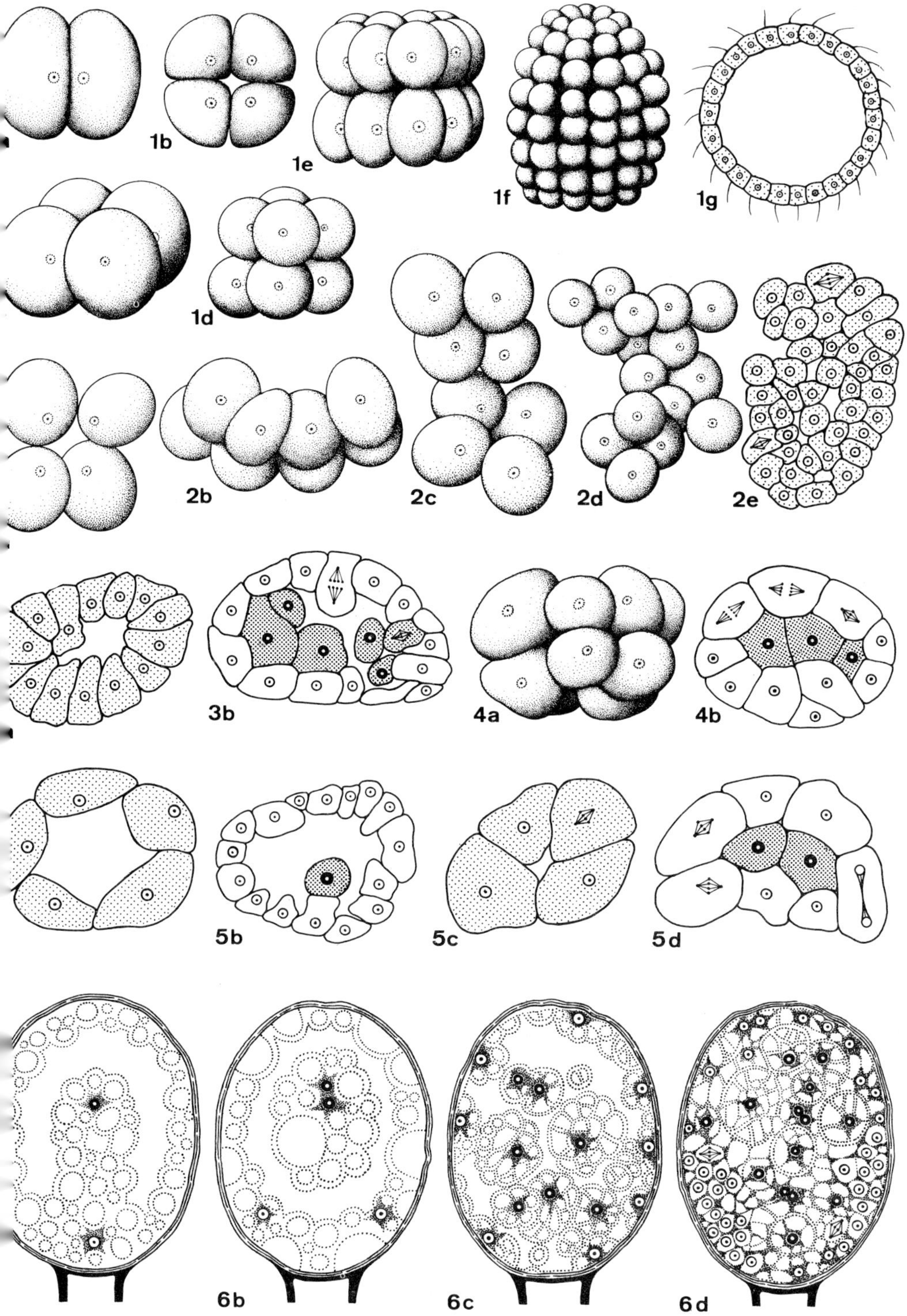

PLATE II

Germ layer formation in Hydrozoa

Fig. 7a–e. Polar ingression in *Aequorea forskalea.* Sections. 7a: Free-swimming blastula larva with highly cylindrical cells at the vegetal pole; 7b: start of inwandering of cells at the vegetal pole; 7c and d: advance of the inwandered cells towards the animal pole; 7d: young sterrogastrula (parenchymella) with blastocoele almost filled; 7e: young planula with germ layers delimited from one another and formation of gastric cavity in the entoderm. (7a–d after Claus 1883; 7e from Mergner 1969, unpublished.)

Fig. 8a–c. Polar ingression in blastula larvae of free medusae. Sections. 8a: start of entoderm development at the vegetal pole by cell division in *Clytia viridicans* as an exception; 8b: continuing entoderm development by ingression of whole blastoderm cells in *Clytia flavidula;* 8c: young sterrogastrula with loosely filled blastocoele in *Mitrocoma annae.* (After Metschnikoff 1886b.)

Fig. 9a, b. Multipolar ingression in *Aeginopsis mediterranea.* Sections. 9a: multipolar ingression of whole blastoderm cells into the blastocoele of the 32-cell blastula; 9b: free-swimming sterrogastrula. (After Metschnikoff 1886b.)

Fig. 10a–d. Entoderm formation by mixed delamination in *Liriope mucronata.* Sections. 10a, b: ingression of whole blastoderm cells and the daughter cells of paratangential divisions into the blastocoele from all sides; 10c: formation of the entoderm vesicle with creation of gastric cavity and excretion of jelly into the residual blastocoele; 10d: displacement of entoderm vesicle to the vegetal pole. (10a–c after Metschnikoff 1886b; 10d after Maas 1905.)

Fig. 11a–d. Mixed delamination in Hydra. Sections. 11a: 128-cell blastula with beginning entoderm ingression; 11b: sterrogastrula; 11c: formation of the interstitial cells at the base of the ectoderm and excretion of the embryonic theca; 11d: formation of the two germ layers and secretion of the mesolamella (After Brauer 1891a.)

Fig. 12a–c. Mixed delamination in *Tubularia mesembryanthemum*, starting at the 30-cell blastula stage (fig. 3b). Sections; in 12b, c, only one half is drawn. 12a: incipient delimitation of the germ layers in the solid 'pseudomorula'; 12b: flattening of the embryo, secretion of the mesolamella and formation of the first pair of tentacles; 12c: growth of the first tentacles in the young actinula and formation of the gastric cavity. (After Brauer 1891b.)

Plate II.

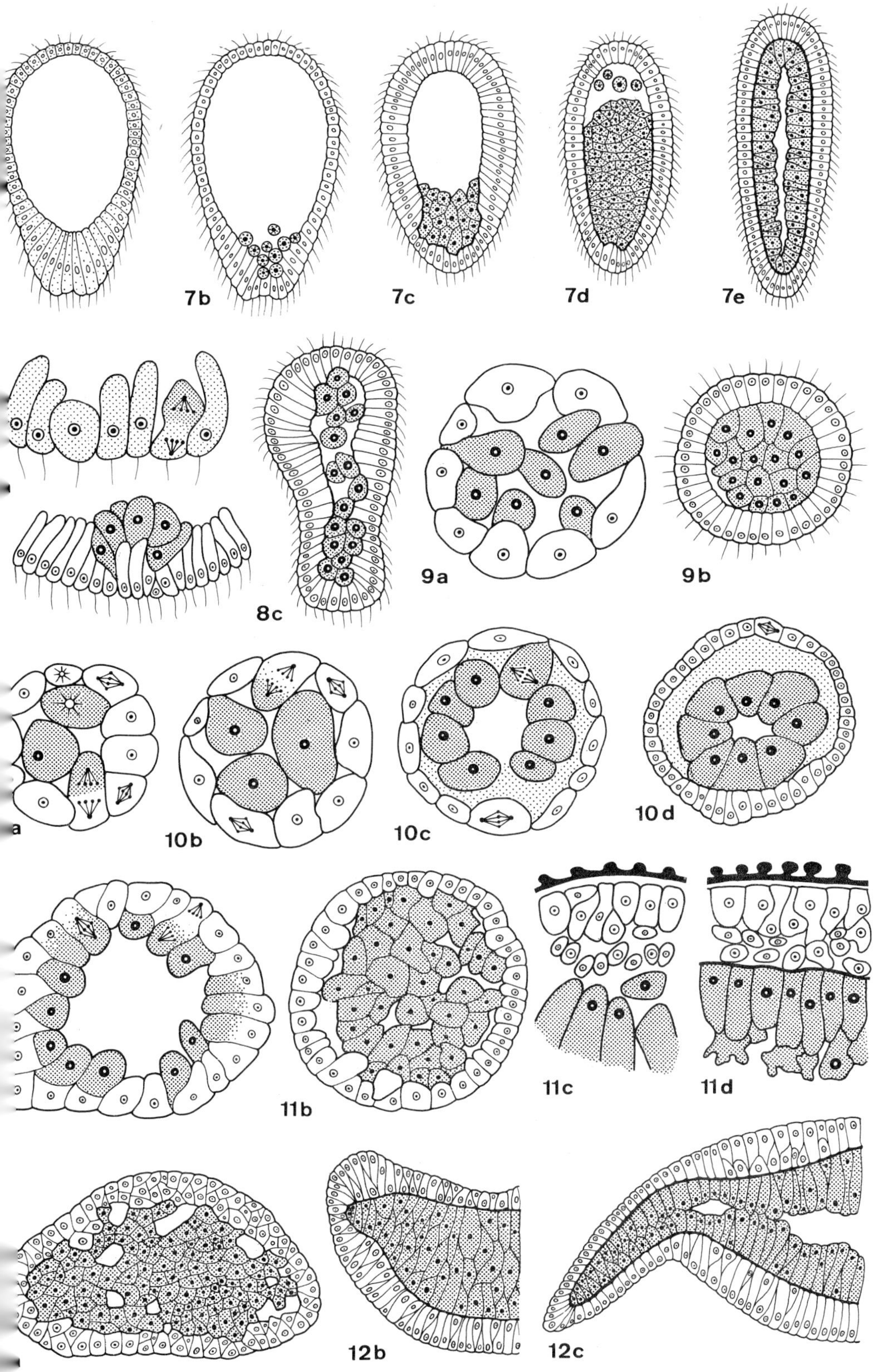

PLATE III

Germ layer formation and postembryonic development of Hydrozoa

Fig. 13a–c. Blastula delamination in *Geryonia fungiformis*. Sections. 13a: few-celled blastula with radial spindle orientation; 13b: simultaneous delamination of daughter cells of paratangential mitoses into the blastocoele; 13c: formation of the closed entoderm vesicle with gastric cavity and excretion of jelly into the residual blastocoele. (After Föl and Metschnikoff.)

Fig. 14a–d. Morula delamination in *Clava squamata*. Sections. 14a: late cleavage stage (morula); the prospective ectoderm and entoderm cells are shown as such; 14b: beginning differentiation of ectoderm and entoderm (delamination); 14c: completed delamination of the two germ layers and secretion of the mesolamella; 14d: planula ready to hatch, showing incipient histological differentiation of ectoderm and formation of gastric cavity in the entoderm. (After Harm 1903.)

Fig. 15a, b. Morula delamination in *Turritopsis nutricola*. Sections. 15a: beginning ectoderm differentiation from the syncytial stage; 15b: young planula with formed ectoderm, but entoderm still disorganised. (After Brooks and Rittenhouse 1907.)

Fig. 16a, b. Syncytial delamination in *Eudendrium racemosum*. Sections. 16a: pregastrula stage; the peripheral nucleus-plasma zones of the preectoderm become delimited from the yolk spheres within the entoderm by formation of the mesolamella, beginning at the oral pole; in the lateral regions the presumptive cnidoblast zone is prominent in the form of numerous nests of I-cell nuclei; 16b: older gastrula with histological differentiation of the sensory nerve cell network, of mucous cells and muscle cells in the ectoderm and of the gastric epithelium and cnidoblasts in the entoderm; disintegrating yolk in the gastrovascular cavity. (After Mergner 1957.)

Fig. 17a–d. Postembryonic development up to the primary polyp in *Eudendrium racemosum*. Sections. 17a: attachment of the fully differentiated planula at the aboral pole; in both body layers same histologically differentiated cell types as within the differentiation gastrula (fig. 16b), but with further increased content; incipient outwandering of the cnidoblasts into the ectoderm; 17b: great flattening of the attached larva as the first stage of metamorphosis; start of periderm formation; further outwandering of cnidoblasts; degeneration of the mucous cells and the aboral sensory cells; 17c: advanced metamorphosis stage; the polyp body rears up from the flattened disc and divides into hydranth and hydrocaulus; great multiplication of I-cells; outwandering of cnidoblasts almost complete; 17d: young primary polyp, showing hydranth and stem; after the mouth aperture has broken through and the cnidocytes have differentiated the polyp is ready for food intake. (From Mergner 1970, unpublished.)

Fig. 18a, b. Postembryonic development of *Clava squamata*. Sections. 18a: Flattening of the attached planula and formation of the periderm theca at the beginning of metamorphosis; 18b: young polyp after completion of metamorphosis; breakthrough of mouth at the oral pole and budding of the first tentacles. (After Harm 1903.)

Plate III.

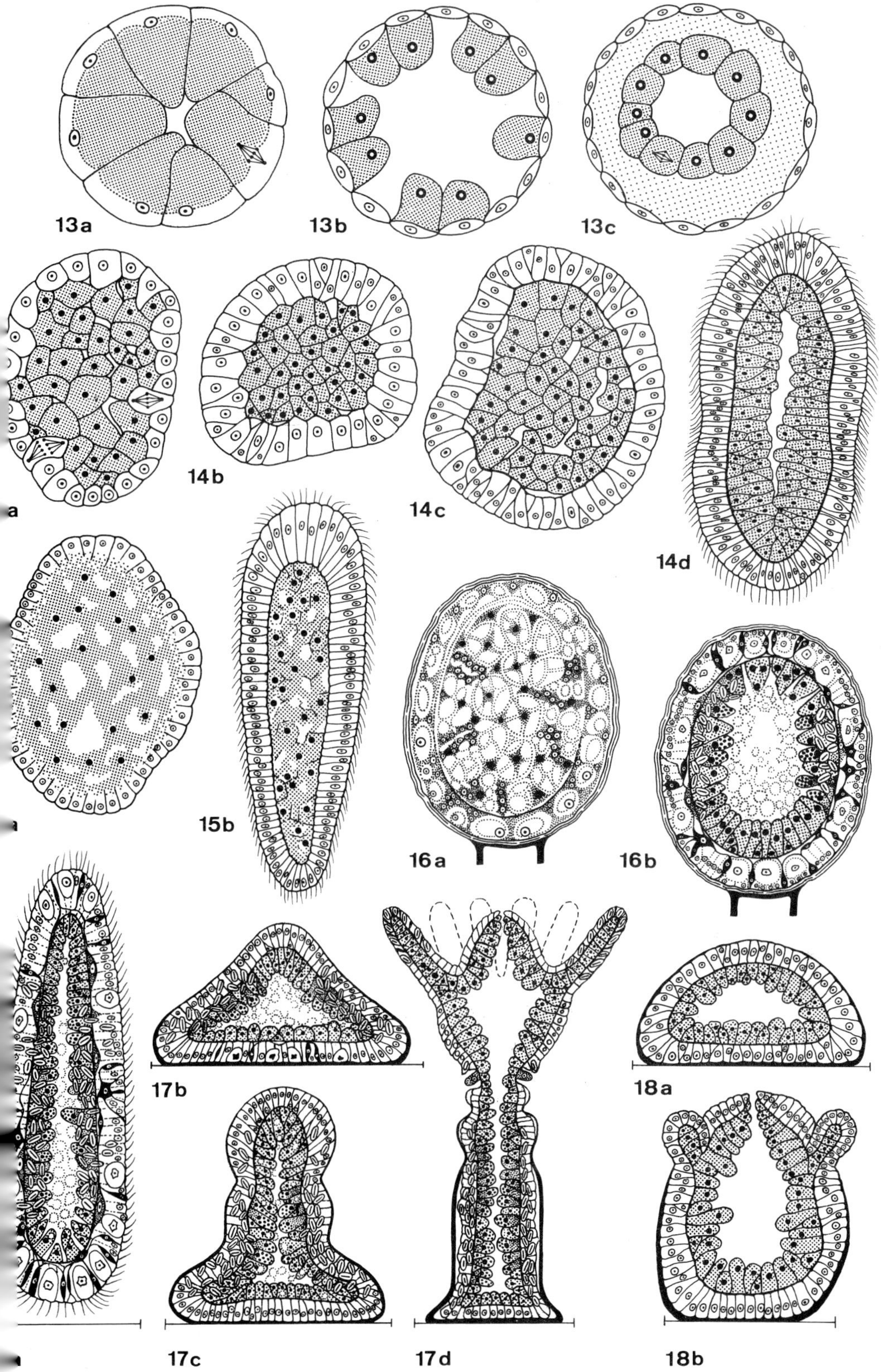

PLATE IV

Postembryonic development of Hydroida

Fig. 19a–c. Postembryonic development of *Pennaria tiarella;* transformation of the whole planula into the primary polyp with early division into hydranth and hydrocaulus. 19a: attached planula with pedal disc, stem and club-shaped terminal swelling, the initial development of the hydranth; 19b: continuing division and growth of the first tentacles; 19c: young primary polyp, separated into hydranth with two circlets of tentacles, annulated hydrocaulus and pedal disk. (After Hargitt 1900.)

Fig. 20a, b. Metamorphosis of the planula of *Turritopsis nutricola,* attached by the side, into the primary polyp. 20a: the whole planula, attached along its length, first becomes the hydrorhiza; only then does the actual polyp primordium arise from its centre; 20b: formation of the polyp with two circlets of tentacles and mouth aperture. (After Brooks and Rittenhouse 1907.)

Fig. 21a–f. Postembryonic development of *Gonothyraea (Laomedea) loveni;* metamorphosis of the whole planula of a thecate hydroid polyp with primary terminal polyp and hydrocaulus. 21a–e are sections, 21f a side view in the optical section. 21a: attachment of the planula at the aboral pole; 21b: histological differentiation of the oral ectoderm with sensory cells, I-cells and nematocysts and accumulation of yolk in the oral entoderm cells; 21c: incipient flattening of the attached larva as first stage of metamorphosis; 21d: flattening of larva complete; 21e: the young hydrocaulus rises up from the flattened disc and already exhibits the typical annulation of the periderm; great accumulation of yolk in the entoderm cells of the tip of the hydrocaulus; 21f: development of the terminal polyp, hydrocaulus and pedal disk. (After Wulfert 1902.)

Fig. 22a–d. Postembryonic development of *Plumularia echinulata;* metamorphosis of the whole planula of a thecate hydroid polyp with axocaulus and terminal growing point. 22a: attached planula; 22b: flattening of the larva and development of the pedal disk; 22c: the young axocaulus rises up from the pedal disk; 22d: formation of the first periderm ring on the axocaulus and growth of the margins of the pedal disk to the hydrorhiza. (After Kühn 1909.)

Fig. 23a, b. Metamorphosis of a planula, attached by the side, of *Mitrocoma annae* with loss of its original polarity; first a part of the hydrorhiza develops from the planula, and only then do the polyp anlagen rise up from it. Sections. 23a: branched hydrorhiza with a polyp anlage at each of the growing ends; 23b: almost complete polyp with hydrotheca at the end of the hydrorhiza. (After Metschnikoff 1886b.)

Fig. 24a, b. Actinula larvae of *Tubularia mesembryanthemum* (fig. 24a) and *Myriothela phrygia* (fig. 24b), the first free stages and outcome of embryonic development. 24a: free actinula with two circlets of tentacles, mouth aperture and gastrovascular cavity; 24b: actinula hatching from the gonophore. (24a after Ciamician 1879; 24b after Allman 1875.)

Fig. 25a, b. Postembryonic development of the hydromedusa *Margelopsis haeckeli* in the process of an actinula as the first free stage. 25a: medusa with embryos and an actinula at the manubrium; 25b: freely suspended polyp with medusa buds, the mouth aperture directed downwards. (After Hartlaub 1899.)

Fig. 26a–c. Postembryonic development of Corymorpha in the process of an actinula-like young stage. 26a and b are sections. 26a: young larva of *Corymorpha palma* with beginning separation into the primordia of the hydranth and stem within the periderm; 26b: a somewhat older larva with growing distal circlet of tentacles in *Corymorpha palma;* 26c: larva of *Corymorpha nutans* with two circlets of tentacles and stem, with rhizostolons sprouting from the latter. (26a, b after Torrey 1907; 26c after Hartlaub 1907.)

Plate IV.

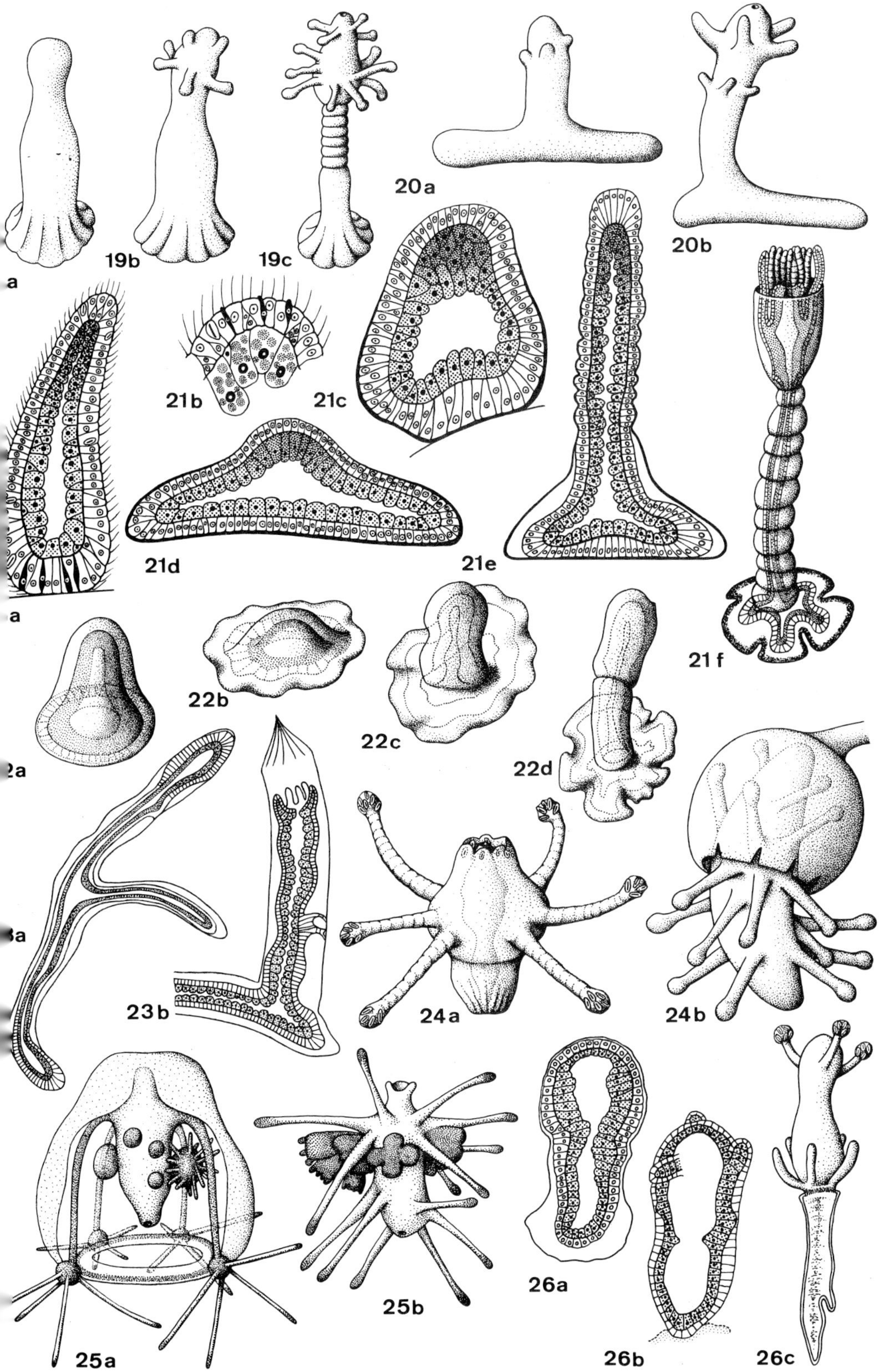

a
19b
19c
20a
20b
21b
21c
a
21d
21e
21 f
a
22b
22c
22d
2a
3a
23b
24a
24b
25a
25b
26a
26b
26c

PLATE V

Postembryonic development of Trachylina (figs. 27–30) and Siphonophora (figs. 31–33)

Fig. 27a–c. Postembryonic development of *Liriope mucronata* (figs. 27a, b) and *Liriope scutigera* (fig. 27c). 27a and b are sections, 27c a meridional section. 27a: formation of the gastrovascular cavity, formation of the tentacles and velum, beginning doming of the exumbrella. 27b: formation of the manubrium, sprouting out of the larval tentacles; 27c older larva: the subumbrella with manubrium and larval tentacles has sunk in; the definitive tentacles and statocysts have formed at the margin of the umbrella; early stages of the ring canal and the radial canals between the interradial sectors of the entoderm already exist. (27a after Maas 1905; 27b after Metschnikoff 1886b; 27c after Brooks 1886.)

Fig. 28a–c. Postembryonic development of Agluara. Optical sections. 28a: older planula with tentacle buds around the mouth cone; 28b: sprouting of tentacles and development of the gastric cavity; 28c: young medusa with open manubrium. (After Metschnikoff 1886b).

Fig. 29. Parasitic development of *Pegantha smaragdina*. Actinula larva from the gastric cavity of the maternal medusa with two daughter actinula larvae on the aboral stolo prolifer. (After Bigelow 1909.)

Fig. 30a–d. Larval development of *Aeginopsis mediterranea*. Sections. 30a: planula; 30b: stretching of the ends of the planula to form the first two tentacles; 30c: formation of the actinula body from the middle piece of the planula; 30d: young actinula with mouth aperture, gastrovascular cavity and the first two tentacles. (After Metschnikoff 1886b.)

Fig. 31a–d. Postembryonic development of the calycophorides Galeolaria (figs. 31a–c) and Muggiaea (fig. 31d); sprouting of the individual persons of the siphonophore colony from a lateral budding zone. 31a–c are sections. 31a: planula; 31b: primordium of the first swimming bell and of the tentacle of the first nutritive polyp; 31c: formation of the gastric cavity, further development of the first swimming bell and first tentacle; 31d: siphonula stage with completed first cormidium composed of swimming bell, nutritive polyp, tentacle and bract; primordium of the second swimming bell on the budding zone, which has remained embryonic. (31a–c after Metschnikoff 1874; 31d after Chun 1886.)

Fig. 32a–c. Postembryonic development of the physophoride *Halistemma picta*. Sprouting of the individual persons from two budding zones begins with the anlage of the pneumatophore and of the first larval tentacle. Sections. 32a: planula with the anlagen of the pneumatophore at the upper end and that of the first tentacle at the side; 32b: growth of the tentacle and development of the gastric cavity; 32c: older larva with two tentacles, nematocyst batteries, pneumatophore and first nutritive polyp. (After Chun 1886.)

Fig. 33a–c. Postembryonic development of the physophoride *Agalma sarsii* begins with the formation of the larval bract. Sections. 33a: larva with the primordia of the primary bract and the pneumatophore, together with the aboral and central liquid cavity; 33b: bract greatly enlarged, pneumatophore sunken in; 33c: after regression of the primary bract the definitive pneumatophore has developed further. (After Woltereck 1905.)

Plate V.

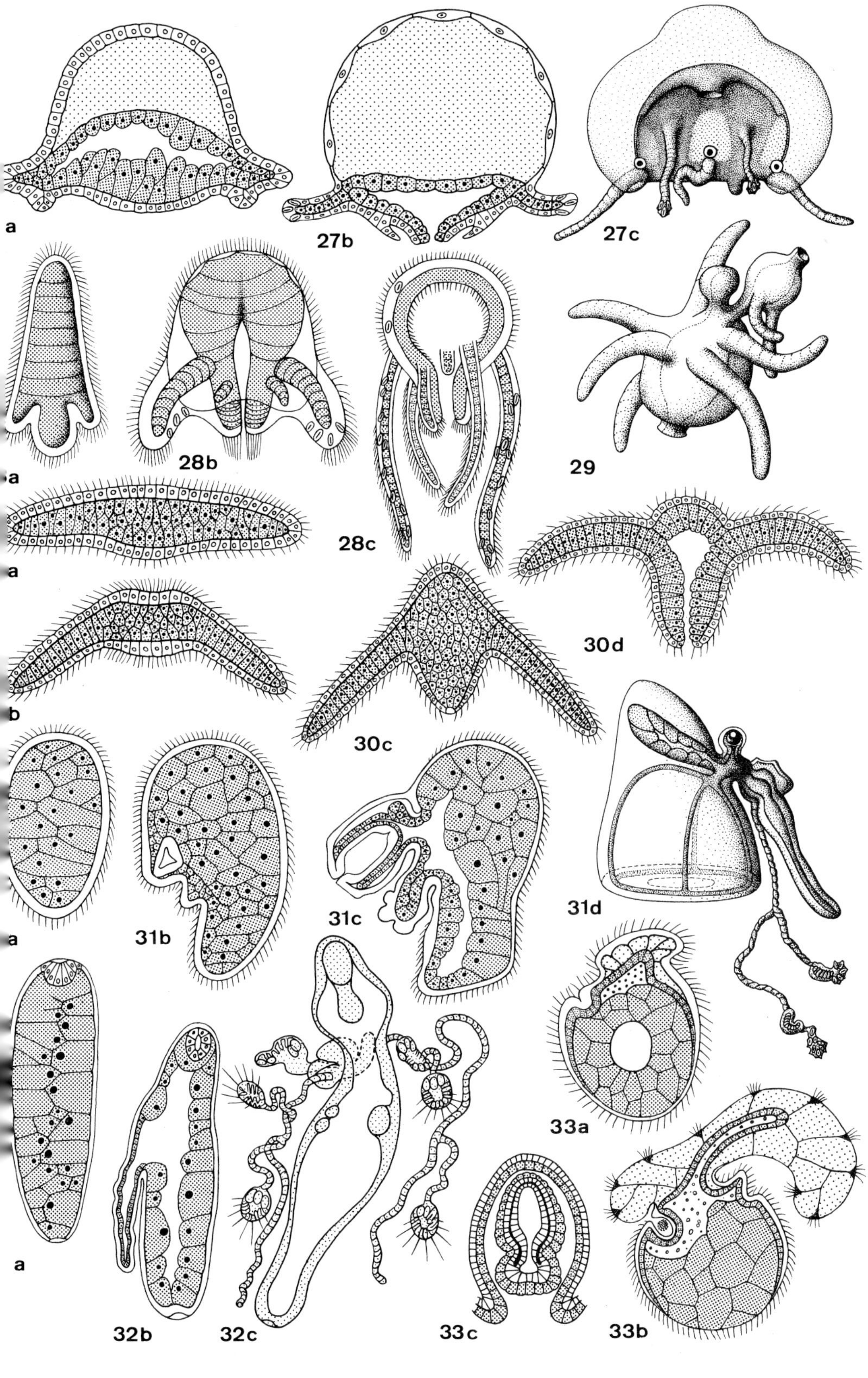
a
27b
27c
a
28b
29
a
28c
a
30d
b
30c
a
31b
31c
31d
33a
a
32b
32c
33c
33b

PLATE VI

Embryonic and postembryonic development of Scyphozoa

Fig. 34a–i. Embryonic and postembryonic development of *Aurelia aurita* up to the ephyra larva. 34a–f are median sections, 34g a cross section, 34h, k side views, 34i, l views from above. 34a: blastula; few single blastoderm cells immigrate into the blastocoele and there degenerate; 34b: invagination gastrula; 34c: young planula with primitive mouth closed; 34d: larva attached with the aboral pole at the beginning of metamorphosis; broadening of the oral end and incipient formation of peristome; 34e: young scyphopolyp with definitive mouth opened, primordia of tentacles and incipient division of the gastrovascular cavity; 34f: somewhat older scyphopolyp with growing tentacles, septal formation and proboscis; 34g: the same stage, cross sectioned, with folding in of the 4 primary interradial septa and formation of the 4 primary perradial gastric pouches; 34h: scyphostoma with constriction of the first circular furrow at the distal end as the start of strobilation; 34i: scyphostoma with 20 tentacles, 4 septal funnels and 4 septa visible through the open mouth; 34k: polydisk strobila with 3 ephyra larvae in different stages of development; 34l: young ephyra, just released, with 4 perradial and 4 interradial marginal lappets each of two wing lappets, with 4 perradial and 4 interradial lappet pouches and 8 anlagen of marginal pouches; the mouth aperture, gastric filaments and sense clubs are shown shaded and the boundary between the gastric cavity and the coronary intestine is shown by a dashed line (34a–g after Hein 1900; 34h, k after Claus 1890/92; 34i after Friedemann 1902; 34l from an original preparation by Mergner 1970.)

Fig. 35a–f. Embryonic and postembryonic development of *Chrysaora hyoscella* until the strobila stage. 35a–e are median sections, 35f a side view. 35a: 4-cell stage; 35b: later cleavage stage with incipient development of blastocoele; 35c: blastula; numerous blastoderm cells enter the blastocoele, where they later disintegrate; 35d: blastula at start of invagination; 35e: invagination gastrula; 35f: polydisk strobila with numerous ephyra larvae at greatly different stages of development. (35a, b, d, e after Teissier 1929; 35c from an original preparation by Mergner 1970; 35f after Hadži 1907.)

Fig. 36. Gastrula of *Linuche unguicula* with entoderm development by invagination, associated with polar ingression from the roof of the archenteron. Median section. (After Conklin.)

Plate VI.

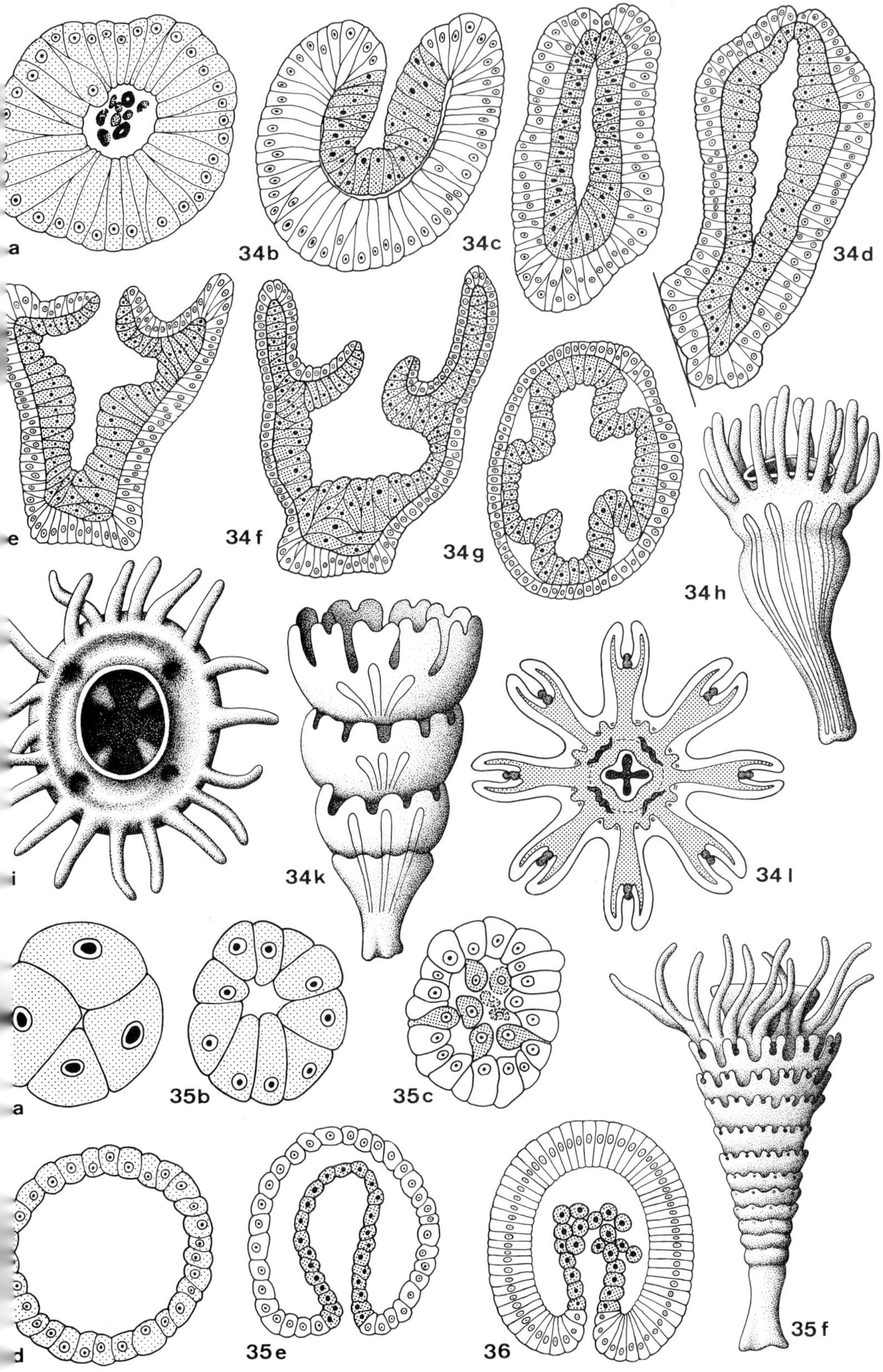

PLATE VII

Embryonic and postembryonic development of Scyphozoa

Fig. 37a–e. Development of *Thaumatoscyphus distinctus* up to the planula. 37c–e are sections. 37a: 8-cell stage; 37b: 16-cell stage; in spite of radial cleavage the animal blastomeres show a slight horizontal twist towards the vegetative blastomeres; 37c: morula; 37d: few-celled polar ingression of the prospective entoderm; 37e: solid planula with column of entoderm cells and flattened ectoderm composed of few cells. (After Hanaoka 1934.)

Fig. 38a–h. Embryonic and postembryonic development of *Haliclystus octoradius;* example of development without a free medusa stage. 38a–d and f are median sections, 38e and h side views, 38g a view from above. 38a: entoderm ingression by polar immigration of a few cells; 38b: sterrogastrula with entoderm development completed; 38c: solid planula with cuboid ectoderm cells and a column of large vacuolated entoderm cells; 38d: attached larva in metamorphosis, with opened mouth, a tentacle anlage, gastric cavity and incipient sinking in of pedal disk; 38e: stage of metamorphosis with 4 tentacles and 4 tentacle primordia; pedal disk deeply sunk in; 38f: 8-tentacle stage; the longitudinal section shows the mouth cone, the septal funnels with the ectodermal cnidoblast forming sites beneath and the ectodermal septal muscles, as well as the pharyngeal tube, gastric cavity and sinking in of pedal disk; 38g: 16-tentacle stage with 4 perradial, 4 interradial and 8 aradial tentacles, 4 septal funnels, 4 rhopaloid thickenings and quadriradial mouth opening; the outlines of the septa and gastric pockets are shown by a dotted line; 38h: young scyphopolyp with incipient development of the tentacle groups, oral cone in 4 parts and the 4 septal funnels. (After Wietrzykowski 1912.)

Fig. 39a–f. Postembryonic development of *Pelagia perla:* example of development without sessile polyps. 39a–e are side views, 39f a view from above. 39a: planula with incipient formation of mouth cone at the oral end; 39b: primordium of the 8 marginal lappets, 39c: growth of the marginal lappets; 39d: incipient shortening of the main axis, flattening of the subumbrella and full development of the wing lappets; 39e: full development of the mouth cone, lappets and sense clubs; after loss of the cilia the larva moves by contraction of the bell; 39f: ephyra larva. (After Delap 1905.)

Fig. 40. Young scyphopolyp of *Charybdaea rastonii* with 2 tentacles. Longitudinal section. (After Okada 1927a.)

Plate VII.

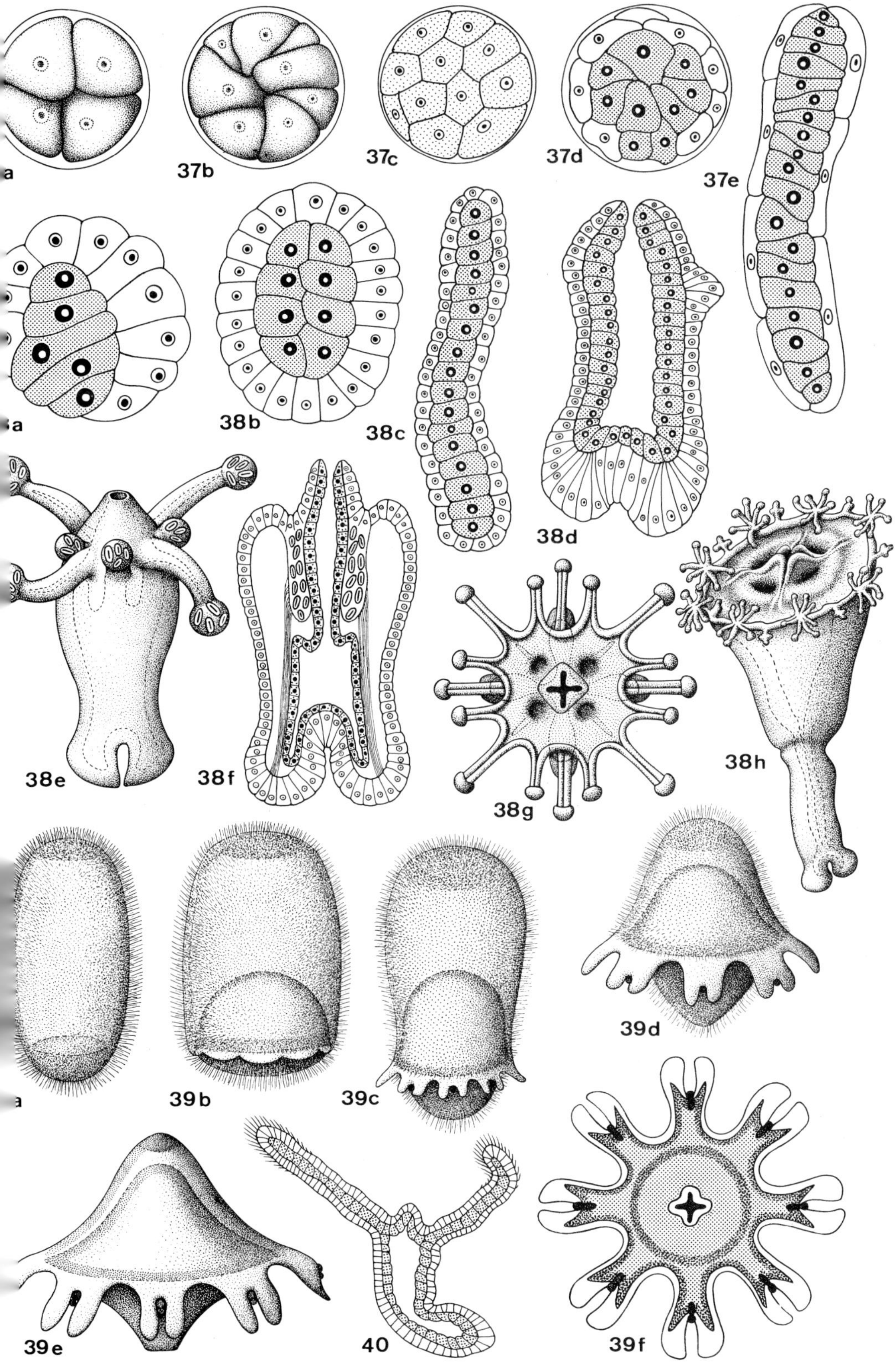

a
37b
37c
37d
37e
a
38b
38c
38d
38e
38f
38g
38h
a
39b
39c
39d
39e
40
39f

PLATE VIII

Embryonic and postembryonic development of Anthozoa

Fig. 41a–m. Embryonic and postembryonic development of *Pachycerianthus multiplicatus;* example of direct development without passing through the Arachnactis stage. 41a–c, i and k are median sections, 41d, h, l and m side views, 41e–g polar views. 41a: first cleavage step, metaphase; 41b: first cleavage step; radial cutting in of first furrow; 41c: early 4-cell stage seen from the pole; 41d: 16-cell stage; 41e–g: pseudospiral cleavage as a variant of the normal radial cleavage shown in figs. 41a–d by subsequent shifting of the daughter blastomeres after their constriction; 41e: 4-cell stage with two blastomeres displaced by approx. 45 degrees; 41f: displacement of a blastomere in the 4-cell stage; 41g: 8-cell stage; displacement of 4 blastomeres into the furrows of the underlying cells; 41h: young blastula as the outcome of normal, as of modified, radial cleavage; 41i: modified course of invagination due to massive filling of the blastocoele with yolk; deformation of the archenteron by cutting in of a circular furrow and simultaneous passage of the yolk into the lumen of the archenteron; 41k: completed invagination; the whole yolk has passed into the cavity of the archenteron through intercellular spaces; 41l young larva with broadened oral end and incipient pharynx formation; 41m: young actinian polyp with 6 tentacles and 6 septa. (After Nyholm 1943.)

Fig. 42a–e. Embryonic and postembryonic development of *Sagartia troglodytes.* 41a, d and e are side views, 42b is a polar view, 42c a median section. 42a: 4-cell stage; 42b: 16-cell stage; 42c: invagination of the archenteron into the yolk-filled blastocoele; 42d: larva with pharyngeal tube sunk in at the oral end, incipient septal formation and aboral ciliary tuft; 42e: older larva, already lengthened, with pharyngeal tube and aboral ciliary tuft. (42a–c, e after Nyholm 1943; 42d after Carlgren 1906.)

Fig. 43. Invagination gastrula of *Urticina crassicornis.* Median section. As a result of the filling of the blastocoele with yolk the archenteron invaginates irregularly. (After Appellöf 1900.)

Fig. 44a–c. Embryonic development of *Metridium marginatum.* Median sections. 44a: blastula; the inner parts of the blastoderm cells with clumps of yolk separate from the definitive blastoderm and decompose in the blastocoele into nutrient substance; 44b: blastula at the start of invagination; 44c: typical invagination gastrula. (After McMurrich 1891.)

Plate VIII.

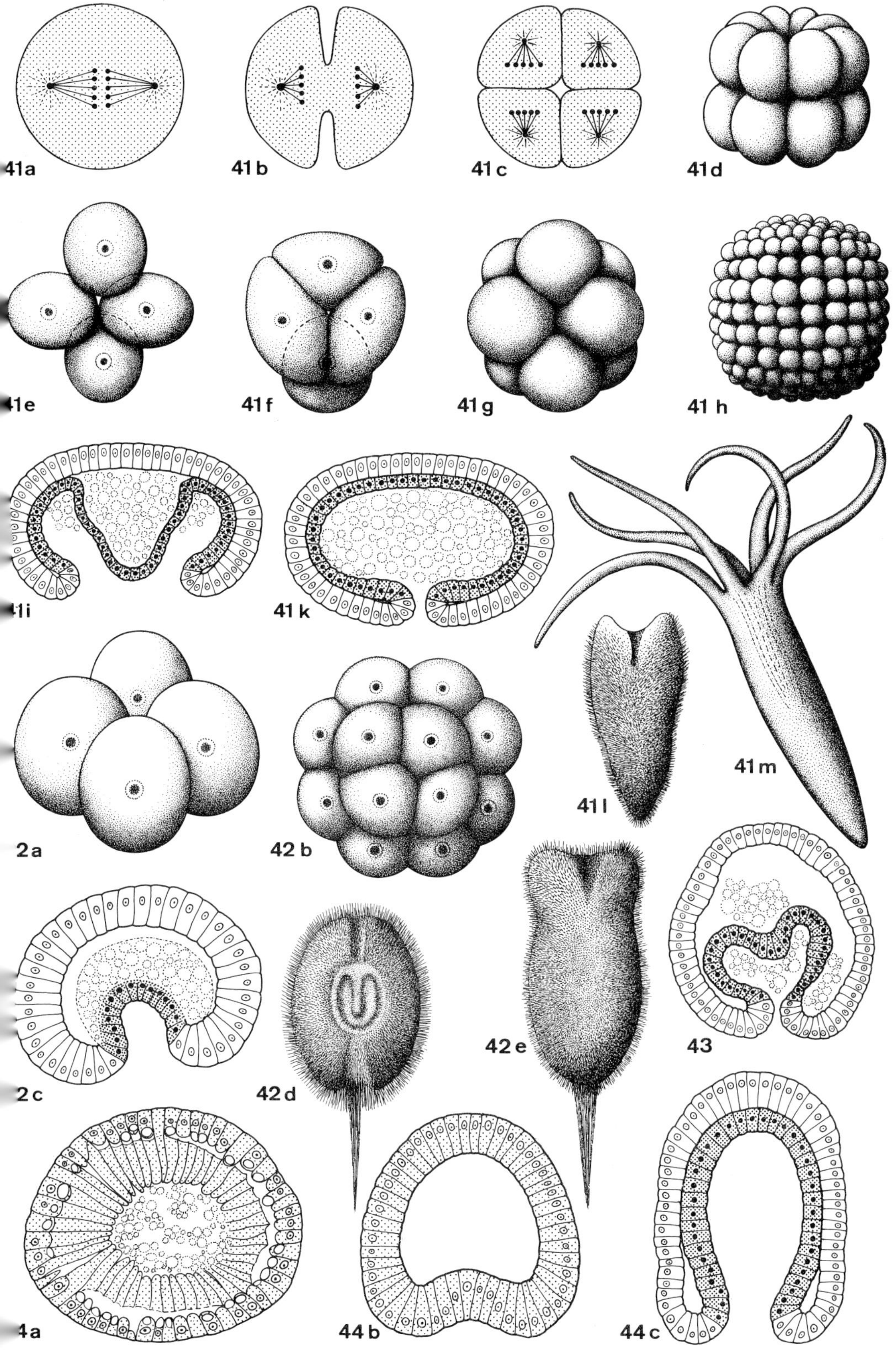
41a
41 b
41 c
41d
1e
41 f
41 g
41 h
1i
41 k
41 m
41 l
2 a
42 b
2 c
42 d
42 e
43
4a
44 b
44 c

PLATE IX

Embryonic and postembryonic development of Anthozoa

Fig. 45a–e. Embryonic and postembryonic development of *Cerianthus lloydii;* example of early development by way of superficial cleavage and invagination. 45a–c: are median sections, 45d a side view, 45e a cross section. 45a: early stage of syncytial cleavage: the first cleavage nuclei are multiplying in the yolk-filled interior; 45b: incipient invagination; all cleavage nuclei have reached the periphery of the embryo; the embryo has somewhat flattened at the vegetal pole; prospective entoderm nuclei accumulate at this location; 45c: completed invagination; development of cell boundaries in both germ layers; 45d: young Arachnactis larva with 4 tentacles, filled with yolk; 45e: same stage sectioned across at the level of the pharyngeal tube shows 2 formed and 2 still incompletely developed septa. (45a–d after Nyholm 1943; 45e after Van Beneden 1891.)

Fig. 46a, b. Embryonic development of *Halcampa duodecimcirrata;* example of embryonic development by way of superficial cleavage and multiphasic morphogenesis of the entoderm; the initial invagination soon passes into ingression. Median sections. 46a: blastula with uncleaved central yolk mass and peripheral blastoderm cells at the end of superficial cleavage; 46b: incipient entoderm development by greatly modified invagination, while the blastomeres eliminate the yolk before their ingression. (After Nyholm 1949.)

Fig. 47a–e. Embryonic and postembryonic development of *Sympodium coralloides;* example of early development by way of holoblastic cleavage and morula delamination. 47a, b, d and e are median and longitudinal sections, 47c is a cross section. 47a: incipient entoderm formation by delimitation of the outer cells as ectoderm; 47b: completed morula delamination; young sterrogastrula; 47c: formation of the gastric cavity and dissolution of the yolk mass; 47d: planula with yolk largely absorbed; 47e: attached larva at beginning of metamorphosis; greatly shortened main axis and sinking in of the pharyngeal tube announce the transformation to the young primary polyp. (After Kowalevsky and Marion 1883.)

Fig. 48a–d. Postembryonic development of *Bolocera turdiae*. 48a and b are median sections, 48c and d cross sections. 48a: older invagination gastrula with incipient sinking in of the pharyngeal tube and abundant yolk in the archenteron; 48b: older larva with deepened pharynx; 48c: formation of the first pair of septa on the attached larva; 48d: order of formation of the first 4 pairs of septa; septal pouches still well filled with yolk. (After Gemmill 1922.)

Plate IX.

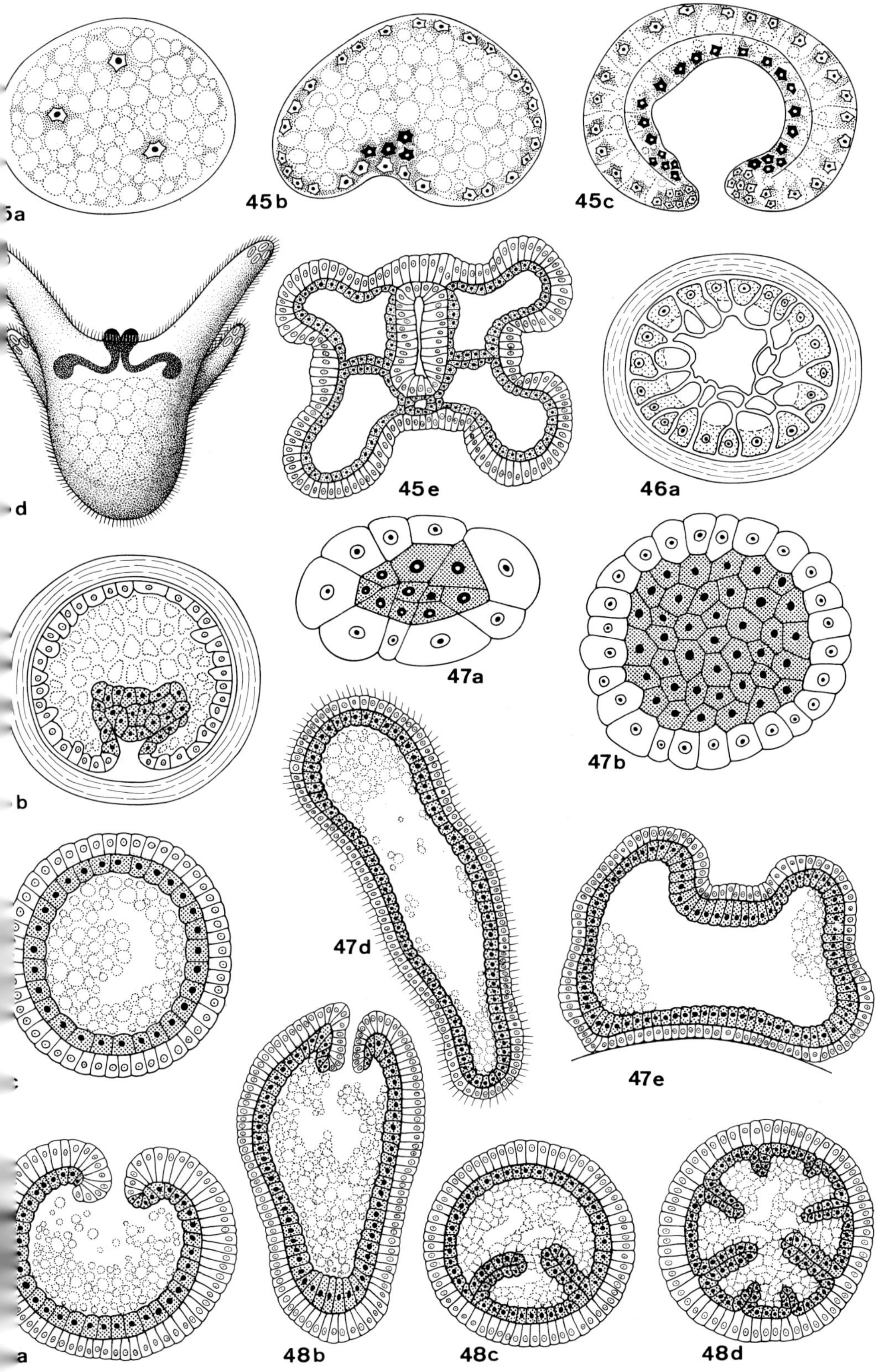

PLATE X

Postembryonic development of Anthozoa

Fig. 49a–d. Development of *Epiactis prolifera*. 49a is a median section, 49b–d are cross sections. 49a: older gastrula with yolk-filled gastrovascular cavity; 49b: older Edwardsia stage; upper part with pharyngeal tube, with 8 complete and 4 incomplete septa and septal pouches filled with yolk; 49c: Halcampula stage; upper part with densely folded pharyngeal tube, with 12 complete and 12 incomplete septa; the yolk has already been absorbed; 49d: 24-tentacle stage; lower part with 8 complete and 16 incomplete septa and formation of muscle bundles on the septa (young actinian). (After Uchida and Iwata 1954.)

Fig. 50a–c. Postembryonic development of Renilla. 50b is a longitudinal section. 50a: older planula with 8 tentacle anlagen, 8 septa, pharyngeal tube sunk in and the first polyp bud; 50b: larva with formation of pharyngeal tube and a septum demonstrated in section; 50c: primary polyp with incipient bud formation; a terminal zoid bud (above) and 4 autozoid buds (below) can be distinguished; one of the latter is shown enlarged at the left. (After Wilson 1884.)

Fig. 51. Primary polyp of *Anthoplexaura dimorpha;* cross section at the level of the pharyngeal tube, 8 septa with muscle bundles. (After Kinoshita 1910.)

Fig. 52a, b. Postembryonic development of *Siderastraea radians*. 52a: older planula with 12 septa, still without tentacles; 52b: young coral polyp with basal plate, 6 tentacles, 12 septa and the primordium of two circlets of 6 sclerosepta each. (After Duerden 1904.)

Fig. 53a–i. Secondary larvae of Ceriantharia (figs. 53a–e), Actinaria, (fig. 53f), and Zoantharia (figs. 53g–i). 53a: *Cerianthula braemi* with 12 septa and 12 tentacles; 53b: *Calpanthula guinensis* with 24 tentacles; 53c: *Anactinia pelagica* with 24 septa, but still without tentacles; 53d: young Arachnactis with 4 tentacles; 53e: *Arachnactis valdiviae* with 12 tentacles in two circlets, 53f: Edwardsia larva of *Urticina felina;* cross section through the upper part with 8 septa, still without tentacles; 53g: *Zoanthella henseni* with ventral row of cilia and aboral porus; 53h: *Zoanthina americana* with 12 septa, aboral ciliary ring and deep constriction behind; 53i: older Zoanthina larva with 12 deeply constricted septa and 12 tentacle buds. (53a–c, e after Carlgren 1924; 53d after Van Beneden 1891; 53f after Appellöf 1900; 53g, h after Conklin 1909; 53i after Carlgren 1906.)

Plate X.

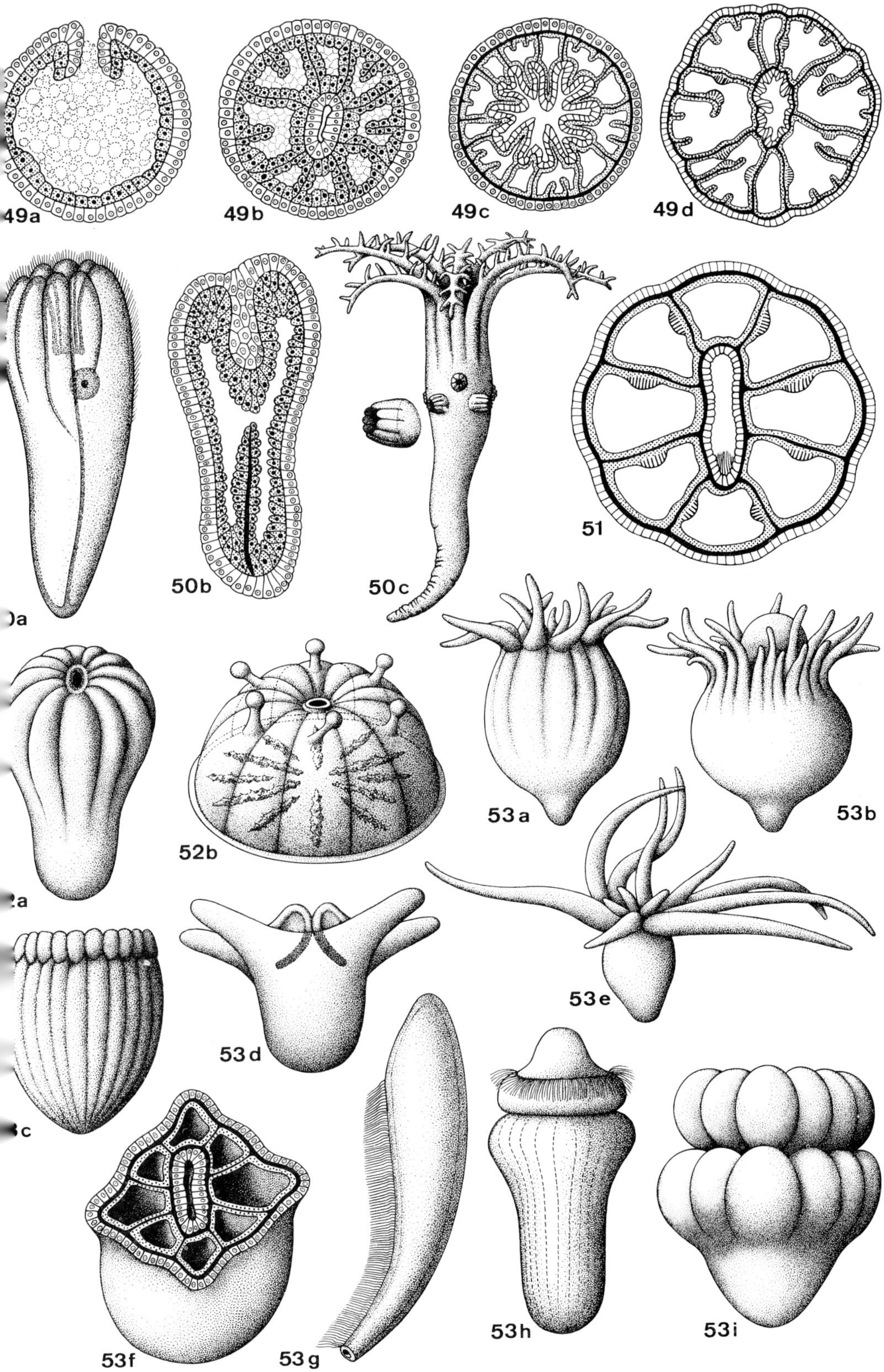

PLATE XI

Experimental analysis in Sertularia operculata

Fig. 54a–k. Normal development of *Sertularia (Amphisbetia) operculata* up to the planula stage. The orange pigment in the free oocyte half (densely dotted), by remaining during cleavage and gastrulation, proves the pigmented cells to be prospective entoderm. 54a: two oocytes in the spadix ectoderm; 54b: shed oocyte; 54c: first cleavage division; 54d: 4-cell stage at the start of the third cleavage step; 54e: 8-cell stage; 54f: late cleavage; 54g: blastula; 54h: start of polar ingression of the presumptive entoderm cells at the vegetal pole of the blastula; 54i: advanced ingression of entoderm cells; 54k: young planula. (After Teissier 1931.)

Fig. 55a–f. Isolation experiments with embryos of *Sertularia (Amphisbetia) operculata* demonstrate the great regulative capacy of blastomeres and incomplete embryos in the course of indeterminative embryogenesis, but also its limitation by the retention of the polarity taken over from the oocyte. 55a: development of isolated half-blastomeres to the planula; 55b: development of isolated quarter–blastomeres to the planula; 55c: separation of the animal from the vegetal half in the 8-cell stage; 55d: separation of the animal from the vegetal half in the blastula stage; 55e: oblique section of the blastula and staining of the cut edges; retention of the polarity of the egg in the planulae (x shows the primary poles); 55f: development of an embryo after displacement of the blastomeres at the 4-cell stage; 55f$_1$: transition from the 4- to the 8-cell stage corresponding to fig. 54d of normal development; 55f$_2$, 55f$_3$: no differentiation into uniform planulae, division into two masses of cells. (After Teissier 1931.)

Plate XI.

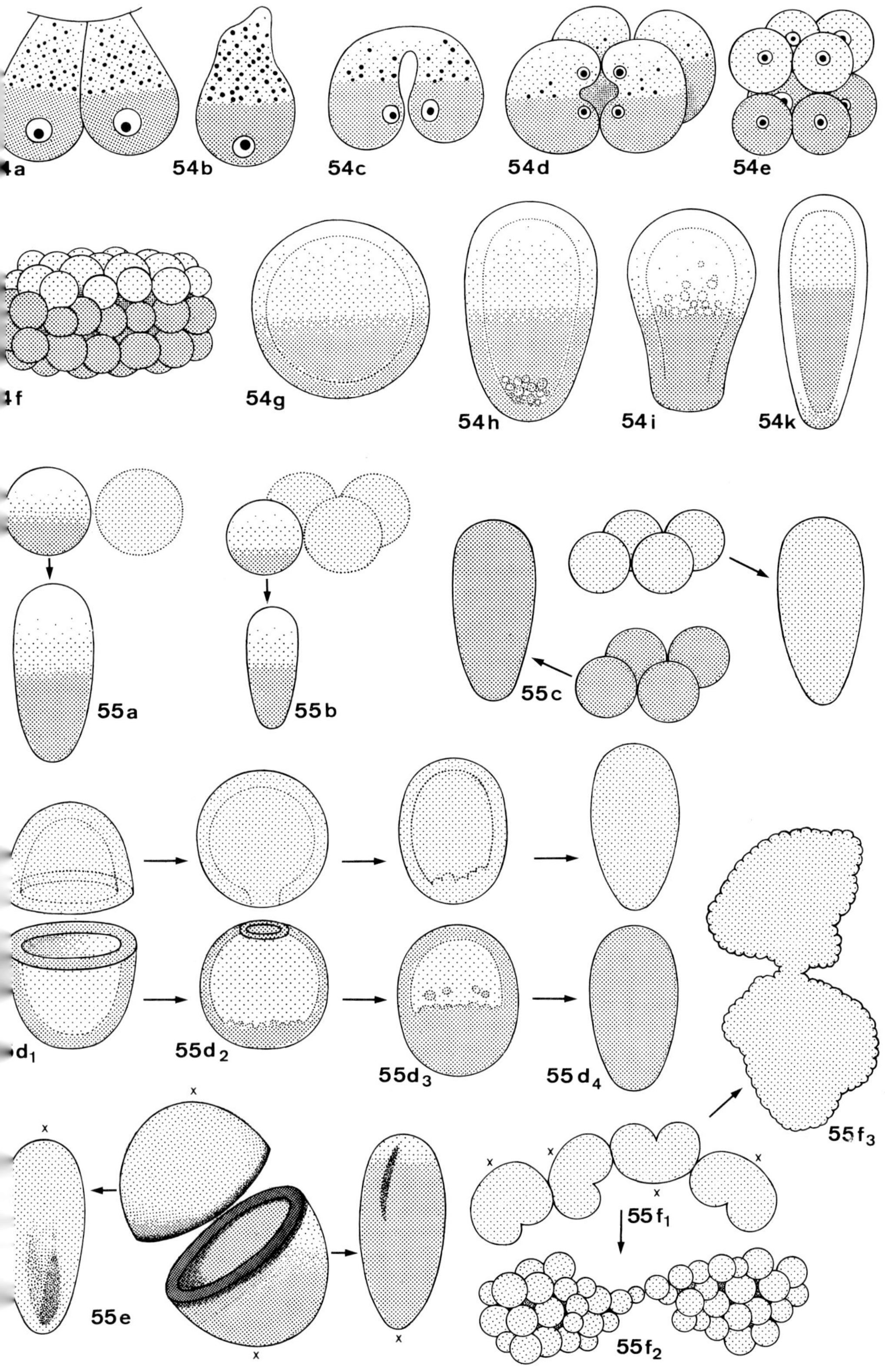
54b
54c
54d
54e
54g
54h
54i
54k
55a
55b
55c
$55d_2$
$55d_3$
$55d_4$
55e
$55f_1$
$55f_2$
$55f_3$

PLATE XII

Experimental analysis in Hydractinia echinata

Fig. 56a–e. Normal embryonic development of *Hydractinia echinata.* 56a: unilateral cutting in of the first furrow at the animal pole; 56b: 2-cell stage; 56c: 4-cell stage at the start of the third cleavage step; 56d: 16-cell blastula with incipient ingression of the entoderm by mixed delamination; 56e: continuing entoderm formation by ingression of whole cells and release of daughter cells of paratangential cleavages of the blastoderm. Multipolar ingression of whole cells and the radial spindle orientation in delamination are indicative of the radial polarity of the egg. Ultimately, however, the primary animal-vegetal axis of the egg determines the main axis of the planula. (After Teissier 1931.)

Fig. 57a–c. Centrifugation experiments with eggs of *Hydractinia echinata* to stratify the egg content. 57a: cleavage and planula formation of a centrifuged egg after separation of the first two unequally sized blastomeres; each of the two half-blastomeres developes to a normal planula of corresponding size; 57b, c: cleavage and planula formation at any angle of the first plane of cleavage to the plane of stratification and at varying sizes of the daughter cells; all half-blastomeres become normally formed planulae with polar differentiation, in spite of the difference in content. Thus despite different distribution of the egg content the primary axis of the egg determines the animal-vegetal main axis of the embryo; thus it must be located in a polar morphogenetic pattern of the cortex, which remains uninfluenced by the centrifugal force. The Hydractinia embryo is also found to be a regulative embryo and its development to be indeterminative. (After Beckwith 1914.)

Fig. 58a–c. Fragmentation experiments in embryos of *Hydractinia echinata* to determine the unilateral cutting in of the first furrow. 58a: position of the telophase nuclei of the first cleavage step in relation to the egg surface during normal cleavage; 50b: design of the operation to isolate the first two cleavage nuclei, each with a quarter of the normal plasma content, the position of the nuclei in relation to the surface of the embryo being changed; the distances are now approximately equal in all directions; 58c: hence at the second cleavage step the furrows cut in regularly from all sides; this proves that the position of the mitosis apparatus is crucial for the mode of cleavage. (After Rappaport and Conrad 1963.)

Plate XII.

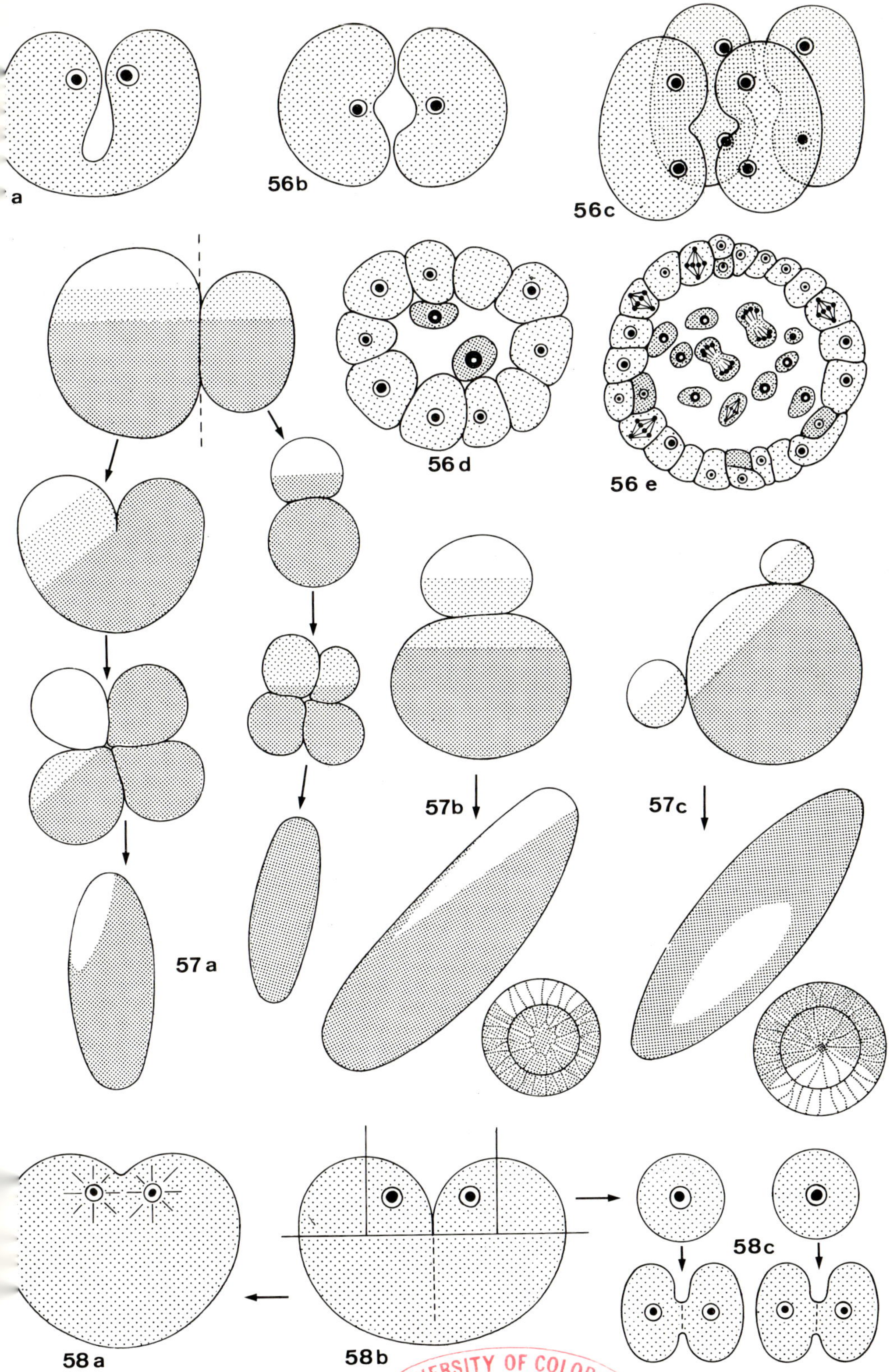

PLATE XIII

Experimental analysis in Chrysaora hyoscella

Fig. 59a–d. Normal development of *Chrysaora hyoscella* up to the planula stage. 59a: 4-cell stage; 59b: late cleavage stage with incipient blastocoele formation; 59c: 120-cell blastula with the final cell size and nucleo-cytoplasm ratio; 59d: as a result of continued nutrition from the maternal organism and continual multiplication of cells the blastula is greatly enlarged; the nucleo-cytoplasm ratio has remained constant, however. (After Teissier 1929.)

Fig. 60a–i. Analysis of the mechanisms underlying termination of cleavage divisions and the initiation of germ layer formation in embryos of *Chrysaora hyoscella;* a so-called 'nature experiment'. Demonstration of the dependence of the larva size on that achieved during blastula growth and of the start of gastrulation by termination of the food supply to the growing blastula. 60a_1, b_1: outlines of blastulae of extremely different sizes as the result of different periods of nutrition; 60a_2, 60a_3: cells of an extremely small blastula, surface view and in section; 60b_2, b_3: the cells of an extremely large blastula have the same dimensions; 60c: start of gastrulation after prolonged growth of the blastula; 60d: start of gastrulation after short growth of the blastula; 60e: planula; 60f_1–f_3: the size of the planulae varies according to that of the blastulae; 60f_1: one of the largest planulae; 60f_2: one of the smallest planulae; 60f_3: planula of average size; 60g–i: cell nuclei from blastulae, gastrulae and planulae, to same scale; in 60h and i the ectoderm nuclei are shown in white and the ectoderm nuclei in black. (After Teissier 1929.)

Plate XIII.

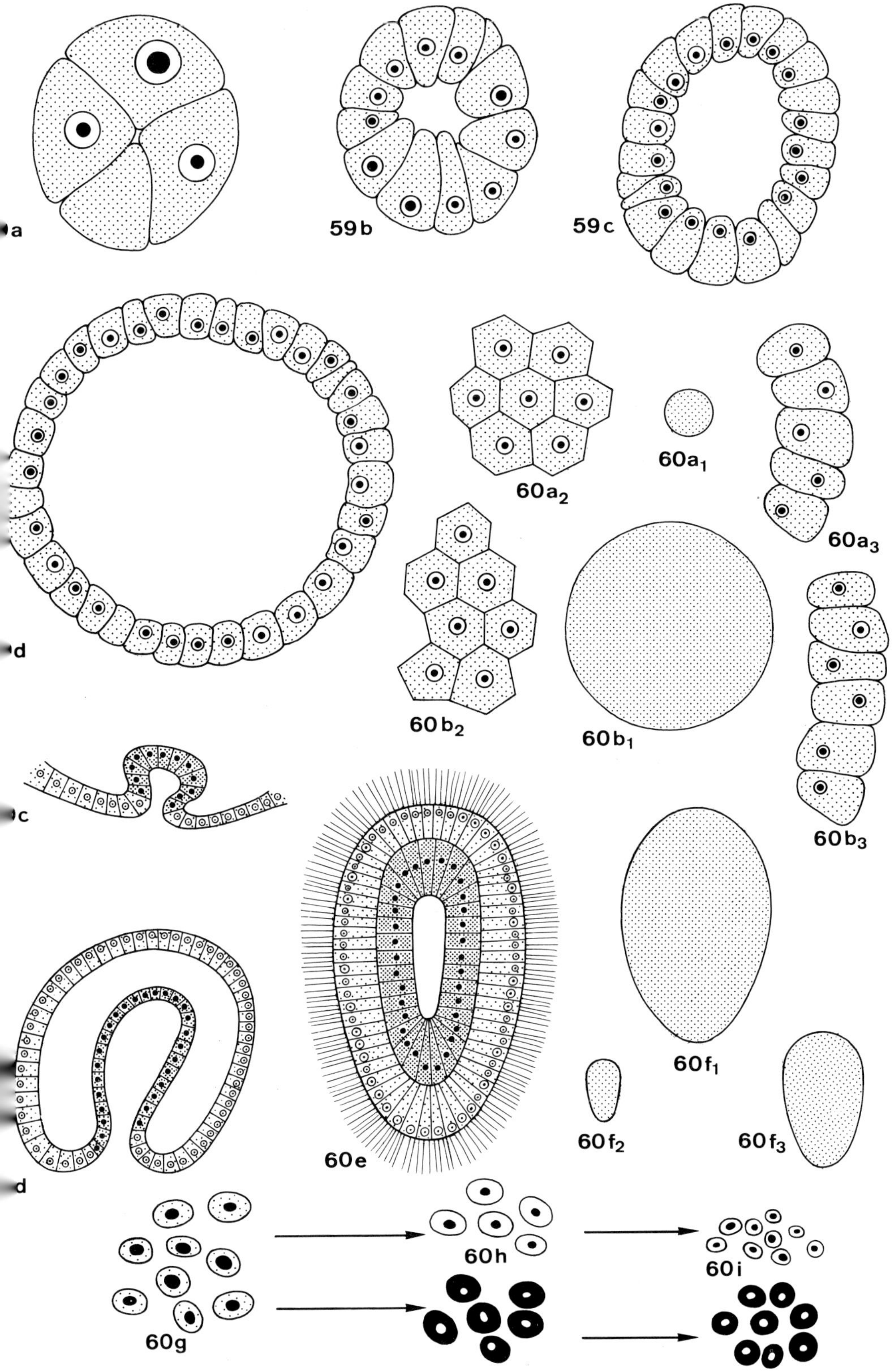
59b
59c
60a1
60a2
60a3
60b1
60b2
60b3
60e
60f1
60f2
60f3
60g
60h
60i

References

AGASSIZ L., 1862. Contributions to the natural history of the United States of America. vol. *4* (Boston).

ALLMAN G. J., 1875. On the structure and development of *Myriothela*. Phil. Trans. Roy. Soc. London *165*.

APPELLÖF A., 1900. Studien über Actinienentwicklung. Bergens Mus. Aarbog.

BECKWITH C. J., 1914. The genesis of the plasma-structure in the egg of *Hydractinia echinata*. J. Morphol. *25*.

BIGELOW H. B., 1909. Reports on the scientific results of the expedition to the Eastern Tropical Pacific. 16. The Medusae. Mem. Mus. Comp. Zool. Harvard Coll. *37*.

BRAUER A., 1891a. Über die Entwicklung von *Hydra*. Z. Wiss. Zool. *52*, 169–216.

BRAUER A., 1891b. Über die Entstehung der Geschlechtsprodukte und die Entwicklung von *Tubularia mesembryanthemum*. Z. Wiss. Zool. *52*, 551–579.

BROOKS W. K., 1886. The life history of the Hydromedusae: a discussion of the origin of the medusae, and of the significance of metagenesis. Mem. Boston Soc. Nat. Hist. *3*.

BROOKS W. K. and S. RITTENHOUSE, 1907. On *Turritopsis nutricola* (Mc Crady). Proc. Boston Soc. Nat. Hist. *33*.

BUNTING M., 1894. The origin of the sex-cells in *Hydractinia* and *Podocoryne*, and the development of *Hydractinia*. J. Morphol. *9*.

CARLGREN O., 1901. Brutpflege bei Actinienlarven. Biol. Zbl. *21*.

CARLGREN O., 1906. Die Actinienlarven. Nord. Plankt. vol. *11*.

CARLGREN O., 1924. Die Larven der Ceriantharien, Zoantharien und Actiniarien der D. Tiefsee-Exp. Ergeb. Tiefsee Exp. *19*.

CARY L. R., 1910. Formation of germ layers in *Actinia bermudensis*. Biol. Bull. *19*.

CHUN C., 1886. Bau und Entwicklung der Siphonophoren. Sitzber. Akad. Wiss. Berlin.

CIAMICIAN J., 1879. Über den feineren Bau und die Entwicklung von *Tubularia mesembryanthemum*. Z. Wiss. Zool. *30*.

CLAUS C., 1883. Untersuchungen über die Organisation und Entwicklung der Medusen. Prag und Leipzig.

CLAUS C., 1890–92. Entwicklung des Scyphostoma von *Cotylorhiza*, *Aurelia* und *Chrysaora*. Arb. Zool. Inst. Wien *9*, *10*.

CONGDON E. D., 1906. Notes on the morphology and development of two species of *Eudendrium*. Biol. Bull. *11*.

CONKLIN E. G., 1909. Two peculiar actinian larvae from Tortugas Florida. Carnegie Inst. Washington *103*.

COWDEN R. R., 1964. A cytochemical study of gonophore and oocyte development in *Pennaria tiarella*. Acta Embryol. Morphol. Exptl. *7*.

DELAP M. J., 1905. Notes on the rearing in an aquarium of *Cyanea*, *Aurelia*, *Pelagia*. Rep. Fish. Ireland *2* and *7*.

DUERDEN J. E., 1904. *Siderastraea radians* and its postlarval development. Carnegie Inst. Washington *20*.

FAUROT L., 1903 and 1907. Développement du pharynx et des cloisons chez les Hexactinies. Arch. Zool. Exptl. Gén. Sér. 4, *1* and *6*.

FRIEDEMANN O., 1902. Die postembryonale Entwicklung von *Aurelia aurita*. Z. Wiss. Zool. *71*.

GEMMILL J. F., 1920. Development of *Metridium* and *Adamsia*. Phil. Trans. Roy. Soc. London *209*.

GEMMILL J. F., 1922. Development of the sea anemone *Bolocera*. Quart. J. Microscop. Sci. *65*.

GERD W., 1892. Zur Frage über die Keimblätterbildung bei den Hydromedusen. Zool. Anz. *15*.

GOETTE A., 1893. Vergleichende Entwicklungsgeschichte von *Pelagia noctiluca*. Z. Wiss. Zool. *55*.

GOETTE A., 1907. Vergleichende Entwicklungsgeschichte der Geschlechtsindividuen der Hydropolypen. Z. Wiss. Zool. *87*.

HADŽI J., 1907a. Über Nesselzellwanderungen bei Hydroidpolypen. Arb. Zool. Inst. Wien *17*.

HADŽI, J., 1907b. Einige Kapitel aus der Entwicklungsgeschichte von *Chrysaora*. Arb. Zool. Inst. Wien *17*.

HADŽI J., 1911. Über die Symbiose von Xanthellen und *Halecium ophiodes*. Biol. Zbl. *31*.

HAECKEL E., 1881. Metagenesis und Hypogenesis von *Aurelia*. Jena.

HAECKER V., 1892. Die Furchung des Eies von *Aequorea forskalea*. Arch. mikr. Anat. *40*.

HAMANN O., 1882. Der Organismus der Hydroidpolypen. Jena. Ztschr. Nat. *15* (N.F. *8*).

HAMANN O., 1883. Beiträge zur Kenntnis der Medusen. Z. Wiss. Zool. *38*.

HANAOKA K., 1934. Notes on the early development of a stalked medusa. Proc. Imp. Acad. *10*.

HANISCH J., 1970. Die Blastostyl- und Spermienentwicklung von *Eudendrium racemosum* Cavolini. Zool. Jahrb. Abt. Anat. Ontog. Tiere *87*.

HARGITT C. W., 1900. A contribution on the natural history and development of *Pennaria tiarella* Mc Cr. Amer. Nat. *34*.

HARGITT C. W., 1904a. The early development of *Pennaria tiarella* Mc Cr. Arch. Entwickl. Mech. Organ. *18*.

HARGITT C. W., 1904b. The early development of *Eudendrium*. Zool. Jahrb. Abt. Anat. Ontog. Tiere *20*.

HARGITT C. W., 1911. Some problems of Coelenterate ontogeny. J. Morphol. *22*.

HARM K., 1903. Die Entwicklungsgeschichte von *Clava squamata*. Z. Wiss. Zool. *73*.

HARTLAUB C., 1899. Zur Kenntnis der Gattungen *Margelopsis* und *Nemopsis*. Nachr. Ges. Wiss. Göttingen.

HARTLAUB C., 1907. Craspedote Medusen, I. Teil, 1. Lfrg.: Codoniden und Cladonemiden. Nord. Plankt. Lief. *12*. Kiel und Leipzig.

HEIN W., 1900. Über die Entwicklung von *Aurelia aurita*. Z. Wiss. Zool. *67*.

HEIN W., 1903. Über die Entwicklung von *Cotylorhiza tuberculata*. Z. Wiss. Zool. *73*.

HÖRSTADIUS S., 1928. Über die Determination des Keimes bei Echinodermen. Acta Zool. (Stockholm) *9*.

HYDE J. H., 1894. Entwicklungsgeschichte einiger Scyphomedusen. Z. Wiss. Zool. *58*.

HYMAN L. H., 1940. The invertebrates: protozoa through Ctenophora (Vol. I). First edition, McGraw-Hill, New York.

KINOSHITA K., 1910. Postembryonale Entwicklung von *Anthoplexaura dimorpha*. J. Coll. Sci. Tokyo *27*.

KLEINENBERG N., 1872. *Hydra*, eine anatomisch-entwicklungsgeschichtliche Untersuchung. Leipzig.

KORSCHELT and HEIDER, 1936. Vergleichende Entwicklungsgeschichte der Tiere. Neu bearbeitet von E. KORSCHELT. Erster Band, Verlag Gustav Fischer, Jena.

KOWALEVSKY A. and A. F. MARION, 1883. Histoire embryogénique des Alcyonaires. Ann. Mus. Hist. Nat. Marseille *1*.

KÜHN A., 1909. Sproßwachstum und Polypenknospung bei den Thecaphoren. Studien zur Ontogenese und Phylogenese der Hydroiden. Zool. Jahrb. Abt. Anat. Ontog. Tiere *28*.

KÜHN A., 1910. Die Entwicklung der Geschlechtsindividuen der Hydromedusen. Studien zur Ontogenese und Phylogenese der Hydroiden II. Zool. Jahrb. Abt. Anat. Ontog. Tiere *30*.

KÜHN A., 1913. Entwicklungsgeschichte und Verwandschaftsbeziehungen der Hydrozoen. I. Die Hydroiden. Ergeb. Fortschr. Zool. *4*.

KÜHN A., 1932. Coelenterata, in: Handwörterbuch der Naturwissenschaften, 2. Auflage. Verlag Gustav Fischer, Jena.

KÜHN A., 1965. Vorlesungen über Entwicklungsphysiologie. 2nd ed. Springer Verlag, Berlin-Heidelberg-New York.

KUMÉ M. and K. DAN, 1968. Invertebrate embryology. Translated from Japanese by J. C. Dan. Nolit, Publishing House, Belgrade.

MAAS O., 1892. Bau und Entwicklung der Cuninenknospen. Zool. Jahrb. Abt. Anat. Ontog. Tiere *5*.

MAAS O., 1905. Experimentelle Beiträge zur Entwicklungsgeschichte der Medusen. Z. Wiss. Zool. *82*.

MCMURRICH J. P., 1891. Development of the Hexactinies. J. Morphol. *4*.

MERGNER H., 1957. Die Ei- und Embryonalentwicklung von *Eudendrium racemosum* Cavolini. Zool. Jahrb. Abt. Anat. Ontog. Tiere *76*.

METSCHNIKOFF E., 1874. Studien über die Entwicklung der Medusen und Siphonophoren. Z. Wiss. Zool. *24*.

METSCHNIKOFF E., 1886a. Medusologische Mitteilungen. Arb. Zool. Inst. Wien *6*.

METSCHNIKOFF E., 1886b. Embryologische Studien an Medusen. Ein Beitrag zur Genealogie der Primitivorgane. Wien.

MORGENSTERN P., 1901. Untersuchungen über die Entwicklung von *Cordylophora lacustris*. Z. Wiss. Zool. *70*.

MOTZ-KOSSOWSKA S., 1905. Contribution à la connaissance des Hydraires de la Méditerranée occidentale. 1. Hydraires gymnoblastiques. Arch. Zool. Exptl. (4) *3*.

MÜLLER-CALÉ K., 1913. Zur Entwicklungsgeschichte einiger Thecaphoren. Zool. Jahrb. Abt. Anat. Ontog. Tiere *37*.

MÜLLER-CALÉ K. and E. KRÜGER, 1913a. Einige biologische Beobachtungen über die Entwicklung von *Aglaophenia helleri*, *Aglaophenia pluma* und *Sertularella polyzonias*. Mitt. Zool. Stat. Neapel *21*, 41–50.

MÜLLER-CALÉ K. and E. KRÜGER, 1913b. Symbiontische Algen bei *Aglaophenia helleri* und *Sertularella polyzonias*. Mitt. Zool. Stat. Neapel *21*, 51–64.

NYHOLM K. G., 1943. Zur Entwicklung und Entwicklungsbiologie der Ceriantharien und Actinien. Zool. Bidr. Uppsala *22*.
NYHOLM K. G., 1949. On the development and dispersal of Athenaria Actinia with special reference to *Halcampa duodecimcirrata* M. Sars. Zool. Bidr. Uppsala *27*.
OKADA Y. K., 1927a. Note sur l'ontogénie de *Charybdaea rastonii*. Bull. Biol. France Belg. *61*, 241–249.
OKADA Y. K., 1927b. Sur l'origine de l'entoderme des Discomeduses. Bull. Biol. France Belg. *61*, 250–262.
PERKINS H. F., 1903. The development of *Gonionemus murbachi*. Proc. Acad. Nat. Sci. Philadelphia *54*.
RAPPAPORT R., 1961. Experiments concerning the cleavage stimulus in sand dollar eggs. J. Exptl. Zool. *148*.
RAPPAPORT R. and G. W. CONRAD, 1963. An experimental analysis of unilateral cleavage in invertebrate eggs. J. Exptl. Zool. *153*.
RITTENHOUSE S., 1910. The embryology of *Stomotoca apicata*, J. Exptl. Zool. *9*.
SALENSKY W., 1911. *Solmundella* und Actinula. Mém. Acad. Sci. St.-Pétersbourg (8) *30*.
SARS M., 1846. Fauna litoralis Norvegiae, 1. Heft: Über die Fortpflanzungsweise der Polypen. Christiania.
SCHNEIDER K. C., 1902. Lehrbuch der vergleichenden Histologie der Tiere. Jena.
SIEWING R., 1969. Lehrbuch der vergleichenden Entwicklungsgeschichte der Tiere. Verlag Paul Parey, Hamburg-Berlin.
SMALLWOOD W. M., 1899. *Pennaria*. Amer. Nat. *33*.
SMITH F., 1891. Gastrulation of *Aurelia flavidula*. Bul. Mus. Harvard Coll. *22*.
STROHL J., 1907. Jugendstadien und 'Vegetationspunkt' von *Antennularia antennina* Johnst. Jena Ztschr. Nat. *42*.
TEISSIER G., 1929. La croissance embryonnaire de *Chrysaora hyoscella*. Arch. Zool. Exptl. *69*.
TEISSIER G., 1931. Étude expérimentale du développement de quelques hydraires. Ann. Sci. Nat., Sér. X, *14*.
TICHOMIROFF A. A., 1887. K istorii ravitija gidroidov (Zur Entwicklungsgeschichte der Hydroiden). Teil 2: *Eudendrium armatum* n.sp. Moscow Soc. Sci. Bull. *50*, Append. 1 (Russian).
TORREY H. B., 1907. Biological studies on *Corymorpha*. 2. The development of *C. palma* from the egg. Calif. Univ. Publ. Zool. *3*.
UCHIDA T., 1926. Anatomy and development of a rhizostome medusa *Mastigias papua*. J. Fac. Sci. Tokyo, Zool. *1*.
UCHIDA T. and F. IWATA, 1954. On the development of a broodcaring actinian. J. Fac. Sci. Hokkaido Univ., Ser. VI, Zool. *12*.
VAN BENEDEN E., 1891. Développement des *Arachnactis*. Arch. Biol. *11*.
VAN BENEDEN E., 1898. Anthozoaires de la Plankton-Expedition. Ergeb. Plankt. Exp. *2*.
VAN BENEDEN P. J., 1849. Recherches sur l'embryologie des Tubularies etc. Nouv. Mém. Acad. Bruxelles.
WASSERTHAL W., 1969. Histologische Untersuchungen zur Ei- und Embryonalentwicklung der Hydroidengattung *Eudendrium* (Ehrenberg). Naturw. Fak. Giessen. Unpublished.
WEISMANN A., 1883. Die Entstehung der Sexualzellen bei den Hydromedusen. Jena.
WERNER B., 1955. Über die Fortpflanzung der Anthomeduse *Margelopsis haeckeli* Hartlaub

durch Subitan- und Dauereier und die Abhängigkeit ihrer Bildung durch äußere Faktoren. Verh. Zool. Ges. Tübingen 1954.

WIETRZYKOWSKI W., 1912. Développement des Lucernaires. Arch. Zool. Exptl. Gén. Sér. 5, *10*.

WIETRZYKOWSKI W., 1914. Développement de l'*Edwardsia*. Bull. Acad. Sci. Krakau, Sci. Nat., Ser. B.

WILSON E. B., 1884. Development of *Renilla*. Phil. Trans. Roy. Soc. London *174*.

WOLTERECK R., 1905. Bemerkungen zur Entwicklung der Narkomedusen und Siphonophoren. Verh. Zool. Ges. *15*.

WULFERT J., 1902. Die Embryonalentwicklung von *Gonothyraea loveni* ALLM. Z. Wiss. Zool. *71*.

ZOJA R., 1895. Sullo sviluppo dei blastomeri isolati delle uove di alcune Meduse. Arch. Entwickl. Mech. Organ. *1*.

G. Reverberi (ed.), Experimental embryology of marine and fresh-water invertebrates

CHAPTER 2

Ctenophores

G. REVERBERI

Zoological Institute, University of Palermo, Italy

2.1. Bibliographic indications

The normal development of the Ctenophores has been described by many authors: first by Kowalewsky (1866, 1873) and later by Agassiz (1874), Metschnikoff (1885), Chun (1880) and Ziegler (1898). These studies, which satisfactorily illustrate the more important problems of development, are considered classical and fundamental. Some additional data have been provided by Yatsu (1911).

In the field of experimental embryology, the names of Chun (1880), Driesch and Morgan (1895), Ziegler (1898), Fischel (1897, 1898, 1903) and Yatsu (1912) should be mentioned. Recently, experimental research has been resumed by Farfaglio (1963a, b), Reverberi and Ortolani (1963) and Ortolani (1963a, b, 1964).

General reviews of normal and experimental development have been published by Morgan (1927), Dawydoff (1928), Schleip (1929), Hyman (1940) and Reverberi (1966).

2.2. The egg

A well-planned review should start with the ovarian egg: however, with the exception of some contributions by Chun employing the usual microscopic techniques, this topic has scarcely been considered. Research with the electron microscope is completely lacking. This situation is probably due to the fact that it is difficult to fix this material properly. In this paper we shall limit our attention to the egg already detached from the ovary, both unfertilized and fertilized.

The Ctenophores are generally planktonic and their eggs are also planktonic. They can, however, be made to lay eggs in the laboratory. The development can then be easily followed and if desired subjected to experimental interventions. The animals lay eggs during most of the year; however, it is not easy to catch them as their occurrence is conditioned by many factors (winds, currents, temperature).

Once brought into the laboratory the animals should be carefully watched. Sometimes long, transparent strands of jelly appear in the glasses in which they are kept; these contain the eggs in the form of many small globules. For experimentation the eggs must be freed from their envelopes; this is easily done by hand with two needles. All operations should be performed on a thin film of agar. The development is very rapid: a motile larva of Bolina is obtained within 8–10 hr. The egg of Bolina is small in comparison with that of *Beroe ovata*, which is 1 mm in diameter. In both species the eggs are transparent, so that development can be followed easily.

2.3. *The normal development*

The egg of *Beroe ovata* has been studied most because of its size. It is enveloped in jelly and protected by a very thin membrane which thickens after fertilization: this peculiarity allows a fertilized egg to be distinguished from an unfertilized one. If examined against a dark background with light from above, the egg appears green-coloured. Some authors maintain that it is fluorescent like a 'piece of uranium glass' and that this fluorescence

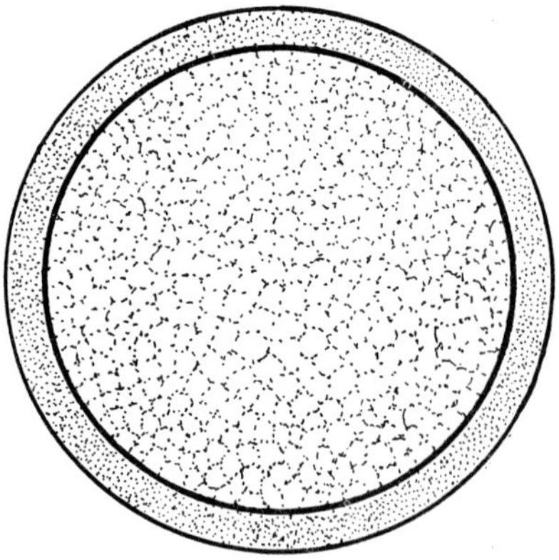

Fig. 1. *Beroe ovata* egg with its three concentric constituents: the thin homogenous outer layer, the fluorescent ectoplasm (fine dots) and the alveolar endoplasm.

increases after stimulation with an electric current. The luminescence is limited to the 'ectoplasm'. According to Yatsu three visible concentric formations should be distinguished in the egg: an extremely thin homogenous outer layer, the ectoplasm, and the endoplasm (fig. 1).

The outer layer is semi-fluid, without granules, and is difficult to observe; the ectoplasm, on the other hand, is a rather thick layer of fine alveolar structure, the endoplasm has a coarse alveolar structure.

An analysis of these structures by electron microscope would be very rewarding. The spermatozoon can enter anywhere and an 'entrance cone' is formed. Sometimes the egg is penetrated by many spermatozoa. Polyspermy is not physiological; however, the supernumerary spermatozoa do not interfere with the normal development and generally degenerate.

The polocytes are given off after fertilization and mark the animal pole of the egg. The first furrow starts at the animal pole, where an accumulation of ectoplasm forms. It is vertical and the furrow is preceded by a downflow of the ectoplasmic thickening.

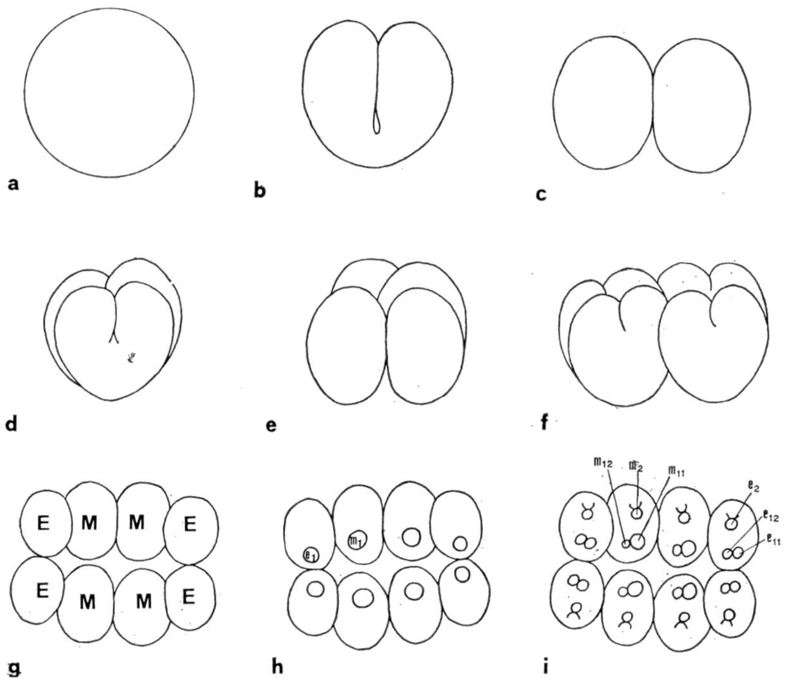

Fig. 2. First stages of development of the Ctenophore egg.

Two blastomeres of equal size and constitution are formed by the first cleavage (fig. 2). At the second segmentation an accumulation of ectoplasm appears again at the animal pole; the cleavage furrow is vertical and 4 equal cells are formed. The third cleavage is oblique, and divides each blastomere into two unequal parts forming 8 cells, the more exterior 4 being smaller than the interior ones. The 4 external cells are indicated by the letter E, the 4 interior ones by the letter M. Together they form a curved plate in which the external cells are situated below. A series of very unequal segmentations follows leading to the formation of many micromeres.

Before describing some of these segmentations in detail it must be remarked that the green ectoplasmic substance is segregated in the micromeres throughout. In the 16–32-cell stage the luminescence is thus strictly limited to the micromeres; this phenomenon has been particularly described by Spek (1926) (fig. 3). The fourth segmentation, as mentioned above, is markedly

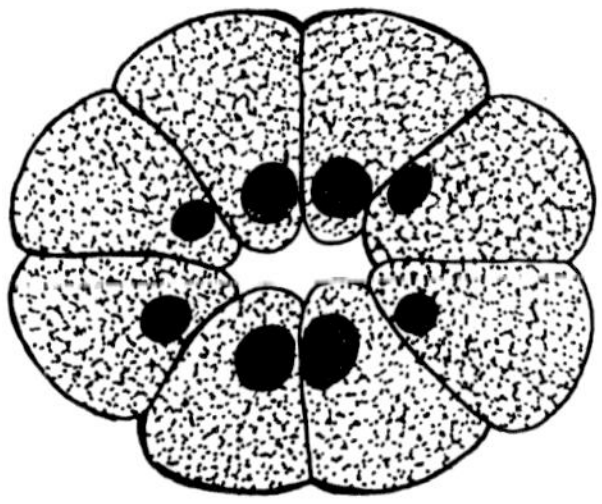

Fig. 3. Egg at 16-cell stage: the fluorescent ectoplasm has been segregated in the micromeres.

unequal. An octet of micromeres is formed: the micromeres which derive from the interior cells are denominated m_1, those deriving from the external cells e_1. These micromeres are budded off at the vegetal pole of the egg; the e_1 are smaller than the m_1, and are emitted somewhat earlier than the latter. A second octet of micromeres (m_2 and e_2) is produced by the fifth segmentation. In the meantime the first octet has divided (giving rise to $m_{1.2}$ and $e_{1.2}$). At this stage (32 cells) each macromere has three micromeres ($m_{1.1}$, $m_{1.2}$, m_2; $e_{1.1}$, $e_{1.2}$, e_2) ($m_{1.1} > m_{1.2}$, while $e_{1.1} = e_{1.2}$).

A new emission of micromeres (e_3) follows, but only from the E-cells. The cell lineage of Bolina is shown in table 1. Through ulterior segmentations

TABLE 1

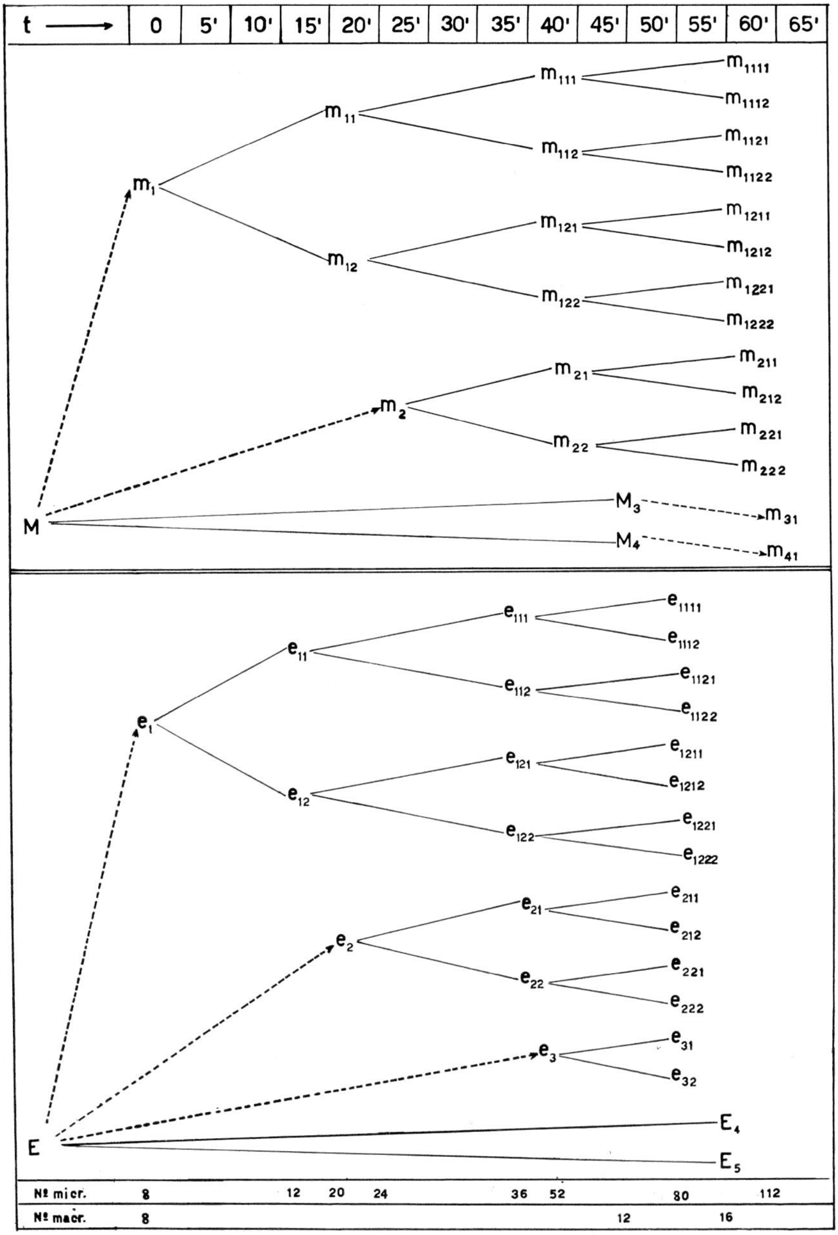
t
0
5'
10'
15'
20'
25'
30'
35'
40'
45'
50'
55'
60'
65'
M
m1
m11
m12
m111
m112
m121
m122
m1111
m1112
m1121
m1122
m1211
m1212
m1221
m1222
m2
m21
m22
m211
m212
m221
m222
M3
M4
m31
m41
E
e1
e11
e12
e111
e112
e121
e122
e1111
e1112
e1121
e1122
e1211
e1212
e1221
e1222
e2
e21
e22
e211
e212
e221
e222
e3
e31
e32
E4
E5
Nº micr. 8 12 20 24 36 52 80 112
Nº macr. 8 12 16

of the micromeres already formed, a crown of micromeres appears at the vegetal pole. A mantle of micromeres, which extends over the macromeres, develops. After these segmentations are completed, the 8 macromeres divide to form 16 macromeres (fig. 4): the direction of the segmentation planes

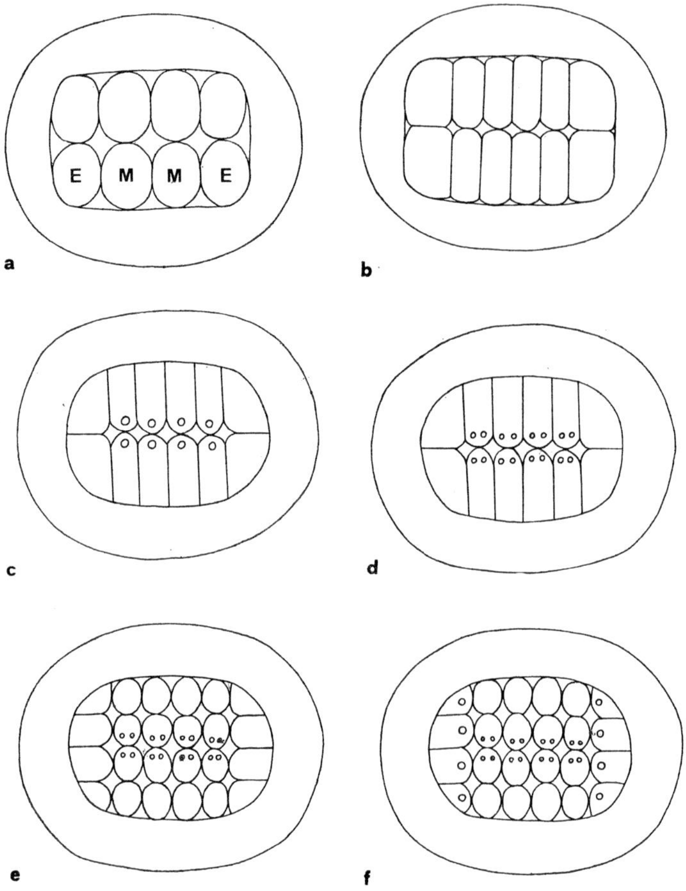

Fig. 4. Formation of the mesodermic micromeres at the animal side of the macromeres.

is different in M and E. The later stage of development is characterized by the fact that the eight derivatives of the M-cells produce two sets of new micromeres, which are formed at the animal pole; the eight derivatives of the E-cells on the other hand (somewhat later) produce only one set of micromeres. If an embryo at this stage of development is observed from the animal pole one notices a sort of crater the bottom of which is made up of the macromeres with apically applied, freshly budded micromeres. The significance of these micromeres will be treated subsequently. A turning of the macromeres produces a cavity, the oesophagus. The larva is rapidly formed; the apical organ appears opposite the mouth and from it radiate four double ectodermic thickenings (the precursors of the comb-plates). The apical organ has a very peculiar structure in the form of a ciliated pit from which emerge four ciliated columns supporting an otholite (fig. 5). These formations are

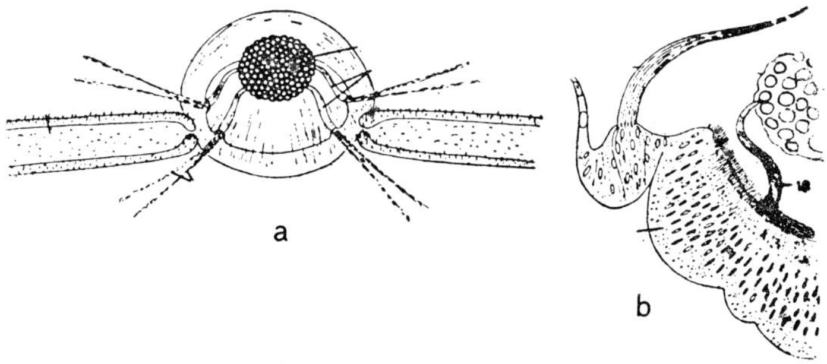

Fig. 5. Apical organ: (a) exterior view; (b) histological structure. (From Hyman 1940.)

protected by a dome consisting of long, transparent cilia. The apical organ controls the equilibrium of the swimming larva. It is connected with the 8 rows of comb-plates by four ciliated furrows.

2.4. The presumptive destiny of the different cells at the 8–16- and 32-cell stage

2.4.1. The destiny of the various cells can be traced in different ways: the most simple method is to mark the blastomeres with coloured granules and to determine the position of these granules in the developed larva.

Let us begin with a granule which has been placed at the animal pole in the immediate vicinity of the polocytes. At the end of development this granule can be seen at the bottom of the gastral cavity in the larva; this indicates that the blastopore, and later the larval mouth (which derives from it), forms at the animal pole.

A granule on the opposite pole does not move much: it remains on the surface and is finally found on the apical organ, which thus forms at the vegetal pole of the egg. A granule which has been put on an E-cell is found at the larval stage in a gastrovascular canal, while a granule on an M-cell is found in the gastric sac. According to these results the E give rise to the longitudinal canals, the M to the gastric pockets.

At the 16-cell stage a granule placed upon an e_1 micromere (fig. 6a) is

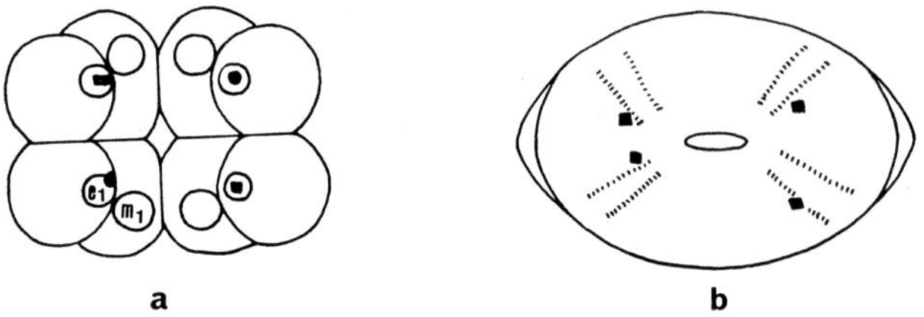

Fig. 6. Egg at the 16-cell stage: (a) the four e_1 micromeres were marked with a coloured chalk granule; (b) the granules are situated on the ciliated combs.

found on the ciliated combs of the larva (fig. 6b). A granule placed on an m_1 micromere (fig. 7a) is found on the apical organ (fig. 7b): this result suggests

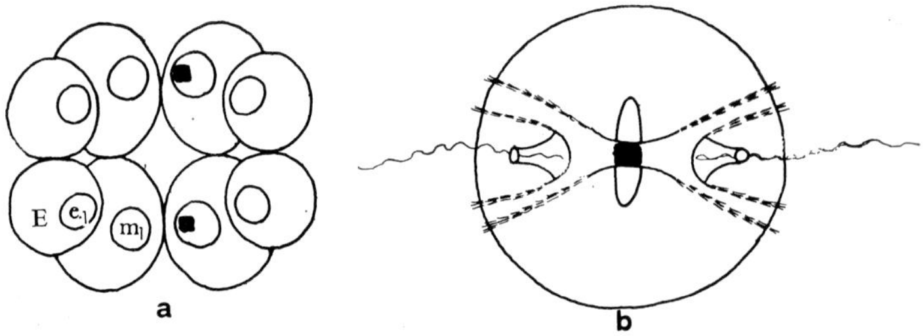

Fig. 7. The four m_1 were marked with chalk granules; the granules are found on the apical organ of the larva.

that the m_1, contrary to the opinion held in the past, have nothing to do with the formation of the ciliated plates. One granule placed on an e_{11} and another on an e_{12} are found in the larva on ciliary plates, but on different rows: this suggests that every e_1 (from which e_{11} and e_{12} derive) is involved in the development of two rows of ciliated plates.

The micromeres emitted from macromeres at gastrulation at the animal pole have a completely different destiny. In fact, they give rise to mesoderm, the derivatives of the M giving rise to the mesenchyme and those of the E to the larval musculature. These conclusions are supported by the results obtained by removing or destroying some blastomeres. Before reporting these experiments (Farfaglio 1963), however, let us examine the results obtained in the past by different authors.

2.4.2 The fragmentation of the unfertilized egg (Beroe) has been performed by Yatsu (1911). He sectioned the egg into two parts and then fertilized the fragments. Each segmented as a miniature egg which gave rise to a small but otherwise normal larva (fig. 8). La Spina (1963) also obtained

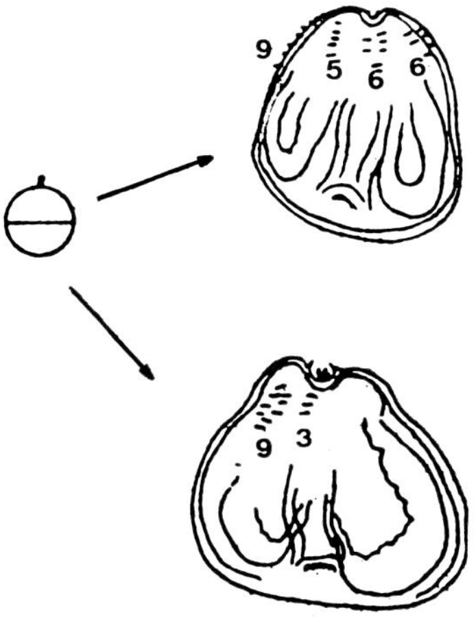

Fig. 8. A fertilized egg of Beroe was cut in two along the equatorial plane: both fragments developed into more or less normal larvae (Yatsu 1912).

corresponding results after breaking the egg by centrifugation. The centrifugal fragment containing the ectoplasm segmented typically, gastrulated and gave rise to a larva having abundant ciliature; the centripetal, larger fragment containing the endoplasm segmented irregularly and gave rise to an undifferentiated cellular mass lacking in cilia (fig. 9). This result emphasizes the importance of the ectoplasm for the development of the ciliary plates.

2.4.3. Experimental isolation of the blastomeres was done by Chun (1880). Studying Eucharis eggs freshly collected from the sea after a storm he noticed that in the same capsule two half-larvae sometimes appeared, each having only 4 rows of ciliary plates instead of the usual eight. Chun supposed that these half larvae had derived from the two mechanically isolated

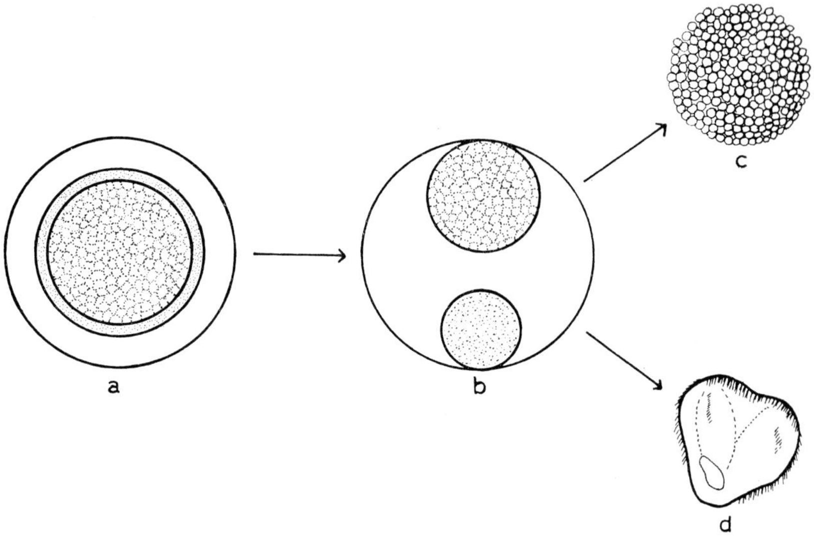

Fig. 9. An egg of Beroe (a) was broken by centrifugation (b); the small centrifugal fragment which contains the green plasm develops into a more or less ciliated larva (d); the centripetal fragment segments irregularly and gives rise to a cellular mass (c) (La Spina 1963).

blastomeres, a hypothesis he confirmed by experiments. A more systematic isolation of the blastomeres at different stages of development of the egg was undertaken by Driesch and Morgan (1895) and later by Fischel (1897, 1898, 1903): they confirmed the strictly mosaic character of the segmented egg.

2.4.4. The isolation of blastomeres from eggs at the 8-cell stage was also done by Yatsu (1912) (fig. 10). From a single E-cell (12 cases) he obtained a larva with only one row of comb-plates. From two isolated E-cells he obtained larvae with two rows of comb-plates. From two M (5 cases) he obtained larvae of various types: 'in one of them two comb-rows were formed . . . in another embryo a very small group of comb-plates was developed. In three no comb-rows were formed at all'. According to Yatsu the developmental potencies of the E and the M, at least with regard to the formation of the comb-rows, are the same: each cell is responsible for one row of combs. The fact that he did not obtain a clear result from isolation of the M was attributed to an 'initial organization or some other physiological condition of the cells'. But Yatsu's assumption 'that each blastomere

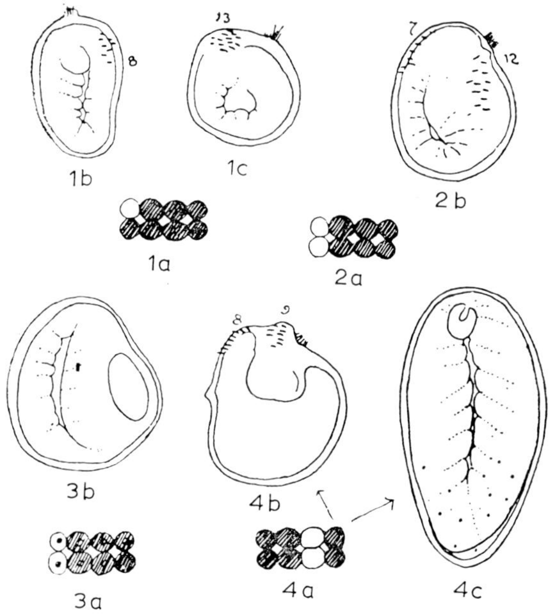

Fig. 10. 1b and 1c: embryos derived from one E (1a); 2b: embryo derived from 2E (2a); 3b: embryo derived from 3a (2E + $2e_1$); 4b and 4c: embryos from 2M (4a). The embryo 4b shows two comb-rows, the embryo 4c is combless (Yatsu 1912).

of the 8-cell stage contains the basis for one comb-row' met insurmountable difficulties in other experiments.

The combination 1E + 2M, according to the hypothesis, should yield larvae with 3 comb-rows; however, this result was not obtained; in fact, in six cases, Yatsu obtained larvae with only two comb-rows. The combination 2E + 1M should also yield larvae with 3 comb-rows, but instead larvae with 2 (one case) or 4 (four cases) comb-rows were formed. The combination 3E + 2M should yield a larva with five rows but Yatsu obtained larvae with six rows (three cases). Finally 2E + 3M should yield larvae with five comb-rows; however, only a larva with four rows appeared (table 2). Yatsu was most perplexed by these results. The smaller number of comb-rows was explained as an arrest of development, but the opposite result was much more difficult to explain. Yatsu was consequently obliged to admit that sometimes the ctenoplasm is unequally distributed among the blastomeres.

TABLE 2

Isolation	Number of experiments	Number of comb-rows formed 1	2	3	4	5	6	7	8	Remarks
1E	12	12	—	—	—	—	—	—	—	
2E	6	—	6	—	—	—	—	—	—	
2M	5	—	1	—	—	—	—	—	—	No comb-rows in 4
1E + 2M	6	—	6	—	—	—	—	—	—	
2E + 1M	5	—	1	—	4	—	—	—	—	
3E + 2M	3	—	—	—	—	—	3	—	—	
2E + 3M	1	—	—	—	1	—	—	—	—	

2.4.5. As Yatsu's explanation seemed unsatisfactory, isolation experiments at the 8-cell stage were resumed (Farfaglio 1963a).

Isolation of the E-cells: at the 8-cell stage all four E of the same egg were separated singly, and their development was followed (fig. 11). The devel-

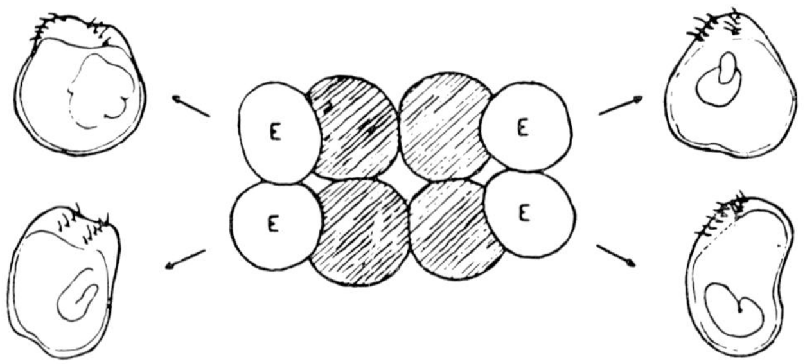

Fig. 11. The four embryos derived from the four isolated E-cells (Farfaglio 1963).

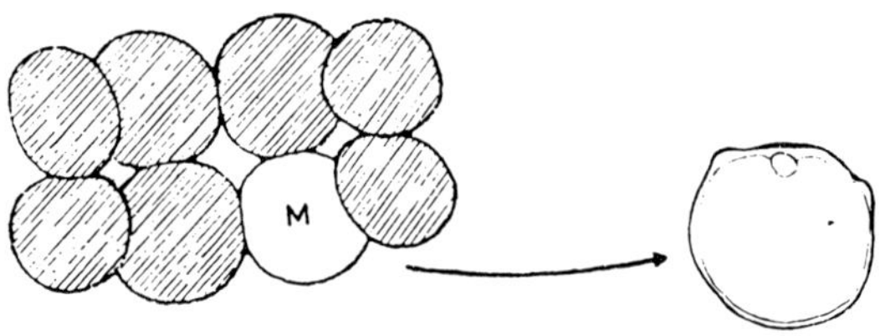

Fig. 12. Larva derived from an isolated M-cell: note the absence of the ciliated plates (Farfaglio 1963).

opment was good; the cells continued to segment as though *in situ* and finally gave rise to 4 partial larvae, each having 2 comb-rows.

Isolation of the M-cells: the isolated M do not develop easily; this is probably due to the fact that the M, as mentioned, produce fewer micromeres than the E and so remain uncovered. In any case, the larvae which developed did not have any combs (fig. 12).

Isolation of 1E + 4M: the larvae had only 2 rows (fig. 13a).

Isolation of 1E + 2M: the larvae obtained from this combination showed only 2 comb-rows (fig. 13b).

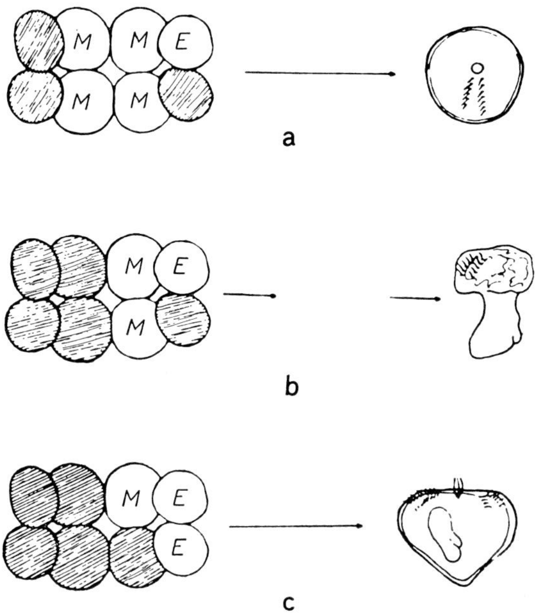

Fig. 13. Embryos derived respectively from 4M + 1E (a); 2M + 1E (b); 1M + 2E (c) (Farfaglio 1963).

Isolation of 2E + 1M: resulted in larvae with 4 comb-rows (fig. 13c).

These results suggest that only the E-cells, and not the M-cells, are involved in the formation of the comb-rows, each of them giving rise to two comb-rows.

2.4.6. This conclusion is confirmed by the results obtained after killing cells of the egg at the 16-cell stage (Farfaglio 1963b).

Destruction of the four e_1: the larvae did not show any ciliated plates (fig. 14a). It is evident from this result that the m_1 are not responsible for the formation of the comb-rows.

Destruction of the four m_1: the larvae which develop from these eggs have all 8 comb-rows (fig. 14b).

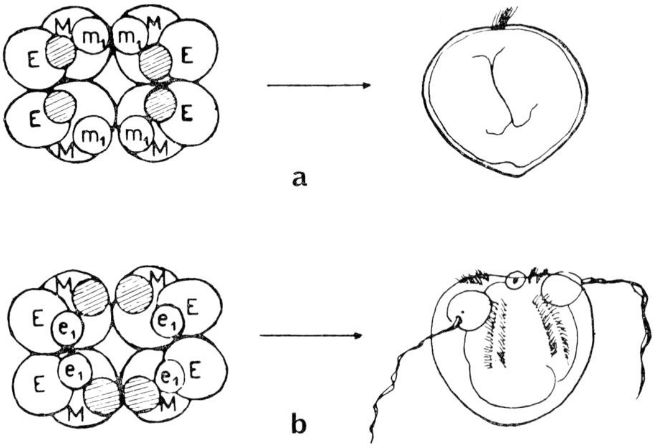

Fig. 14. (a) Embryo derived from an egg in which the four e_1 were killed; (b) embryo from an egg in which the four m_1 were killed (Farfaglio 1963).

Development of four singly isolated m_1: larvae totally lacking in ciliary plates are obtained (fig. 15a).

Development of four singly isolated e_1: small larvae (ectodermic) with some ciliated plates which are however, not arranged in regular rows, are obtained (fig. 15b).

These data again lead to the conclusion that only the e_1 are responsible for the ciliated plates.

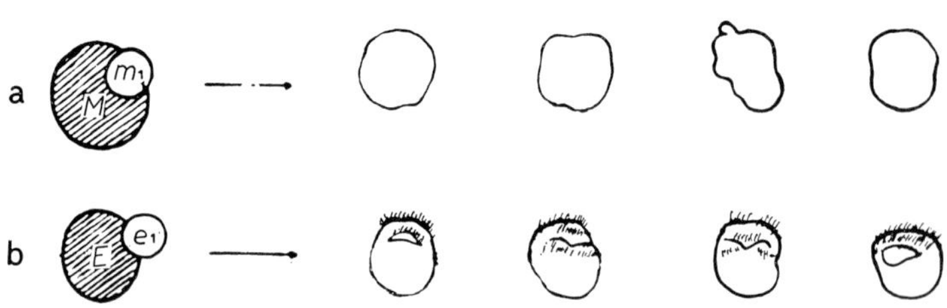

Fig. 15. (a) Embryos derived from m_1; (b) embryos developed from e_1 (Farfaglio 1963).

2.4.7. As shown above, each e_1 is responsible for 2 comb-rows. However, the cleavage of e_1 forms e_{11} and e_{12}: should not each of these two cells be responsible for one comb-row? The experiments give an affirmative answer. Ortolani (1964) destroyed respectively the four e_{11} or the four e_{12} and obtained larvae with only 4 comb-rows.

2.4.8. If we now return to the results obtained by Yatsu (table 2) we realize that all contradictions can be explained by attributing to the E-cells the capacity of forming the ciliary plates and in particular attributing to a single E-cell the responsibility for 2 comb-rows.

2.4.9. Since the m_1 are not involved in the genesis of the comb-rows, what is their function? The problem was approached by Ortolani (1964). As remarked earlier, a granule placed on the m_1 is found on the apical organ in the larva: this result suggests that the m_1 are involved in its formation. The killing or removal of the m_1 confirms this hypothesis. Ortolani removed all the m_1 and obtained larvae with an apical organ, defective in respect of some structures; the long cilia forming the protective dome were still present, but the four pillars supporting the otholite and the short cilia of the pit were wanting. Probably the m_1 are responsible for the formation of the ciliary pillars: each m_1 would be responsible for one pillar.

2.4.10. The destiny of the e_2, m_2 and e_3 was also investigated. In table 3 the development of the single cells up to 32–64-cell stage is indicated. The results of the investigations reported earlier show that before segmentation, the Ctenophore egg is equipotent. At the 2-cell stage, however, it is already a mosaic of parts: the left blastomere, if isolated, gives rise to a left half-larva, the right to a right half-larva. This is true in respect of the comb-rows (although the situation of the gastric sacs is not clear); inductions or interrelations between the different territories were never observed.

2.4.11. The relation between green plasm (in Beroe) and the formation of the comb-rows seems particularly important: this relationship is so close that the green plasm has sometimes been termed ‘ctenoplasm’. Is it actually responsible for the formation of the combs?

It should be recalled that the green plasm is segregated equally at the 4th segmentation into the 8 micromeres; only four, however, (the e_1), give rise to the ciliary rows. The chemical constitution of the green plasm is not known: is it really fluorescent, as asserted? Does it fluoresce more intensively

TABLE 3

Cell-lineage of the egg of the Ctenophores (Ortolani 1964).

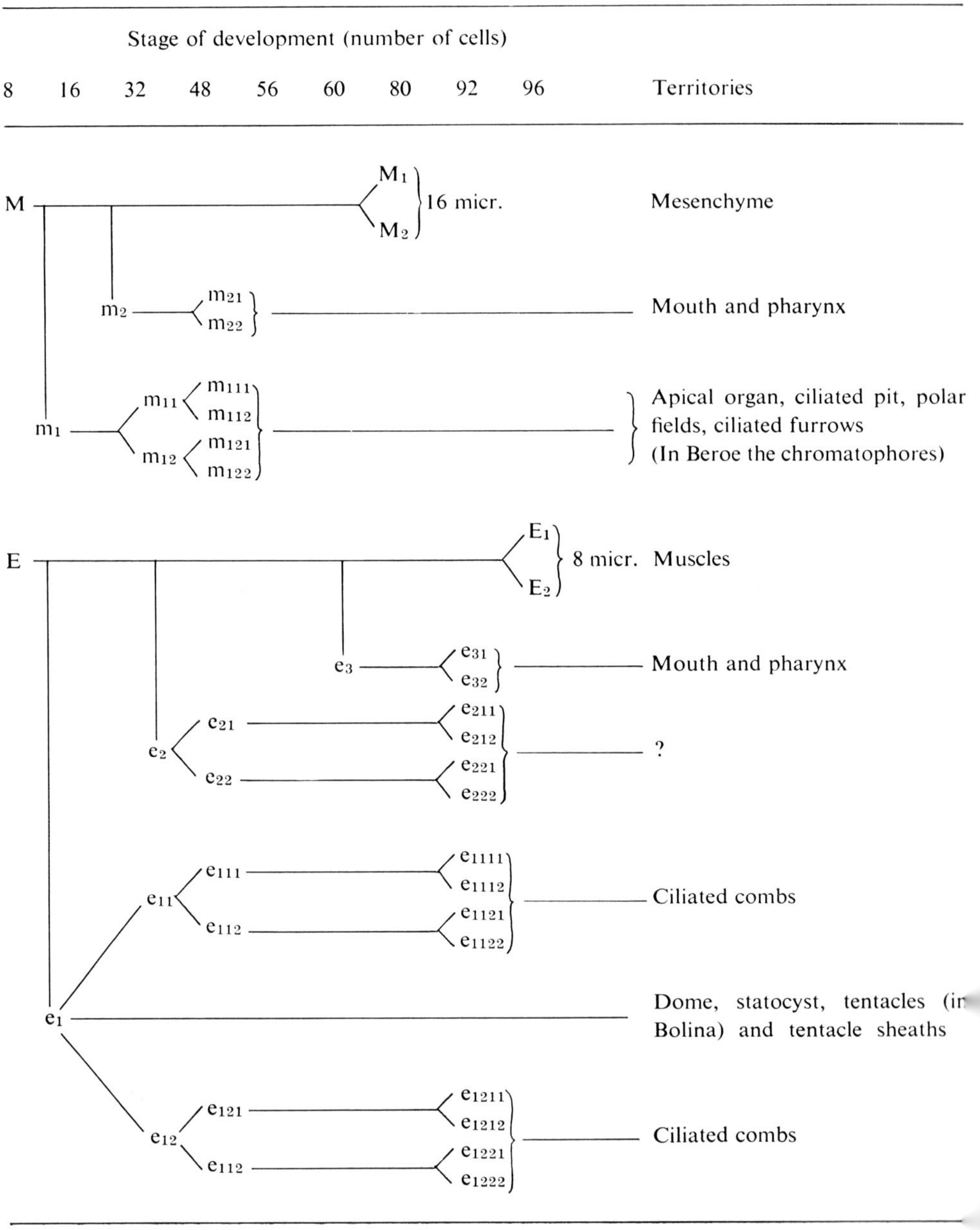

when stimulated electrically? This phenomenon should be investigated; it seems rather that the green luminescence is only a physical process, due to absorption of light by a particular cortical structure. What is this structure? Some authors speak of a granular plasm. This seems possible, because with the Nadi reaction and vital staining with Janus green the cortex stains blue (fig. 16). This suggests that it is rich in mitochondria. If so, its segregation

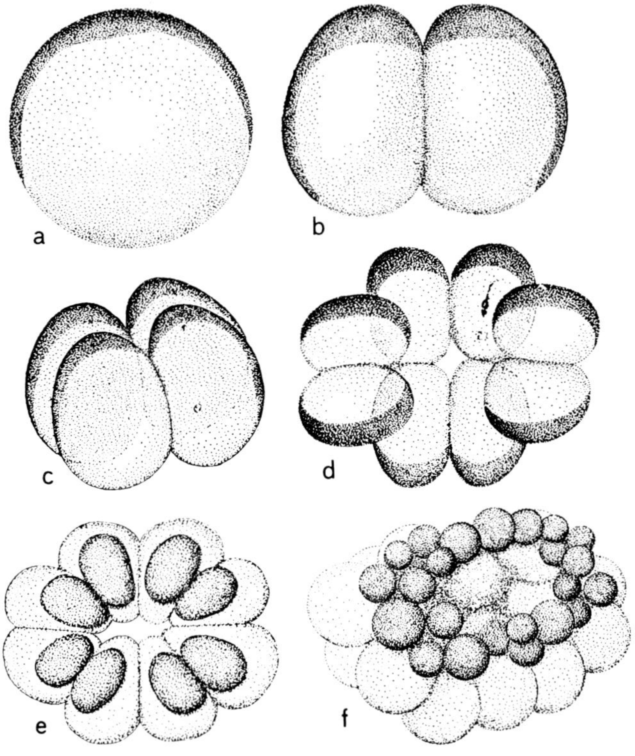

Fig. 16. Eggs of Beroe treated with the Nadi reaction at different stages of development; the green substance which is gradually segregated into the micromeres shows a positive Nadi reaction indicative of the presence of large amounts of mitochondria (Reverberi and Pitotti 1940).

in the cells from which the ciliary combs (e_1) and the pillars of the apical organ (m_1) derive is understandable. The mitochondria, in fact, are the organs that produce energy for the cells. Now the cells which give rise to the combs are exceptional in that they require much energy both for construction of the ciliary proteins and afterwards, in the larva, for the movement of the ciliary plates. This does not necessarily indicate that the mitochondria are plasmatic structures responsible for the ctenoblasts; differentiation of a ciliary cell is certainly dependent upon a substance which represses or activates the genes. What is this substance? It must be present in the green plasm but it is not the mitochondria: these are necessary for differentiation but only because without energy no construction is possible.

References

AGASSIZ A., 1874. Embryology of the Ctenophorae. Mem. Am. Acad. Arts Sci. *10*, 357–398.

CHUN C., 1880. Die Ctenophoren des Golfes von Neapel. Fauna und Flora des Golfes von Neapel. Monographie I. Engelman, Leipzig.

DAWYDOFF C., 1928. Traité d'embryologie comparée des invertébrés. Masson & Cie., Paris.

DRIESCH H. and T. H. MORGAN, 1895. Zur Analyse der ersten Entwicklungsstadien des Ctenophoreneies. I. Von der Entwicklung einzelner Ctenophorenblastomeren. Arch. Entwicklungsmech. Organ. *2*, 204–215.

FARFAGLIO G., 1963a. Experiments on the formation of combs in the Ctenophores. Experientia *19*, 303–304.

FARFAGLIO G., 1963b. Experiments on the formation of the ciliated plates in Ctenophores. Acta Embryol. Morphol. Exptl. *6*, 191–203.

FISCHEL A., 1897. Experimentelle Untersuchungen am Ctenophorenei. I. Von der Entwicklung isolierter Eiteile. Arch. Entwicklungsmech. Organ. *6*, 109–130.

FISCHEL A., 1898. Von der künstlichen Erzeugung (halber) Doppel- und Missbildungen. III. Über Regulation der Entwicklung. IV. Über den Entwicklungsgang und die Organisationsstufe des Ctenophoreneies. Arch. Entwicklungsmech. Organ. *7*, 557–630.

FISCHEL A., 1903. Entwicklung und Organdifferenzierung. Arch. Entwicklungsmech. Organ. *15*, 679–750.

HYMAN L. H., 1940. The invertebrates: protozoa through Ctenophora. Vol. 1. McGraw-Hill, New York.

KOWALEWSKY A., 1966. Entwicklungsgeschichte der Rippenquallen. Mém. Acad. Sci. St. Pétersbourgh *10*, Sér. 7, n. 4.

LA SPINA R., 1963. Development of fragments of the fertilized egg of Ctenophores and their ability to form ciliated plates. Acta Embryol. Morphol. Exptl. *6*, 204–211.

METSCHNIKOFF E., 1885. Vergleichend-embryologische Studien. IV. Über die Gastrulation und Mesodermbildung der Ctenophoren. Z. Wiss. Zool. *42*, 648–656.

MORGAN T. H., 1927. Experimental embryology. Columbia Univ. Press.

ORTOLANI G., 1963a. Ricerche sui territori organo-formativi nell'uovo dei Ctenofori. Boll. Zool. *30*, 25–31.

ORTOLANI G., 1963b. Sulla origine del mesoderma nei Ctenofori. Rend. Accad. Naz. Lincei *34*, s.8, 434–435.

ORTOLANI G., 1964. Origine dell'organo apicale e dei derivati mesodermici nello sviluppo embrionale di Ctenofori. Acta Embryol. Morphol. Exptl. *7*, 191–200.

REVERBERI G., 1966. Quelques nouvelles recherches expérimentales sur le développement des Cténophores. Année Biol. *5*, 375–390.

REVERBERI G. and G. ORTOLANI, 1963. On the origin of the ciliated plates and of the mesoderm in the Ctenophores. Acta Embryol. Morphol. Exptl. *6*, 175–190.

SCHLEIP W., 1929. Die Determination der Primitiventwicklung. Akademie Verlag, Leipzig, pp. 41–74.

SPEK J., 1926. Über gesetzmässige Substanzverteilung bei der Furchung des Ctenophoreneies und ihre Beziehung zu den Determinationsproblemen. Arch. Entwicklungsmech. Organ. *107*, 54–73.

YATSU N., 1911. Observations and experiments on the Ctenophore egg. II. Notes on the early cleavage stages and experiments on cleavage. Ann. Zool. Japon. *7*, 333–346.

YATSU N., 1912. Observations and experiments on the Ctenophore egg. III. Experiments on germinal localization of the egg of *Beroe ovata*. Ann. Zool. Japon. *8*, 5–13.

ZIEGLER H. E., 1898. Experimentelle Studien über die Zellteilung. III. Die Furchungszellen von *Beroe ovata*. Arch. Entwicklungsmech. Organ. *7*, 34–64.

G. Reverberi (ed.), Experimental embryology of marine and fresh-water invertebrates

CHAPTER 3

Planarians

R. J. SKAER

Zoological Laboratory, Cambridge, England

3.1. Introduction

The embryology of planarians differs markedly from that of every other group; cleavage produces an irregular shaped group of loosely associated cells that may well not touch each other, and development proceeds without the formation of germ layers and without gastrulation. Most of the yolk, moreover, is external to the embryo in yolk-cells. The embryology of other orders within the class Turbellaria, however, especially that of the Polycladida, the Archoophora, the Eulecithophora and the Perilecithophora (Beauchamp 1961) enables one to relate, albeit with difficulty, the type of development found in triclads to that in animals with spiral cleavage.

Since the power of regeneration of adult planarians is so great and has been so much worked on, the experimental embryology of planarians is of considerable interest. Very little experimental embryology of planarians has, however, been carried out, and none has so far been done on other members of the phylum Platyhelminthes. In the case of planarians this neglect may be due to the considerable technical difficulties of working with planarian embryos, for the eggs are laid within cocoons with an extremely tough shell. This shell, that contains chitin (Acconci 1919) and quinone-tanned proteins (Nurse 1950), is highly impermeable to almost all fixatives and is dark brown and nearly opaque. Furthermore, the early stages of development are surrounded by yolk-cells. Thus it is very difficult without killing the embryo to see the effects of experimental procedures. The majority of work on both normal and experimental embryology of planarians has, in fact, to be done using sectioned material. The embryos have to be fixed, usually after opening the cocoon – a procedure that may cause structural modifications. It is difficult to pierce the cocoon without causing damage, since, until all the yolk-cells have been ingested by the embryos, the contents of the cocoons appear to be under pressure, for no matter how carefully they are opened, yolk-cells spurt out at the first incision. The slow penetration of fixatives into intact capsules may also result in artifacts.

Early embryos after removal from the cocoon have not been cultured in a way that yields normal embryos. Present methods for culturing early embryos of triclads, themselves produce extreme abnormalities (Seilern-Aspang 1957) and this would complicate the interpretation of further experiments.

Another possible complication is that each cocoon usually contains several embryos. Reynoldson (1960) found that the number of young that hatch from each cocoon of *Polycelis tenuis* varied from none to six, with an average of 2.7, *SD* 1.3, and the average number of young per cocoon varied slightly with the season. In *Dugesia lugubris* he found (Reynoldson 1961) that the numbers were from 1 to 4 with an average of 2.8, *SD* 0.70 and the average number did not vary with the season. Similar results were produced by Balazs and Burg (1962). Jenkins and Brown (1963) calculated that the average number of young in each cocoon of *Dugesia dorotocephala* is as high as 16.5. The land planarian *Geoplana notocelis* lays cocoons that contain from five to eight eggs (Carlé 1935).

Early embryos of planarians are in an apparently haphazard orientation with reference to the shell of the cocoon. Thus it is difficult to apply a centrifugal force in a direction known with reference to a particular embryo within the cocoon.

Although there are accounts of the embryology of planarians in textbooks (Bresslau 1933; Hyman 1951; Beauchamp 1961), I shall begin this chapter with a description of the normal course of development since it differs so much from that in other phyla, and since the features of development that have been investigated experimentally are barely mentioned in these accounts. This description of early development is based on the work of Acconci (1919), and is amplified by the work of Seilern-Aspang (1956, 1958) and of Le Moigne (1963); for the account of later development reference is made to the work of Seilern-Aspang (1958), Le Moigne (1963), and of Skaer (1962, 1965).

3.2. *Normal development*

For convenience I shall number the stages of development as in Le Moigne (1963): the timing of each stage was obtained from capsules of *P. tenuis* kept at 19.4 °C (Skaer 1965).

The extensive researches of Benazzi and his group, summarised in Benazzi (1963) have shown that there is great variation in life cycle from one race* of planarians to another. Some races are diploid, have normal meiosis in both male and female cells to produce gametes, and both parents contribute chromosomal material to the fertilised egg (amphimixis). Other races are, however, polyploid, and undergo chromosomal doubling before meiosis, resulting in polyploid eggs; others again may perform only a mitotic division to produce the egg. There is a similar variation in the manner of formation of the sperm. In some races doubling of the chromosomal complement precedes meiosis, in others elimination of chromosomes occurs before the normal reduction divisions. The sperms in some races disintegrate after activating the egg, in others they are eliminated with the first polar body (Benazzi-Lentati 1966). Thus in many races, especially of polyploid animals, sperms do not contribute chromosomal material to the zygote. This is known as pseudogamy.

A wide variation in chromosomal cycle may occur within the gonads of one individual; a single planarian may have both diploid and tetraploid oocytes which may be amphimictic or pseudogamous, and may show normal or abnormal meiosis (Benazzi-Lentati 1966). Thus by crossing two races of *Dugesia lugubris*, Benazzi-Lentati (1962) found five types of individual were produced. Some had either exclusively amphimictic oocytes or exclusively pseudogamous ones irrespective of the degree of ploidy. Other individuals had two types of oocyte, diploid and tetraploid. In some of these animals the diploid oocytes were amphimictic and the tetraploid pseudogamous, in others the tetraploid were amphimictic and the diploid pseudogamous. Benazzi-Lentati also found that in some individuals with diploid and tetraploid oocytes, both types of egg could be either amphimictic or pseudogamous.

Benazzi (1963) points out that the type of life cycle in which pseudogamy occurs with polyploidy is common in planaria and he claims that it must, therefore, confer an evolutionary advantage over other types of life cycle. The importance of these studies from the point of view of our understanding of speciation, however, cannot be considered here. On the other hand the experimental cross-breeding of planarians that Benazzi's group has carried out reveals the genetic control of some aspects of fertilisation.

3.2.1. Fertilisation (stage 1)

Benazzi and his group have shown that one can cross diploid planaria with

* The term 'genodeme' is more accurate than the term 'race' in this context (Gilmour and Heslop-Harrison 1954).

polyploid and obtain normal offspring; animals showing amphimixis in the life cycle can be crossed with those that are pseudogamous. This has revealed that both sides contribute genetic material to the zygote if the eggs are from diploid amphimictic individuals, and the sperm come from pseudogamous individuals, whereas in the reverse type of cross, sperm from diploid amphimictic animals only rarely contribute genetic material to the zygote if the eggs are from a pseudogamous line. Thus almost all sperm have the capacity to form fully functional pronuclei, but whether or not they do so is controlled by the egg.

Benazzi infers that pseudogamy is a character that can be transmitted by the nucleus of the sperm, and he points out that despite the correlation between polyploidy and pseudogamy in natural populations, in experimentally produced individuals at least, the level of ploidy does not control whether or not the life cycle is pseudogamous.

Fertilisation occurs in the oviduct before the cocoon is built up. The eggs, which may contain yolk-granules (Benazzi-Lentati and Gremigni 1966) are fertilised at metaphase of the first maturation division. Lepori (1949) regards it as normal for cocoons to be built up at the time of formation of the polar bodies. The polar bodies, which in some races may contain sperm, are given off in the normal way – unlike those of the Archoophoran Polychoerus (Costello 1960) which disintegrate intracellularly.

3.2.2. Laying

The time the cocoon is laid does not correspond to a particular stage of the nuclear cycle of the zygote (Melander 1963). The eggs in the cocoon may still be completing their meiotic divisions (Le Moigne 1963) or may be in interphase before the first cleavage (Lanfranchi 1964). During meiosis there is no visible influence of the egg-cells on the surrounding yolk-cells; the 20–40 egg-cells lie irregularly scattered among the 80–90,000 yolk-cells present in cocoons of *Dendrocoelum lacteum*. It is worth mentioning that the number of egg-cells in each cocoon is larger than the number of young that hatch from cocoons of that species. Thus the maximum number of embryos found by Acconci (1919) in cocoons of *D. lacteum* was 13, and Le Moigne (1963) found 5–15 egg-cells in cocoons of *Polycelis nigra* (or *tenuis*) – a number that is significantly higher than the number of young that hatch. Benazzi and Benazzi-Lentati (1948) have attempted to investigate the genetics of the prenatal mortality involved in such a reduction in number of embryos during development. They cross-bred two strains of *Dugesia gonocephala* each of which had approximately 30 egg-cells in each cocoon.

From one strain an average of 2.5 young hatched, in the other strain an average of one hatched. Unfortunately the situation proved too complex for classical genetical investigation.

3.2.3. Clumping of yolk-cells

Up to six hours from the time of laying, neighbouring yolk-cells clump round each egg-cell, and, especially in *D. lacteum* become elongate in such a way that they are arranged like radiating spokes on a hub, the egg-cell. These oriented yolk-cells stain more strongly with both eosin and basic dyes than before, and are polarised in such a way that the yolk-granules are nearest the egg-cell, and the lipid droplets and the nuclei at the opposite pole (fig. 1).

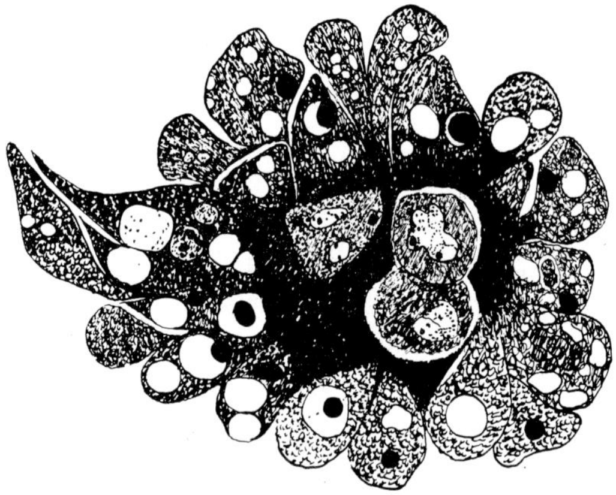

Fig. 1. Embryo of *Pl. polychroa* at the 5 blastomere stage showing the clump of elongate yolk-cells that are beginning to form the yolk-syncytium. (Redrawn from Acconci 1919.)

Both Acconci (1919) and Seilern-Aspang (1958) ascribe this change in the yolk-cells to a chemotactic substance emitted from the egg. Seilern-Aspang found that certain darkly-staining yolk-cells in the cocoons of *Procerodes lobata* also acted as centres of aggregation for the normal yolk-cells.

Le Moigne (1963), on the other hand, claimed that in cocoons of *Polycelis nigra*, two distinct types of yolk-cell are present from the time of laying. The most abundant are polygonal cells with large PAS positive granules, and interspersed with these are fusiform yolk-cells rich in RNA and with smaller PAS positive granules. He held it is these scattered, spindle-shaped cells that move to the egg-cell and fuse to form the yolk-syncytium. Fusiform yolk-cells with cytoplasmic basiphilia are certainly visible among the normal yolk-cells in the region of the egg-cell.

Although the clumping of the yolk-cells around the egg-cells has been described by many authors, and although Seilern-Aspang (1958) has investigated the phenomenon experimentally, its cause is far from clear. We do not know if it is primarily a question of the migration of certain yolk-cells (perhaps due to a chemotactic substance – compared by Seilern-Aspang to acrasin in the cellular slime moulds), or whether it is a change in shape of the neighbouring yolk-cells that is induced and results in a change in their packing around the egg. Until electron micrographs of unopened cocoons are taken, it is not possible to say what the normal intercellular space between the yolk-cells is, nor whether it is decreased during clumping. Investigations on opened cocoons suggest that the strength of adhesion of yolk-cells to the egg-cell is increased at this stage, for though the majority of yolk-cells can be washed away with a jet of water, the egg-cells remain coated by them. The importance of this clumping reaction is shown by the experimental results of Seilern-Aspang (1958), which will be described later.

3.2.4. Cleavage

Cleavage of the egg is total and begins about 12–18 hr after laying. Even in a single species the plane of the second cleavage is irregular in orientation; it may be either parallel or normal to the plane of first cleavage (Acconci 1919). Thus, in the first case, a row of four blastomeres is produced. There is no sign of synchrony in the divisions. The blastomeres produced are spherical, equal in size, and often do not touch each other; their arrangement appears to be haphazard.

At the 5–20 blastomere stage, 24–164 hr after laying, cell boundaries of those yolk-cells that lie in the immediate surroundings of the blastomeres become invisible in the light microscope (fig. 1). Seilern-Aspang (1958) claimed that this is due to a chemical given out by the blastomeres. The 'yolk-syncytium' formed in this way extends progressively until it has a diameter of 200–300 μ – approximately double the diameter of the original clump of yolk-cells – and is visibly differentiated into an inner and an outer region. The inner syncytium is rich in mitochondria and ribonucleic acid, whereas the outer syncytium appears frothy in sections and contains nuclei of yolk-cells, much glycogen, and fat (Kościelski 1964). The yolk-syncytium is normally spherical, but Kościelski has published pictures that suggest it is sometimes dumb-bell shaped in *D. lacteum*. His conclusion, however, that this is evidence for natural polyembryony must be balanced against the fact that the number of embryos that hatch is less than the number of egg-cells in each cocoon.

3.2.5. Migration of the blastomeres

This stage is characterised by the enclosure of the inner syncytium within a wall of flattened blastomeres to form the primary epidermis (Mattiesen 1904). The cells that behave in this way can be recognized at the stage of the early blastomere group in *Procerodes lobata*, for they are smaller, more basiphil, and have a higher frequency of division than the other blastomeres (Seilern-Aspang 1956). It is not clear how these cells form a sphere at a certain level within the syncytium. The explanation in Seilern-Aspang (1958) that it is due to the mutual repulsion of negatively cytotaxic blastomeres within a monocentric embryonal field, is unhelpful, since it merely redescribes the result in hypothetical dynamic terms, and he alone claims to have observed a membrane delimiting the inner syncytium from the outer, on which negatively-cytotaxic blastomeres could flatten to form the epidermis.

A more satisfactory suggestion is given by Acconci (1919). From her work it would appear that the whole blastomere group moves to one side of the inner syncytium, and some of the blastomeres begin to differentiate into the embryonic pharynx. From the outermost part of this group of cells, certain blastomeres with strong cytoplasmic basiphilia flatten and grow round the basiphilic inner syncytium. It seems likely that the primary epidermis extends round the inner syncytium by more and more blastomeres becoming flattened, rather than by mitosis of those blastomeres that are already flattened. The pictures given in Le Moigne (1963) support this account. He claims, however, that it is at the periphery of the yolk-syncytium that the cells flatten. Thus it would seem that the spherical shell of the primary epidermis develops by the gradual extension of a hemisphere, rather than as individual cells that move to a fixed distance from the centre of the syncytium, flatten, and join up – as postulated by Seilern-Aspang (1958).

It is interesting that in the winter eggs of the Perilecithophoran, *Prorhyncus*

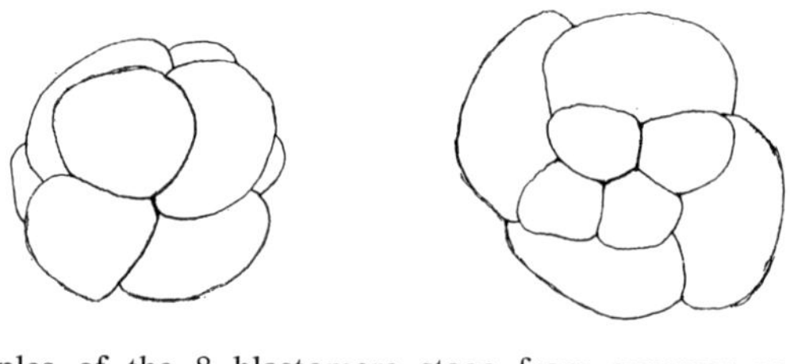

Fig. 2. Two examples of the 8 blastomere stage from summer eggs of *Prorhynchus stagnalis*. (From Steinböck and Ausserhofer 1950.)

stagnalis the blastomeres move to the boundary of the egg, flatten, and enclose the yolk-cell mass in a thin layer of cells, but in summer eggs of the same species, the blastomeres remain adherent to each other as an aggregate within the yolk-cells. The cleavage of these summer eggs may be radial or may be spiral (fig. 2) (Steinböck and Ausserhofer 1950).

3.2.6. Embryonic pharynx and gut (stage 2)

This occurs in embryos at the 60–70 blastomere stage and occurs 2–4 days after laying. Approximately 10 blastomeres form the hollow sphere that is the primary epidermis; this sphere contains the inner yolk-syncytium (together with a few nuclei from the original yolk-cells). Attached to the inside of the sphere are the remainder of the blastomeres, approximately two dozen of which will form the embryonic pharynx and gut.

Fuliński (1938) and Le Moigne (1963) found that these organs are built up of the following cells:

(a) 4 small cells that are continuous with the epidermis and that form the mouth;

(b) 4 large cells that form the internal cells of the pharynx;

(c) 4 elongate cells that grow round the large cells and form the external envelope of the pharynx;

(d) a certain number of small cells migrate between the last two groups of cells and form a ramifying, contractile tissue in the embryonic pharynx;

(e) 4 round cells that lie on the side of the internal cells nearest the gut;

(f) 4 cells that delimit the embryonic gut (see fig. 3).

These organs have no cavity at first, but develop one in the centre of tetrads b, e and f.

The presence of an embryonic pharynx during development is by no means peculiar to planarians. The distribution of such structures throughout the animal kingdom has been described by Weygoldt (1966).

3.2.7. The swallowing of the external yolk-cells (stage 3)

The fully differentiated embryonic pharynx ingests, apparently actively, the external yolk-cells. It is possible that this process is assisted by the waves of contraction that pass over the outer epithelial cells (Kölliker 1846). As the embryonic gut becomes filled with yolk-cells, it enlarges and its epithelial cells increase in number. When most of the yolk-cells have been taken into the gut – about 8 days after laying – the embryo is still spherical and the cells of the primary epidermis have become extremely flattened (to as little as 1 μ thick in most places) (Skaer 1965). Thus the embryo now consists of a

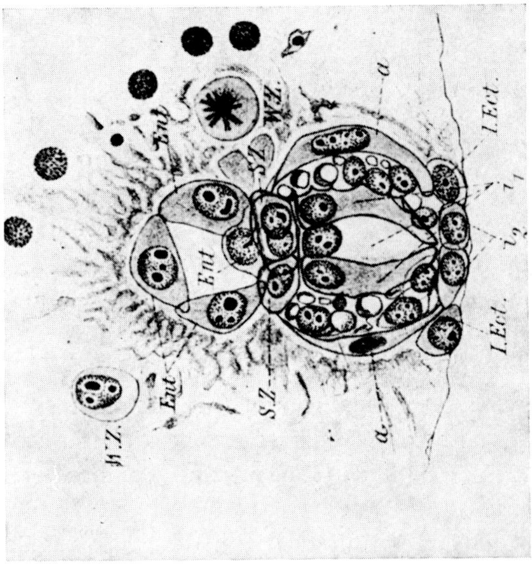

Fig. 3b.

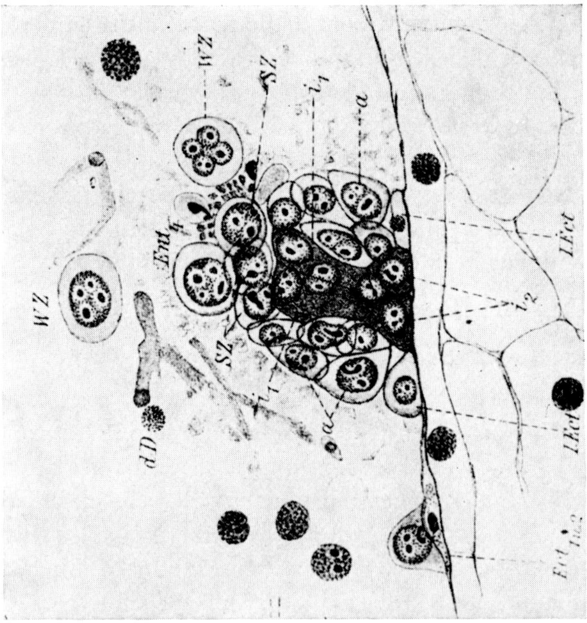

Fig. 3a.

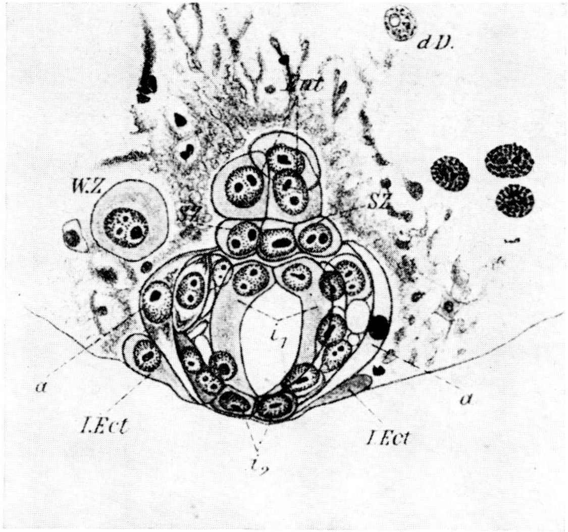

Fig. 3c.

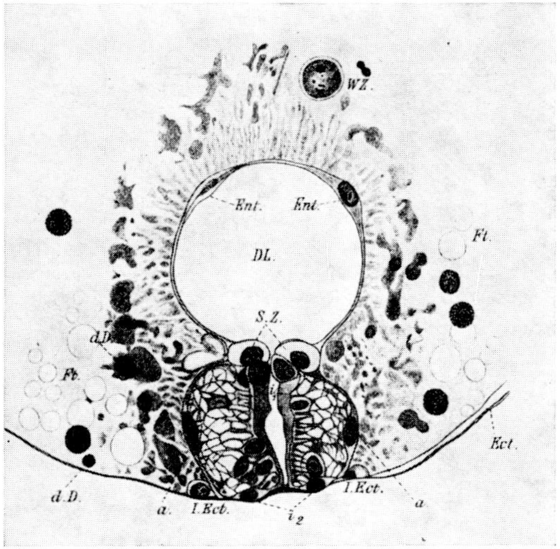

Fig. 3d.

Fig. 3. Four stages in the development of the embryonic pharynx. (From Mattiesen 1904).

syncytium that contains a few blastomeres and is bounded on the outside by flattened blastomeres and on the inside by the gut. The blastomeres in the syncytium have a high frequency of division at this stage and are more abundant in the hemisphere that contains the embryonic pharynx than in the other hemisphere.

Towards the end of this stage when the embryo is still spherical, the smooth contour of the primary epidermis becomes distorted by large cells that emerge as 'secondary' epidermal cells (Skaer 1965). These cells flatten very little on the surface of the embryo, so they appear as pimples in surface view, and are ciliated. As more and more of these large cells enter the epidermis, the primary epidermal cells increase in height and diminish in width. Muscle cells also differentiate at this stage. They are spindle-shaped cells that may be seen with the electron microscope to contain much glycogen and thick and thin myofilaments in rather irregular arrays. Secondary epidermal cells entering the epidermis are often deformed by these muscle cells (Skaer 1965), for the circular muscles lie immediately beneath the epidermis and with the electron microscope can be seen to be closely packed – as are many of the blastomeres in the hemisphere that contains the embryonic pharynx. Electron micrographs show, however, that large spaces are present beneath the epidermis and in the aboral hemisphere of the embryo; these are presumably parts of the original syncytium (Skaer unpublished).

The fact that the number of embryos that hatch is smaller than the number of egg-cells in each cocoon may be due to developmental failure at this stage, for Le Moigne (1963) reports that many embryos of *P. torva* absorb no yolk-cells and develop no further. Benazzi-Lentati (1948) has studied the anatomy of embryos that die at various stages before hatching. She found that the highest mortality occurred early in development – that is up to the stage when the yolk has been ingested.

3.2.8. Transitory pharynx (stage 4)

The transitory pharynx consists of a crown-shaped group of blastomeres that is concentric with the embryonic pharynx and is situated between this and the gut. It is peculiar to *Polycelis nigra* (Le Moigne 1963), and its presence here may perhaps be correlated with the fact that in this species the adult pharynx has a lumen that coincides with the lumen of the embryonic pharynx.

This stage, which begins when all the external yolk-cells have been ingested, $8\frac{1}{2}$ days after laying, may be generally recognized by the fact that the embryo has become elongate and slightly flattened dorsoventrally. Seilern-Aspang (1958) claimed that the axis that passes through the embryonic pharynx is the future dorsoventral axis of the embryo, but in *P. nigra* this is not so: here the embryonic pharynx opens onto the ventral surface and is at an acute angle to it. Nor is it true of *Planaria maculata* or *P. simplissima* where the embryo flattens in such a way that the embryonic pharynx is on, and at an acute angle to, the dorsal surface (Curtis 1902 1905; Stevens 1904).

In the gut of embryos of this age, the yolk-cells form a smoothly contoured ovoid mass. Posteriorly this mass is divided by a narrow cleft, thus initiating the trifid gut.

3.2.9. Adult pharynx (stage 5)

It is customary to distinguish three blastemas ventrally – an anterior, a middle and a hind blastema; the distinction is, however, by no means clear cut. Le Moigne (1963) found that the brain differentiates at this stage in the anterior blastema, and that part of the middle blastema forms the adult pharynx. This develops as a clump of cells caudal to the embryonic pharynx, which degenerates. The lumen of the adult pharynx appears as a split in the blastema lined by flattened cells. In *P. nigra* and *P. tenuis*, the adult pharynx develops on virtually the same site as the embryonic pharynx – which then degenerates. Le Moigne (1966) claimed to have proved experimentally the existence of, and to have found histologically, neoblasts at this stage.

Since a great deal occurs at this stage it will be further subdivided:

9-day embryos. Embryos of this age and older will develop normally if the capsule is opened and left in pond water. The lateral margins of the trifid gut develop a scalloped edge at this stage.

Certain darkly-staining cells in the ventral regions of embryos of this age only, contain vesicles lined by cilia and are held to be replacement cells for the epidermis (Skaer 1965). Rhabdite-forming cells also differentiate at this stage. They form first in the parenchyma of the lateral margins of the body and only later appear in the epidermis (Metschnikoff 1883; Skaer 1962, 1965). It has been suggested (Skaer 1961, 1962, 1965) that rhabdites may be used as intracellular markers that enable one to follow the centrifugal migration of these cells into the epidermis and the continuous replacement of the epidermis by these cells (Skaer 1961). Lentz (1967) has also suggested this.

9¼-day embryos. Rhabdite-containing cells are found in the epidermis of embryos of this age. These cells increase in number as development proceeds until the entire epidermis is made of them and may be called a 'tertiary' epidermis (Skaer 1965). With the electron microscope, junctional complexes may be seen between epidermal cells in $9\frac{1}{4}$-day embryos. The embryos move sporadically by ciliary activity, but no muscular responses can be detected, even though the electron microscope shows that muscle cells are differentiated.

A large proportion of cells in these embryos make up what is apparently the developing flame-cell system (Skaer 1962). Some flame cells are already recognisable and have osmiophilic cytoplasm and 'flames' of cilia, but in more dorsal regions of the embryo cells can be observed that are essentially similar but that lack cilia and a smooth lumen. Nevertheless, they contain clear, cytoplasmic vesicles which in places are confluent from cell to cell. Certain groups of these cells have a continuous, though irregular-shaped lumen. After this stage, the number of protonephridial cells relative to other types of cell decreases, and each tubular cell becomes more elongate. It is suggested (Skaer 1962), that by cellular rearrangement and the development of cilia these cells will form part of the flame-cell system.

$9\frac{3}{4}$-day embryos. The bud of the adult pharynx enlarges and moves more anteriorly – thus enhancing the trifid nature of the gut. Some sections of embryos of this age suggest that rhabdite-containing cells penetrate and destroy the secondary cells (Skaer 1965). Whether or not this is the cause of their disappearance, no secondary epidermal cells remain in 11-day embryos.

3.2.10. Development of peripheral nerves (stage 6)

Le Moigne (1966) gives a diagram showing that peripheral nerves have developed from the brain, three quarters of the way to the bud of the adult pharynx by this stage. These embryos are 10 days old. When prodded they show a muscular contraction on the side touched, but the embryos are not able to stick to the substratum (Skaer 1962). At this stage, rhabdites are longer than the cells of the epidermis are tall, and they come to be spread like spokes of a wheel in the flattened epidermal cell. The process of formation of this pattern has been watched in living embryos of this age (Skaer 1965).

Embryos 11 days old may also be grouped under this stage. They are able to adhere quite strongly to glass and are not dislodged by a gentle current of water. This property may perhaps be correlated with the development, at this stage, of acidophil gland cells. The epidermis contains only primary epidermal cells and tertiary epidermal cells; all secondary epidermal cells have disappeared. A basement membrane to the epidermis first develops in 11-day embryos.

3.2.11. Differentiation of the eyes (stage 7)

Although embryos 15 days old have no general body pigmentation, the eye cups are black. A further flattening and elongation of the embryo has occurred by this stage, and the pharynx has moved to the centre of the animal and has itself become elongate. These embryos have well-developed powers of locomotion. It may be that this can be correlated with the differentiation of mucous cells, to be found first in embryos of this age (Skaer 1962).

The epidermis is an entirely new structure – a tertiary epidermis, that is, it consists of rhabdite-containing cells only. These are continuously intercalated into the epidermis and cause this to increase in height until some time after hatching (Skaer 1965). This change in shape of the epidermal cells is correlated with the disappearance of the radiating pattern of the rhabdites, which become aligned normal to the surface of the epidermis. Mitoses are absent from the epidermis, and this bears out the suggestion (Skaer 1961) that the epidermis is not autogenous, but is renewed in both embryos and adults by centrifugal migration of cells from the parenchyma.

Le Moigne (1963) found that nerve trunks are well developed throughout the body at this stage.

The embryos normally hatch within 12 hours of 21 days (Skaer 1962); Le Moigne (1963), however, found a greater variability than this in the time of hatching.

3.3. Experimental embryology

3.3.1. Irradiation experiments

Benazzi (1963a) has used X-ray irradiation to investigate the function of sperm in fertilisation. He irradiated whole *Dugesia lugubris* with 1000 to 70,000 r and allowed sperm from these animals to fertilise amphimictic or pseudogamous eggs. The irradiated sperm were found to activate both types of egg. In the case of pseudogamous eggs, development proceeded normally, but in amphimictic eggs development was normal only if the sperm had received less than 2000 r. If the sperm had had between 2000 and 3000 r, cleavage of the egg occurred normally, but the embryo died at the stage when it takes in the yolk cells.

Since X-rays are assumed to act largely on nuclear material, this example illustrates, as Benazzi points out, two of the different functions of the sperm in fertilisation: activation of the egg and the contribution of nuclear material to the zygote.

Benazzi-Lentati (1964) found that amphimictic eggs of *D. lugubris* fertilised with sperms irradiated with 55,000–60,000 r could develop pseudogamously.

The longevity of irradiated sperm is striking. Benazzi (1963a) found that sperm from the seminal receptacle retain for several weeks the power to activate eggs.

Seilern-Aspang (1958) has employed two experimental techniques to study the embryology of planarians; he has centrifuged cocoons of *D. lacteum* and *P. torva* of a whole range of ages at 4000 rpm for 10 min, and has grown early embryos in coverslip culture in such a way that the embryos are abnormally flattened.

3.3.2. Centrifugation experiments

He finds that when cocoons that have just been laid are centrifuged, even for long periods of time, they develop normally. Centrifugation of cocoons 2–6 hr after they are laid, however, may result in abnormal embryos – especially symmetrical twins (see table 1). No animals hatch if centrifugation is for 30 min or longer at this stage. Seilern-Aspang points out that this period of sensitivity to centrifugation is very clear-cut and corresponds to the time when yolk-cells are clumping around the egg-cell, and ends when

TABLE 1

Time of centrifugation of cocoons after laying, in hours at 18-20 °C	Normal embryos	Symmetrical twinning	Asymmetrical twinning	Cyclopia	3 and 4 eyes	Abnormal pharynx	Half of body thickened	Quarter of body thickened	Lacking hind end	Laterally reversed blastemas
0–10	80%	9%	2%	1%	6%				2%	
10–72	100%									
72–96	80%						20%			
72–120	80%							20%		
120–168	70%					30%				
168–240	95%									5%
240–hatched	100%									

(From Seilern-Aspang 1958.)

the yolk-cell clump is complete. He has demonstrated histologically that in such centrifuged cocoons the centre of the clump of yolk-cells is displaced centrifugally relative to the egg-cell, or that the egg-cell is centripetal. In some cases, the egg-cell had influenced the yolk-cells before displacement and the elongate yolk-cells can be seen to radiate out from a point some distance from the egg. After centrifugation, the egg-cell may have no influence at all on the yolk-cells that now surround it, but often the egg-cell still retains the ability to influence yolk-cells to some extent. Seilern-Aspang claims that it is in the latter case that normal embryos are produced and suggests this is through regulation. He stresses, moreover, the morphogenetic as well as the nutritional importance of extraembryonic yolk-cells in the production of normal embryos.

It seems likely that centrifugation of cocoons when the yolk-cells are clumping results in symmetrical 'Siamese' twins or triplets by causing two or even three egg-cells to lie so close together that their spheres of influence

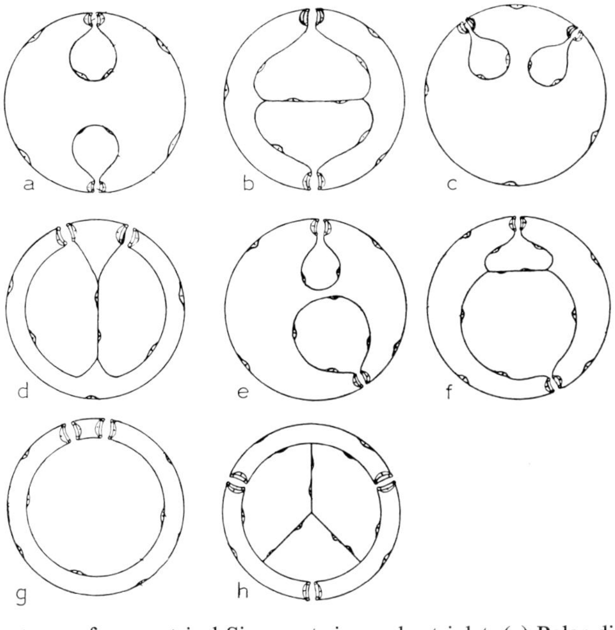

Fig. 4. Early stages of symmetrical Siamese twins and a triplet. (a) Polar distribution of embryonic pharynges. (b) Later stage of the same. (c) Two embryonic pharynges in the same hemisphere. (d) Later stage of the same. (e) Embryonic guts of different size (? age). (f) Later stage of the same. (g) Twins with a common gut. (h) Siamese triplet. (From Seilern-Aspang 1958.)

on the yolk-cells overlap. It may also be possible for such twins or triplets to be produced if the yolk-syncytia of two egg-cells become confluent and rounded off (Seilern-Aspang 1958). Neither of these suggestions has been conclusively proved, nor do we know the reason for the ending of this period of sensitivity to centrifugation – whether it is just that an egg-cell surrounded by its full complement of clumped yolk-cells does not move in the centrifugal field, or whether some more fundamental morphogenetic principle is at work. The fact that embryos develop normally if cocoons are centrifuged immediately after they are laid suggests that there is more to twinning than merely egg-cells developing in abnormally close proximity (if indeed that is the effect of centrifugation at that early age), but Seilern-Aspang does not give the number of cocoons he used for this experiment, so the result may not be significant.

A wide variety of types of Siamese twins (or double embryos) is undoubtedly produced by centrifugation at the stage of yolk-cell clumping; twins that were not joined in some way have not been distinguished from normal by Seilern-Aspang. He does, however, describe four classes of symmetrical Siamese twins, the early stages of development of which are shown in fig. 4. In the first type the embryos are joined back-to-back – thus the pharynx of the upper twin is uppermost (fig. 5a). In the second type, the

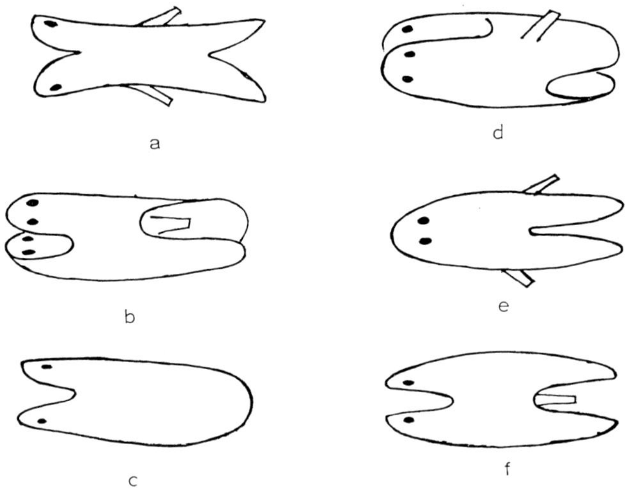

Fig. 5. Four types of symmetrical Siamese twins. (a) Type 1 with twins back-to-back, side-view. (b) Type 2 oblique view. (c) Type 2 side-view. (d) Type 3 oblique view. (e) Type 3 side-view. (f) Type 4 with animals belly-to-belly, side-view. (From Seilern-Aspang 1958.)

Siamese twins have the two front ends arranged back-to-back; relative to the front ends, the hind ends are positioned as if they were rotated through 90 degrees about the longitudinal axis, and have the two ventral surfaces facing each other (fig. 5b, c). The pharynges may have no external opening, and sometimes only one large pharynx develops. In the third type of symmetrical Siamese twin, the head ends have their ventral surfaces facing each other and the hind ends are positioned as in the second type but with their ventral surfaces outwards (fig. 5d, e). In the fourth type the Siamese twins lie belly-to-belly (fig. 5f). Seilern-Aspang uses the evidence of these Siamese twins to support his idea that the planarian embryo is at one stage made up of four blastemas; the head and tail blastemas of the right and left side – and that these can be made to fuse with their neighbours in a variety of ways.

Some twins are found to be fused side-by-side so that the animal produced is, to varying degrees, double laterally (fig. 6). Embryos with three or four

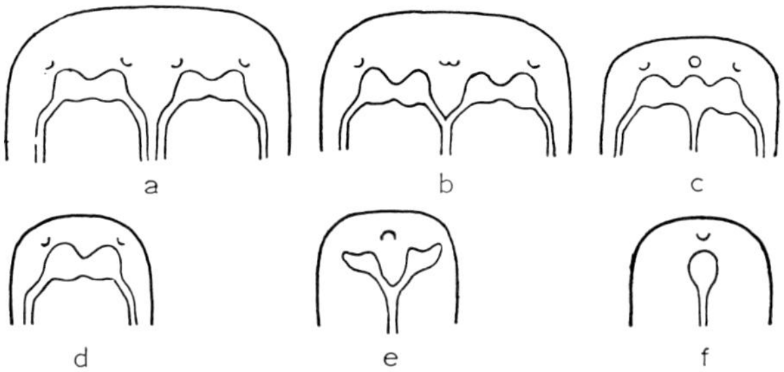

Fig. 6. The nervous system in some abnormal embryos of planarians. (a) Complete lateral doubling, (b) and (c) various degrees of fusion, (d) normal embryo, (e) laterally reversed nervous system, (f) embryo in which nervous system has not become double. (From Seilern-Aspang 1958.)

eyes may be produced, and the nervous system, for instance, may be completely double (fig. 6a), or may be fused to the extent that it consists of three, instead of four, cerebral ganglia and these give off a median as well as two lateral nerves (fig. 6c). The median nerve would seem to correspond to two lateral nerves that have fused in the mid-line. The existence of these partially double organs is, according to Seilern-Aspang, evidence that differentiation at the time of organogenesis depends on the location of cells, not on their cell lineage.

Formation of the yolk-syncytium is not normal in embryos centrifuged at the time of the clumping reaction of the yolk-cells; a syncytium forms around each individual blastomere and the shape of the whole syncytial mass is related to the grouping of the blastomeres. Seilern-Aspang (1958) has observed a similar relationship in embryos cultured under coverslips. There seems little point in discussing the influence of centrifugation on the hypothetical substance that is said to induce the formation of the syncytium until the existence of both have been more definitely proved.

Seilern-Aspang has shown that further aberrant types of development occur in these centrifuged cocoons; in some cases the flattened blastomeres that form the primary epidermis grow round regions of the outer syncytium as well as enclosing the inner syncytium. He claims that this enlarged embryonal area then fragments, and he implies that this gives two or more smaller embryos in the place of the one large one. A similar fragmentation occurs when young embryos are highly flattened in culture under a coverslip (Seilern-Aspang 1957).

If cocoons are centrifuged uninterruptedly from the time of laying for three or four days, the development of the embryonic pharynx is inhibited. Seilern-Aspang (1958) claimed that separate cells of the pharynx do not differentiate in these circumstances. He suggested that the cells of the embryonic pharynx are determined only after they have clumped together as a blastema. The evidence for such a later determination of these cells is based on the fact that separate, but differentiated cells of the embryonic pharynx have not been found; these, however, would be very difficult to recognise and identify in normal histological preparations. Seilern-Aspang (1958) has found, nevertheless, that highly abnormal embryonic pharynges may be present in embryos from cocoons that have been centrifuged.

The thickening of the right or left body halves in some embryos following centrifugation of the cocoon 72–96 hr after laying is ascribed by Seilern-Aspang (1958) to an interference with the doubling of the blastema that is normally associated with the formation of a bilaterally symmetrical embryo. Seilern-Aspang suggests that the fact that one side of the head end may alone be thickened by centrifugation a little later on (see table 1) is evidence that front and back blastemas are differentiated after right and left blastemas. In other cases, especially when the embryo has been flattened against the wall of the cocoon by the centrifugal force, some organs, for example the cerebral ganglia and eyes, that are double in the normal embryo, develop as single median structures (fig. 6f). How this inhibition of doubling is achieved is not known.

Approximately 5% of embryos that had been centrifuged 7–10 days after laying show an extremely odd abnormality in which structures that are normally lateral come to lie in the midline, and normally median organs are doubled and lie peripherally (fig. 6e). These embryos have anteriorly a single, median, backwardly directed eye and one median nerve tract, but two lobes of the gut. After a short while these embryos show a regulation that tends to make them more normal, and they develop, for example, lateral nerve tracts (Seilern-Aspang 1958).

3.3.3. Regulation and regeneration

Le Moigne (1965a, b, c) has studied the ability of planarian embryos to regulate; he amputated the anterior end of embryos at various stages of development to see if the hind end regenerated the lost part. He was particularly interested in whether undifferentiated tissues are able to induce the formation of blastema, but found that the tissues have to be somewhat differentiated before any regeneration occurred. The earliest stage that will form a blastema is stage 5, and the formation of a blastema in these embryos is slower than in embryos that were amputated at stages 6 or 7. He points out that the embryonic tissues differentiate before the blastema is formed; nerve trunks are always present in embryos that have started to regenerate, they are also present in some embryos that have not regenerated. These results show, moreover, that the presence of the brain is not necessary for the differentiation of nerve cells.

The question whether regeneration in embryos is carried out by neoblasts is a matter of definition, but Le Moigne (1965c) has shown that a dose of 750–3000 r of X-rays, sufficient to kill neoblasts in the adult animal, when applied at stage three of embryology results in hatchlings that are apparently normal but that fail to regenerate when the head end is amputated. Stages earlier than stage 3 are more sensitive to X-rays and are killed even by 500 r. It is still not clear whether the lack of regeneration in early stages of embryology is due to a lack of appropriate cells, or whether it is due to an inadequate stimulus to regeneration.

Experimental embryology of planarians thus does not yet answer many questions; evidence is still insufficient for one to say if the egg is a mosaic or whether it is a regulation egg; no induction process in the embryo has been proved conclusively. Present experiments have shown that regulation occurs even at an early stage of development, but that embryos regenerate only when the cells have undergone some differentiation.

References

ACCONCI C., 1919. Osservazioni sullo sviluppa delle planarie d'acqua dolce. Boll. Ist. Zool. R. Univ. Palermo *1*, 49–76.

BALAZS A. and M. BURG, 1962. Quantitative data to the changes of propagation according to age. II. Fertility and number of embryos in *Dugesia lugubris*. Acta Biol. Hung. *12*, 297–304.

BENAZZI M., 1963. Genetics of reproductive mechanisms and chromosome behaviour in some fresh-water triclads. In: The lower metazoa comparative biology and phylogeny, E. C. Dougherty, Z. N. Brown, E. D. Hanson and W. D. Hartman, eds., pp. 405–422, University of California Press.

BENAZZI M., 1963. L'azione dello spermio irradiato nello sviluppo anfimittico e pseudogamico delle planarie. Atti Ass. Genet. Ital. *8*, 255–256.

BENAZZI M. and G. BENAZZI-LENTATI, 1948. Sulla mortalità prenatale dei Tricladi. Boll. Soc. Ital. Biol. Sper. *24*, 1–2.

BENAZZI-LENTATI G., 1948. Sui processi istologici della mortalità prenatale dei Tricladi d'acqua dolce. Mem. Soc. Tosc. Sci. Nat. B *55*, 1–13.

BENAZZI-LENTATI G., 1962. Due modalita' di sviluppo dello stesso tipo di uova in ibridi interrazziali di planarie. Acta Embryol. Morphol. Exptl. *5*, 145–160.

BENAZZI-LENTATI G., 1964. Sulla regolazione dei processi maturativi di ovociti di planarie attivati da spermi normali ed irradiati. Boll. Zool. *31*, 963–971.

BENAZZI-LENTATI G., 1966. Amphimixis and pseudogamy in fresh-water triclads: experimental reconstruction of polyploid pseudogamic biotypes. Chromosoma *20*, 1–14.

BENAZZI-LENTATI G. and V. GREMIGNI, 1966. Ulteriori studi morfologici ed istochimici sugli ovociti della planaria *Dugesia lugubris*. Mem. Soc. Tosc. Sci. Nat. B *73*, 101–111.

BRESSLAU E., 1933. Handbuch der Zoologie, W. Kükenthal and T. Krumbach, eds., vol. 2, pp. 52–304, W. de Gruyter and Co., Berlin and Leipzig.

CARLÉ R., 1935. Beiträge zur Embryologie der Landplanarien. Z. Morphol. Ökol. Tiere *29*, 527–558.

COSTELLO D. P., 1960. The internal polocytes of the egg of *Polychoerus carmelensis*. Anat. Rec. *137*, 346–347.

CURTIS W. C., 1902. The life history, the normal fission, and the reproductive organs of *Planaria maculata*. Proc. Boston Soc. Nat. Hist. *30*, 515–559.

CURTIS W. C., 1905. The location of the permanent pharynx in the planarian embryo. Zool. Anz. *29*, 169–175.

DE BEAUCHAMP P., 1961. Traité de Zoologie, P. – P. Grassé, ed., Tome 4, vol. 1, pp. 35–212, Masson & Cie., Paris.

FULIŃSKI B., 1938. Zur Embryonalpharynxfrage der Tricladen. Zool. Polon. *2*, 185–207.

GILMOUR J.S.L. and J. HESLOP-HARRISON, 1954. The deme terminology and the units of micro-evolutionary change. Genetica *27*, 147–161.

HYMAN L. H., 1951. The invertebrates. II. Platyhelminthes and rhynchocoela. The acoelomate bilateria, pp. 176–179, McGraw-Hill Book Company, Inc., New York.

JENKINS M. M. and H. P. BROWN, 1963. Cocoon production in *Dugesia dorotocephala* (Woodworth). Trans. Am. Microscop. Soc. *82*, 167–177.

KÖLLIKER A., 1846. Ueber die contractilen Zellen der Planarienembryonen. Arch. Naturgesch. *12*, 291–295.

KOŚCIELSKI B., 1964. Polyembryony in *Dendrocoelum lacteum* O.F. Muller, J. Embryol. Exptl. Morphol. *12*, 633–635.

LANFRANCHI A., 1964. Aspetti cito-morfologici in embryoni irradiati della planaria *Dugesia lugubris* (O. Schmidt). Boll. Zool. *31*, 555–565.

LE MOIGNE A., 1962. Étude de formules chromosomiques de quelques *Polycelis* (Turbellariés, Triclades) de la région parisienne. Bull. Soc. Zool. Fr. *87*, 259–270.

LE MOIGNE A., 1963. Étude du développement embryonnaire de *Polycelis nigra* (Turbellarié, Triclade). Bull. Soc. Zool. Fr. *88*, 403–422.

LE MOIGNE A., 1965a. Mise en évidence d'un pouvoir de régénération chez l'embryon de *Polycelis nigra* (Turbellarié, Triclade). Bull. Soc. Zool. Fr. *90*, 355–360.

LE MOIGNE A., 1965b. Sur la régénération et différentiation de fragments d'embryons de *Polycelis nigra – tenuis* (Turbellarié, Triclade). Compt. Rend. Soc. Biol. *159*, 54–57.

LE MOIGNE A., 1965c. Effet des irradiations aux rayons X sur le développement embryonnaire et le pouvoir de régénération à l'éclosion de *Polycelis nigra* (Turbellarié, Triclade). Compt. Rend. *260*, 4627–4629.

LE MOIGNE A., 1966. Étude du développement embryonnaire et recherches sur les cellules de régénération chez l'embryon de la Planaire *Polycelis nigra* (Turbellarié, Triclade). J. Embryol. Exptl. Morphol. *15*, 39–60.

LENTZ T., 1967. Rhabdite formation in planaria: the role of microtubules. J. Ultrastruct. Res. *17*, 114–126.

LEPORI N. G., 1949. Ricerche sulla ovogenesi e sulla fecondazione nella Planaria *Polycelis nigra* Ehr. con particolare riguardo all'ufficio del nucleo spermatico. Caryologia *1*, 280–295.

MELANDER Y., 1963. Cytogenetic aspects of embryogenesis in Paludicola, Tricladida. Hereditas *49*, 119–166.

METSCHNIKOFF E., 1883. Die Embryologie von *Planaria polychroa*. Z. Wiss. Zool. *38*, 331–354.

NURSE F. R., 1950. Quinone tanning in the cocoon-shell of *Dendrocoelum lacteum*. Nature *165*, 570.

REYNOLDSON T. B., 1960. A quantitative study of the population biology of *Polycelis tenuis* (Iijima) (Turbellaria, Tricladida). Oikos *11*, 125–141.

REYNOLDSON T. B., 1961. A quantitative study of the population biology of *Dugesia lugubris* (O. Schmidt) (Turbellaria, Tricladida). Oikos *12*, 111–125.

SEILERN-ASPANG F., 1956. Frühentwicklung einer marinen Triclade (*Procerodes lobata*, O. Schmidt). Arch. Entwicklungsmech. Organ. *148*, 589–595.

SEILERN-ASPANG F., 1957. Polyembryonie als abnorme Entwicklung bei *Procerodes lobata* O. Schmidt (Turbellaria). Zool. Anz. *159*, 187–193.

SEILERN-ASPANG F., 1958. Entwicklungsgeschichtliche Studien an Paludicolen Tricladen. Arch. Entwicklungsmech. Organ. *150*, 425–480.

SKAER R. J., 1961. Some aspects of the cytology of *Polycelis nigra*. Quart. J. Microscop. Sci. *102*, 295–317.

SKAER R. J., 1962. Aspects of the cytology of triclads, Ph.D. Thesis, University of Cambridge.

SKAER R. J., 1965. The origin and continuous replacement of epidermal cells in the planarian *Polycelis tenuis* (Iijima). J. Embryol. Exptl. Morphol. *13*, 129–139.

STEINBÖCK O. and B. AUSSERHOFER, 1950. Zwei grundverschiedene Entwicklungsabläufe bei einer Art *Prorhynchus stagnalis* M. Sch. (Turbellaria). Arch. Entwicklungsmech. Organ. *144*, 155–177.

STEVENS N. M., 1904. On the germ cells and the embryology of *Planaria simplissima*. Proc. Acad. Nat. Sci. Philad. *56*, 208–220.

WEYGOLDT P., 1966. Die Ausbildung transitorischer Pharynxapparate bei Embryonen. Zool. Anz. *176*, 147–160.

G. Reverberi (ed.), Experimental embryology of marine and fresh-water invertebrates

CHAPTER 4

Annelids

G. REVERBERI

Zoological Institute, University of Palermo, Italy

4.1. Introduction and bibliographical notes

4.1.1. The eggs of Annelids are highly appropriate for embryological research, because of both the facility with which they are obtained and the fact that they can be fertilized artificially and their development followed up to the late stages. In fact, as documented by many publications, they have been widely used.

In a book the scope of which is to present, in a rapid review, the principal acquisitions in a particular embryological field, it does not seem necessary to report the complete bibliography: the reader can, however, refer to the many sources which exist and which will soon be indicated.

4.1.2. The first embryological research on the Annelids, like that in other fields, was descriptive. Nearly all the representatives of this important phylum were investigated; some of these investigations are generally considered 'classic', in particular those of Kowalewsky (1871), Hatschek (1878), Whitman (1878), Kleinenberg (1879), Salensky (1882) and Goette (1882), which are the oldest.

In 1892 two important papers by E. B. Wilson appeared: the first on 'The embryology of the earthworm', and the other on 'The cell lineage of Nereis'. These two papers can be regarded as models for all ulterior research. A new impetus came with the papers of Mead, 'The early development of marine Annelids' (1897); Nelson, 'The early development of Dinophilus: a study in cell lineage' (1905); and Child, 'The early development of Arenicola and Sternaspis' (1900).

4.1.3. These last papers closed the era of descriptive embryology, but in the meantime experimental research had developed. It is not possible to mention here even the milestones along this road which has been followed by embryologists such as Wilson, Morgan and Lillie. The reader will find valuable bibliographic indications in Schleip's well-known book 'Die Determination der Primitiventwicklung' (1929) or, still better, in 'Experimental embryology' by Morgan (1927).

For the more recent literature the reader may profitably consult volumes 9 and 16 of 'Fortschritte der Zoologie'.

Finally for information on the eggs and their manipulation the booklet on 'Methods for obtaining and handling marine eggs and embryos' of Costello et al. (1957) will be of great use.

4.2. The organization of the egg

To obtain a knowledge of the intimate structure of the egg has been, of course, one of the first and principal tasks of researchers. To mention only one of these researchers we shall refer to the work of Lillie (1902, 1906, 1909) on Chaetopterus which still constitutes an inexhaustible mine of information. However, the remarkable paper by Raven et al. (1950) on *Sabellaria alveolata* (to mention one of the moderns) should also be indicated. As it is not possible to enter this labyrinthine topic we shall draw the attention of the reader to only a few questions connected with the structure of the egg.

4.2.1. The first point is the importance and activity of the nucleolus during oogenesis. This problem, which had long been attentively studied, has been recently reconsidered by Martoja and Martoja-Pierson (1959) (Pomatocerus and Hirudo) and by Srivastava (1952) (Lumbricus). The former authors noted the large amount of SH-proteins in the nucleolus. These would be passed continuously to the cytoplasm during the period of yolk synthesis. Srivastava, on the other hand, reported the transfer of granular nucleolar material through the nuclear membrane: the products of the disintegration of these granules can be displaced by centrifugation and collected in a layer.

4.2.2. Another old question of great importance is that concerning the origin, the fate and the function of the 'yolk nucleus' (or 'Balbiani body'); in the past this question was much debated. The situation in Lumbricus is of some interest. Calkins (1895) had noticed that the oogonia do not have a 'yolk nucleus'; this first appears only in the young oocytes as a heap of granules sticking to the nuclear membrane. In older oocytes the mass begins to disintegrate and move away from the nucleus; the disintegrated mass would produce the 'yolk plates'. According to Calkins, when still compact the 'yolk nucleus' always stains like the chromatine; when disintegrated, it stains like the cytoplasm. The origin of the 'yolk plates' from the disintegration of the yolk nucleus is, however, contested by Gatenby and Nath (1926). A reconsideration of the matter, possibly using the electron microscope, would certainly be useful. Another point which should also be verified with the electron microscope is whether, as Gatenby and Nath maintain, the already mature oocyte which contains many Golgi spherules and mitochondria, does or does not have true yolk granules.

4.2.3. Of great importance is the localization of some chemical substances in the egg. In this regard the papers of Fauré-Frémiet (1924) and Raven et al. (1950) (on Sabellaria) and Allen (1961, 1967) (on Diopatra and Autolytus) are relevant. This research was initiated by Spek (1930, 1934) who used for his analysis some vital stains having the properties of pH-indicators. In the unsegmented egg of Nereis and Chaetopterus he observed that the animal cytoplasm shows an 'alkaline' reaction, whereas the vegetal cytoplasm shows an acid one. According to the author this 'bipolar differentiation' derives from a cataphoretic process at fertilization, as a result of which the positively and negatively charged colloidal particles are separated and accumulate at the poles. Later the ectoderm arises from the cells containing the alkaline cytoplasm and the entoderm from the cells which contain the acid cytoplasm.

Raven (1938), confirming the bipolar distribution of the ooplasmic substances, considered it to be due to a respective accumulation of the proteinic yolk in the vegetal region of the egg, and of hyaloplasm at the animal pole. Raven further studied the distribution and amount of some particular chemical substances, especially those to which a morphogenetic role is generally attributed, such as those included in the polar lobes. In the unfertilized egg RNA is present in the peripheral hyaline plasm, around the nuclear membrane and in the nucleolus. Glutathione is contained in the germinal vesicle. Glycogen appears to be evenly distributed in the cytoplasm. After the germinal vesicle has broken, glutathione spreads through the cytoplasm: an accumulation of it is observed around the spindle. With the cytochemical methods used, Raven was unable to show that the substances segregated in the polar lobes were different from the rest of the cytoplasm: he concluded that the properties of these morphogenetic plasms must reside in some structural peculiarities of their proteins. Alternatively one can attribute the morphogenetic value of the polar lobes not to the substances they contain, but to a particular cortical structure.

In the egg of Diopatra the RNA is distributed along an animal-vegetal gradient. Glycogen is scarce in *D. cuprea*, but is found in large amounts in *D. neapolitana*. Lipids are abundant. An important localization is that of the proteins and mitochondria at the animal pole: this localization is probably related to meiosis and initiation of the first cleavage. The young oocytes of Autolytus incorporate uridine-H^3 very actively: the incorporation is intense where large amounts of RNA are present, particularly in the nucleolus. Fully-grown oocytes no longer synthesize RNA: it is as if the machinery for its synthesis is turned off.

Phenylalanine-H^3 is incorporated everywhere by the young oocytes, which means that proteins are also synthesized everywhere. Fully-grown oocytes do not synthesize proteins. The segmenting egg, on the other hand, synthesizes them again and this synthesis is particularly active in the region of the spindle.

4.2.4. The polar plasms

Important features, probably present in all eggs but more evident in some types, are the polar plasms, so called because they are accumulated at the poles of the egg. They are particularly evident in the eggs of Clepsine (Schleip 1913, 1914, 1929) and Tubifex (Penners 1924a, b, c).

The Clepsine egg is released from the ovary in the first metaphasic stage: at this time, however, no polar organization of the plasms can be observed. When the first polocyte is emitted, the egg undergoes a contraction which redistributes the plasms. The contraction begins at the vegetal pole and gradually extends in the direction of the animal pole. After this contraction

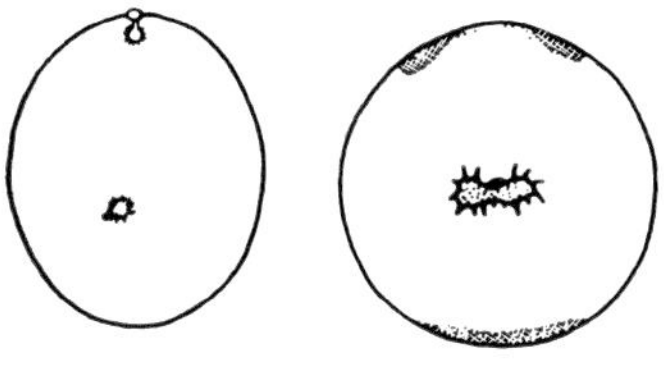

a **b**

Fig. 1. Meridional sections of an egg of Clepsine: in (a) the meiotic spindle at the animal pole, and the male pronucleus are shown; in (b) the distribution of the polar plasms can be seen. (Redrawn from Schleip 1914b.)

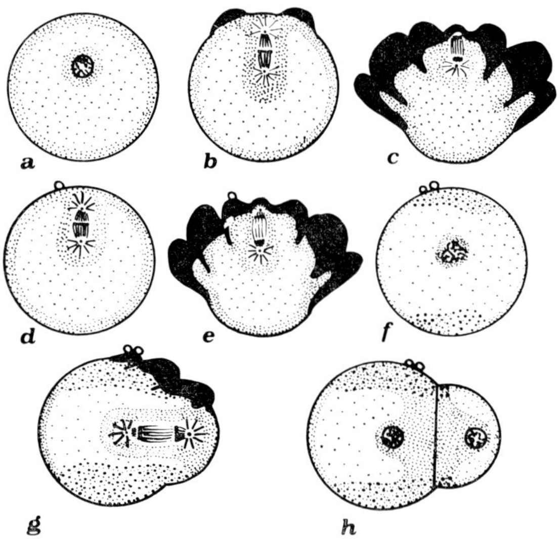

Fig. 2. Egg of Tubifex in different periods of maturation and at the first segmentation: a peculiar accumulation of some unsaturated lipids in the lobopodes is shown. (Redrawn from Hess 1959.)

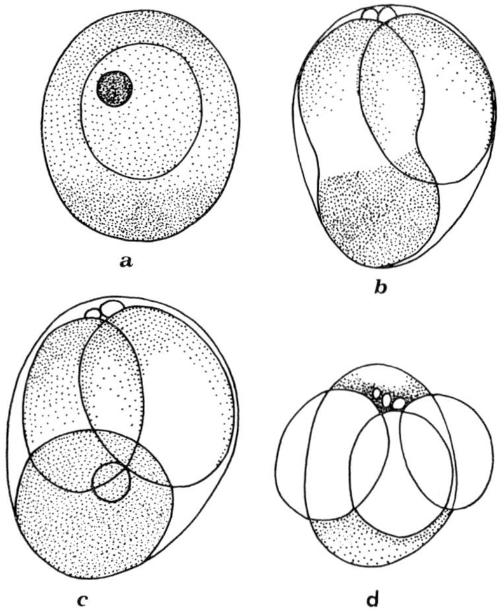

Fig. 3. Egg of Myzostoma before and after segmentation: the green plasm situated at the vegetal pole is segregated, at the 2-cell stage, in the polar lobe; at the 4-cell stage it is segregated in the D-cell. (Redrawn from Carazzi 1904.)

the egg resumes its spherical shape and orients itself with the animal pole uppermost. At this moment two caps can be distinguished at the poles. The animal plasm has a circular configuration with the first polocyte at the centre: the 'vegetal plasm', on the other hand, forms a compact cap (fig. 1). In Tubifex this distribution also follows a peristaltic contraction (at meiosis) which is so strong that it gives rise to the formation of lobes. It is of interest that some lipids become segregated in these lobes (Hess 1959): the significance of this fact is unknown (fig. 2).

A very peculiar plasma segregation occurs in the eggs of Chaetopterus, Sabellaria and Myzostoma. Like some molluscan eggs these eggs form, before cleaving, a cytoplasmic lobe at the vegetal pole: it is usually indicated as the yolk lobe or polar lobe. In Myzostoma (fig. 3) the polar lobe is green; a red plasm is also present at the animal pole (Carazzi 1904).

4.2.5. The ultrastructure of the egg

Naturally we should like to know more about the composition of at least some of the plasms described above. Unfortunately the electron microscope has been very little used for study of the eggs of Annelids. The few in-

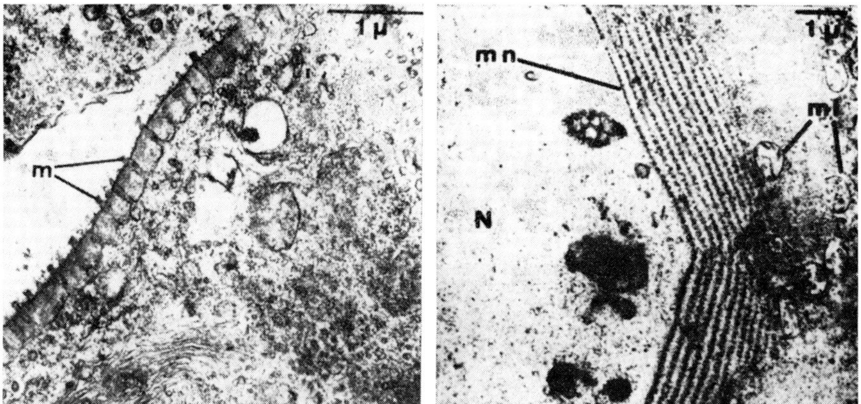

Fig. 4. Egg of Nereis: (a) cortical region of a normal egg; (b) egg which was deprived of the action of the neural hormone. Note the lamellae annulatae in the proximity of the nucleus: m = microvilli; G = Golgian body; N = nucleus; l = lamellae annulatae; mn = nuclear membrane; mt = mitochondria (Durchon and Boilly 1964).

vestigations that have been made concern the eggs of Nereis (Pasteels 1966; Fallon and Austin 1967; Durchon and Dhainaut 1964; Durchon and Boilly 1964; Dhainaut 1966, 1968, 1969), Hydroides (Colwin and Colwin 1961), Sabellaria (Pasteels 1965), and Merceriella (Sichel 1966). In very young

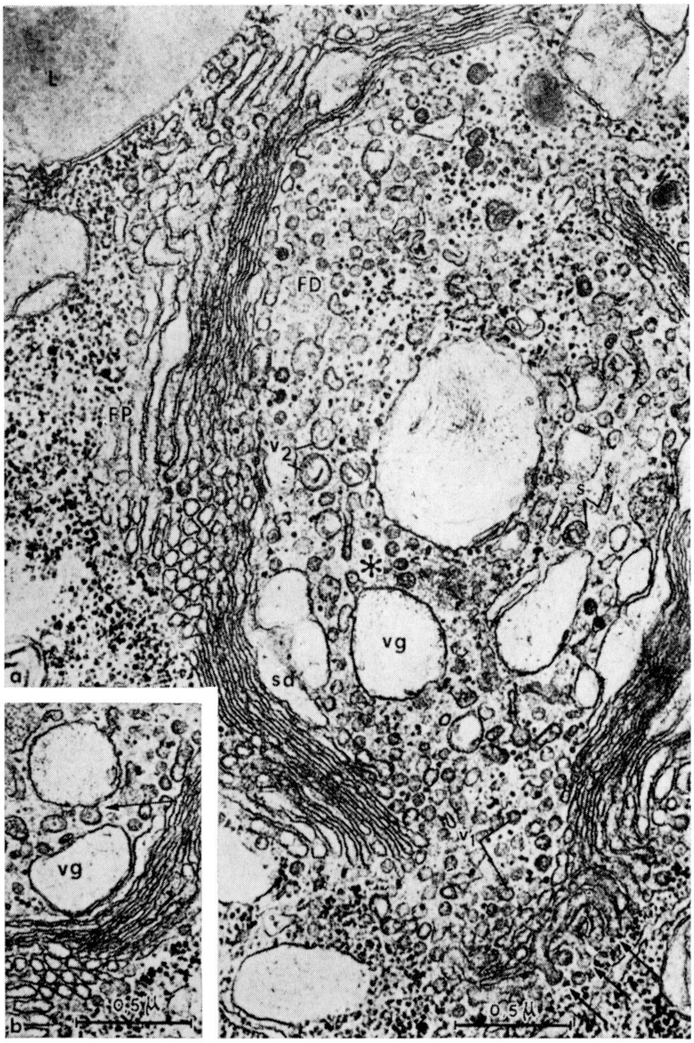

Fig. 5. The Golgian complex in an egg of Nereis: the complex elaborates a fibrous material which is collected in the vacuoles. vg = golgian vesicles; sd = distal sacs; FD = distal face; FP = proximal face (Dhainaut 1969).

oocytes of Nereis, Dhainaut demonstrated a large amount of so-called multivesicular bodies, which seem to be implicated in yolk synthesis. In the vitellogenetic oocytes (40–90 μ), on the other hand, he showed many rosettes situated in the proximity of the microvilli or connected with the lipid granules; these are probably glycogen rosettes.

Very interesting structures were also found in oocytes which, in consequence of the removal of the parapods in which they lay, were for 10–20 days deprived of the action of the cerebral hormone (Durchon and Boilly 1964): large quantities of 'lamellae annulatae' are observed (fig. 4). Of much interest is the Golgian body which elaborates a fibrous material which is finally collected in the vacuoles (fig. 5).

The egg of Nereis has also been investigated by Pasteels and by Fallon and Austin. According to these authors a differentiated cortex with many large alveoli is present; the alveoli contain a fibrous material, probably a precursor of the jelly coat. The cortex is limited by the plasma membrane from which many thick microvilli depart; these traverse the perivitelline space and

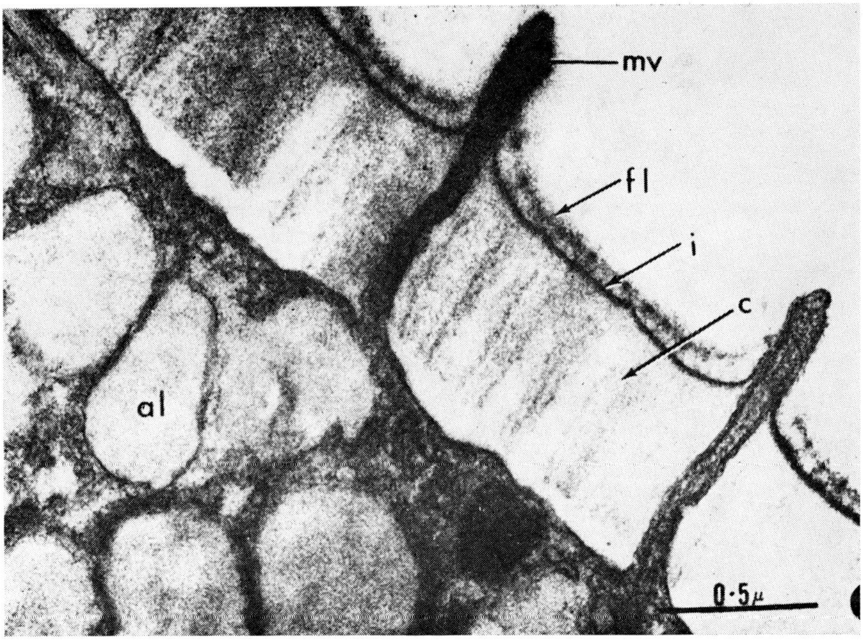

Fig. 6. Cortical region of an egg of *Nereis limbata:* note the alveoli (al) and the microvilli (mv) which project through the chorion; fl = fibrous layer; i = intermediate layer; c = canal layer (Fallon and Austin 1967).

reach the chorion. Along the microvilli the jelly (or its precursor) which is found peripherally to the egg, seems to be poured out (fig. 6). A thick chorion made up of three layers of different structures surrounds the egg.

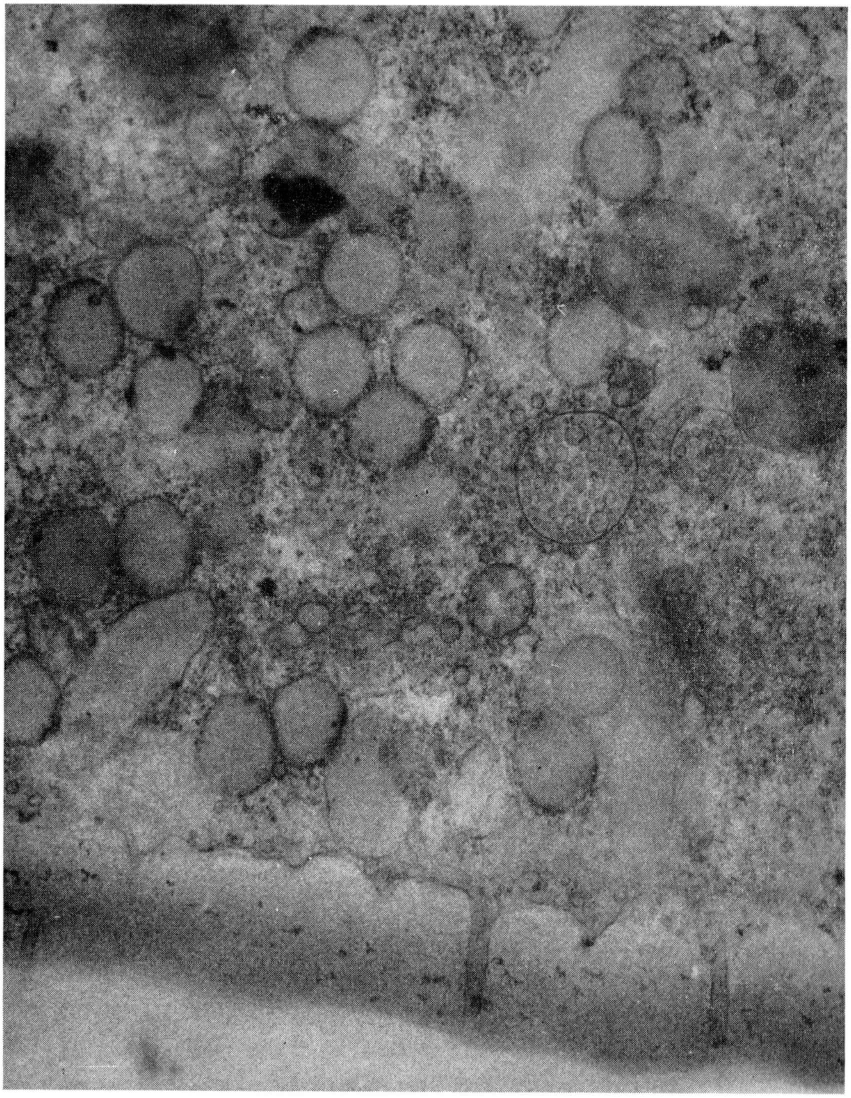

Fig. 7. Cortical region of an egg of Hydroides: notice the microvilli, the chorion and the multivesicular bodies (original). × 22.000.

The egg of Hydroides has also been investigated by Colwin and Colwin (1961), who, however, limited their research to the membranes. Their

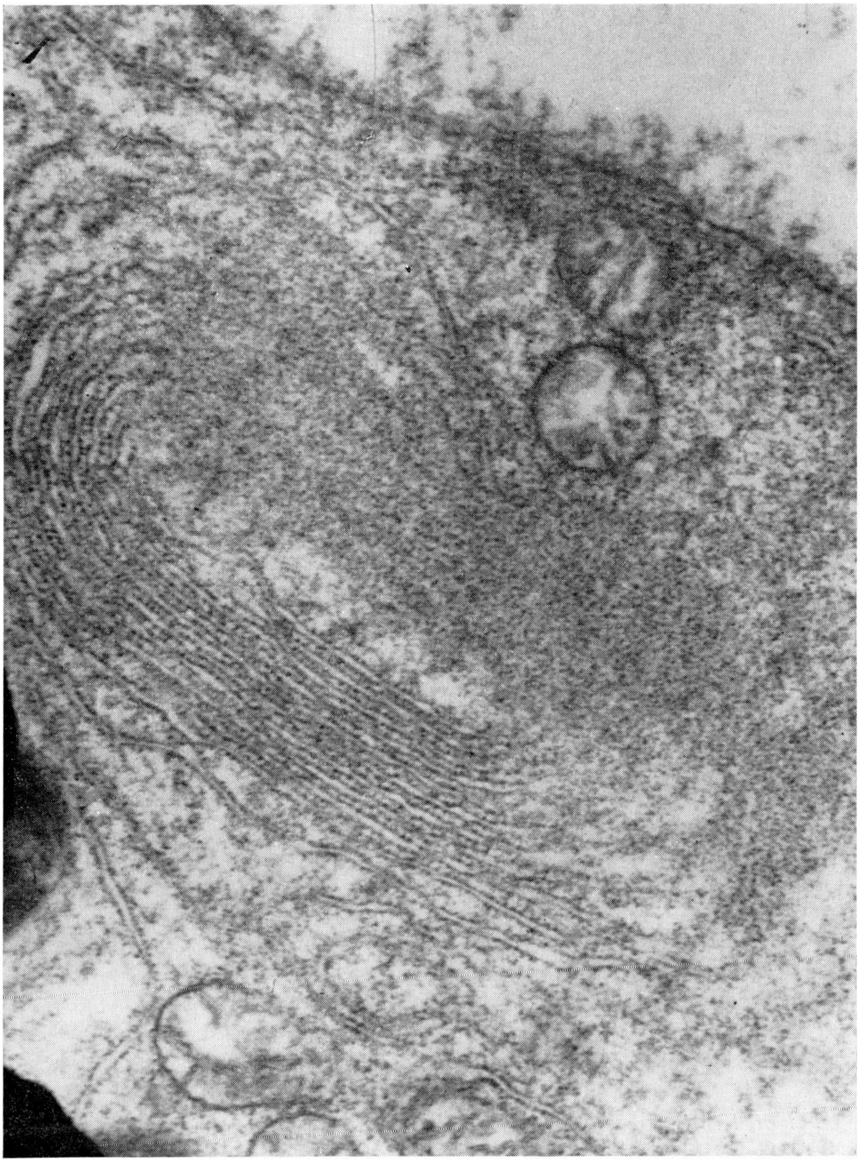

Fig. 8. A peculiar ergastoplasmic whorl situated in the proximity of the nucleus in the egg of Hydroides (original). × 35.000

findings can be completed by other data which can be derived from some of our photographs (figs. 7, 8, 9).

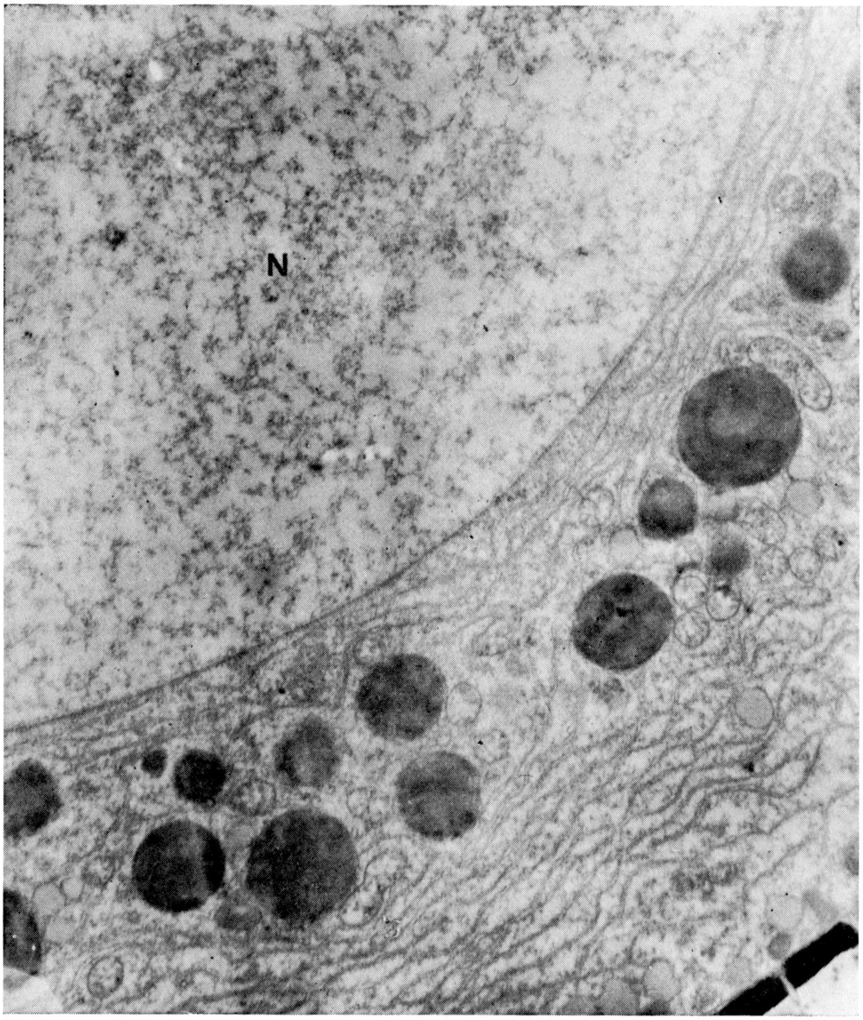

Fig. 9. A peculiar aspect of an egg of Hydroides: note the nucleus (N) and the dense lamellar endoplasmic reticulum (original). × 10.000

4.3. *Fertilization*

4.3.1. The structure of the egg, as well as of its membranes, changes profoundly at fertilization. These changes were described accurately in the past but only recently have they been checked with the electron microscope. The early research of Lillie (1902, 1906, 1909, 1911) on the Nereis egg is still interesting. The unfertilized egg shows a distinct radial striation at the periphery due to a layer of alveoli (fig. 10). At fertilization the alveolar layer

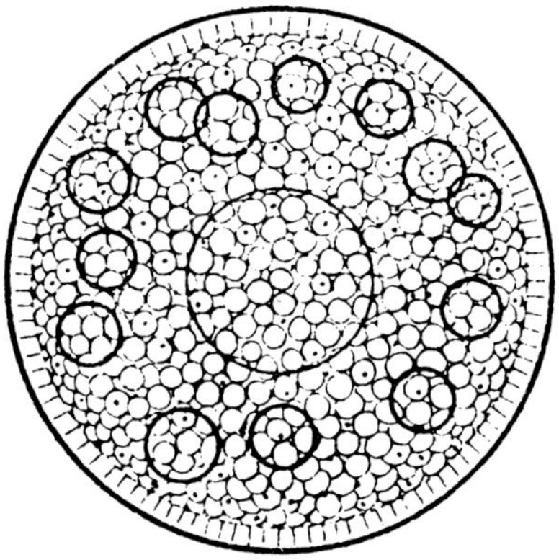

Fig. 10. Diagram of an unfertilized Nereis egg: the periphery is occupied by a sheet of vacuoles (Costello 1940).

disappears and a great amount of jelly is produced (fig. 11). According to Lillie, the alveoli are situated inside the unfertilized egg; this view is supported by the observation of Novikoff (1939a) that eggs from which the membranes were removed still possessed the alveolar layer. Costello (1949) also noticed

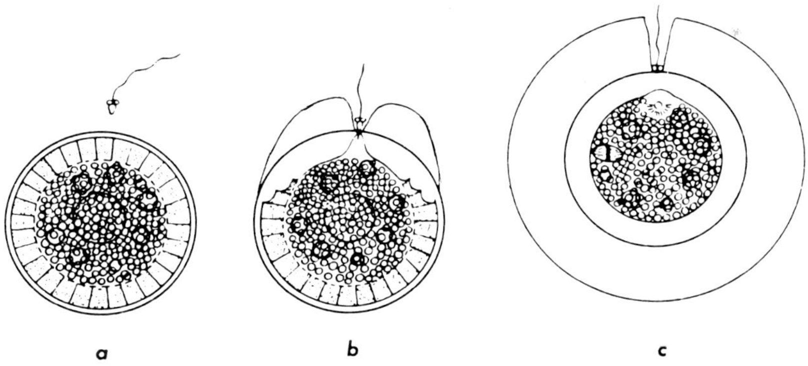

Fig. 11. Aspects of the egg of Nereis before (a) and after (b), (c) fertilization. At fertilization the alveoli break down (b) and their content is poured out: hydrated jelly is formed around the egg (c) (Austin 1965).

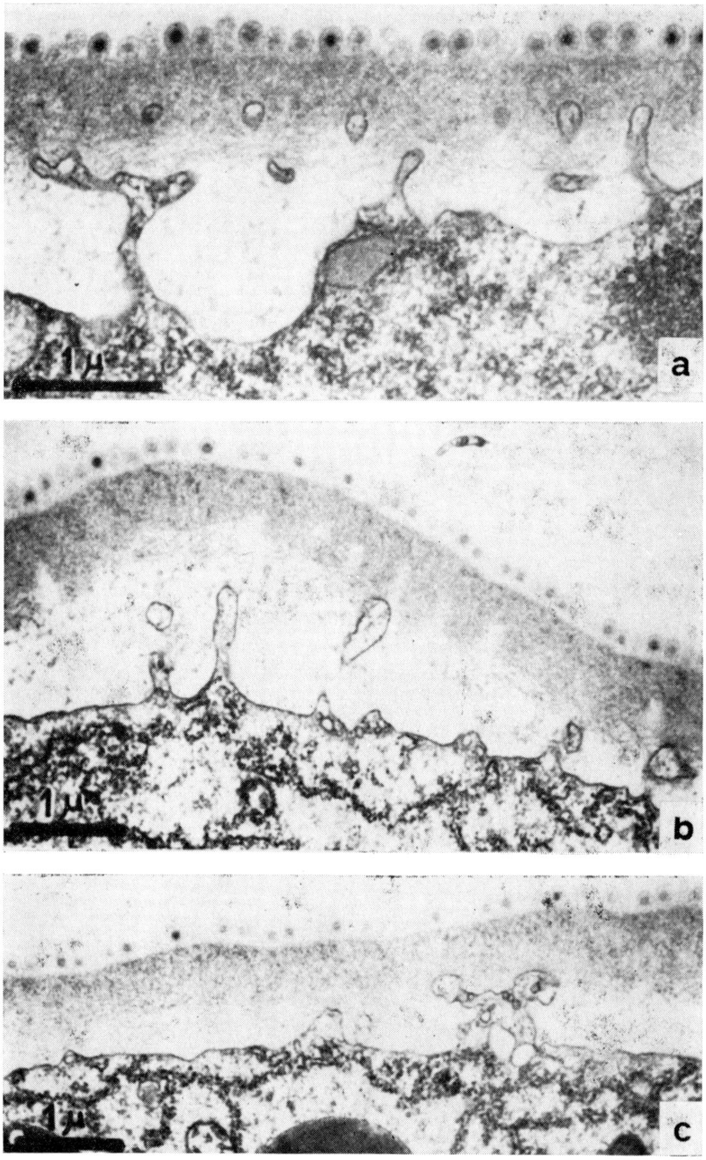

Fig. 12. Egg of Sabellaria: (a), (b) cortex and membranes in an unfertilized egg which was in contact with sea water for 30 sec and 45 sec; (c) the cortex and the membranes in a fertilized egg (Pasteels 1965).

that the alveoli can be displaced to the centrifugal pole by centrifugation.

Examination by electron microscope shows the situation to be the following (Pasteels 1966; Fallon and Austin 1967): around the unfertilized egg (fig. 6) a thick envelope (chorion), composed of many heterogenous layers, is seen; the most interior of these layers has numerous prismatic lodges and lies directly on the plasma membrane, which is composed of short microvilli filled with a densely fibrous substance. The microvilli are continuous with the interior of the cytoplasm and are interspersed among the cortical alveoli. At fertilization the layer of cortical granules is expelled in toto, the perivitelline space enlarges to make room for a jelly-like substance which is produced, and a new plasmalemma is formed. Thus observation with the electron microscope confirms the opinion of Lillie that the layer of the alveoli is situated inside the egg.

Another difference between the unfertilized and fertilized egg is the appearance of the microvilli, which in the unfertilized egg are thick and short and in the fertilized egg are long and sinuous. The total surface of the fertilized egg is consequently greatly increased; at the same time, as the sinuous microvilli reach the more external layer of jelly, a connection between the sea water and the interior of the egg is established. Modifications at fertilization have also been observed in the egg of Sabellaria: described earlier by Horst (1881), Fauré-Frémiet (1924), Waterman (1934), Novikoff (1939b) and Raven et al. (1950) they have also been investigated by electron microscope (Pasteels 1965).

The chorion of the mature egg is made up of three distinct layers; the exterior one with spherical globules; the plasmalemma from which depart short, cylindric microvilli, and a layer of alveoli containing a granular or fibrous substance. The unfertilized egg which remained for a while in sea water, no longer exhibits the alveoli, all the perivitelline space being occupied by their remains (fig. 12).

The fertilization does not change anything, only the microvilli are retracted. It can thus be affirmed that the egg of Sabellaria does not change significantly at fertilization, and that the elevation of the vitelline membrane is not a typical feature of fertilization.

4.3.2. The modifications of the Annelid spermatozoon when it comes into contact with the egg have also been followed by electron microscope: all investigations concern either the spermatozoon of Nereis (Fallon and Austin 1967) or Hydroides (Colwin and Colwin 1961).

Fig. 13 shows some features of the acrosome of the spermatozoon of

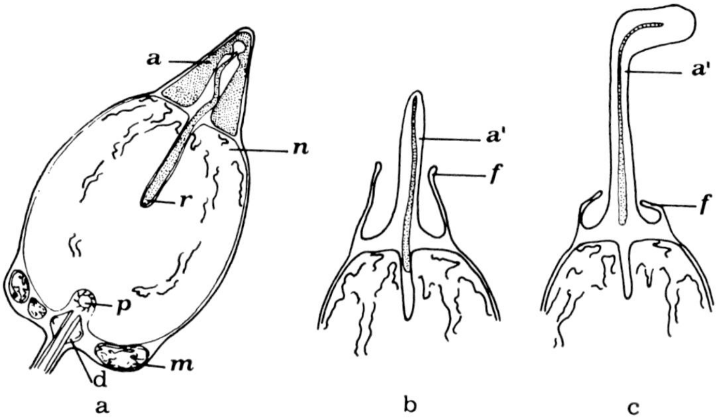

Fig. 13. Diagrams showing some features of the acrosome of *Nereis limbata* in an unreacted (a) and reacted (b) (c) spermatozoon; r = axial rod; a = acrosome; n = nucleus; p = proximal centriole; m = mitrochondrion; d = distal centriole; a′ = acrosome filament; f = folds. (Redrawn from Fallon and Austin 1967.)

Nereis, before and after the reaction. As illustrated by the figures, the acrosome has the shape of a hollow cone, which encloses an axial chamber ending anteriorly in a small cavity. In the axial chamber there is a rod-like structure, which fits posteriorly into an invagination of the nucleus. When the spermatozoon touches the egg the acrosome opens, discharging its contents: an acrosomial filament is also produced, which, through one of the cylindrical lodges of the chorion, reaches the plasma membrane and fuses with it.

These modifications were first described by Colwin and Colwin (1961) in Hydroides. Since the mechanism of the penetration of the vitelline membrane by the spermatozoon and its incorporation into the egg is generally described in all books concerned with fertilization, we shall dispense with further discussion of this subject.

4.4. Segmentation

4.4.1. The Annelid egg has always been regarded as choice material for cell-lineage studies. The influence which studies of cell-lineage have had on the whole development of embryology is well documented. Unfortunately we cannot report such investigations in detail here, and prefer instead to suggest the reading of the papers by Wilson (1892) on Nereis and Lumbricus; Lillie (1906) on Chaetopterus; Nelson (1905) on Dinophilus; and Kleinenberg (1886) on Lepadorhynchus. These papers have always been regarded as a fount of valuable information and every student of embryology should be acquainted with them. Let us here report some evaluations.

Concerning Kleinenberg's paper on Lepadorhynchus, Davydoff in his 'Traité d'embryologie comparée des invertebrés' notes: 'tant pour la richesse des documents que par l'originalité des vues, reste toujours l'un des plus beaux de la bibliographie des Polichetes'; while of Wilson's paper on Nereis he writes: 'Ce fut E. Wilson qui apporta dans son fameux travail sur le développement de la Nereis (1892) les méthodes désirées, ouvrant ainsi une voie nouvelle aux recherches embryologiques. C'est Wilson qui a montré le parti qu'on pouvait tirer de l'étude de la généalogie des blastomères . . . Le travail de Wilson a servi de modèle à toutes les recherches récentes . . .'

4.4.2. The segmentation of the Annelid egg is typically 'spiral': slight

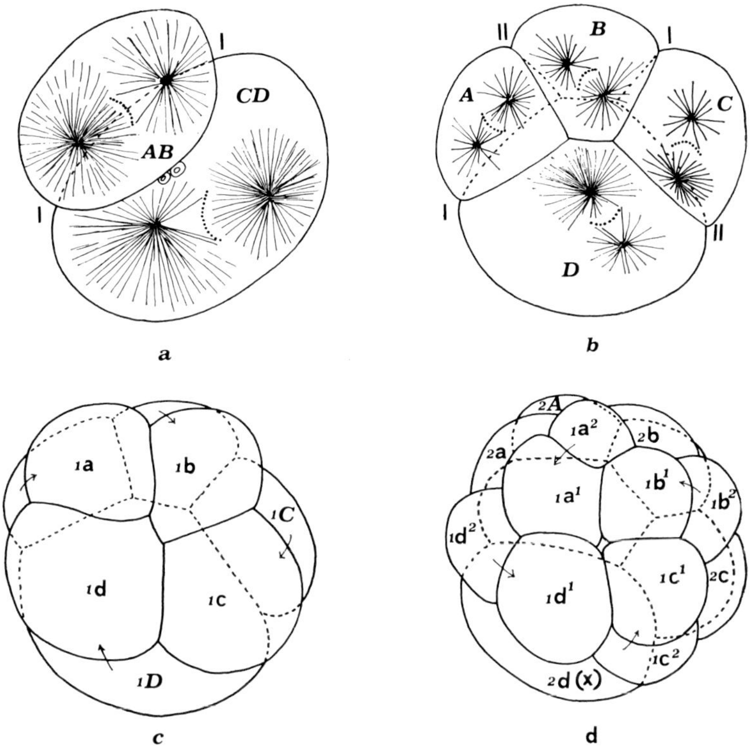

Fig. 14. First stages of development of the egg of *Arenicola cristata;* the trochoblasts are indicated in (d) as $1a^2$, $1b^2$, $1c^2$, $1d^2$. (Redrawn from Child 1900.)

variations of this pattern occur in the different classes, however. We cannot describe these variations, and shall only report the general lines of this segmentation, selecting in particular the eggs of polichaetes and oligochaetes. The polichaete egg cleaves unequally from the beginning. By the 1st cleavage two unequal blastomeres, AB and CD, are formed, the latter being the larger (fig. 14). The 2nd segmentation plane cleaves the egg into 4 cells which can be readily individuated: D, the largest, has a special role in ontogenesis. The 3rd segmentation plane divides the egg into 8 cells, 4 macromeres and 4 micromeres: the latter are indicated as 1a, 1b, 1c, 1d and are rotated dexiotropically by 45° with respect to the macromeres. The second quartet of micromeres, indicated as 2a, 2b, 2c, 2d, is laeotropically rotated (stage of 12 cells); by two new segmentations two other quartets of micromeres are produced (they are rotated dexio- and laeotropically respectively) indicated as 3a–3d; 4a–4d. Segmentation of the two first quartets produces two new quartets: $1a^1$, $1b^1$, $1c^1$, $1d^1$; $1a^2$, $1b^2$, $1c^2$, $1d^2$. The destiny of the 64 cells so formed is known: 1a–1d; 2a, 2b, 2c, and 3a–3d, give rise to the ectoderm; $1a^1$–$1d^1$ give rise to the (cephalic) ectoderm and the brain; $1a^2$–$1d^2$ give rise to the primary trochoblasts. The 2d cell, which is larger than its sister cell (and can thus be easily individuated), is the first somatoblast, responsible for the formation of the trunk ectoderm. Cell 4d (= second somatoblast), on the other hand, is responsible for the formation of the mesoderm; 4a, 4b and 4c, together with the original macromeres, give rise to the entoderm.

The gastrula is formed by epiboly and invagination: 4d migrates interiorly

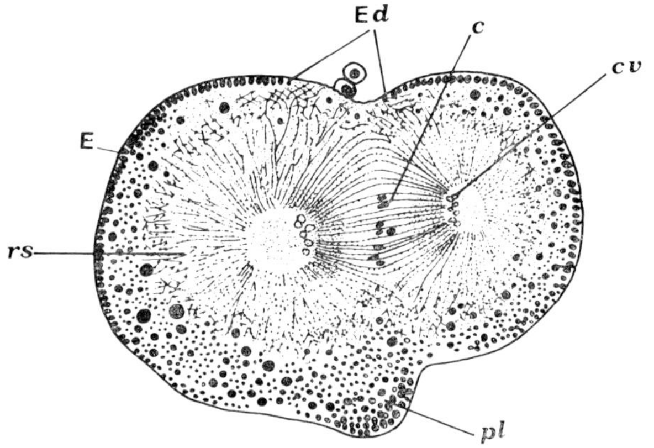

Fig. 15. Section of an egg of Chaetopterus at first cleavage (anaphase); pl = polar lobe; rs = residual substance of the germinal vesicle; E = ectoplasm; Ed = ectoplasmic defect; c = chromatin masses; cv = chromosomal vesicles. (Redrawn from Lillie 1906.)

and divides into $4d^1$ and $4d^2$, which are symmetrically situated in relation to the medial sagittal plane of the embryo. These two cells are the mesodermic teleoblasts which, however, as Wilson has shown, are not purely mesodermic, since they also give rise to some small cells which participate in the formation of the intestine (so that they should be designated as entomesodermic entoblasts). The larva is a trochophore with a trochus and a long ciliary tuft at the apex.

The segmentation described above is a little more complicated in eggs forming a polar lobe, such as those of Sabellaria. The segmentation of this egg has been followed by Fauré-Frémiet (1924), Hatt (1932), and Novikoff

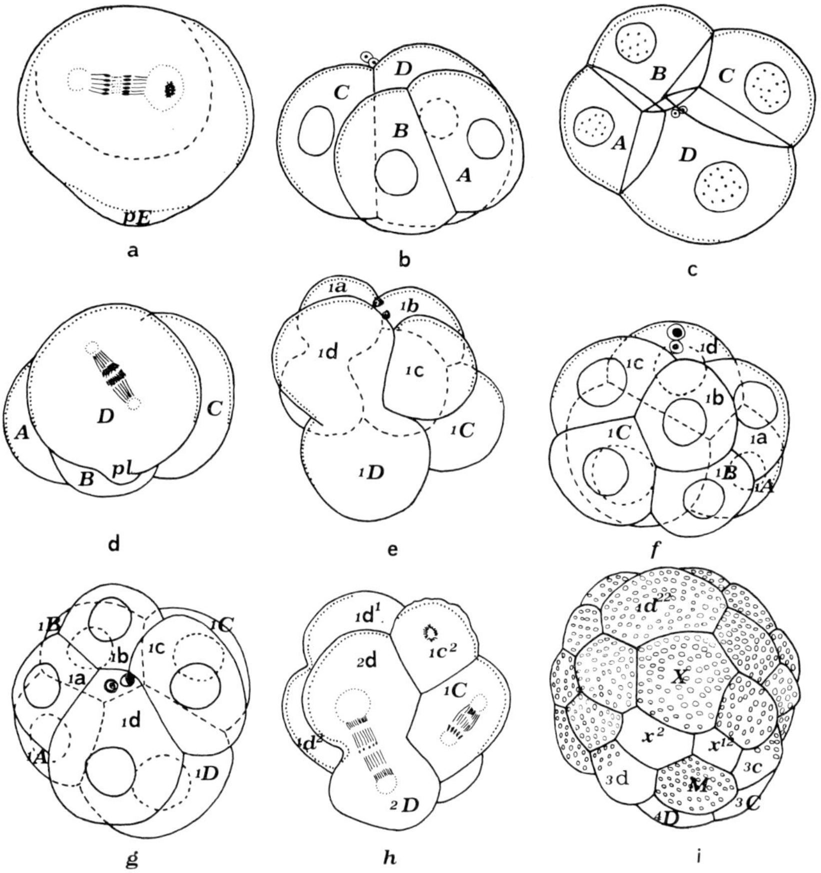

Fig. 16. First stages of the development of the egg of Chaetopterus; pl = polar lobe; X = 1st somatoblast; M = 2nd somatoblast. (Redrawn from Lillie 1906.)

(1938b). According to their descriptions, before the egg cleaves it produces, at the vegetal pole, a large cytoplasmic protrusion which persists for a while (trefoil stage) but is soon reabsorbed in one of the two already formed cells (CD). The same protrusion reappears before the 2nd cleavage. It is again reabsorbed in one cell (D) of the 4-cell stage. A third polar lobe is formed before the 3rd segmentation which flows definitely into the D-macromere and its derivatives.

A similar behaviour has also been described for the eggs of Chaetopterus (figs. 15 and 16) and Myzostoma (fig. 3). Our knowledge of the chemical constitution of the substances included in the polar lobe is very poor. Raven et al. (1950) studied this problem and concluded their investigation with the following words: 'In connexion with the importance of the lobes for morphogenesis, and the fact that they contain the determining factors of the apical tuft and post-trochal region of the larva (Hatt 1932; Novikoff 1938b) their cytochemical composition has been studied with great care, in order to determine if any differences with the rest of the egg could be observed. The results of this study are mainly negative: no clear differences have been observed. In particular, it must be emphasized that no special accumulation of ribonucleic acid or sulphydril compounds in the region of the lobes occurs, and that the oxidative enzymes studied cannot be demonstrated in the Sabellaria egg

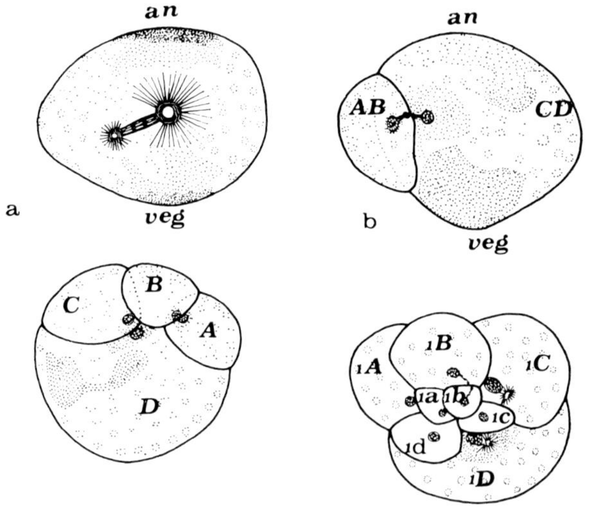

Fig. 17. First stages of development of the egg of Tubifex: the animal and vegetal polar plasms are indicated by a fine granulation. (Redrawn from Penners 1922.)

at all . . . The only observation which might be explained as pointing to a different physicochemical composition of the lobes is their somewhat greater stainability with brilliant cresyl violet and the occurrence of dark granules with this staining. The interpretation of this fact remains difficult, however.' On the other hand, Dalcq and Pasteels (1963) have shown that the polar lobes of Chaetopterus are rich in enzymes of the mononucleotide phosphohydrolase type.

A plasm segregation is also observed in the egg of some Oligochaetes. We shall describe it in Tubifex according to a recent publication (Inase 1967). The egg is laid at the metaphase of the first maturation division. The first polar body is extruded 30–40 min after spawning together with a marked change in the shape of the egg (fig. 2): this change occurs again before the emission of the 2nd polar body, and before the first segmentation. The egg cleaves into two blastomeres of unequal size (CD being larger than AB) (fig. 17), CD retaining all the polar plasms. At the 2nd cleavage which occurs 2 hr after the first, CD divides earlier than AB: an intermediate stage of 3 cells is thus formed. By division of AB into two equal cells the egg reaches the 4-cell stage: the polar plasms become totally segregated in D. At the 3rd segmentation D and C divide earlier than A and B and thus a 6-cell stage is formed. At the end of the 3rd segmentation the egg has 8 cells, 4 macromeres (1A–1D) and 4 micromeres (1a–1d): the polar plasms are all in the 1D cell. The 4th cleavage is initiated by the 1D cell; the 1C soon

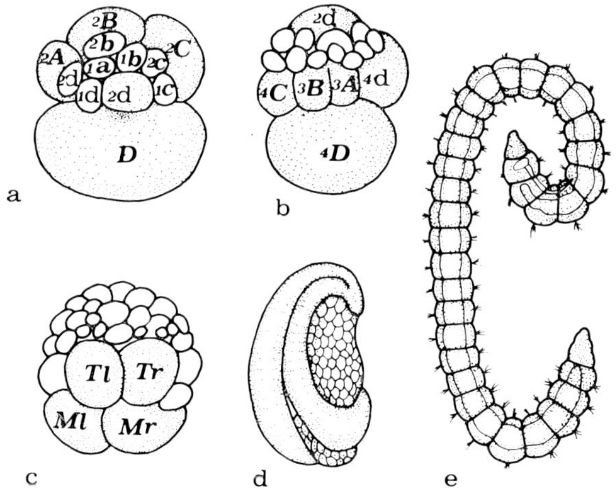

Fig. 18. Egg of Tubifex at different stages of development (4th, 5th cleavage, blastula stage, young embryo); at the extreme right the embryo after hatching. (Redrawn from Inase 1967.)

divides and a 10-cell stage is formed. The 4th cleavage is completed by the segmentation of 1A and 1B and the formation of the second quartet of micromeres (2a–2d) (fig. 18). The 'animal plasm', at this stage (12 cells) is in the 2d-cell while the 'vegetal plasm' is in 2D. Soon the micromeres of the first quartet cleave, and the 16-cell stage is reached. The ulterior destiny of the 'polar plasms' is as follows: the animal plasm enters $2d^1$, and the vegetal plasm 3D; later (22-cell stage) it enters the $2d^{11}$- and 4d-cells (derived respectively from $2d^1$ and 3D). This last is the mother cell of the mesodermal bands (second somatoblast).

At the blastula stage the $2d^{11}$ cell, containing the animal plasm (first

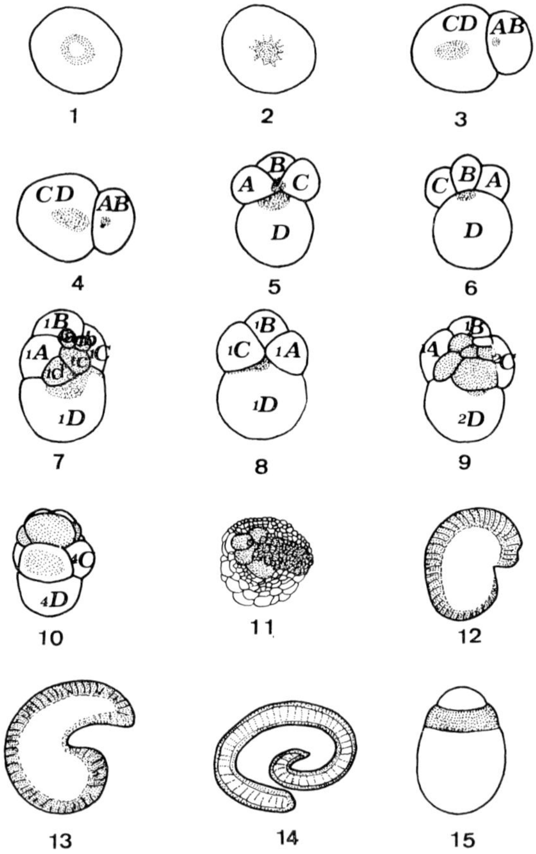

Fig. 19. Egg of Tubifex at different stages of development after treatment with Nadi: the fine granulation indicates the regions which are Nadi positive: in (15) an egg is shown which was strongly centrifuged. (Redrawn from Carrano and Palazzo 1955.)

somatoblast), divides into Tr and Tl (the teloblasts) from which the ectodermal bands derive; 4d, for its part, gives rise to Mr and Ml from which derive the mesodermal bands.

The precise distribution of the polar plasms suggests that they may have some morphogenetic role. This hypothesis is soundly based on the results of certain experiments which will be reported later. Before dealing with this topic we should examine the problem of the chemical constitution of the polar plasms. Lehmann (1941, 1948) first showed that the polar plasms are very rich in indophenol-oxidases. Carrano and Palazzo (1955) subsequently showed that they are rich in oxidative mitochondrial enzymes (cytochrome-oxidase and succino-dehydrogenases). This precocious mitochondrial segregation is probably inherent in the function of the cells deriving from 2d and 4d, and possibly in their differentiation, as is attested by the results of the treatment of the egg with specific enzyme inhibitors. In any case, the presence of large amounts of mitochondria in the polar plasms is well documented by investigations with the electron microscope (Lehmann and Mancuso 1957; Weber 1958).

4.5. The developmental potentials of the unsegmented and segmented egg

The developmental potentials of the egg can only be investigated by experimentation. To this end the Annelid eggs have long been fused, fragmented, centrifuged, compressed or disarticulated, the principal aim being that of discovering the morphogenetic value of the visible or invisible plasms. To report all these investigations would take more space than is permitted, so we shall mention only that research which in our opinion provides the best view of the potentialities of this egg.

4.5.1. Fusion

The fusion of two eggs of Sabellaria before segmentation was performed by Hatt (1931). The result was not the formation of a single gigantic embryo but of double or multiple embryos (fig. 20).

4.5.2. Fragmentation

The egg can be fragmented into two parts. This has been done, for instance, with the eggs of Nereis (Costello 1940) shortly before the first cleavage (fig. 21). If the egg is cut across the equator one obtains an animal fragment which contains the cleavage spindle and a vegetal fragment which contains

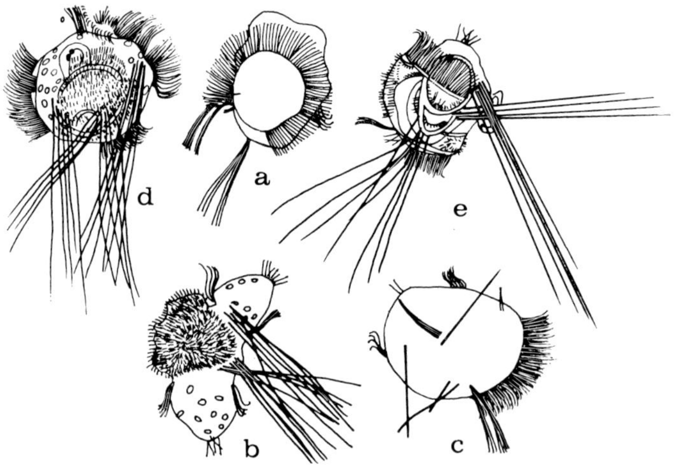

Fig. 20. Larvae of Sabellaria arising from the fusion of two eggs. (a) and (b) larvae of 23 and 52 hr respectively, (c) larva of 49 hr, (d) another larva of 49 hr with one internal gut, (e) larva of 55 hr with two internal guts. (Redrawn from Novikoff 1938b.)

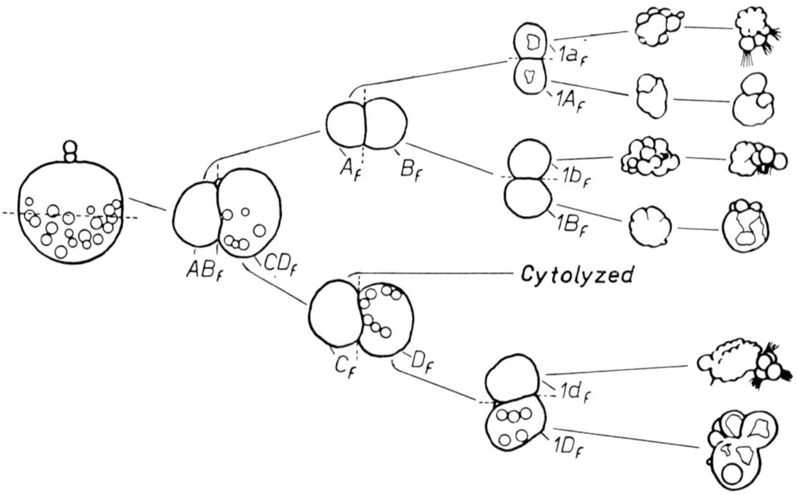

Fig. 21. Final products of differentiation (row at extreme right) of the progressive blastomere separation of a segmenting animal fragment of a fertilized egg of Nereis (Costello 1945).

no nuclear material. Of course the latter does not develop at all: the animal fragment, however, cleaves and can also give rise to an essentially normal embryo. This means that before segmentation the egg is omnipotent.

4.5.3. Centrifugation

The plasms can be completely displaced or disarranged by centrifugation. Afterwards, when the egg cleaves, the plasms can be segregated in unusual blastomeres so that their formative value can be tested. These experiments have been carried out on many kinds of eggs: in Chaetopterus by Lillie (1906, 1909), Morgan (1910) and Raven (1938).

The eggs fall at random in the centrifuge and their materials become stratified in any position with regard to the primary axis. The first cleavage which falls through the poles gives rise to two unequal blastomeres, the smaller of which contains every possible kind of substance (oils, yolk, etc.); the polar lobe, which appears at the usual times, may also contain all these materials (Raven 1938). It is really unexpected to obtain normal larvae from such disarranged eggs; Morgan (1927) comments: 'it is not without interest to note that despite the variety of visible materials a normal embryo may develop after the interior materials have been redistributed and arranged in zones. Again, although this egg has the typical spiral cleavage of other Annelids, it may develop normally after centrifuging. It is evident that this result is due to the retention of the normal type of cleavage despite the displacement of its visible materials. The evidence furnishes a very convincing argument in favour of the view that these materials are not formative stuffs'.

Furthermore, by submitting the egg to stronger centrifugation it can be broken into fragments, each containing single and different plasms. The experiments undertaken by Wilson (1929, 1930) and E. Harvey (1939) have yielded very interesting results. In the cases in which the egg was split into two pieces these were nearly of the same size; one was white and contained the polar spindle, the oil cap, and the hyaline plasm; the other was yellow and contained yolk granules. After fertilization the former cleaved like a miniature egg, and in some cases developed into a trochophore differing from a normal one in size and transparency; the latter showed only amoeboid movements and only in few cases divided into 2–4 cells. After parthenogenetic activation with KCl (parthenogenetic merogony) this fragment also showed amoeboid modifications (not a polar lobe), and sometimes cleaved into two cells, but did not develop further, thus behaving differently from the parthenogenetic merogonies of the sea urchin egg. 'Probably the unfertilized egg of Chaetopterus is more highly organized than that of sea urchin.' Similar results were also obtained with the eggs of Nereis (Costello 1940). These were centrifuged at high speed, but the two parts did not separate, as is shown in fig. 22, but remained in the thick membrane connected by a

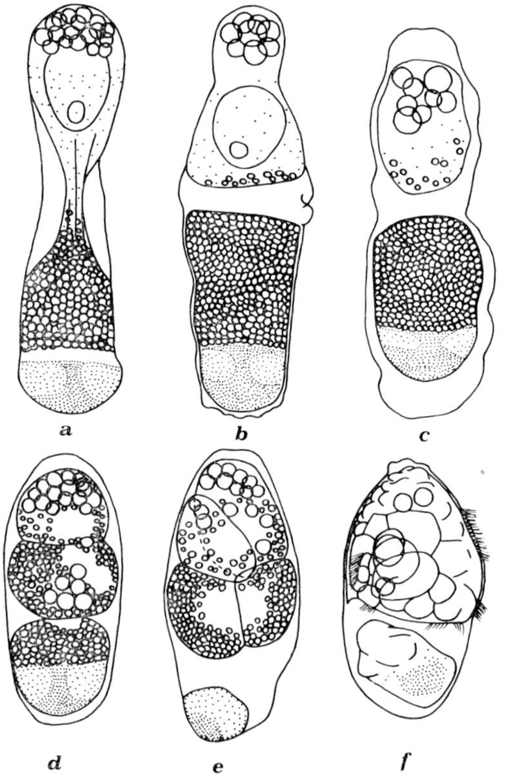

Fig. 22. (a) Nereis egg after centrifugation; (b), (c) the centrifuged egg was cut transversely and fertilized; (d), (e) the centripetal fragment has cleaved; (f) a ciliated larva has developed from the centripetal fragment. (Redrawn from Costello 1940a.)

bridge. However, the bridge was afterwards broken by needles and the two pieces separated. The centrifugal fragment, which contains yolk, the material of the cortical granules and some erythrophilic granules, does not cleave at all after fertilization: the spermatozoon is present but does not seem able to stimulate the formation of an aster in the cytoplasm. Costello maintains that it is unable to do so because it requires conditioning by the nuclear sap. The nucleated fragment (containing the oils and the clear plasm), on the other hand, develops in some way. If the anucleated fragment is activated parthenogenetically with KCl (parthenogenetic merogony) it does not show any development, but only amoeboid movements. In other words the parthenogenetic merogonies cannot develop at all.

Centrifugation has also been carried out on the eggs of Tubifex (Parseval

1922) and Sabellaria (Fauré-Frémiet 1924; Harris 1935; Raven et al. 1950). In Tubifex the polar plasms can be completely displaced, with the result that these eggs never develop into normal larvae. After centrifugation of the unfertilized Sabellaria egg the ooplasms are distributed as follows: the strongly refringent oil droplets at the animal pole; the heavy yolk granules at the vegetal pole; and the hyaline cytoplasm in the intermediate region. Once fertilized, the eggs develop, and the ooplasm is distributed in an arbitrary way in the different regions; sometimes the oils were found even in the polar lobe, which means that its formation is independent of its content.

4.5.4. Compression

No less interesting are the results obtained from the development of eggs compressed between two plates. If Chaetopterus eggs are submitted to this compression (Titlebaum 1928; Tyler 1930) they divide equally instead of cleaving into two unequal cells and, as a consequence, the vegetal plasm is distributed equally in the cells AB and CD. These eggs develop into double larvae (fig. 23) (Tyler 1930). These are perfectly normal if the blastomeres AB and CD were previously separated. Similar results have also been obtained with the eggs of Nereis (Wilson 1896; Morgan 1910): the embryos that develop from eggs compressed before the first segmentation are generally double.

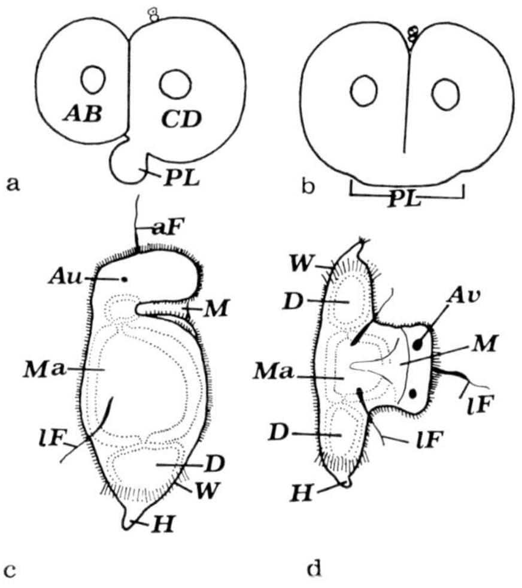

Fig. 23. Egg of Chaetopterus: (a) normal 2-cell stage with the polar lobe; (c) normal trochophore; (b) egg which segmented into two equal blastomeres; (d) trochophore derived from (b): duplicitas cruciata. (Redrawn from Titlebaum 1928.)

4.5.5. *Physical or chemical treatments*

Double embryos from a single egg have also been obtained in Tubifex following treatment of the egg by heat, lack of oxygen, etc. (Penners 1922, 1924a, b, c, 1926, 1936, 1938). This result is due to the fact that the polar plasms are equally distributed into the two blastomeres which are formed at the first segmentation.

4.6. *Differentiation without cleavage*

Another important result deriving from the experimental investigations on the Annelid egg is that concerning the possibility of differentiation without cleavage. This result has been obtained by Lillie (1902) with the egg of Chaetopterus. The egg, whether unfertilized or fertilized, if exposed for a certain time to treatment with KCl and afterwards returned to normal sea water, fails to cleave. The egg forms the polar lobes normally, the ectoplasm extends over the endoplasm as in a normal larva and a unicellular gastrula is formed. Later the ectoplasm becomes vacuolated as in a normal larva and develops cilia; so an acellular larva, similar to a normal one, is formed. It does not have the long apical ciliary tuft, nor the trochus, but moves actively in the water (fig. 24). The extraordinary fact made evident by Lillie (1906) (and confirmed by Pasteels 1934, 1965, 1966 and Brachet 1937) 'is that the capacity to develop the specific ground plane of the embryo is

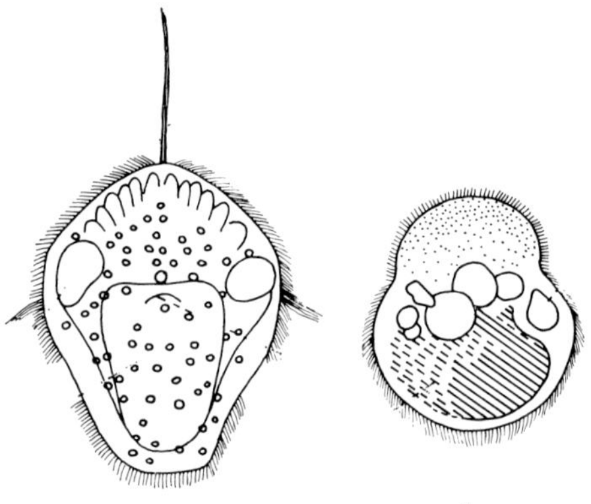

Fig. 24. (a) Normal trochophore of Chaetopterus; (b) acellular trochophore derived from an activated but not segmented egg (differentiation without segmentation). (Redrawn from Lillie 1902.)

present in the egg before segmentation . . . and that this capacity can materialize up to a certain plane without the intervention of segmentation' (Weiss 1939, p. 224).

4.7. The developmental capacities of the isolated blastomeres

A pioneering investigation along these lines was done by Wilson (1904) on the Lanice egg. The blastomeres, when isolated, cleave as if still part of the entire egg: CD gives rise to a rather normal trochophore (with proto-trochus, apical organ, eye, and post-trochal region); AB, on the other hand, yields only a partial larva (which has prototrochus and apical organ but neither eye nor post-trochal region). Similar results have been obtained with the eggs of Sabellaria (Hatt 1932; Novikoff 1938) (fig. 25). AB cleaves without forming polar lobes, finally giving rise to a spherical larva which lacks the apical tuft and the post-trochal bristles, but which has, however, the

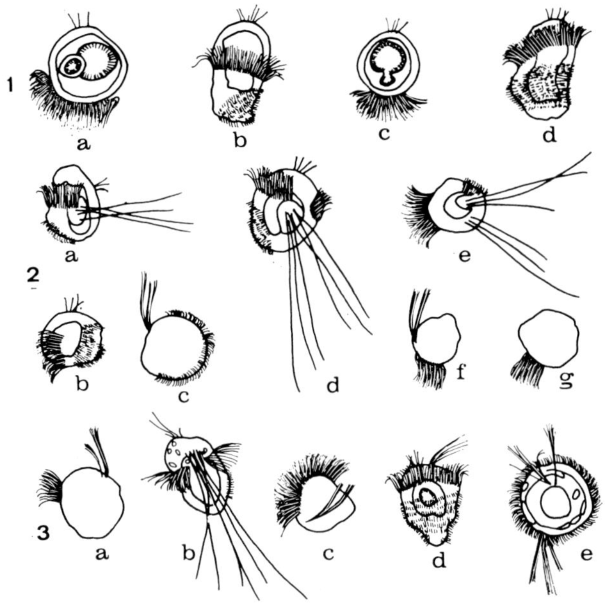

Fig. 25. Larvae from isolated blastomeres (1a) = E-Pl_1 larva, 71 hr, with internal gut; (1b) = E-Pl_1 larva, 50 hr with exogastrulated entoderm; (1c) = AB larva; (1d) = AB larva; (2a) = AD larva; (2b) = BC larva; (2c) = ABC larva; (2d) = ABD larva; (2e) = D larva; (2f) = C larva; (2g) = D-Pl_3 larva; (3a) = CD larva; (3b) = CD larva; (3c) = CD-Pl_2 larva; (3d) = E-Pl_2 larva; (3e) = E-Pl_2 larva. (Redrawn from Novikoff 1938.)

prototrochal and the apical cilia. CD, on the other hand, forms the polar lobes and gives rise to a normal larva (although the cilia and tuft are disproportionately large).

According to Novikoff: 'Isolation experiments . . . demonstrate that the formation of the apical tuft in partial larvae is dependent upon the presence of the first polar lobe and the C cell; that the post-trochal region develops only when the three polar lobes and the 1D cell are present'. The results obtained by Novikoff in different isolation experiments are reported in table 1.

TABLE 1

Summary of larvae obtained from isolation experiments.

	Number operated	Number surviving	Apical tufts			Post-trochal bristles		Prototrochal cilia			Apical cilia		
			+	−	?	+	−	+	−	?	+	−	?
E-PL1	30	21	0	21		0	11	21	0		14	0	
AB	36	25	0	24		0	18	23	0	1	11	6	1
CD	56	47	37	1	9	32	1	42	0	5	1	26	6
CD-PL2	8	5	5	0		0	4	5	0		0	3	1
E-PL2	3	3	3	0		0	2	3	0		2	0	
BC	10	8	6	2		1	5	8	0		6	0	
AD	6	3	0	3		3	0	3	0		2	0	1
ABC	10	8	7	0	1	0	5	8	0		4	1	
ABD	7	5	0	5		4	0	5	0		4	0	
C	17	11	7	3	1	0	7	11	0		0	7	
D	23	15	0	12	3	9	4	15	0		0	9	4
D-PL3	7	5	0	5		0	3	5	0		0	3	
	213	156											

No less interesting are the results obtained by Costello (1945) with the eggs of Nereis. The blastomeres were isolated after the eggs had been deprived of their membranes (with alkaline NaCl). They continued to cleave as if still forming part of the embryo, as is shown very clearly in fig. 26, where the results of isolation of single cells are shown at the 16-cell stage. The development is a true 'mosaic work', with neither interaction between the blastomeres nor induction ever being observed.

The results obtained by Penners (1922, 1924a, b, c, 1926, 1936, 1938) with

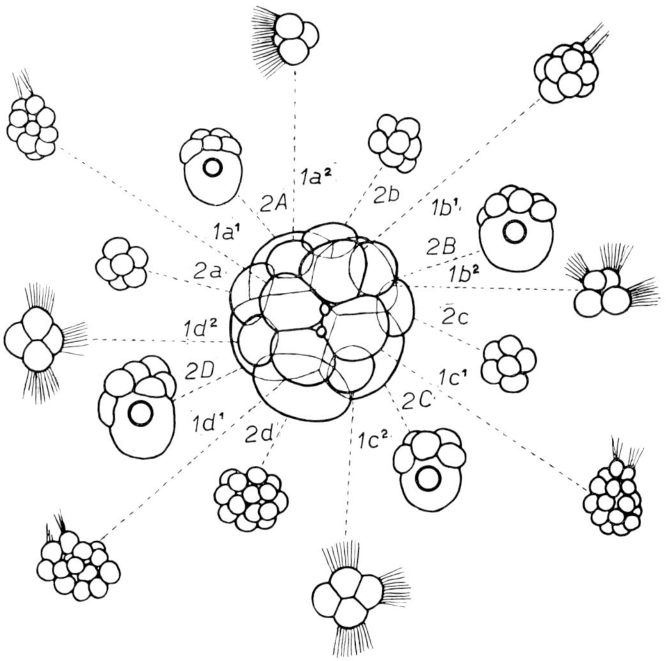

Fig. 26. Egg of Nereis at the 16-cell stage: results of development after disarticulation of the blastomeres (Costello 1945).

the egg of Tubifex are also very interesting. In these experiments the blastomeres were not isolated, but some were killed with ultraviolet radiation and the development of the remainder was studied (fig. 27). The results were as follows:

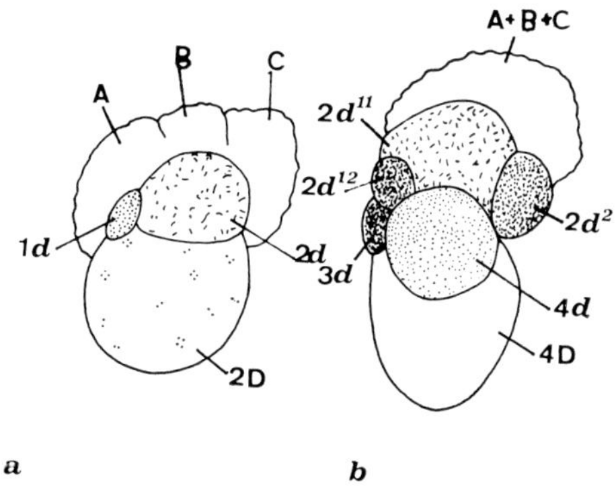

Fig. 27. Development of an egg of Tubifex after removal (a) of the A, B, C, cells. (Redrawn from Penners 1925.)

(a) development of the isolated D-cell (A, B, C, killed): it continues to develop as if still in the usual conditions, giving rise to the 1st and 2nd somatoblasts, and finally to a small perfectly proportioned animal. This result shows that the D-cell, which ordinarily would have formed only a part of the embryo, is capable by 'regulation' of giving rise to an entire embryo;

(b) development of A + B + C: only formless and undifferentiated masses of ectoderm and entoderm result from the development of these blastomeres;

(c) development of the egg from which the 2d- and 4d-cells were removed: the germ bands do not form and the embryo is similar to that obtained in (b);

(d) development of the egg without the 2d-cell: the remaining cells continue to divide and give rise to the second somatoblast (4d), Myr and Myl and the mesodermic bands. The embryo as finally formed, lacks the organs deriving from the ectodermic bands. A regulation however is observed; embryos reared for a long time do form the nervous system. It can be shown that its formation is due to a regulation arising from the mesodermic derivatives (Penners 1936);

(e) development without 4d: the remaining cells continue to develop typically; they give rise to the first somatoblasts ($2d^{111}$) and afterwards to its derivatives (Nr and Mr). At the end of development, however, the embryo lacks all the organs which derive from the mesodermic bands; the organs which derive from the ectodermic bands are also irregularly disposed. That means that the territories deriving from 4d exert an influence on the territories deriving from 2d.

The importance of the animal and vegetal plasms for normal development has also been shown in Clepsine (Mori 1932); their elimination cannot be compensated and their dislocation by centrifugation is followed by abnormal morphogenesis.

An interaction between the embryonic layers, as shown in Tubifex, is not,

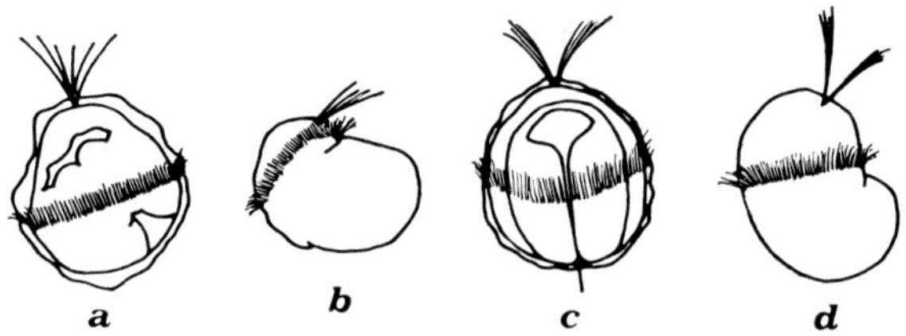

Fig. 28. (a) and (c) normal larvae of Sabellaria; (b) and (d) exogastrulae obtained by treatment with NaCl. (Redrawn from Novikoff 1938.)

however, the rule in the Annelids. In the Nereis egg such mutual influences are not present; according to Novikoff (1938) the same is true of the egg of Sabellaria. The treatment of this egg with isotonic NaCl at pH 9.6 produces exogastrulation so that the entoderm remains on the exterior and the ectoderm does not enter into contact with it (fig. 28). In this condition, however, they both differentiate normally. It will be recalled that the situation is completely different in Amphibians.

The conclusion that in Sabellaria the different territories do not exert any influence on each other is strengthened by the results of transplanting the polar lobe (Novikoff 1938). The transplanted lobe does not induce any particular structure in the host; consequently 'since the polar lobe does not affect the differentiation of any cell through contact with that cell, it is not possible to consider the polar lobe as an organizer in the sense of Spemann' (Novikoff 1938, p. 231).

4.8. Biochemical investigations

This field, which should constitute the ultimate step in embryological research, has been scarcely cultivated, probably for technical reasons: a few investigations may be mentioned.

One is that by Weber (1958), who used the egg of Tubifex, which is relatively large and so sufficiently adaptable for biochemical research. Weber proposed to ascertain whether the different blastomeres of the develop-

TABLE 2

Regional distribution of cytochrome oxidase and cathepsin in the somatoblast embryo (Weber 1958).

Enzyme	Experiment	Nadi-positive region somatoblasts 2d, 4d, and micromeres: N^+	Nadi-negative region endoderm-forming cells: N^-	Ratio N^-/N^+
Cytochrome oxidase	1	40.10	19.10	0.47
$\times 10^{-3}$ μ 10_2h μg TN	2	79.80	23.70	0.30
	3	0.135	0.157	1.16
Cathepsin CU/μg TN	4	0.151	0.152	1.01
	5	0.108	0.098	0.91

ing egg have a different metabolism, and eventually whether there is a causative relation between metabolism and differentiation.

The activity of cytochrome oxidase and cathepsin in 2d–4d-cells (somatoblasts) and macromeres was considered: the results are reported in table 2. As shown there the activity of cytochrome oxidase in 2d and 4d (Nadi-positive somatoblasts) is three times higher than in the Nadi-negative macromeres: this result agrees with the data obtained by cytochemistry and electron microscopy concerning the distribution of the mitochondria. The activity of the cathepsin, on the other hand, is lower in 2d and 4d than in the macromeres (if the data are related to the amount of yolk present in the macromeres).

A new line of research is that of Winesdorfer (1965b) who studied the amino acid's incorporation in the post-mitochondrial supernatants of the unfertilized egg of Sabellaria.

References

ALLEN M. J., 1961. A cytochemical study on the developing oocytes and attached nurse cells of the polychaetous annelid *Diopatra cuprea* (Bosc). Acta Embryol. Morphol. Exptl. *4*, 219–239.

ALLEN M. J., 1967. Nucleic acid and protein synthesis in the developing oocytes of the budding form of the Syllid, *Autolitus Edwarsi* (class Polichaeta). Biol. Bull. *133*, 287–302.

BRACHET J., 1937. La différenciation sans clivage dans l'œuf de Chétoptère, envisagée aux points de vue cytologique et métabolique. Arch. Biol. *48*, 561–589.

CALKINS G., 1895. Observations on the yolk nucleus in the eggs of Lumbricus. Trans. N.Y. Acad. Sci. *13*, 222–230.

CARAZZI D., 1904. Ricerche embriologiche e citologiche sull'uovo di *Myzostoma glabrum* Leuchart. Mon. Zool. Ital. *15*, 62–78, 87–100.

CARRANO F. and F. PALAZZO, 1955. Localizzazione di alcuni enzimi nello sviluppo dell'uovo di *Tubifex rivulorum*. Riv. Biologia (Rome) *47*, 193–202.

CHILD C. M., 1900. The early development of Arenicola and Sternaspis. Arch. Entwicklungsmech. Organ. *9*, 587–723.

COLWIN H. L. and A. L. COLWIN, 1961. Changes in the spermatozoon during fertilization in *Hydroides hexagonus* (Annelida). I. Passage of the acrosomal region through the vitelline membrane. J. Biophys. Biochem. Cytol. *10*, 231–254.

COSTELLO D. P., 1938. The studies on fragments of centrifuged Nereis eggs. Science *88*, 436.

COSTELLO D. P., 1940a. The fertilizability of nucleated and non-nucleated fragments of centrifuged Nereis egg. J. Morphol. *66*, 99–114.

COSTELLO D. P., 1940b. The cell origin of the protroch of *Nereis limbata*. Biol. Bull. *79*, 369–370.

COSTELLO D. P., 1940c. Development of fragments of Nereis egg. Anat. Record *78*, Suppl. 133.

COSTELLO D. P., 1940d. The development of isolated blastomeres of Nereis egg fragments. Anat. Record *78,* Suppl. 133.

COSTELLO D. P., 1945. Experimental studies of germinal localization in Nereis. I. The development of isolated blastomeres. J. Exptl. Zool. *100,* 19–66.

COSTELLO D. P., 1948. Ooplasmic segregation in relation to differentiation. Ann. N.Y. Acad. Sci. *49,* 663–683.

COSTELLO D. P., 1949. The relations of the plasma membrane, vitelline membrane, and jelly in the egg of *Nereis limbata.* J. Gen. Physiol. *32,* 351–366.

COSTELLO D. P., M. E. DAVIDSON, A. EGGERS, M. H. FOX and C. HENLEY, 1957. Methods for obtaining and handling marine eggs and embryos. Marine Biol. Lab., Woods Hole, Mass.

DALCQ A. M. and J. PASTEELS, 1963. La localization d'enzymes de déphosphorylation dans les œufs de quelques invertébrés. Develop. Biol. *7,* 457.

DHAINAUT A., 1966. Étude ultrastructurale de l'évolution cytoplasmique au cours des premiers stades de l'ovogenèse chez *Nereis pelagica* L. (Annélide polychète). Compt. Rend. *262,* 2616.

DHAINAUT A., 1968. Étude par autoradiographie à haute résolution de l'élaboration des mucopolysaccharides acides au cours de l'ovogenèse de *Nereis pelagica* L. (Annélide Polychète). J. Microscopie, *7,* 1075–1080.

DHAINAUT A., 1969. Origine et structure des formations mucopolysaccharidiques de la zone corticale de l'ovocite de *Nereis diversicolor* O. F. Müller (Annélide polychaete). J. Microscopie *8,* 69–86.

DURCHON M. and B. BOILLY, 1964. Étude ultrastructurale de l'influence de l'hormone cérébrale des Néréidiens sur le développement des ovocytes de *Nereis diversicolor* O. F. Müller (Annélide polychète) en culture organotypique. Compt. Rend. *259,* 1245.

DURCHON M. and A. DHAINAUT, 1964. Influence de l'hormone cérébrale des Néréidiens sur la croissance des ovocytes. Étude en culture organotypique. Compt. Rend. *259,* 917–919.

FALLON J. and C. R. AUSTIN, 1967. Fine structure of gametes of *Nereis limbata* (Annelida) before and after interaction. J. Exptl. Zool. *166,* 225.

FAURÉ-FRÉMIET E., 1924. L'œuf de *Sabellaria alveolata* L. Arch. Anat. Microscop. Morphol. Exptl. *20,* 211.

GATENBY J. B., 1922. The gametogenesis of Saccocirrus. Quart. J. Microscop. Sci. *66,* 1–48.

GATENBY J. B. and V. NATH, 1926. The oogenesis of certain invertebrata with special reference to Lumbricus. Quart. J. Microscop. Sci. *70,* 371–390.

GOETTE A., 1882. Abhandlungen zur Entwicklungsgeschichte der Tiere. III. Ueber die Entwicklung der Anneliden. Leipzig.

HARRIS J. E., 1935. Studies on living protoplasm. I. Streaming movements in the protoplasm of the egg of *Sabellaria alveolata* L. J. Exptl. Biol. *12,* 65–79.

HARVEY E. B., 1939. Development of half-eggs of *Chaetopterus pergamentaceus* with special reference to parthenogenetic merogony. Biol. Bull. *76,* 385–404.

HARVEY L., 1925. On the relation of the mitochondria and Golgi apparatus to yolk-formation in the egg-cells of the common earthworm. Quart. J. Microscop. Sci. *69,* 291–316.

HATSCHEK B., 1878. Studien über Entwicklungsgeschichte der Anneliden. Arb. Zool. Inst. Wien *1,* 277–404.

HATT P., 1931. La fusion expérimentale d'œufs de *Sabellaria alveolata* L. et leur développement. Arch. Biol. (Liège) *42*, 303–323.

HATT P., 1932. Essais expérimentaux sur les localizations germinales dans l'œuf d'un Annélide *(Sabellaria alveolata)*. Arch. Anat. Microscop. Morphol. Exptl. *28*, 81–98.

HESS O., 1959. Phasenspezifische Änderung im Gehalt an ungesättigten Fettsäuren, beim Ei von Tubifex während der Meiosis und der ersten Furchung. Z. Naturforsch. *14*, 342, 345.

HESS O., 1963. Entwicklungsphysiologie der Anneliden. Fortschr. Zool. *16*, 347–379.

HORST R., 1881. Sur la fécondation et le développement de l'*Hermella alveolata*. Bull. Sci. Nord *4*, 1–4.

INASE M., 1960a. On the double embryo of the aquatic worm, *Tubifex hattai*. Sci. Rept. Tohuku Univ. Fourth Ser. *26*, 59–64.

INASE M., 1960b. The culture solution of the eggs of Tubifex. Sci. Rept. Tohuku Univ. Fourth Ser. *26*, 65–67.

INASE M., 1967. Behavior of the pole plasm in the early development of the aquatic worm *Tubifex hattai*. Sci. Rept. Tohoku Univ. Fourth Ser. *33*, 223–231.

KLEINENBERG N., 1879. The development of the earthworm, *Lumbricus trapezoides*, Dugès. Quart. J. Microscop. Sci. *19*, 1–41.

KLEINENBERG N., 1886. Die Entstehung des Annelids an der Larve von Lepadorhyncus. Z. Wiss. Zool. *44*.

KOWALEWSKY A., 1871. Embryologische Studien an Würmern und Arthropoden. Mém. Acad. Sci. St. Pétersbourgh *16*, 70.

LEHMANN F. E., 1941a. Die Indophenolreaktion der Polplasmen von Tubifex. Naturwissenschaften *29*, 101.

LEHMANN F. E., 1941b. Die Zucht von Tubifex für Laboratoriumswerke. Rev. Suisse Zool. *48*, 559.

LEHMANN F. E., 1948. Zur Entwicklungsphysiologie der Polplasmen des Eies von Tubifex. Rev. Suisse Zool. *55*, 1.

LEHMAN F. E. and V. MANCUSO, 1957. Verschiedenheiten in der submikroskopischen Struktur der Somatoblasten des Embryos von Tubifex. Arch. Julius Klaus-Stift. Vererbungsforsch. Sozialanthropol. Rassenhyg. *32*, 482–493.

LILLIE F. R., 1902. Differentiation without cleavage in the egg of the Annelid *Chaetopterus pergamentaceus*. Arch. Entwicklungsmech. Organ. *14*, 477–499.

LILLIE F. R., 1906. Observations and experiments concerning the elementary phenomena of embryonic development in Chaetopterus. J. Exptl. Zool. *3*, 153–268.

LILLIE F. R., 1909. Polarity and bilaterality of the annelid egg. Experiments with centrifugal force. Biol. Bull. *16*, 54–79.

LILLIE F. R., 1911. Studies on fertilization in Nereis. I. The cortical changes in the egg. II. Partial fertilization. J. Morphol. *22*, 361–393.

MARTOJA R. and M. MARTOJA-PIERSON, 1959. Sur quelques caractères histochimiques du nucléole de l'ovocyte au cours de l'interphase de grand accroissement. Bull. Biol. France Belg. *93*, 335–354.

MEAD A. D., 1895. Some observations on maturation and fecondation in *Chaetopterus pergamentaceus*. Cuv. J. Morphol. *10*, 313–317.

MEAD A. D., 1897. The early development of marine Annelids. J. Morphol. *13*, 227.

MORGAN T. H., 1910. The effects of altering the position of the cleavage planes in eggs with precocious specification. Arch. Entwicklungsmech. Organ. *29*, 205–224.

MORGAN T. H., 1927. Experimental embryology. Columbia University Press, New York.

MORI Y., 1932. Entwicklung isolierter Blastomeren und teilweise abgetöteter älterer Keime von *Clepsine sexoculata*. Z. Wiss. Zool. *141*, 399–431.

NATH V., B. L. GUPTA and S. L. MANOCHA, 1958. Histochemical and morphological studies of lipids in the oogenesis of *Pheretima posthuma*. Quart. J. Microscop. Sci. *99*, 475, 484.

NELSON J. A., 1905. The early development of Dinophilus: a study in cell lineage. Proc. Acad. Nat. Sci. Phila. *56*, 687–737.

NOVIKOFF A. B., 1938a. Embryonic determination in the annelid *Sabellaria vulgaris*. I. The differentiation of ectoderm and endoderm when separated through induced exogastrulation. Biol. Bull. *74*, 198.

NOVIKOFF A. B., 1938b. Embryonic determination in the annelid *Sabellaria vulgaris*. II. Transplantation of polar lobes and blastomeres as a test of their inducing capacities. Biol. Bull. *74*, 211.

NOVIKOFF A. B., 1939a. Changes at the surface of *Nereis limbata* eggs after insemination. J. Exptl. Biol. *16*, 403.

NOVIKOFF A. B., 1939b. Surface changes in unfertilized and fertilized eggs of *Sabellaria vulgaris*. J. Exptl. Zool. *82*, 217.

NOVIKOFF A. B., 1940. Morphogenetic substances or organizers in Annelid development. J. Exptl. Zool. *85*, 127.

PARSEVAL M., 1922. Die Entwicklung zentrifugierter Eier von *Tubifex rivulorum* Lam. Arch. Entwicklungsmech. Organ. *50*, 468.

PASTEELS J., 1934. Recherches sur la morphogenèse et le déterminisme des segmentations inégales chez les Spiralia, Aplysia, Myzostoma, Chaetopterus. Arch. Anat. Microscop. *30*, 161–197.

PASTEELS J., 1965. Étude au microscope électronique de la réaction corticale. J. Embryol. Exptl. Morphol. *13*, 327.

PASTEELS J., 1966. La réaction corticale de fécondation de l'œuf de *Nereis diversicolor*, étudiée au microscope électronique. Acta Embryol. Morphol. Exptl. *9*, 155.

PENNERS A., 1922. Die Furchung von *Tubifex rivulorum*. Zool. Jahrb. Abt. Anat. Ontog. Tiere *43*, 323.

PENNERS A., 1924a. Experimentelle Untersuchungen zum Determinationsproblem am Keim von *Tubifex rivulorum* Lam. I. Die duplicitas cruciata und organbildende Keimbezirke. Arch. Mikr. Anat. *102*, 51–100.

PENNERS A., 1924b. Ueber die Entwicklung teilweise abgetöteter Eier von *Tubifex rivulorum*. Verhandl. Zool. Ges. Berlin *29*, 69–73.

PENNERS A., 1924c. Doppelbildungen bei *Tubifex rivulorum* Lam. Zool. Jahrb. Abt. Allgem. Zool. *41*, 91–120.

PENNERS A., 1926. Experimentelle Untersuchungen zum Determinationsproblem am Keim von *Tubifex rivulorum* Lam. Z. Wiss. Zool. *127*, 1–140.

PENNERS A., 1937. Regulation am Keim von *Tubifex rivulorum* Lam. nach Ausschaltung des ektodermalen Keimstreifs. Z. Wiss. Zool. *149*, 86–130.

PENNERS A., 1938. Abhängigkeit der Formbildung vom Mesoderm im Tubifex-Embryo. Z. Wiss. Zool. *150*, 305–357.

RAVEN C. P., 1938. Experimentelle Untersuchungen über die 'bipolare Differenzierung' der Polychaeten und Molluskeneies. Acta Neerl. Morphol. *1*, 337.

RAVEN C. P., J. M. BRINK, J. C. van KAMER and J. A. van de MILTENBRIRG, 1950. Cytological and cytochemical investigations on the development of *Sabellaria alveolata* L. Verhandel. Koninkl. Ned. Akad. Wetenschap. *47*, 1.

SALENSKY W., 1882. Études sur le développement des Annélides. Arch. Biol. (Liège) *3*, 345–378.

SALENSKY W., 1885. Études sur le développement des Annélides. Développement de Branchiobdella. Arch. Biol. (Liège) *6*, 1–64.

SCHLEIP N., 1913. Die Furchung des Eies von Clepsine und ihre Beziehungen zur Furchung des Polychäteneis. Freiburg B. Ber. Naturforsch. Ges. *20*, 177–188.

SCHLEIP W., 1914a. Die Furchung des Eies der Russelegel. Zool. Jahrb. Anat. *37*, 313–368.

SCHLEIP W., 1914b. Die Entwicklung zentrifugierter Eier von *Clepsine sexoculata*. Verhandl. Deutsch. Zool. Ges.

SCHLEIP N., 1929. Die Determination der Primitiventwicklung. Akademie Verlag, Leipzig.

SICHEL G., 1966. Modificazioni ultrastrutturali dell'ooplasma in rapporto alla vitellogenesi, in *Merceriella enigmatica*, Fauvel (Annelida Polychaeta). Atti Accad. Gioenia Sci. Nat. Catania *18*, 21–32.

SPEK J., 1930. Zustandsänderungen der Plasmakolloide bei Befruchtung und Entwicklung des Nereis Eies. Protoplasma *9*, 370–427.

SPEK J., 1934. Über die bipolare Differenzierung der Eizellen von *Nereis limbata* und *Chaetopterus pergamentaceus*. Protoplasma *21*, 394.

SRIVASTAVA D. S., 1952. Nucleolus and nucleolar extrusion in the oocytes of *Lumbricus terrestris* L. Cellule Rec. Cytol. Histol. *65*, 131–135.

SRIVASTAVA D. S., 1953. Cytological studies on the egg cells of *Lumbricus terrestris*. Cellule Rec. Cytol. Histol. *55*, 187–200.

TITLEBAUM A., 1928. Artificial production of Janus embryos of Chaetopterus. Proc. Nat. Acad. Sci. US *14*, 245–247.

TYLER A., 1930. Experimental production of double embryos in Annelids and Molluscs. J. Exptl. Zool. *57*, 347–407.

VEIDOVSKY F., 1881. Untersuchungen über die Anatomie, Physiologie und Entwicklung von Sternaspis. Denkschr. Akad. Wiss. Wien *18*, 33–90.

WATERMAN A. J., 1934. Observations on reproduction, prematuration, and fertilization in *Sabellaria vulgaris*. Biol. Bull. *67*, 97.

WEBER R., 1958. Über die submikroskopische Organisation und die biochemische Kennzeichnung embryonaler Entwicklungstadien von Tubifex. Arch. Entwicklungsmech. Organ. *150*, 542–580.

WEISS P., 1939. Principles of development. Holt & Co., N.Y.

WHITMAN C. O., 1878. The embryology of Clepsine. Quart. J. Microscop. Sci. *18*, 215–315.

WILSON E. B., 1889. The embryology of the earthworm. J. Morphol. *3*, 388–442.

WILSON E. B., 1892. The cell lineage of Nereis. J. Morphol. *6*, 361–470.

WILSON E. B., 1896. On cleavage and mosaic work. Arch. Entwicklungsmech. Organ. *3*, 19–26.

WILSON E. B., 1904. Mosaic development in Annelid egg. Science *20*, 748–750.

WILSON E. B., 1929. The development of egg fragments in Annelids. Arch. Entwicklungsmech. Organ. *117*, 179.

WILSON E. B., 1930. Notes on the development of fragments of the fertilized Chaetopterus egg. Biol. Bull. *59*, 7.

WINESDORFER J. E., 1965a. Marine Annelids: Sabellaria. in 'Methods of developmental biology' ed. by Wilt and Wessel. Th. Y. Crowell Comp., New York.

WINESDORFER J. E., 1965b. Demonstration of polyribosomal aggregates and amino acid incorporation in post-mitochondrial supernatants from unfertilized eggs of *Sabellaria cementarium*. Am. Zool. *5*, 635.

G. Reverberi (ed.), Experimental embryology of marine and fresh-water invertebrates

CHAPTER 5

Nemertinae

SVEN HÖRSTADIUS

Zoological Institute, Uppsala, Sweden

In the group of Metanemertini the development is direct. In the Heteronemertini we find a metamorphosis, either from a swimming larva, the pilidium (Cerebratulus), or from a larva with a shortened development inside a membrane, Desor's larva (Lineus). In both cases the body wall of the worm is formed by the inner wall, the imaginal discs, of amniotic invaginations of the larval ectoderm. In the following pages the egg of Cerebratulus, which was the only one used for embryological experiments, will be considered.

5.1. The egg

A piece of ripe female will shed its eggs when brought into sea water. For experimental work it is suitable to envelope the piece of the female in cheesecloth which retains the slime but allows the eggs to fall through. Coe (1899) has shown that the germinal vesicle gradually fades away when the eggs come out in the sea water. The first polar spindle remains until fertilization occurs. Zeleny (1904) recommends adding sperms only half an hour after shedding to secure uniform developing stages. The eggs form no fertilization membrane but are surrounded by a jelly and soft membrane; they are easily removed by gentle shaking. The cleavage stages exhibit a remarkable tolerance of operation. Blastomeres can be isolated by shaking (C. B. Wilson 1900) or cut apart by a fine scalpel (E. B. Wilson 1903) or a glass needle (Hörstadius 1937).

A further advantage of the Cerebratulus egg as object for experiments is the fact that the egg-axis can be accurately determined from the moment of discharge from the ovary. The liberated egg has a conical protuberance at the pole diametrically opposite the point at which the polar bodies afterwards are formed (fig. 1). C. B. Wilson (1900) states that the protuberance represents the remains of the pedicle of attachment of the egg in the ovary.

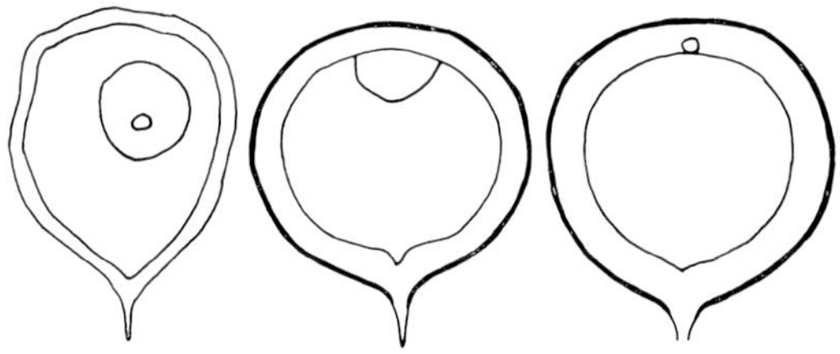

A B C

Fig. 1. Eggs of *Cerebratulus lacteus*. A: Ovocyte immediately after liberation from the ovary, with vegetal protuberance. B: Maturing egg with a vaguely defined clearer area where the first polar spindle is formed. C: Egg with first polar body, vegetal protuberance disappearing. (From E. B. Wilson 1903.)

The germinal vesicle usually lies eccentrically on the side opposite the protuberance. After being in the sea water for 10–20 min the vesicle fades away forming a clear area which moves towards the animal pole where the polar bodies will be formed (fig. 1B, C). In this way an orientation of the egg according to the egg axis is possible from the ovocyte to the first cleavage stages. The polar bodies give a good landmark even in cleavage stages.

5.2. Segmentation and development

The cleavage of the nemertine egg is total and spiral. The animal blastomeres in the eight-cell stage, which in the spiral type usually are called micromeres, are, however, in the nemerteans of equal size with the macromeres or even larger than these. The latter is the case in Cerebratulus (Coe 1899). The first two furrows are meridional and give four blastomeres of equal size. We therefore have no predominance of a D-quadrant. The third cleavage is dexiotropic, forming four micromeres (1a–1d) and four smaller macromeres (1A–1D; fig. 2A). At the next (laeotropic) cleavage the micromeres are divided into two layers each of four cells ($1a^1$–$1d^1$ and $1a^2$–$1d^2$) and the macromeres bud off a second quartet of micromeres (2a–2d). The macromeres are now called 2A–2D (fig. 2B).

The use of the words micro- and macromeres in this case and of the terms

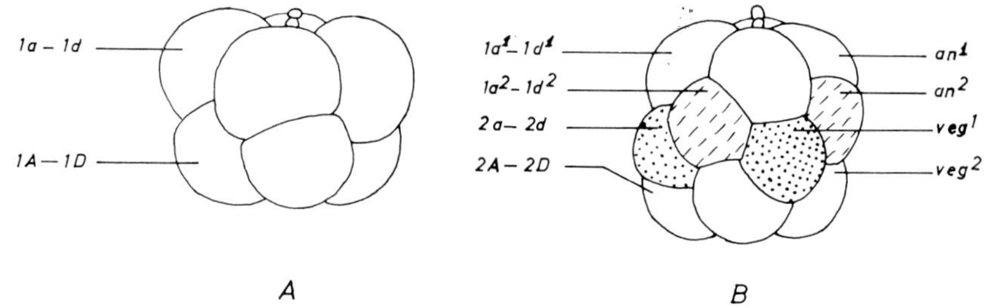

Fig. 2. Eight-cell stage (A) and 16-cell stage (B) of Cerebratulus. (From Hörstadius 1937.)

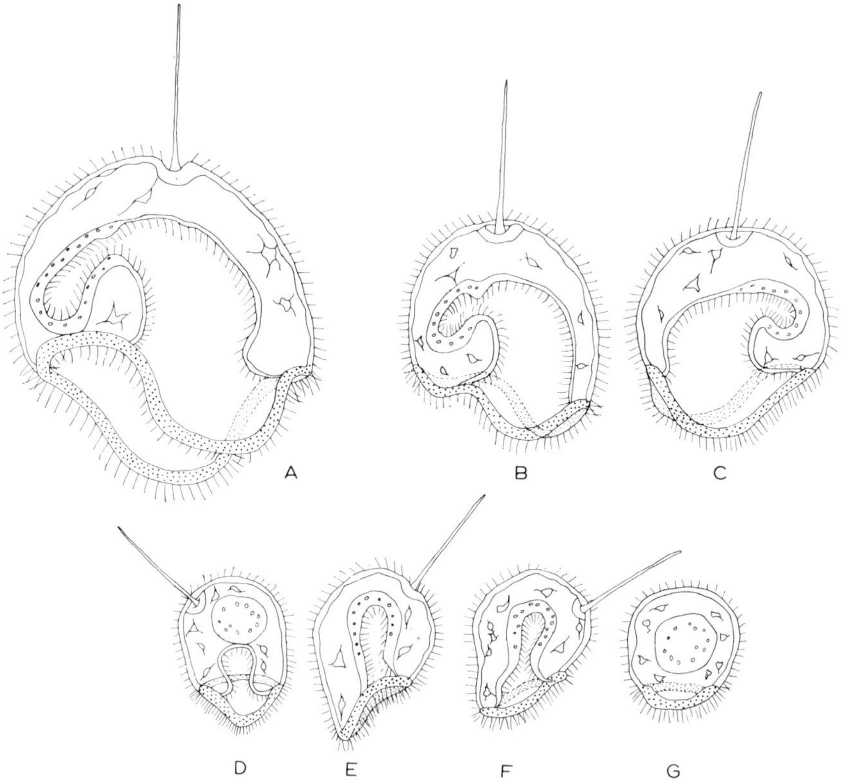

Fig. 3. A: Normal pilidium. B, C: Pilidia from the isolated two half-blastomeres of one egg. D-G: Four larvae from the isolated quarter-blastomeres of one egg. (From Hörstadius 1937.)

$1a^1$ etc. is rather confusing. To simplify the matter we designate the four layers as an_1, an_2, veg_1 and veg_2 (fig. 2B), as has been done with the four layers in the sea urchin egg (Hörstadius 1935). The composition of a larva may be then expressed by a formula; e.g. $an_1 + an_2 + veg_1$ denotes a larva from which veg_2 has been removed, etc.

The gastrula is still rather opaque but the pilidium larva is quite transparent. The blastopore forms the mouth, leading into a wide oesophagus and a stomach. There is no intestine or anus (fig. 3A). Two lappets grow out of the left and right sides of the mouth. The ectoderm is ciliated but the lappets are bordered by a ciliated band where the cilia are longer, more concentrated and beat in another way than those of the rest of the ectoderm. There is an apical organ at the animal pole, consisting of a long flagellum in a thickened pit of the ectoderm. The flagellum is composed of a bundle of fine threads. As there is no anus it is difficult to say what corresponds to the dorsal and what to the ventral side. As the ventral side presumably is very short, the dorsoventral axis may form an acute angle with the egg (animal–vegetal) axis. After metamorphosis the egg axis corresponds to the dorsoventral axis of the worm.

5.3. The prospective significance of egg regions

One of the blastomeres in 2-cell stage was stained by leaning it against a piece of agar saturated with Nile-blue sulphate (Hörstadius 1937): the staining could not be quite exact but it seems that the first furrow is not confined to a certain plane. Isolated half-blastomeres therefore probably represent right or left, dorsal or ventral, or oblique meridional halves. The boundary between the layers of the 8- and 16-cell stages could not always be determined with complete accuracy with the same method.

An_1 forms the greater part of the pretrochal ectoderm including a small anterior piece of the ciliated band. This was confirmed by the following experiments: an_1 was isolated, stained, and transplanted back on the an_2-cells: the fragment was evidently rotated 180° at the transplantation as a patch of ciliated band appeared too high up on the posterior side of the pilidium. An_2 gives rise to the rest of the outside ectoderm and a great part of the ciliated band. Veg_1 also contributes to the ciliated band and moreover represents the material for the insides of the lappets, and the oesophagus. Veg_2 forms the stomach and perhaps a small part of the oesophagus as well.

5.4. Isolation experiments

5.4.1. Fragments from any part of the unfertilized egg may be fertilized and develop. E. B. Wilson (1903), Zeleny (1904), and Yatsu (1904, 1910a) found that such fragments may segment as whole eggs. Zeleny and Yatsu divided eggs during the stages between the fertilization and the completion of the first cleavage and stated that the fragments segment as halves when isolated just before the first cleavage.

Yatsu (1910b) and Hörstadius (1937) followed the further development of fragments from unfertilized eggs and found that they, if of sufficient size, may give dwarf pilidia. Thus animal fragments are able to gastrulate and vegetal ones are able to form apical organs. These results are in great contrast to the differentiation of the animal and vegetal cell layers of 8- and 16-cell stages.

5.4.2. C. B. Wilson (1900) was the first to obtain two pilidia from one egg, isolating the half-blastomeres by shaking. The other four authors mentioned above also studied the development of twins from half-blastomeres. Some produced typical dwarf pilidia (figs. 3B, C); in other cases the larvae were asymmetrical or cylindrical, the apical organ was doubled or displaced towards the anterior end. Yatsu (1910b) concluded that bilaterality of the egg substances cannot be detected at the 2- or 4-cell stage. Hörstadius (1937) reared the half-larvae derived from one egg together. He was not able to refer the half-larvae to certain parts of the egg. Like Yatsu he found that the egg in the 2-cell stage is not so markedly bilaterally organized that half-larvae show deficiences on the side where material has been removed.

5.4.3. Isolated quarter-blastomeres undergo a fractional cleavage (Wilson 1903; Zeleny 1904). The dwarf larvae never attain a quite typical pilidium shape although they possess the organs of such a larva, an apical organ (often displaced), an ectoderm with ciliated band and a stomach (figs. 3D-G) (Wilson 1903; Yatsu 1910b; Hörstadius 1937).

5.4.4. From isolation of cell layers perpendicular to the egg axis we no longer meet an harmonic development of the fragments. Their differentiation instead reveals a pure mosaic, without tendency of regulation towards a whole. Larvae developing from the animal quartet of the 8-cell stage have an apical organ, sometimes two with separated ectodermal pits, a pretrochal pavement epithelium and a ciliated band, but no archenteron (figs. 4A–D). With the exception of the frequent duplication of the apical organ this differentiation corresponds to the prospective significance of the material (Zeleny 1904; Yatsu 1910b; Hörstadius 1937). The vegetal quartet demonstrates the same rigid mosaic development (fig. 4E–G). An archenteron invaginates but it

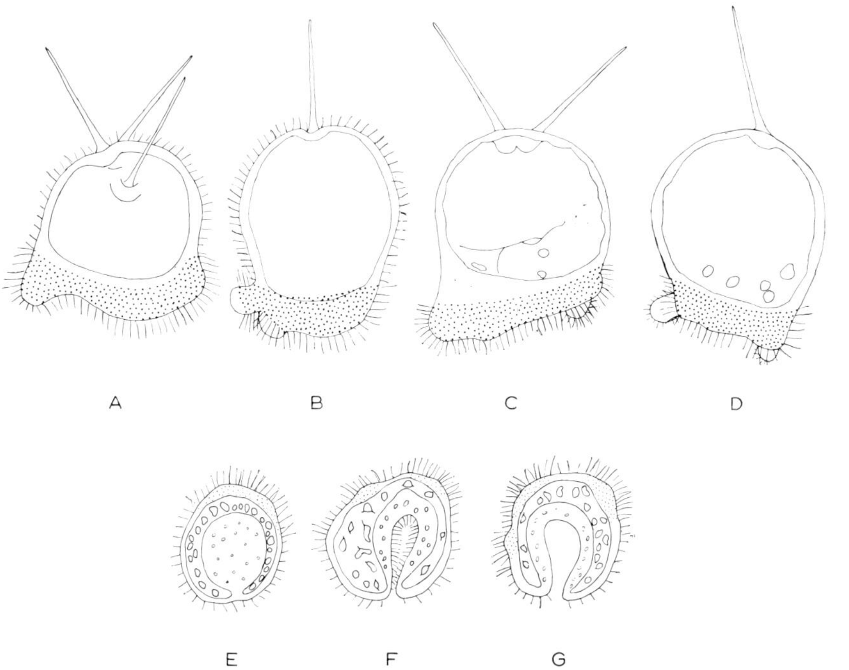

Fig. 4. A-D: Larvae from the isolated four animal cells of the 8-cell stage. E-G: Larvae from the isolated four vegetal cells of the 8-cell stage. (From Hörstadius 1937.)

evidently corresponds only to the stomach of the pilidium, as it differentiates in the same way, containing the small crystals characteristic of the stomach. The 'ectoderm' of the gastrula corresponds to the oesophagus and the insides of the lappets, and at the animal pole of the vegetal fragment we find a ciliated field or patches, presumably derived from the material that normally contributes to the formation of the band.

Already Zeleny succeeded in isolating the four an_1 cells of the 16-cell stage but they were not reared into full differentiation. A further analysis has shown that such fragments form blastulae with none, one or two apical organs and at the opposite pole a small field of ciliated band tissue (figs. 5A–D). This patch evidently represents that small part of the ciliated band

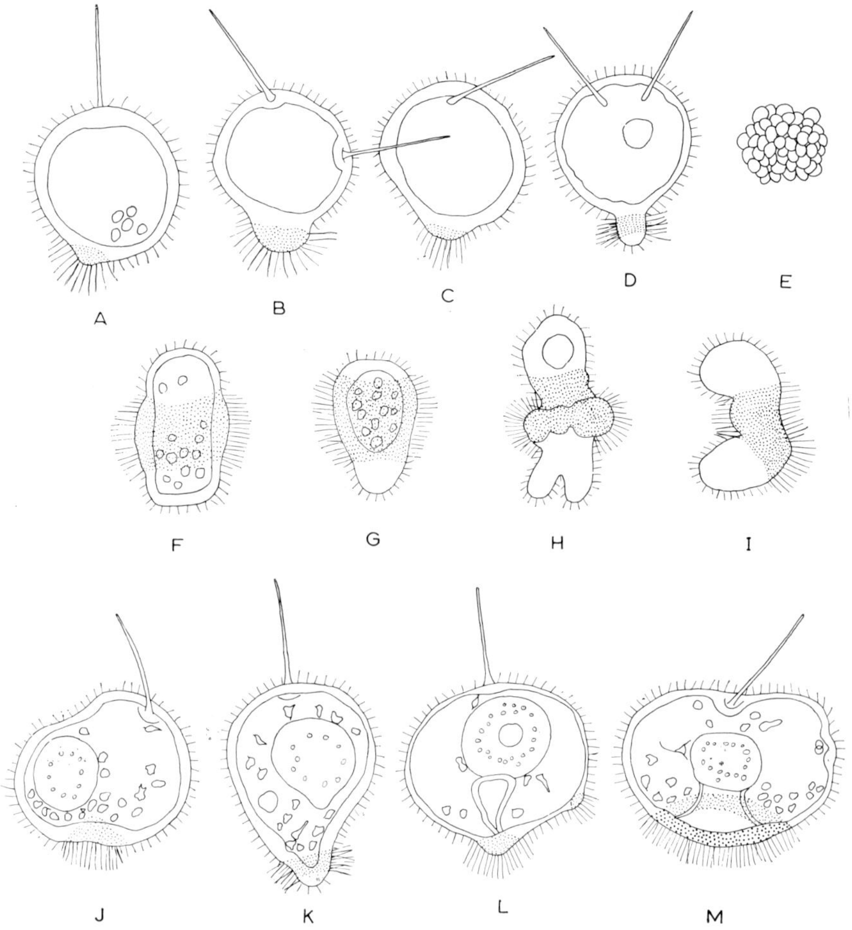

Fig. 5. A-D: Larvae from the isolated most animal cells of the 16-cell stage: an_1. E: Isolated veg_2 (the most vegetal cells of the 16-cell stage). F-I: Larvae from the two middle layers of the 16-cell stage: $an_2 + veg_1$. J-L: Larvae composed of the most animal and the most vegetal layers of the 16-cell stage: $an_1 + veg_2$. M: $an_1 + (an_2/2) + veg_2$. (From Hörstadius 1937.)

which on the basis of staining and transplantation experiments (see above) was ascribed to an_1 cells.

Veg_2 constitutes the material for the stomach and perhaps also a small part of the oesophagus. When isolated the cells of this region do not live long. Without an ectoderm cover they only formed a mass of cells (fig. 5E). It has not been possible to isolate each of an_2 and veg_1. But the fate of

$an_2 + veg_1$ could be studied after removal of an_1 and veg_2 (figs. 5F–I). As the ectoderm does not expand the larvae are rather small with a thick and broad equatorial ciliated band. No apical organ nor any archenteron appear. This picture confirms the impression of mosaic development. When larvae consist of three of the four 16-cell layers the tissues differentiate in a more typical way. As would be predicted $an_2 + veg_1 + veg_2$ has a ciliated

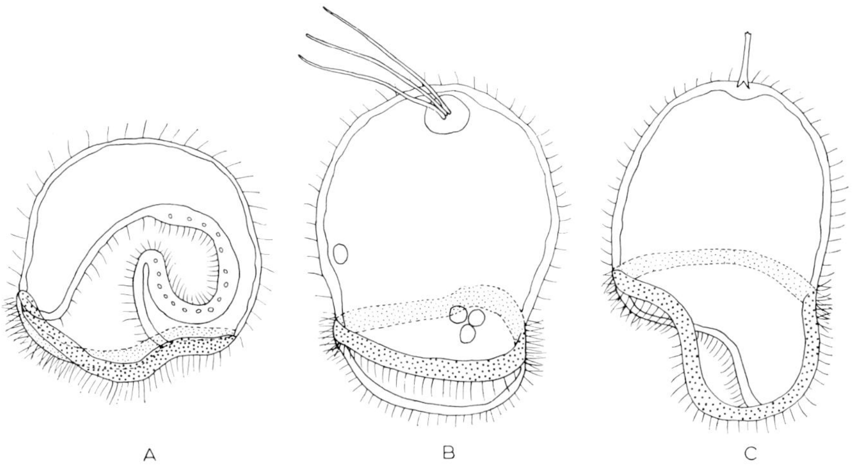

Fig. 6. A: $an_2 + veg_1 + veg_2$. B, C: $an_1 + an_2 + veg_1$. (From Hörstadius 1937.)

band, oesophagus and stomach, but no apical organ (fig. 6A). On the other hand $an_1 + an_2 + veg_1$ possesses this organ and a well-developed ectoderm with a ciliated band and lappets but is devoid of a digestive tract (figs. 6B, C).

5.5. Transplantation experiments

It could be expected that a combination of the two extreme layers, an_1 and veg_2, would give a rather harmonic larva, with all organs more or less represented. Such fragments were fused by the method worked out for transplantation of parts of sea urchin eggs (Hörstadius 1928). Larvae of this composition could as matter of fact at their best produce apical organ, ectoderm with ciliated band tissue, oesophagus and stomach; of course they did not present the full pilidium shape (figs. 5J–L). If two an_2-cells were added a real band appeared (fig. 5M).

Animal halves of 8-cell stages were vitally stained and fused with meridional

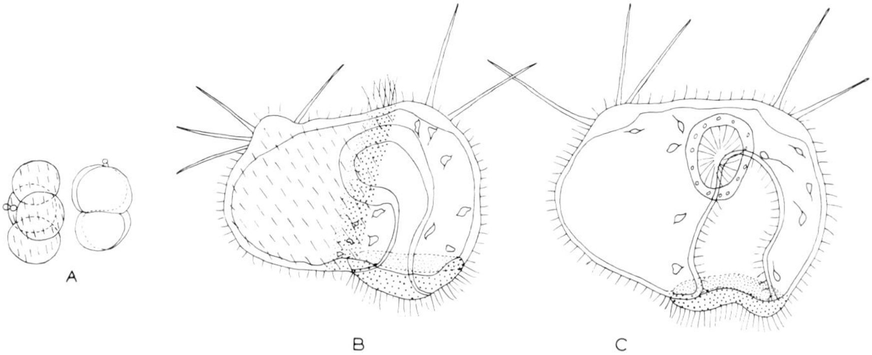

Fig. 7. A: Fusion of a meridional half and a vitally stained (stippled) animal half of the 8-cell stage. B: The larva has differentiated conformably to the prospective significance of the halves. C: Commencing regulation by redifferentiation. (From Hörstadius 1937.)

halves. The polarity of the animal fragment will stand at right angles to that of the meridional one (fig. 7A).

The problem to test was whether, as in sea urchins (Hörstadius 1928), the entoderm material of the meridional half could bring about any entodermization of the adjacent presumptive ectoderm and a regulation to a complete larva could ensue. This was not the case. Each component differentiated in the same way as it would have done normally. There seems to be no interaction between them whatsoever (fig. 7B). But it is highly interesting to note that once the embryological differentiation had taken place according to a mosaic scheme a regulative interaction occurred in the following development. The greater part of the ciliated band of the animal component gradually disappeared, only a short part remaining to form a complete ring together with the horseshoe of the meridional half (fig. 7C). This means a step towards individualization of this heterogeneous larva by means of redifferentiation.

5.6. General remarks

Experiments by several authors mentioned above have shown that any fragment of sufficient size of unfertilized Cerebratulus eggs may develop into a pilidium. On the other hand isolated layers of 8- and 16-cell stages demonstrate a mosaic differentiation. Yatsu (1910b) observed that an animal half of even a mature egg could form a pilidium, and also that removal of most of the animal part of the blastomeres of a 2-cell stage had no effect. But when most of the vegetal material was cut off the archenteron was missing or too small. Another experiment with compressed cleavage stages lead him to the conclusion that at the 4-cell stage we have the same animal–vegetal localization as in the 8-cell stage. Thus the transition from a regulation to a mosaic type takes place between the maturation and the 8-cell stage. It has often been pointed out that there is no fundamental difference between the mosaic and regulation eggs. In some eggs the determination sets in at an earlier stage than in others.

It is of interest in this respect to compare the Cerebratulus egg with that of the sea urchin. Even after isolation at 32- or 64-cell stages animal and vegetal halves do not differentiate in conformity to the significance of the material. Animal halves as a rule do not form a ciliated band and stomodeum whereas vegetal halves may produce these organs. Vegetal material implanted in animal fragments may induce an archenteron and cause development of a complete dwarf pluteus. An animal half combined with a meridional half also forms an harmonic larva (cf. Cerebratulus above, figs. 7A, B). In the sea urchin we may furthermore obtain two plutei from one egg after cutting at right angles to the egg-axis, i.e. if we cut twice and put the polar parts together. The middle part will thus form more animal and vegetal differentiations than it would have done normally, and the animal and vegetal fragments will form organs characteristic of the excised middle part of the egg. This power of regulation has to be compared to the mosaic development we have seen in fig. 5. Conversely, a bilateral organisation in the sea urchin egg is already demonstrable in the 2-cell stage (Hörstadius 1928, 1935, 1936; Hörstadius and Wolsky 1936).

In the eggs of annelids and molluscs with the typical spiral cleavage we can often distinguish precociously differentiated 'organ-forming' substances, the polar plasms. These are distributed only into the D-quadrant which gives rise to the ectodermal and entomesodermal germ bands. In Cerebratulus we also have an early rearrangement of substances, namely at the time of the break-down of the germinal vesicle (fig. 1). But these substances are not, as far as we know, unevenly distributed to the quadrants. The authors differ on the mode of origin of the mesenchyme in Cerebratulus (cf. Hörstadius 1937).

Wilson (1903), Zeleny (1904) and Yatsu (1910b) have reported rather startling results of rearing of fragments of blastulae. Animal halves could gastrulate and vegetal ones could be provided with an apical organ. These results are in conflict with our knowledge of gradual restriction of potencies in the course of the development. As it is extremely difficult to cut a blastula exactly in the desired plane I am, like Wilson, inclined to explain the results on the grounds of oblique sections.

It is possible that the apical organ does not follow a strict mosaic design. The fact that all four quadrants of an egg form such an organ does not speak against a mosaic development as each fragment may contain a part of the rudiment. But we have also seen that some animal and meridional fragments may have more than one, even with separate pits, or be without apical organ (figs. 3G, 4A, C, 5B, D, 6B, 7B). Such observations lead Yatsu (1904) to the suggestion, that the factors for the apical organ in the uncleaved egg are not localized in the animal pole but in a broad zone above the equator. In the cleavage stages this can hardly be the case as most fragments with an_1-material have produced apical organs whereas larvae without an_1 lack this organ (figs. 8A, 9). On the other hand Yatsu (1910b) found that the apical organ regenerates when it has been removed from late gastrula or young pilidia (cf. the power of redifferentiation in the larva, fig. 7).

References

COE W. R., 1899. The maturation and fertilization of the egg of Cerebratulus. Zool. Jahrb. Abt. Anat. Ontog. Tiere *12*, 425.

HÖRSTADIUS S., 1928. Über die Determination des Keimes bei Echinodermen. Acta Zool. (Stockholm) *23*, 1.

HÖRSTADIUS S., 1935. Über die Determination im Verlaufe der Eiachse bei Seeigeln. Pubbl. Staz. Zool. Napoli *14*, 251.

HÖRSTADIUS S., 1936. Weitere Studien über die Determination im Verlaufe der Eiachse bei Seeigeln. Arch. Entwicklungsmech. Organ. *135*, 40.

HÖRSTADIUS S., 1937. Experiments on determination in the early development of Cerebratulus lacteus. Biol. Bull. *73*, 317.

HÖRSTADIUS S., and A. WOLSKY, 1936. Studien über die Determination der Bilateralsymmetrie des jungen Seeigelkeimes. Arch. Entwicklungsmech. Organ. *135*, 1.

WILSON C. B., 1900. The habits and early development of Cerebratulus lacteus (Verrill). Quart. J. Microscop. Sci. *43*, 97.

WILSON E. B., 1903. Experiments on cleavage and localization in the nemertine egg. Arch. Entwicklungsmech. Organ. *16*, 411.

YATSU N., 1904. Experiments on the development of egg fragments in Cerebratulus. Biol. Bull. *6*, 123.

YATSU N., 1910a. Experiments on cleavage in the egg of Cerebratulus. J. Coll. Sci. Tokyo *27*, 10.

YATSU N., 1910b. Experiments on germinal localization in the egg of Cerebratulus. J. Coll. Sci. Tokyo *27*, 17.

ZELENY C., 1904. Experiments on the localization of development factors in the nemertine egg. J. Exptl. Zool. *1*, 293.

G. Reverberi (ed.), Experimental embryology of marine and fresh-water invertebrates

CHAPTER 6

Mytilus

G. REVERBERI

Zoological Institute, University of Palermo, Italy

6.1. *Short bibliographical notes*

The Mytilus egg is very useful for experimentation, particularly because, like the eggs of Ilyanassa and Dentalium, it forms the polar lobe.

6.1.1. The normal development has been described by Wilson (1886), Matthews (1913), Field (1921, 1922) and more recently by Rattenbury and Berg (1954) and Abd-El-Wahab (1957). The ovarian egg has been studied by Worley (1944), Urbani-Mistruzzi and Scollo-Lavizzari (1954) and Reverberi (1967), while the mature egg has been investigated cytochemically by Pucci (1961). It has also been studied by electron microscope (Reverberi and Mancuso 1961; Dan 1962; Humphreys 1962, 1964, 1967; Reverberi 1967).

6.1.2. Experimental analysis of the early development has been done by Rattenbury and Berg (1954).

6.1.3. A biochemical investigation of isolated blastomeres and the polar lobe has been undertaken by Berg and Kutsky (1951), Abd-El-Wahab and Pantelouris (1957) and Berg and Prescott (1958).

6.2. *The egg*

6.2.1. *The ovarian egg*

The ovarian egg has been investigated with the light microscope by Worley (1944), Urbani-Mistruzzi and Scollo-Lavizzari (1954) and Reverberi (1967); and with the electron microscope by Reverberi (1967).

Worley devoted much attention to Golgi bodies and mitochondria; Urbani-Mistruzzi and Scollo-Lavizzari, for their part, laid great emphasis on some RNA-rich bodies (probably of nucleolar origin) interpreted as yolk nuclei and held to be responsible for the synthesis of the yolk granules. Some features shown by the light microscope are illustrated in fig. 1. The

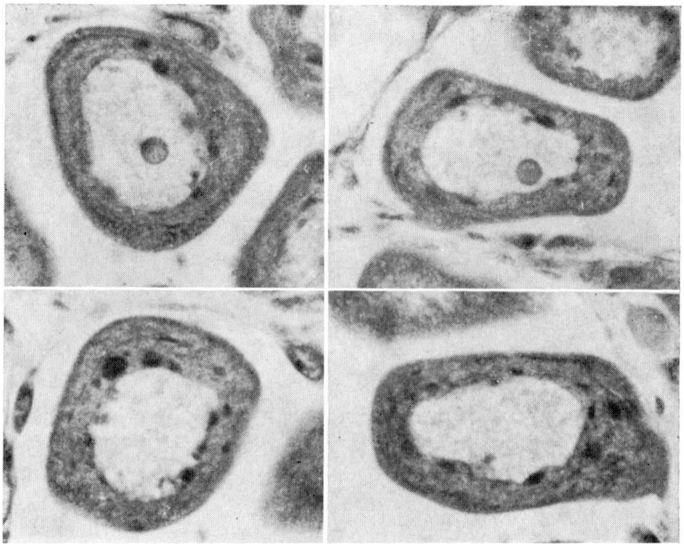

Fig. 1. Oocytes of Mytilus: note the large, clear germinal vesicle; the nucleolus, and the numerous cytoplasmic bodies once interpreted as yolk nuclei or Golgi bodies (original).

picture obtained with the electron microscope is shown in fig. 2: a very large germinal vesicle occupies nearly the whole egg; its shape is not regular, but rather amoeboid, and probably changes according to its activity. The surrounding membrane, as usual, is double and is perforated by numerous pores which allow a communication with the cytoplasm. Yolk and lipid granules are very numerous and are dispersed in an endoplasmic reticulum composed of small vesicles crowded with ribosomes. A striking figure in the cytoplasm is the appearance of numerous plurilamellar formations in the form of whorls, crowded with ribosomes and in the centre delimiting a space occupied by an amorphous substance or a few mitochondria (fig. 3): no doubt these plurilamellar bodies are the same as the Golgi bodies noticed by Worley, or the basophilic structures (interpreted as pallial formations or yolk nuclei) described by Urbani-Mistruzzi and Scollo-Lavizzari. The

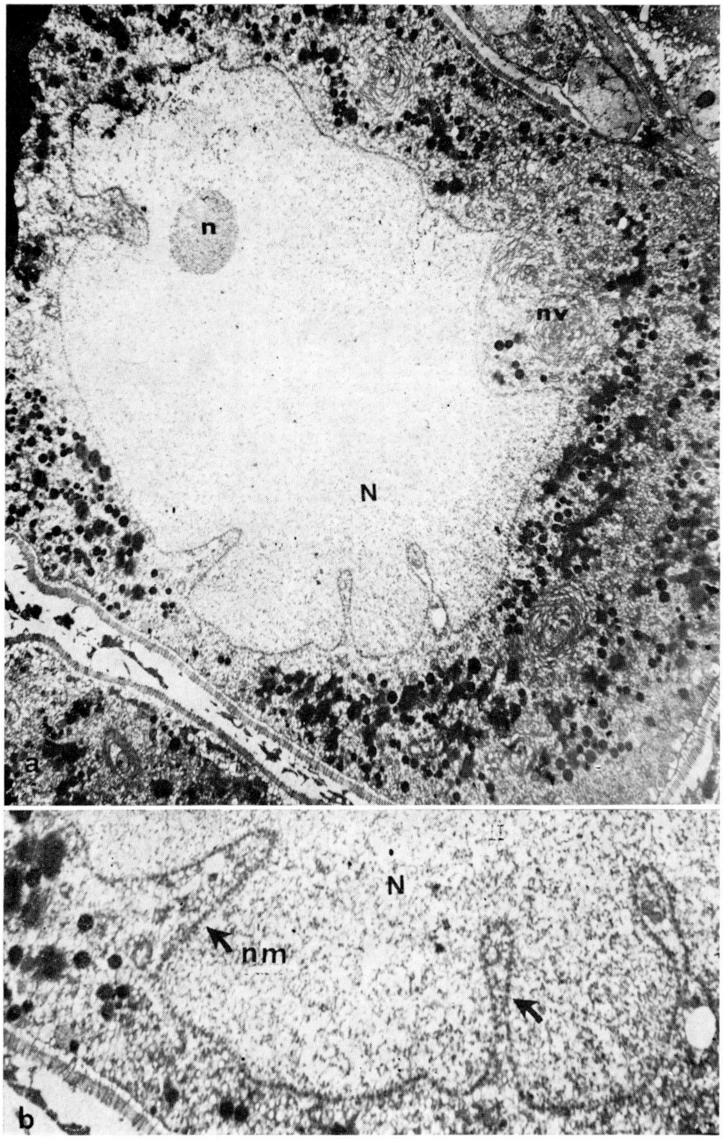

Fig. 2. Electron micrograph of vitellogenetic oocyte. (a) The germinal vesicle has an amoeboid shape, the nuclear membrane, on the other hand, shows numerous indentations; the nucleolus and some concentric lamellar formations are evident (nv), the vitelline membrane is traversed by microvilli. N = nucleus; n = nucleolus; nm = nuclear membrane. (b) The indentations of the nuclear membrane at higher magnification (Reverberi 1967).

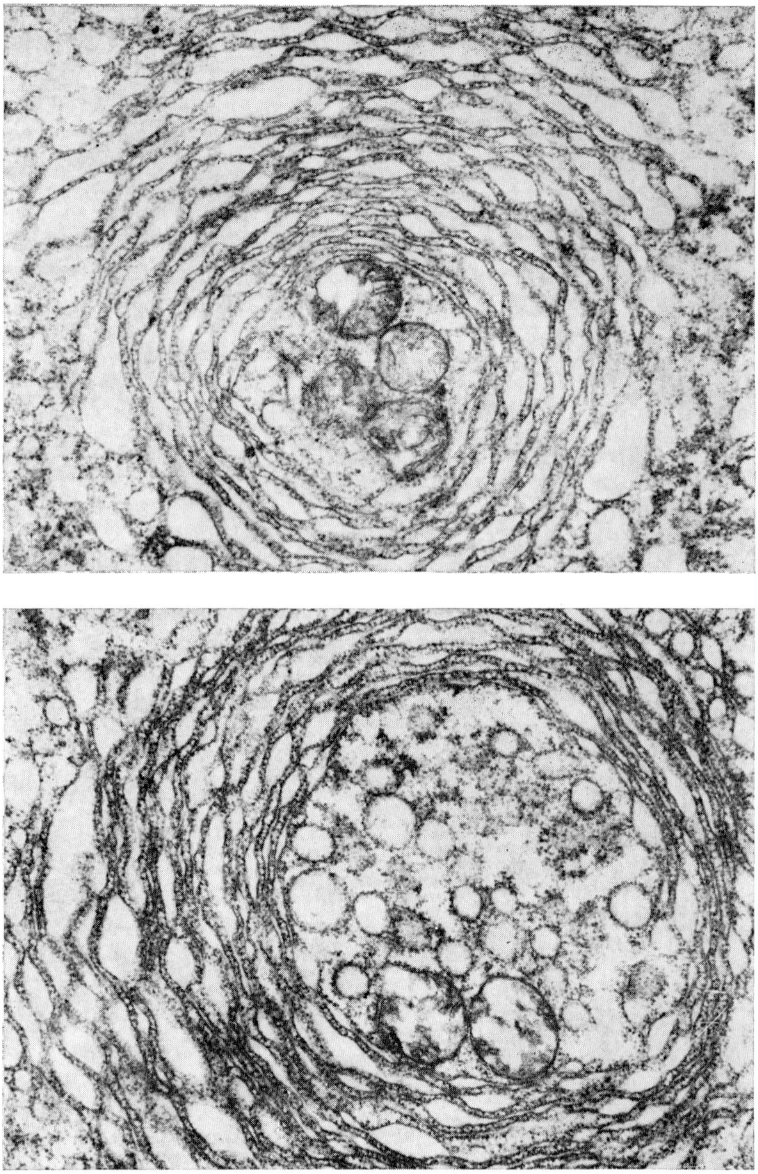

Fig. 3. An aspect of the cytoplasmic bodies shown by electron microscope: they appear as plurilamellar formations forming whorls. The lamellae are crowded with granules and a central space is occupied by vesicles of endoplasmic reticulum and mitochondria (Reverberi 1967).

periphery of the egg is completely occupied by the closely packed cortical granules. The plasma membrane is raised in a large number of slender, regularly disposed microvilli, which traverse the perivitelline space, reach and cross the vitelline membrane, emerge on its exterior surface and send fine processes into the jelly (to the formation of which they probably also contribute with their secretion). It seems that Worley identified these microvilli as mitochondria; the latter have a completely different distribution, however.

6.2.2. The mature eggs

The mature eggs are shed spontaneously from the female at the time of maturity extending from September to February (in Palermo). Deposition of the eggs can also be induced artificially (it is recommended to keep the animals dry, in a refrigerator, for some days, and then to put them into sea water at 20 °C; electric stimulation also seems to be effective) (cf. Breese et al. 1963; Bayne 1965; Iwate 1949).

Just after being spawned the eggs have an irregular shape and a large germinal vesicle is clearly visible. After some time, however, the eggs become perfectly spherical, the germinal vesicle disrupts and the first meiotic figure is completed. Both polar globules are emitted after fertilization. The eggs (about 70 μ in diameter) are opaque and are enveloped in a thick (1 μ) vitelline membrane and some jelly (which can be made evident by dissolving India ink in the sea water). (For the technical treatment of the eggs consult Berg 1967.) The ripe and just spawned egg has been investigated cytochemically by Pucci (1961).

As noted above, the egg has been studied with the electron microscope by Reverberi and Mancuso (1961), Dan (1962), Humphreys (1962, 1964, 1967) and Reverberi (1967). It is surrounded by a mass of jelly and a thick vitelline membrane; this is traversed by numerous evenly spaced microvilli, the presence of which probably renders it more consistent (fig. 4). The microvilli very likely not only have the function of reinforcing the vitelline membrane; it seems that they also contribute to increasing the mass of the peripheral jelly-coat: possibly this is done by conveying the materials contained in the cortical granules. These are very numerous (fig. 4b), forming a continuous sheet under the plasma membrane and are not displaced by centrifugation; they disappear after treatment of the egg with sperm extract. Their fine structure has recently been described by Humphreys (fig. 5).

The endoplasm is mostly made up of vesicles 20–400 mμ studded with granules 15 mμ in diameter: mitochondria, yolk-granules, lipidic drops and

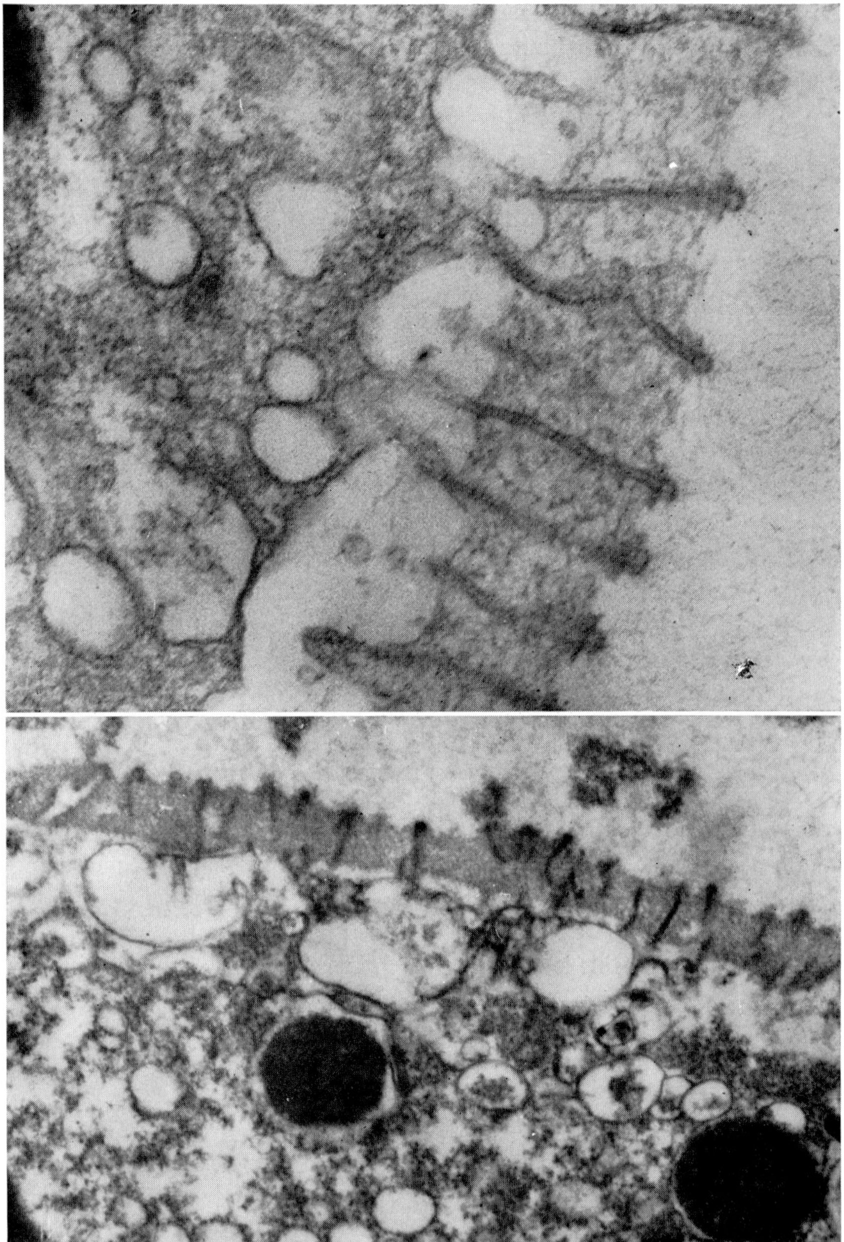

Fig. 4. Cortical region of an egg: cortical granules, microvilli, perivitelline space, vitelline membrane traversed by microvilli (Reverberi 1967).

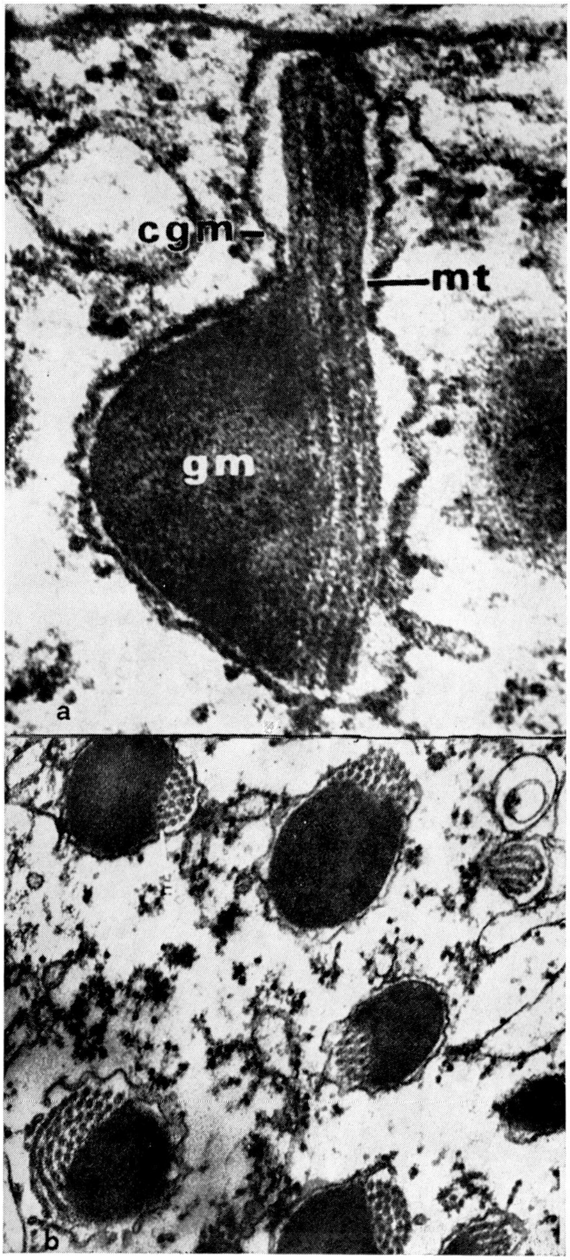

Fig. 5. (a) Structure of a cortical granule of an unfertilized egg: note the granular mass (gm) and the bundle of microtubules (mt); these are enclosed in a membrane (cgm). (b) Cortical granules in a subtangential section: the microtubules are cut in cross-section: normally they are oriented perpendicularly to the egg surface (Humphreys 1967).

multivesicular bodies contribute to the constitution of the endoplasm. The mitochondria are rod-shaped: some seem to be transitional between yolk granules and true mitochondria. No differential distribution of these endoplasmic constituents was noticed.

6.3. Normal development

As mentioned earlier, before cleaving the egg forms a lobe at the vegetal pole: at this stage is it pear-shaped. After cleavage the egg has a trefoil aspect; the polar lobe is connected to the CD-cell. As it subsequently fuses with this cell, CD will be larger than AB. A second polar lobe is formed before the second cleavage: it is reabsorbed by the D-cell. At the third cleavage, which is spiral, A, B and C divide equally; D however, divides unequally: 1D and 1d are formed. At the 4th cleavage 1D cleaves equally: the two resulting cells, 2d and 2D, are rather large. The subsequent cell-lineage has not been followed in detail: after some time a cap of micromeres is found at the animal pole, tending to envelope the vegetal macromeres. After 6 hr of development the embryo has formed cilia, the micromeres have covered the vegetal macromeres and invagination of the latter has begun. The area of invagination is very narrow: a small tubular archen-

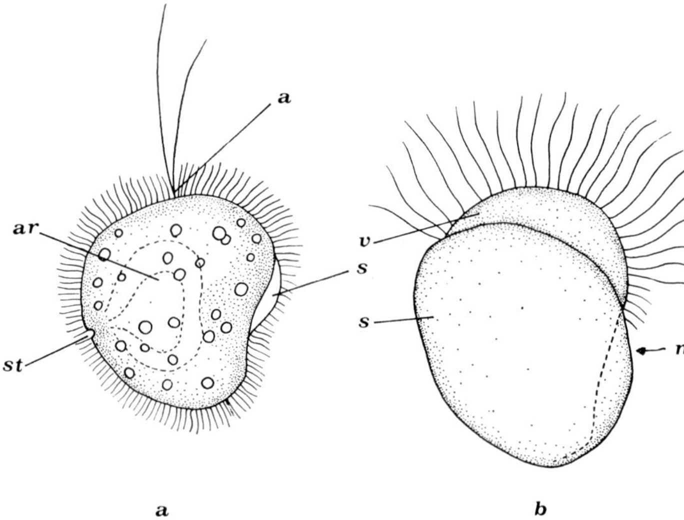

Fig. 6. Trochophore at 24 hr: ar = archenteron; st = stomodeum; s = shell; a = apical tuft; a normal veliger at 48 hr is shown in (b); v = velum; m = mouth. (Redrawn from Rattenbury and Berg 1954.)

teron is formed communicating exteriorly with a circular blastopore. The mesoderm probably originates from the 4d-cell and its derivatives. The larva (fig. 6) is referred to as a trochophore although it has no prototroch: the surface is ciliated, and a long tuft is present apically; a stomodeum, a thickening of the dorsal ectoderm, the shell gland and a small transparent shell are also present. After 48 hr the larva (fig. 6b) has a large velar lobe and a bivalve shell into which the velar lobe can be withdrawn. After 72 hr the larva has a foot (Rattenbury and Berg 1954).

6.4. *The ultrastructure of the fertilized egg, blastomeres AB and CD and the polar lobe*

The ultrastructure of the fertilized egg (Humphreys 1964) does not greatly differ from that of the unfertilized one: 'subtle but consistent differences' are, however, found.

The cytoplasm of the polar lobe does not possess a particular differential ultrastructure: mitochondria are not more abundant in the polar lobe than elsewhere: fewer cortical granules and multivesicular bodies are however present; microvilli are also numerous. The cytoplasm of the polar lobe is separated from the cytoplasm of blastomeres AB and CD by a sheet of vesicles, in a similar manner to when the egg divides into two cells.

6.5. *Experimental research*

Experimental investigation of the Mytilus egg is still rare and in any case is not comparable with that done on Ilyanassa. The development of isolated AB- and CD-blastomeres of the egg deprived of the polar lobe has been followed by Rattenbury and Berg (1954).

6.5.1. Development of CD (figs. 7a, b)

CD, if isolated, continues its normal development, dividing 'partially'. Before segmenting, it forms the second polar lobe, which is subsequently reabsorbed in the D-cell. A trochophore which is very similar to a normal trochophore is formed. The 48-hr embryo does not show the bivalve shell but possesses a normal velum.

6.5.2. Development of AB (figs. 7c, d)

The isolated AB-blastomere also cleaves partially. The 24-hr AB-embryo is uniformly ciliated and bears no apical tuft; shell gland, shell and stomo-

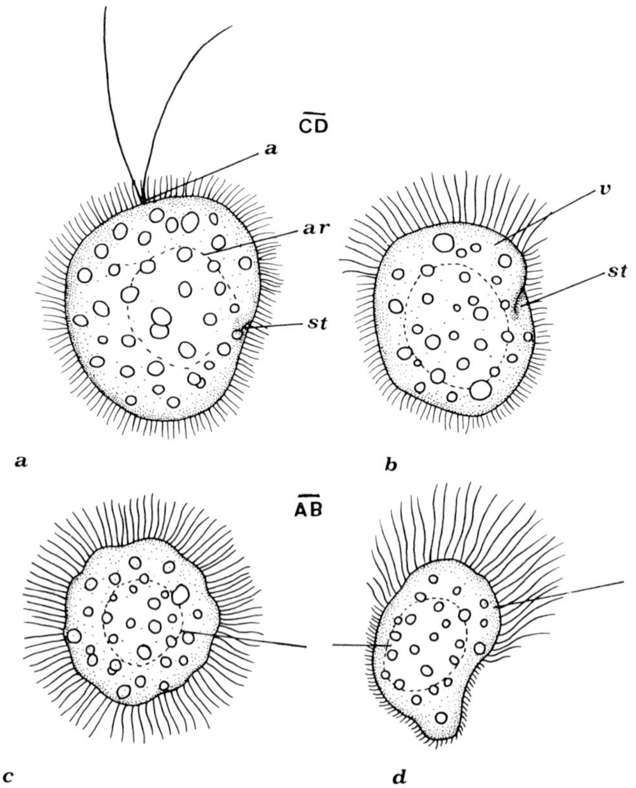

Fig. 7. Twenty-four-hour embryo (a) and 48 hr embryo (b) from an isolated CD-blastomere; 24-hr embryo (c) and 48-hr embryo (d) from an isolated AB-blastomere. (Redrawn from Rattenbury and Berg 1954.)

deal invagination are also absent. The 48-hr embryo is slightly elongated and the anterior end bears a developed velum.

6.5.3. Development of isolated A, B, C, and D (at the 4-cell stage)

Only the larvae that derive from D bear an apical tuft; the 24-hr larvae deriving from A, B and C are indistinguishable from one another. They are spherical in shape and have a fine ciliature. The 48-hr D-larvae have a velum but no shell gland; shell or mesoderm are present.

6.5.4. Development of the egg without the polar lobe (figs. 8a, b)

The larvae that develop from these eggs have no apical tuft; a reduced shell gland, a stomodeum and an archenteron are however formed. No mesoderm

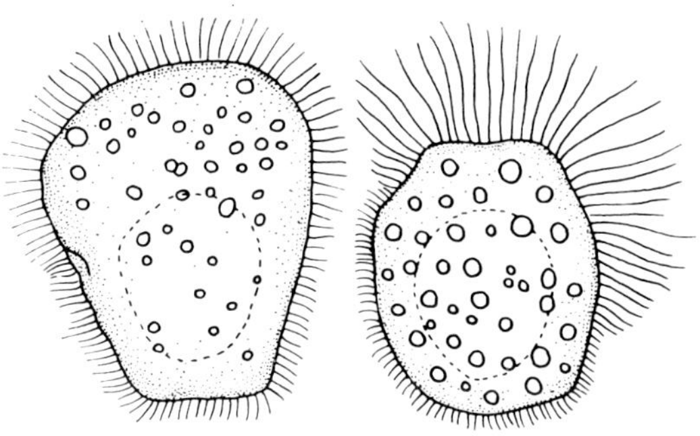

Fig. 8. Twenty-four-hour (a) and 48-hr (b) embryos from an egg from which the first polar lobe was removed. (Redrawn from Rattenbury and Berg 1954.)

is formed, and after 48 hr they develop a velum. The above-mentioned results allow a number of conclusions to be drawn. (a) The material responsible for the formation of the apical tuft of the trochophore is located in the 1st polar lobe; from there it passes to the CD-blastomere but does not enter the 2nd polar lobe. (b) The material responsible for the formation of the shell gland and shell is localized in the CD-cell. (c) The capacity for forming stomodeum is localized in the CD-blastomere. (d) The material for the formation of mesoderm and generally for the formation of the post-trochal structures seems to be segregated into the CD-blastomere, probably via the first polar lobe. The material for the formation of the mantle and the foot is not clearly localized: neither mantle nor foot were seen in partial embryos. Their development is possibly dependent upon an interaction between AB- and CD-blastomeres. The failure of the partial embryos to develop features which are typical of the adult is probably not due to a lack of competence but to arrested development.

6.6. *The chemistry of development*

The cellular differentiation is certainly dependent on, or correlated with, the metabolism. This should be different in the different blastomeres of a segmenting egg and an investigation into the metabolism of isolated blastomeres is therefore likely to be rewarding.

6.6.1. The oxygen consumption in AB- and CD-blastomeres and isolated polar lobes has been determined by Berg and Kutsky (1951) with the

Cartesian diver. According to their data the isolated polar lobe has a very low oxidative metabolism. The average oxygen uptake of the developing AB-embryo is 0.90 $\pm$ 0.032 μl $O_2/hr/mm^3$ cytoplasm; that of CD-embryo is 0.78 $\pm$ 0.27 and that of the polar lobe is 0.42 $\pm$ 0.25.

Interpretation of the data is difficult in view of the fact that, as noted above, the mitochondria, which are responsible for the processes of respiration, are uniformly distributed in the AB- and CD-blastomeres and the polar lobe.

6.6.2. The rate of incorporation of amino acids or bases (adenine) has also been determined (Abd-el-Wahab and Pantelouris 1957). AB and CD incorporate at the same rate: the isolated polar lobe, on the contrary, incorporates very poorly. This finding has been attributed to the absence of a nucleus in the polar lobe, but other factors could be responsible.

6.6.3. The uptake of phosphate has been determined by Berg and Prescott (1958). The isolated polar lobe accumulates phosphate at a lower rate than AB and CD. The explanation of this result is not easy: it has been related (by Humphreys) to the fact that the polar lobe has no microvilli, that is, that its surface area of absorption is reduced.

6.6.4. The peptidase content has been determined by Berg (1954).

References

ABD-EL-WAHAB A., 1957. Influence of the antimitotic naphthoquinone on the cleavage of the eggs of the mollusc, *Mytilus edulis* L. Arch. Entwicklungsmech. Organ. *149*, 613–623.

ABD-EL-WAHAB A. and E. M. PANTELOURIS, 1957. Synthetic processes in nucleated and non-nucleated parts of Mytilus eggs. Exptl. Cell Res. *13*, 78–82.

BAYNE B. L., 1965. Growth and the delay of metamorphosis of the larvae of *Mytilus edulis* (L.). Ophelia *2*, 1–47.

BERG W. E., 1950. Lytic effects of sperm extracts on the eggs of *Mytilus edulis*. Biol. Bull. *98*, 128–138.

BERG W. E., 1954. Large scale constriction and segregation of polar lobes from the eggs of *Mytilus edulis*. Exptl. Cell Res. *6*, 162–171.

BERG W. E., 1967. Some experimental techniques for eggs and embryos of marine invertebrates, in 'Methods in developmental biology' ed. by Wilt and Wessels. Th. Y. Crowell Comp., New York.

BERG W. E. and P. B. KUTSKY, 1951. Physiological studies of differentiation in *Mytilus edulis*. I. The oxygen uptake of isolated blastomeres and polar lobes. Biol. Bull. *101*, 47–61.

BERG W. E. and D. M. PRESCOTT, 1958. Physiological studies of differentiation in *Mytilus edulis*. II. Accumulation of phosphate in isolated blastomeres and polar lobe. Exptl. Cell Res. *14*, 402–407.

BREESE W. P., R. E. MILLEMANN and R. E. DIMICK, 1963. Stimulation of spawning in the mussels, *Mytilus edulis* Linneaeus and *Mytilus californianus* Conread, by kraft mill effluent. Biol. Bull. *125*, 197–205.

DAN J. C., 1962. The vitelline coat of the Mytilus egg. I. Normal structure and effect of acrosomal lysine. Biol. Bull. *123*, 531–541.

FIELD I. A., 1921–1922. The biology and economic value of the sea mussel, *Mytilus edulis*. Bull. Bur. Fish. US *38*, 127–259.

HUMPHREYS W. J., 1962. Electron microscope studies on eggs of *Mytilus edulis*. J. Ultrastruct. Res. *7*, 467–487.

HUMPHREYS W. J., 1964. Electron microscope studies of the fertilized egg and the two-cell stage of *Mytilus edulis*. J. Ultrastruct. Res. *10*, 244–262.

HUMPHREYS W. J., 1967. The fine structure of cortical granules in eggs and gastrulae of *Mytilus edulis*. J. Ultrastruct. Res. *17*, 314–326.

IWATA K. S., 1949. Spawning of *Mytilus edulis*. II. Discharge by electrical stimulation. Bull. Japon. Soc. Sci. Fisheries *15*, 443–446.

MANCUSO V., 1960. La membrana ovulare di *Mytilus edulis* studiata al microscopio elettronico. Rend. Ist. Super. Sanita *23*, 793–796.

MATTHEWS A., 1913. Notes on the development of *Mytilus edulis* and *Alcyonium digitatum*. J. Marine Biol. Assoc. UK *9*, 557–560.

MEVES F., 1915. Über den Befruchtungsvorgang bei der Miesmuschel (*Mytilus edulis*). Arch. Mikroskop. Anat. *87*, 47–62.

PUCCI I., 1961. Cytochemical investigations on the egg of Mytilus. Acta Embryol. Morphol. Exptl. *4*, 96–101.

RATTENBURY J. C. and W. E. BERG, 1954. Embryonic segregation during early development of *Mytilus edulis*. J. Morphol. *95*, 393–414.

REVERBERI G., 1966. Electron microscopy of some cytoplasmic structures of the oocytes of Mytilus. Exptl. Cell Res. *42*, 392–394.

REVERBERI G., 1967. Some observations on the ultrastructure of the ovarian Mytilus egg. Acta Embryol. Morphol. Exptl. *10*, 1–14.

REVERBERI G. and V. MANCUSO, 1961. The constituents of the egg of Mytilus as seen at the electron microscope. Acta Embryol. Morphol. Exptl. *4*, 102–121.

SUGIURA Y., 1962. Electrical induction of spawning in two marine invertebrates (*Urechis unicinctus*, hermaphroditic *Mytilus edulis*). Biol. Bull. *123*, 203–206.

URBANI-MISTRUZZI L. and G. SCOLLO-LAVIZZARI, 1954. Osservazioni citologiche e citochimiche sulla oogenesi di *Mytilus edulis*. Rend. Accad. Naz. Lincei, *16*, Suppl 8, 786–791.

WADA S. K., J. R. COLLIER and J. C. DAN, 1956. Studies on the acrosome. V. An egg-membrane lysin from the acrosome of *Mytilus edulis* spermatozoa. Exptl. Cell Res. *10*, 168–180.

WILSON E. B., 1886. On the development of the common mussel (*Mytilus edulis*). 5th Rep. Fish. Board Scot. App. F. VI, 247–256.

WORLEY L. G., 1944. Studies of the vitally stained Golgi apparatus. II. Yolk formation and pigment concentration in the mussel, *Mytilus californianus* Conrad. J. Morphol. *75*, 77–101.

G. Reverberi (ed.), Experimental embryology of marine and fresh-water invertebrates

CHAPTER 7

Ilyanassa

A. C. CLEMENT

Department of Biology, Emory University, Atlanta, Ga. 30322, U.S.A.

7.1. Introduction

There are two features in particular which make eggs of the marine snail *Ilyanassa obsoleta* unusually attractive material for developmental studies. One is the fact that during first cleavage the egg forms a large cytoplasmic protrusion at the vegetal pole known as the polar lobe, which experiments have shown to be highly influential in morphogenesis. The fact that the lobe is easily detached has permitted a variety of types of analysis, both of the properties of the isolated lobe, and of the alteration in properties of the lobeless egg as compared with the normal. Most of the experimental work on *Ilyanassa* has in fact been directed toward the role of the lobe cytoplasm in developmental processes, either at the biochemical or structural level. The second feature which makes the eggs of *Ilyanassa* favorable material for embryological work is the fact that they can be obtained in abundance throughout most of the year from snails maintained in small salt water aquaria in inland laboratories. Culture requirements are simple, both for the adult snails and for eggs and embryos. The period required for development of a well-formed veliger larva is about one week at room temperature, and hence sufficiently short to make a variety of experiments feasible.

Ilyanassa is in present usage a subgeneric name. The taxonomic designation currently used for the species is *Nassarius obsoletus* (Say). *Nassa* and other generic terms have been used in the past. Since *Ilyanassa* is the name generally used by embryologists for *N. obsoletus*, it will be employed in the present account in reference to this species.

Reference will be made occasionally to another marine prosobranch gastropod, *Crepidula*. Generally speaking, *Ilyanassa* is the more favorable form for experimental work on problems of embryonic development. On the other hand, the monumental descriptive works of Conklin (1897, 1902) on the development and cytology of *Crepidula* provide a wealth of detailed information which is not available for *Ilyanassa*, and which may be of interest in connection with experimental work on the latter form.

7.2. Methods for obtaining, handling and observing eggs

I. obsoleta is commonly found on mud flats in shallow water along much of the Atlantic coast of the United States, and may often be seen exposed in large numbers at low tide.

Although the breeding season normally begins with warm weather in the spring and lasts on into the summer, snails collected in the vicinity of Woods Hole, Massachusetts and brought into the laboratory at any time between about mid-November and midsummer will produce eggs. Freshly collected snails packed in moist sea weed are easily shipped for long distances by air parcel post. Although the natural breeding season at Woods Hole probably ends about midsummer, snails collected in June or earlier will continue to lay fertile eggs under laboratory conditions until September at least.

At a marine station where flowing sea water is available, 50–100 snails may be kept in an aquarium of around 15 gallons capacity. For steady egg production, the snails should be fed regularly, preferably at the end of each day. If an irregular feeding schedule is followed, eggs are not likely to be obtained on the day after the snails are fed. A freshly chopped up clam is a convenient food for use at a marine laboratory; uneaten fragments may be removed from the aquarium the next day.

In inland laboratories snails may be successfully maintained in circulating sea water systems of small volume, using either natural or artificial sea water. In our laboratory we have had good success for several years with an artificial sea water made from Instant Ocean salts* and tapwater. Thirty or forty snails may be kept in a system containing no more than 15 or 20 gallons of sea water. For convenience in collecting eggs we keep the snails in a small glass aquarium (around five gallons capacity), in which the water is kept well aerated. From the aquarium the water is moved with a simple air-lift pump into a filter unit which contains sand and crushed oyster shell. After passing through the filter the water flows by siphon tubes into a reservoir and then back into the aquarium. A part of the sea water is replaced occa-

* Supplied by Aquarium Systems, Inc., Wickliffe, Ohio.

sionally, and the specific gravity is maintained at a normal level. We feed the snails daily on slices of commercial frozen shrimp. The quantity is adjusted so that none is left unconsumed. Snails will usually not begin to lay eggs until they have been in the laboratory for several days or a week. Egg production will probably fall off after several weeks, and the snails may then be replaced with a new batch. Reserve snails may be stored in a refrigerator for several weeks in a relatively small amount of water, if the latter is changed occasionally.

The sexes are separate in *Ilyanassa*. Fertilization is internal, and the eggs are deposited in capsules which, in the laboratory, are attached to the walls or floor of the aquarium. Each capsule usually contains from 50 to 200 eggs at approximately the same stage of development. Females in good condition may deposit a row of several capsules.

To start the day with a good supply of recently deposited eggs, all capsules should have been removed at the end of the previous day. The morning's collection of capsules may be scanned rapidly under a dissecting microscope, and those containing eggs of the desired stage selected for further attention. If eggs in the very earliest stages are desired, capsules may be collected as deposited; females in the process of depositing capsules are easily observed through the glass walls of the aquarium.

Egg capsules are easily removed from the aquarium walls with a scraper made by attaching a single-edge razor blade to a wooden handle long enough to avoid the necessity of placing the hands in the aquarium. A capsule should not be torn or compressed during removal. Detached capsules may be sucked up in a syringe made from glass tubing and a large rubber bulb.

Freshly laid eggs will usually not have extruded the first polar body, and at room temperature the first cleavage will not take place for about three hours.

A simple technique for removing eggs from the capsule involves snipping off one end of the capsule with curved cuticle scissors, and gently squeezing the eggs from the capsule with a pair of sharpened steel needles. The eggs within the capsule are embedded in a jelly mass which will retract into the capsule if the pressure on the latter is released; to prevent this the capsule may be torn; the jelly soon dissolves and the eggs are freed. During spherical stages the liberated eggs orient in the dish with the animal pole uppermost. To observe the polar bodies during stages when the uncleaved eggs are spherical, it is necessary to roll the eggs around by agitating the dish or by the use of gentle spurts of water from a pipette.

If eggs are to be reared *in vitro*, after removal from the capsule, precautions are necessary to prevent harm from microorganisms, or toxic chemicals. To avoid the possibility of chemical contamination, it is advisable to use glassware which has been scrupulously washed and rinsed, and which has not previously been used for fixatives or other potentially toxic substances. Disposable plastic dishes may be used instead of glassware. To reduce the hazard of infection by microorganisms, capsules may be freed of adherent jelly and debris before opening by rolling them around gently on a paper towel, then placing them into pasteurized sea water for opening. Once removed from the capsule, the eggs should be transferred through several baths (four or five) of pasteurized sea water before placing them in the culture dish. Small stender dishes (38 mm diameter) make convenient culture vessels. Such a dish about half filled with pasteurized sea water will accommodate up to 20 eggs, without further change of the sea water. The dish should be covered and placed inside a moist chamber to prevent loss of water by evaporation. It may be left undisturbed until a week has elapsed, at which time normal development will have produced well developed veliger larvae, at room temperature. Pasteurized sea water is obtained by heating filtered sea water to 70–75 °C for around 15 min; it should be well aerated by shaking after it has cooled to room temperature.

Many structural details of the larvae can be observed in living specimens. Larvae may be anesthetized with chloretone and mounted under supported coverslips for examination with the compound microscope. An equal mixture of sea water and a saturated solution of chloretone in sea water is a satisfactory anesthetic solution. Excellent permanent whole mounts of veliger larvae may be obtained by first anesthetizing the larvae with the chloretone solution mentioned, then fixing them in a solution containing about 4% formaldehyde in sea water, and mounting them according to the method described by Clement and Cather (1957). This involves dehydration in alcohol, faint staining with Orange G, and mounting under supported coverslips in euparal or diaphane. With this method the velum may be preserved in an expanded condition, cilia, shell, and statoliths are preserved, and in many ways the larvae are comparable in appearance with live material. When examined under polarized light, the shell, statoliths and operculum of either living larvae or whole mounts prepared by the method just described are birefringent.

Many details of *Ilyanassa* eggs and embryos are well observed after preservation in full strength picrosulfuric acid fixative (Mayer's formula) and staining in Mayer's acid hemalum diluted with nine times its volume of 0.5% glacial acetic acid in distilled water; the eggs or embryos are mounted whole under supported coverslips. Cather (1958) has described a technique involving Gomori's chromalum hematoxylin which may be used for detailed study of the chromosomes. Bouin's fixative and Mayer's acid hemalum may be used for sections. Embedding of larvae in agar blocks after fixation is an aid to orientation and handling of material to be sectioned.

Cather (1967) has used marking with carbon particles or localized staining with Nile blue sulfate to follow the derivatives of individual cells during development.

To observe the stratification pattern of cellular inclusions produced by centrifugal force, uncleaved eggs may be suspended in sea water over a very thick solution of gum arabic in sea water. The interface between gum arabic solution and sea water may be mixed slightly to establish a gradient. With forces of around 2000 *g* applied for a few minutes, three cytoplasmic zones are obtained. Additional zones may be separated with a proper balance between centrifugal force and length of exposure. It is also rather easy to break the egg into fragments with centrifugal force. For further details concerning various centrifugation media, time and speed of centrifugation, patterns of fragmentation, etc., the papers of Morgan (1933, 1935a, 1936, 1937), Butros (1956), and Clement (1935, 1968) may be consulted.

By centrifuging eggs held in a more or less reverse orientation in a chilled gelatin-sea water gel, Morgan (1933, 1935b) was able in some cases to dislocate yolk from the region of the vegetal pole and to replace it with other materials. In an elaboration of this technique, Clement (1968) first dislocated the yolk, then freed the eggs from the gelatin, recentrifuged them in a raffinose-sea water solution and pulled them into halves. In this way it was possible to obtain nucleated halves derived partly or wholly from the vegetal region of the egg, but which had been deprived of most of the yolk.

7.3. Structure of the egg

The eggs of *Ilyanassa*, once removed from the capsule and jelly mass in which they are normally enclosed, appear perfectly naked; that is, there is no indication of any membrane other than the egg surface itself. The upper part of the egg is occupied by a protoplasmic cap which is largely free of yolk, and in which the nucleus or division figure is to be found. The remainder of the egg is occupied by yolk granules of widely varying diameter. The protoplasmic cap is rich in both mitochondria and lipid droplets, whereas these constituents are relatively sparse in the more yolky region (Clement and Lehmann 1956a). When the polar lobe appears, it contains much yolk.

With centrifugal force of around 2000 *g*, the egg contents are readily stratified into three zones: (1) a centripetal oil cap, (2) a rather granular 'clear' zone containing the nucleus, and (3) the centrifugal yolky zone. It is possible to achieve a further separation into five zones: (1) the oil cap; (2) a hyaline zone which contains the nucleus and which appears homogeneous and structureless in the living egg; (3) a mitochondrial band; (4) a second clear zone which contains a few small yolk granules and appears in fixed preparations to be different in nature from the upper clear zone; and (5) the yolk zone (Clement 1935). It would be of considerable interest to compare the upper and lower clear zones with the electron microscope, but this has not been done.

The fine structure of the polar lobe of *Ilyanassa* at the trefoil stage was described by Crowell (1964). The most prominent inclusions of the lobe were the yolk particles. Found among them were elements of endoplasmic reticulum, double membrane vesicles, ribonucleoprotein particles, and mitochondria. Yolk particles were found to consist of a central homogeneous core, an outer granular zone, and a limiting membrane. The double membrane vesicles, of unknown composition and significance, were found in the polar lobe at first cleavage, or before the trefoil stage of first cleavage, in the vegetal region of the egg, but not in the animal region. Mitochondria were less abundant in the polar lobe than in other regions. The cortical zone of the polar lobe was found to be narrower than the corresponding zone of the animal region of the egg, and its cytoplasm was denser. Sparse, short microvilli were described for the polar lobe surface, in contrast to numerous long microvilli for the more animal regions. The cortex of the egg and polar lobe contained many RNP particles.

7.4. Development from deposition of the egg to veliger larva

7.4.1. Nuclear events and polar lobe formation during maturation and first cleavage

Both Morgan (1933) and Cather (1963) have reported observations on the *Ilyanassa* egg between the time of laying and first cleavage, and the latter author has constructed a detailed time schedule of the sequence of nuclear events and associated form changes of the egg during the period of meiosis and early cleavage. Much of the following account is taken from Cather's study; the time schedule refers to eggs maintained at 23 ± 1 °C.

Although the freshly laid egg may sometimes show an intact germinal vesicle, usually the first polar spindle is forming deep in the cytoplasm of the animal third of the egg, and a sperm has entered. In other cases the spindle is already at the surface. Under adverse culture conditions females may retain the eggs until they have reached much later stages. Hence careful examination is necessary to establish the stage of the eggs within a particular capsule. The eggs within a capsule will usually be at approximately the same stage. Dispermic eggs are occasionally encountered, and also unfertilized eggs. The latter occur in increasing numbers toward the end of the breeding season. The germinal vesicle of the unfertilized egg has a large and prominent nucleolus. The diameter of the spherical egg is approximately 166 μ.

As the first meiotic spindle of the fertilized egg moves to the surface at the animal pole, a brief and rather inconspicuous bulge at the vegetal pole comprises the first polar lobe. The first polar body appears, and within a few minutes the egg is again spherical and remains so for about 15 min. The second polar lobe then appears, just prior to the metaphase stage of the second meiotic division; it becomes large and conspicuous and persists altogether for about 50 min. The second polar body is given off approximately 20 min after the appearance of the second polar lobe, and about this time the lobe shows irregularities of outline not seen at other stages. The irregularities disappear and the resorption of the lobe is followed by another spherical period.

Following the second meiotic division the sperm nucleus, previously small and dense, enlarges and assumes a characteristic interphase condition. It then migrates to a position beneath the egg pronucleus, having meanwhile entered the prophase condition, as has the egg pronucleus. The third polar lobe makes its appearance during late prophase, while the asters are enlarging,

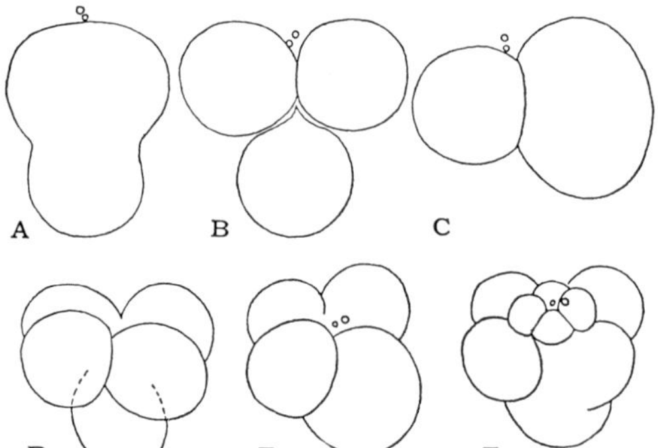

Fig. 1. Early cleavage of the *Ilyanassa* egg, showing the polar lobe. (From Clement, A. C., 1952, J. Exptl. Zool. *121*, 597, figs. 1–6.)

and before the pronuclear membranes have disappeared. The interval between the disappearance of the second polar lobe and the appearance of the third is about 55 min. The third polar lobe (fig. 1A) enlarges to a prominent condition during metaphase. During anaphase the animal pole flattens and the egg elongates in the plane of the cleavage spindle. As the cleavage furrow forms the lobe is also further constricted from the egg, until the trefoil stage is reached (fig. 1B). At trefoil the egg consists of three spheres of roughly the same size, namely, the two blastomeres and the polar lobe. A fine cytoplasmic filament may be seen extending from the upper end of the polar lobe to the notch at the bottom of the two blastomeres, but it is not evident at this time with which of the blastomeres the lobe will eventually fuse. The relative size of the third polar lobe varies in different lots of egss from about one-quarter to one-third the volume of the whole egg. It is easy to distinguish the polar lobe from a blastomere, even after they have been separated from one another. The polar lobe is very yolky in appearance, whereas the blastomeres are recognizable by a lighter upper polar area, where the nucleus and the largely yolk-free cytoplasmic cap occur. The trefoil stage occurs about three hours after the time of laying and the beginning of meiosis. It soon becomes apparent that the lobe is attached to one of the blastomeres; the connection broadens and the lobe is gradually incorporated into the blastomere. In this way an unequal first cleavage is achieved; the larger blastomere which has received the lobe is known as CD, and the smaller one as AB (fig. 1C).

For a detailed history of karyokinesis and cytokinesis during the period of meiosis and cleavage, the elegant monograph of Conklin (1902) on *Crepidula* should be consulted. The details are similar in *Ilyanassa*, according to Cather (1963), except for differences related to polar lobe formation. In *Crepidula* the first cleavage appears to be equal, although Conklin (1902) has described a small cytoplasmic lobe, probably homologous with the polar lobe of *Ilyanassa*, which attaches to and is resorbed into one of the first two blastomeres.

7.4.2. Development from first cleavage to larva

The early cleavage of *I. obsoleta*, through the initial subdivisions of the mesentoblast cell, was described in detail by Clement (1952). No comprehensive account of later development has been published. The veliger larva has been figured by Clement (1952, 1962, 1967) and others. Collier (1965a) has outlined the stages of development from the recently fertilized egg to the veliger larva. Cather (1967) has dealt with the development of the shell gland and shell. The development of two related European forms has been described in a general way (see Bobretzky 1877, on *Nassa mutabilis*, and Pelseneer 1911, on *N. reticulata*). There is sufficient similarity in both the early cleavage history and the larval anatomy of *Ilyanassa* and *Crepidula* to make Conklin's (1897) classic work on the latter form a useful aid in the study of *Ilyanassa*.

As mentioned, the polar lobe passes into the CD cell of *Ilyanassa* at first cleavage, and accounts for the larger size of this blastomere. The lobe reappears for the fourth time prior to the second cleavage, and passes into the D blastomere of the 4-cell stage (figs. 1D, E); the fourth lobe is of about the same size as the third.

The first quartet of micromeres (1a–1d) results from a dexiotropic cleavage (figs. 1F, 2A). The 1d micromere is noticeably smaller than the other three, even though 1D is the largest of the four macromeres (fig. 2A). The second

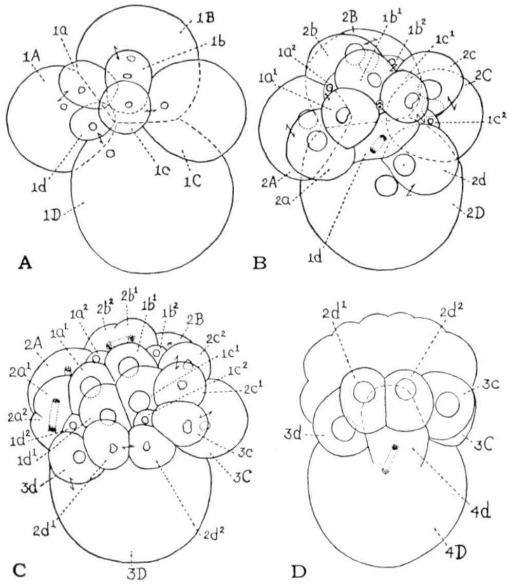

Fig. 2. Normal cleavage stages in *Ilyanassa*. (From Clement, A. C., 1952, J. Exptl. Zool. *121*, 599, figs. 7–10.)

quartet is formed by a leotropic division, and all cells of this group (2a–2d) are of similar size and appearance (fig. 2B). Soon after the appearance of the second quartet cells, those of the first quartet subdivide to produce the turret cells ($1a^2$–$1d^2$), which are much smaller than their sister cells ($1a^1$–$1d^1$)

lying nearer the animal pole. The smaller 1d micromere subdivides later than the other three (fig. 2B). The third quartet cells (3a–3d) arise from a dexiotropic division of the macromeres, and are uniform in appearance; at about the same time the second quartet cells undergo a division (fig. 2C). There are now 24 cells present. The next cells to divide are the apical cells $1a^1$–$1c^1$. Soon thereafter the mesentoblast cell, 4d, makes its appearance (fig. 2D). The interval between first cleavage and the appearance of 4d is about 6–6½ hr at 25–26 °C. The entoblastic members of the fourth quartet (4a–4c) appear about 3 hr after the formation of 4d. Meanwhile 4d will have divided equally in the plane of bilateral symmetry of the embryo, and the resulting cells will have begun the subdivisions which eventually segregate some entoblastic cells and form the primary mesoblast bands (fig. 3A–D).

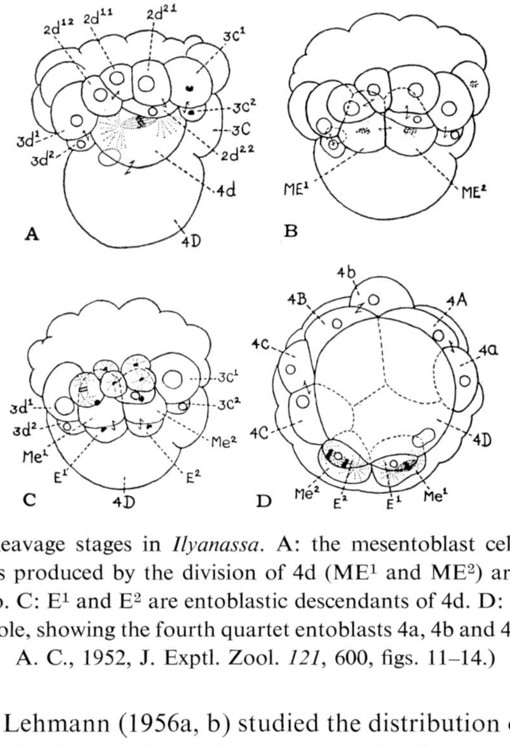

Fig. 3. Normal cleavage stages in *Ilyanassa*. A: the mesentoblast cell, 4d, is about to divide. B: the cells produced by the division of 4d (ME^1 and ME^2) are on the posterior side of the embryo. C: E^1 and E^2 are entoblastic descendants of 4d. D: later stage, viewed from the vegetal pole, showing the fourth quartet entoblasts 4a, 4b and 4c. (From Clement, A. C., 1952, J. Exptl. Zool. *121*, 600, figs. 11–14.)

Clement and Lehmann (1956a, b) studied the distribution of lipid droplets and mitochondria during the early cleavages. In the uncleaved egg both types of inclusion were found mainly in a broad cytoplasmic cap in the animal hemisphere. There appeared to be an equal distribution of lipid and mitochondria to the first four blastomeres. Numerous mitochondria were distributed to the cells of the micromere quartets, but the macromeres 4A–4D appeared to receive relatively few. The distribution of lipid droplets follows a somewhat different pattern. The first quartet micromeres receive relatively few, whereas the second and third quartets receive increasing quantities. Most of the lipid of the 3D macromere was found to go into the 4d cell, but at the subdivision of 3A–3C the distribution was more even. A few yolk particles are found in 4d, but not as many as in 4a–4c.

While the precise relationships of the early cleavage furrows to the future embryonic axes have not been worked out in *Ilyanassa*, we may assume that the general relationships are the same as in other spirally cleaving mollusk eggs. Thus 2d and 4d lie in the median plane, and 4d is at the posterior end. As viewed from the animal pole of the cleaving egg, the A quadrant would lie to the left of the median plane, and the C quadrant to the right. Raven (1958) has discussed the axial relationships of the molluscan egg and embryo in some detail.

Gastrulation is by epibole. At summer temperatures, the embryo is ciliated and rotating at the end of the second day, an early veliger stage has been reached at five days, and hatching takes place about seven to nine days after laying. The relationship of temperature to larval development has been studied by Scheltema (1967).

The development of the shell gland has been studied in detail by Cather (1967). In embryos reared at $20 \pm 1\,^{\circ}\mathrm{C}$, the cells which are to form the shell gland have become columnar by 76 hr and form a flat plate lying along and to the left of the dorsal midline of the embryo, at its posterior end. Invagination of the shell gland begins at about 80 hr, but evagination later takes place. The deposition of calcium carbonate starts just before or during the first stage of torsion, at about 120 hr. By carbon marking experiments, Cather determined that the shell gland normally arises from derivatives of the 2d micromere, but 2c derivatives are later incorporated into the mantle. Collier and McCann-Collier (1964) have described the position of the developing shell gland with respect to the entodermal cells.

By the time of hatching of the veliger, the yolk has been largely or entirely absorbed, and various organs can be clearly seen in the living larva. Among the prominent parts are the paired velar lobes and eyes, the foot with a pair of statocysts and an operculum, the larval shell, the digestive tract, and a beating heart. Some of these details are shown in fig. 4D.

Scheltema (1962) has reared the larvae of *I. obsoleta* through metamorphosis in laboratory culture, feeding them on phytoplankton; according to him the velum is cast off *in toto* at metamorphosis.

7.5. Experimental and biochemical studies

7.5.1. Embryological studies

In the early days of experimental embryology, H. E. Crampton (1896) discovered that removal of the polar lobe of *Ilyanassa* at the trefoil stage of first cleavage prevents the appearance of a characteristic mesoblastic pole cell and the formation of the mesoblast bands. After detachment of the polar lobe, the blastomeres of the 2- and 4-cell stages were of equal size. Successive quartets of micromeres were produced, but the fourth derivative of the D quadrant (4d) lacked the characteristics of the typical mesoblast cell, and instead resembled the 4a, 4b and 4c entoblastic cells of the other quadrants. The lobeless embryos lived long enough to develop cilia and swim around, but they showed little differentiation under the culture conditions employed.

Crampton's result emphasized the importance of cytoplasmic factors in differentiation, although he was careful not to say that the polar lobe contained prelocalized mesoblast material; he suggested on the contrary that the presence of the yolk mass in the D macromere may be the stimulus which causes this cell to act differently from the other three macromeres. As we shall see later, however, the yolk mass cannot be regarded as the significant morphogenetic factor.

It was not until many years later that the morphogenetic value of the polar lobe of *Ilyanassa* was more fully analyzed (Clement 1952). Through the use of pasteurized sea water as a culture medium, it became possible to keep lobeless embryos alive for 10 or 12 days, whereas only five or six days are necessary for whole eggs to produce veliger larvae under the same conditions. It was thus possible to learn which larval features are lobe-dependent, and which are able to develop in the absence of the polar lobe.

Before discussing the later stages of development, it should be noted that the polar lobe plays a more extensive role in determination of the cleavage pattern than Crampton realized. A careful, step-by-step comparison of cleavage in the lobeless and the normal embryo revealed a further difference between the two as early as the first quartet of micromeres (Clement 1952). In the normal egg the first micromere derivative of the D quadrant (1d) is smaller than the other three, whereas in the lobeless egg all first quartet micromeres are of the same size. Thus the presence of the lobe material in the D blastomere in some unknown manner causes this cell to produce a smaller micromere than would otherwise be the case. In a number of other ways the presence of the lobe material was shown to modify the cleavage pattern in the D quadrant. It was found, for example, that the rate of subdivision in the 1d line lags, in the normal egg, as compared with that of the comparable cells in other quadrants; and that the tip cell of the molluscan cross in the D quadrant ($2d^{11}$) is much larger than the corresponding cells ($2a^{11}$, $2b^{11}$, $2c^{11}$) of other quadrants. All of these distinguishing features of the D quadrant disappear in the lobeless egg, and the four quadrants cleave in the same manner. In the case of the fourth quartet, the 4d cell (properly designated the mesentoblast cell) is not only distinctive in appearance and history in the normal egg, it also arises about three hours ahead of the entoblastic fourth quartet cells (4a, 4b, 4c) of the other quadrants (at $\pm$ 25°C). In the lobeless embryo all of the fourth quartet cells are yolky and similar in appearance, as Crampton discovered, and they also appear synchronously. It is thus evident that the presence of the lobe material confers special characteristics on the 4d cell, and causes the cleavage which produces it to occur precociously by about three hours. It is very difficult to imagine what properties of the lobe material could result in such a variety of cleavage effects.

The mesentoblast cell (4d) normally marks the posterior pole of the future embryo. Since a typical mesentoblast cell is not produced if the polar lobe is removed at first cleavage, and the lobeless embryo does not show any clear indication of an antero-posterior axis in other respects, it appears that the lobe is associated, either as cause or effect, with the establishment of the posterior pole.

It would be of great interest to know whether the posterior end of the future embryo is predetermined before the first cleavage, or is established at that time. In the first case, it would already be predetermined with which blastomere the polar lobe should fuse at the close of the first cleavage; in the second case the posterior pole of the embryo would be established by the chance fusion of the lobe with one or the other of the blastomeres. There is at present a lack of sufficient evidence on which to judge this question in *Ilyanassa*. Morgan (1936) sought an answer by separating one blastomere from the trefoil figure with a glass thread, or by pricking and destroying it with a glass needle; if it is already predetermined which cell the lobe is to fuse with, one would expect the lobe and the remaining blastomere to fuse in about half of the cases. Both blastomeres usually disintegrated in Morgan's experiments, however, and he was left with too few cases for a definite conclusion.

Under suitable culture conditions, the lobeless embryo differentiates certain larval tissues and parts, but fails to develop a great many others

(Clement 1952, 1962). The parts which typically fail to develop may be spoken of as lobe-dependent structures. These include eyes, foot, statocysts, operculum, external shell, heart and intestine. Present are velar cilia, small cilia, an everted stomodeal structure, some differentiation of entodermal tissue, pigment, and active muscle fibers. Frequently present also are one or more internal masses of birefringent shell material. Thus, while shell-secreting tissue develops, it does not form a typical, external shell. In general appearance, the lobeless larva bears little resemblance to the normal veliger (fig. 4). The velar tissue commonly occurs in several masses and lacks the

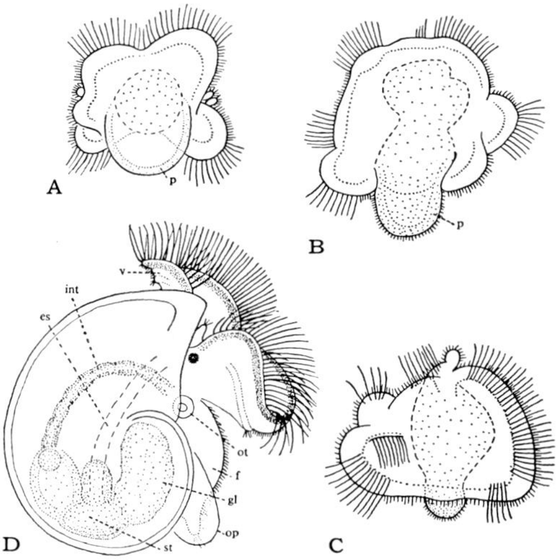

Fig. 4. Comparison of lobeless and normal larvae of *Ilyanassa*. A, B and C: lobeless larvae. p: everted stomodeum. The broken line outlines entoderm in the lobeless larvae. D: normal veliger. es: esophagus; f: foot; gl: digestive gland; int: intestine; op: operculum; ot: statocyst; st: stomach; v: velum. (From Clement, A. C., 1952, J. Exptl. Zool. *121*, 611, figs. 27–30.)

organization it assumes in the normal veliger. The stomodeal evagination appears to mark approximately the original vegetal pole of the egg. It would seem that both the morphogenetic movements and the differentiation of parts which serve to delineate the antero-posterior axis of the normal larva fail to occur. Since lobeless larvae develop at about the same rate as the controls, attain a rather uniform structural pattern, and remain healthy and vigorous for some days after the controls are fully developed veligers, it is apparent that the loss of the polar lobe material results in the failure of certain developmental processes and not of others.* In the absence of the lobe material, much of the genetic potential remains unrealized.

The development of isolated blastomeres of the 2- and 4-cell stages demonstrates that the morphogenetic influence of the polar lobe is allocated to the CD blastomere at first cleavage, and to the D blastomere at second cleavage (Clement 1956). Isolated CD blastomeres develop into larvae which, although incomplete in some respects, may show such parts as eyes, foot and external shell (fig. 5B), whereas AB blastomeres give partial larvae lacking the lobe-dependent features (fig. 5A). Only the isolated D blastomere of the

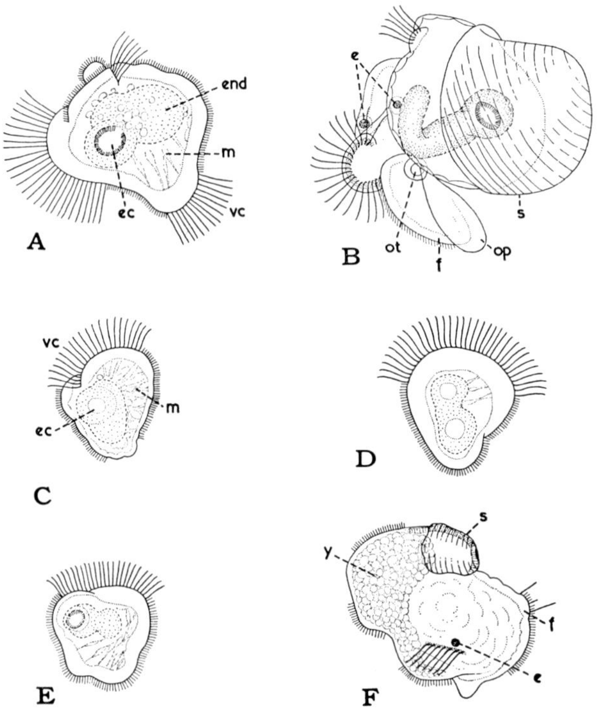

Fig. 5. Partial larvae of *Ilyanassa*, from isolated blastomeres. e: eye; ec: enteric cavity; end: entoderm; f: foot; m: muscle; op: operculum; ot: statocyst; s: shell; vc: velar cilia; y: yolk. A: larva from AB half. B: larva from CD half. C: larva from A blastomere. D: larva from B blastomere. E: larva from C blastomere. F: larva from D blastomere. (From Clement, A. C., 1956, J. Exptl. Zool. *132*. A, C, D, E, F from figs. 1, 2, 3, 4, 6, p. 431; B from fig. 7, p. 433.)

* A more detailed analysis of the structure of the lobeless larva has recently been completed by Atkinson (1968).

4-cell stage differentiates lobe-dependent features. Although D blastomeres in general tend to develop poorly, rare cases may differentiate an eye and a small external shell (fig. 5F). All classes of isolated blastomeres from the 4-cell stage may differentiate velar cilia, pigment and entoderm. The general appearance of the partial larvae from A, B and C quarters is similar (fig. 5C, D, E). The development of isolated one-quarter blastomeres thus indicates that the morphogenetic influence of the polar lobe accompanies the physical bulk of the structure into the D macromere at the second cleavage.

It is largely during the period of micromere formation that the morphogenetic influence of the polar lobe region is exerted (Clement 1962). If the large 4D macromere, which contains the bulk of the yolky material originally contained in the polar lobe, is removed from the embryo following the stage of formation of the 4d micromere, a small but well-formed veliger develops. By this stage the D macromere is of nutritive significance only. A series of deletion experiments in which the D macromere was removed at successive cleavage stages gave evidence that the lobe-dependent structures are determined mainly during the period of the appearance of the third micromere quartet and the mesentoblast cell, 4d. Some structures may be partially determined in earlier stages. When the D blastomere was removed at the four-cell stage, the ABC embryos produced deficient partial larvae, but there was a very low incidence of cases which may have shown eye differentiation. Larvae of about the same type developed from ABC + 1d combinations (obtained by destroying the 1D macromere); in one case the identification of a lens made positive recognition of an eye possible. Most of the larvae arising from ABC + 1d + 2d combinations (obtained by destroying the 2D macromere) resembled those of the two preceding groups, but occasional cases developed a small external shell. In these cases there must already have occurred a weak or partial determination of external shell. When removal of the D macromere was delayed until the third quartet micromeres were present, the operated embryos (ABC + 1d + 2d + 3d) frequently showed a great gain in differentiation capacity as compared with earlier members of the series. There was considerable individual variation, but in the best cases there was good bilateral organization of the velum, a foot with one or two statocysts, a pair of eyes, and a well-formed shell. It is apparent therefore that a number of important larval features are determined during or subsequent to the formation of the third quartet, but before the appearance of 4d. There was some indication in the experiments that better differentiation was obtained if the removal of 3D was delayed as long as possible – that is, until shortly before the time for the appearance of 4d. This suggested the possibility that the gain in differentiation capacity of ABC + 1d + 2d + 3d embryos, as compared with the ABC + 1d + 2d type, might be less a matter of the presence of the extra cell, 3d, than of the association for a longer period of time of the other cells with the 3D macromere. That is, the determination of eyes, velum, foot and shell might be effected through an inductive action of the D macromere which was in progress at this time. This idea is supported by the consequences of deleting the 3d micromere: this operation results in absence of the left statocyst and the left half of the foot, but eyes, velum, shell, and other parts may develop well. A comparable deficiency, but affecting the right statocyst and right half of the foot, is produced by deleting the 3c micromere.

The best ABC + 1d + 2d + 3d embryos failed to develop a heart and intestine; these structures appear to require the presence of the 4d micromere.

Judging from deletion experiments, most of the micromeres have rather specific embryonic values (Clement 1963, 1967). Each of the micromeres of the first, second and third quartets, and also 4d, has been removed by destroying it with a glass needle.

In the case of the first quartet, removal of 1a led to absence of the left eye and a slight reduction in size of the left velar lobe, whereas removal of 1c led to comparable deficiencies of the right eye and right velar lobe. Removal of 1b tended to reduce the distance between the eyes. Evidently this micromere contributes to the ectoderm which normally lies between the eyes. The removal of 1d did not appear to produce any specific defect.

Loss of each of the second quartet micromeres was followed by a different result. Removal of 2a affected the development of the left velar lobe, and frequently, also, the development of the left eye and left statocyst. The loss of 2b resulted in little noticeable effect. Removal of 2c resulted in severe defects, including reduction in size of the shell, absence of the heart, and eversion of the stomodeum. Loss of 2d led to a reduction in size of the shell, or its absence.

As mentioned earlier, Cather (1967) determined through carbon-marking experiments that 2d micromere derivatives form the shell gland, and 2c derivatives are incorporated into the mantle. Removal of either 2d or 2c was found by Cather to interfere with development of the shell, and removal of both resulted in its absence.

In the case of the third quartet micromeres, removal of 3a or 3b produced relatively slight effects (Clement 1963). Loss of 3a resulted usually in a small reduction in size of the left velar lobe, and loss of 3b tended to reduce slightly the size of the right velar lobe. As already mentioned, the 3d and 3c micromeres are concerned with the development of the statocysts and foot, 3d being necessary for development of the left statocyst and left half of the foot, and 3c for the right statocyst and right half of the foot.

Removal of the mesentoblast cell 4d was found to result in visceral abnormalities, including absence of the intestine and heart (Clement 1960). However, the velum, eyes, foot, statocysts, stomodeum, esophagus and shell could show normal or nearly normal development. These results thus provide no indication that the mesoblast bands, derived from 4d, play a role as an inductor or organizer of ectodermal derivatives. The absence of the heart in *Ilyanassa* after removal of either 2c or 4d suggests an important contribution by each of these cells to this structure.

The results of blastomere isolation and micromere deletion in *Ilyanassa* indicate an early specification of cell values, with some limited capacity of the embryo for regulation. It is likely, for example, that the left eye is normally derived from the A quadrant and the right eye from the C quadrant. Yet isolated D blastomeres may develop an eye, and isolated CD blastomeres sometimes develop a pair of eyes. In isolation, therefore, these cells would appear to be capable of exceeding their prospective fate. Although the eyes probably arise normally from the A and C quadrants, an ABC blastomere combination will not usually differentiate eyes. For an embryo to develop eyes the large D cell with its polar lobe component must usually be present through the time of origin of the third quartet of micromeres. Removal of the 1d, the 2d, or the 3d micromere does not prevent development of the eyes, however. These observations suggest an inductive role for the polar lobe material in the development of eyes. That is, the polar lobe, although segregated to the D quadrant at second cleavage, appears to provide some essential factor or influence for the differentiation of the eyes, even though the evidence indicates that these structures arise in other quadrants. Occasional instances of eye differentiation in ABC, ABC + 1d, or ABC + 1d + 2d larvae might be indicative of an unusually early inductive effect of the D quadrant in those cases.

The differentiation of the shell in *Ilyanassa* involves complex relationships. Removal of the polar lobe at first cleavage prevents the formation of an external shell; yet atypical, internal masses of birefringent shell material may be secreted by the lobeless embryo. These have been observed also in isolated AB halves and A, B and C quarters. When it is recalled that an isolated D quarter may produce an external shell, it is seen that every quadrant has the capacity to secrete shell material. Yet Cather's (1967) evidence indicates that the shell normally arises from the 2d and the 2c micromeres. Hence under normal circumstances the A and B quadrants are presumably inhibited from differentiating shell material, as Cather has pointed out. In further experiments, Cather finds that the isolated ectoblastic component of an embryo does not differentiate shell material, either externally or internally; however, if any one of the macromeres 3A, 3B, 3C, or 3D is left in combination with the ectoblast, the isolate is able to form shell material. Also, if a detached polar lobe is held in contact with the underside of the isolated ectoblast of an embryo for six or seven days, the ectoblast may develop shell material. The nature of the influence exerted on the ectoblast by the macromere, or by the polar lobe held in contact with it, is unknown.

A polar lobe detached from the remainder of the egg at the trefoil stage does not cleave (it has no nucleus), but it does undergo rhythmic changes in form (Morgan 1933, 1935c). These form changes, during which the lobe may become constricted, were found by Morgan to occur three or more times, with an intervening spherical phase. The form changes occurred more slowly than the cleavages of the controls. Morgan (1935a) noted also that vegetal fragments of the centrifuged egg which lacked chromatin, or contained chromatin of the egg nucleus but not of the sperm nucleus, formed a polar lobe even though they did not cleave. Animal hemisphere fragments, however, formed no polar lobe even though a full complement of chromatin was present, and they showed equal first and second cleavages. Thus lobe formation is a property of the vegetal region of the egg, and does not require the presence of a nucleus or division figure. While lobe formation normally takes place at the time of certain cell divisions, it appears to be to some extent an independent phenomenon.

By preventing eggs from orienting as usual under centrifugal force, Morgan (1933, 1935b) was able in some cases to displace yolk from the vegetal polar region, and to move other materials into the area. The polar lobe continued to form at or near the vegetal pole however, even though its interior contained unusual materials, or an unusual combination of yolk and other materials. This suggested to Morgan that the cause of lobe formation was to be sought in surface layers not disturbed by centrifuging, rather than in the interior.

The analysis was carried further by Clement (1968), who first centrifuged eggs in an inverted position (in a chilled gelatin-sea water gel), and then broke them into light and heavy fragments by recentrifuging in a raffinose-sea water solution. The light halves of inverted eggs would contain the vegetal polar region, the nuclear material, oil cap, and cytoplasmic zone, but very little yolk. Partly inverted eggs might contain only part of the vegetal polar area. The light halves of inverted eggs were found to form a polar lobe of approximately the correct relative size, to cleave unequally in the usual pattern, and when reared to the larval level, to produce, in about 48 % of the cases, lobe-dependent structures; about 17 % developed into recognizable veliger larvae. When eggs were broken into fragments by centrifuging in raffinose without a prior inversion treatment, the nucleated light fragments represented the animal hemisphere of the egg, since *Ilyanassa* eggs, if permitted to do so, will orient under centrifugal force with the animal pole toward the centripetal end. Such nucleated animal halves had approximately the same visible composition (oil, yolk, etc.) as the nucleated halves of inverted eggs, but as a rule they cleaved equally without forming a polar lobe; about 96 % of the larvae arising from them lacked lobe-dependent structures. The exceptional cases may have contained a portion of the vegetal hemisphere.

From these results it appears most unlikely that the yolk particles, which normally occupy most of the volume of the polar lobe, have anything to do with the lobe's morphogenetic role. The latter must depend on some property of the vegetal polar area which can withstand centrifugal force sufficient to displace the coarser cytoplasmic inclusions, and it very probably resides in the cortical region.

If *Ilyanassa* eggs are compressed before first cleavage they may be made to divide equally (Morgan 1936). Styron (1967), through the use of a special apparatus, was able to exert pole-to-pole compression. The compression was begun before the appearance of the lobe which precedes first cleavage, and was continued until the polar lobe of compressed eggs began to pass to the CD blastomere. Aside from cases which cytolyzed or did not cleave, three classes of compressed eggs were recognized: (1) those which formed a polar lobe and cleaved unequally; (2) those which cleaved unequally without forming a polar lobe; and (3) those which formed no lobe and cleaved equally. The first two classes developed normally, in a majority of instances. In the third class, where first cleavage was equal, a polar lobe was observed on both blastomeres at second cleavage, and the division resulted in two large and two small blastomeres. It appears probable that in these cases each of the first two cells behaved, in effect, as a CD blastomere at the second cleavage. The later development of such eggs was greatly inhibited; in general, their products resembled lobeless larvae, but were even less well developed. From Styron's experiments it appears that dividing the lobe material between the first two blastomeres results in a greater disturbance of development than removal of the lobe entirely. Styron considers the inhibition of development following equal first cleavage to result from a derangement of the cortical pattern of the egg and of ooplasmic segregation. It would be interesting to isolate the first two blastomeres of eggs which had undergone equal first cleavage, to see whether their capacity for differentiation was improved by their separation.

The effect of chemical agents on the development of *Ilyanassa* has been the subject of a few preliminary reports. Morrill (1961a) treated eggs and embryos with cobaltous chloride and produced eye malformations and other defects. The same author (Morrill 1961b) induced eye anomalies with lithium chloride treatment; these included absence of an eye on one side or the presence of supernumerary eyes on one or both sides of the larva. Supernumerary eyes were found twice as frequently on the right side as on the left.

Feigenbaum and Goldberg (1965) found that actinomycin D applied during early stages of development permitted cleavage and the formation of a ciliated embryo, but inhibited later development. Nuclei of treated embryos showed irregular distribution of chromatin and atypical nucleolar features.

Collier (1966), in an effort to relate the time of gene transcription to the differentiation of particular structures in *Ilyanassa* has studied the effect of exposing embryos of different ages to actinomycin D. So far, only a preliminary account of this work has appeared. Treatment at any stage prior to and including the third day of development blocks organ formation. If embryos older than three days are treated with concentrations of 25 μg/ml for 5 or 6 hr and then returned to sea water for further development, the degree of differentiation varies according to the stage of treatment. Thus embryos treated when 4 days old formed a nearly normal foot, an operculum, and a much reduced velum; other structures were largely suppressed. If treatment occurred 24 hr later, there was in addition partial differentiation of statocysts, shell, esophagus and intestine, and complete differentiation of eyes; but certain structures (stomach, digestive gland, heart) failed to appear. On the basis of these results, Collier has set up a time schedule for gene transcription in *Ilyanassa*. In the case of eyes, for example, treatment with actinomycin D at 4 days virtually suppressed eye development, whereas embryos treated at 5 days of age developed a normal pair of eyes; the eyes did not make their appearance until 6½–7 days of development, however. It is not clear how these results are to be related to those obtained by Clement (1952, 1962), who found that most of the principal organs of the *Ilyanassa* veliger (including the eyes) are dependent upon the presence of the polar lobe region, but only through certain rather early cleavage stages. Perhaps we have evidence here of a step-wise process of embryonic determination: an initial step represented by the action of the polar lobe influence during the first day, and a later step involving DNA-dependent RNA synthesis a day or so preceding appearance of the organ in question. The concentration of actinomycin D used by Collier in the experiments mentioned (25 μg/ml) appeared to decrease DNA and protein synthesis, as well as RNA synthesis, but to a lesser extent. However, a concentration (10 μg/ml) which decreased RNA synthesis, but not DNA and protein synthesis, was reported to affect the differentiation process. While the evidence does not demonstrate that the inhibiting effects of actinomycin D on morphogenesis can be attributed principally to a suppression of messenger RNA production, it would appear to be consistent with this interpretation.

By centrifuging eggs prior to the second maturation division Clement (1935) was able to force the meiotic spindle away from the animal pole. After removal from the centrifuge some of these eggs formed a giant second polar body. In some cases the giant polar body, rather than the egg proper, later underwent cleavage, presumably as a result of the inclusion of the sperm nucleus in the polar body. Morgan (1937) concluded, from similar experiments, that stretching of an egg, or of an egg fragment, is a contributing factor to the division brought about by the displaced maturation spindle.

The behavior of the cell surface during the second maturation division and the first cleavage was studied by Dan and Dan (1942) by observing the distance between attached kaolin particles. Their observations indicated that during the second maturation division the animal polar region shrinks while the polar lobe region stretches, and that these changes are maintained during the succeeding spherical stage. There thus appeared to be a shift of surface material from the vegetal area of the egg to the animal area during the second maturation division. Whether a similar shift occurs during the first maturation division was not investigated.

The delay in cleavage induced by X-radiation administered at different stages during the mitotic cycle was studied by Cather (1959). The effect of several metabolic inhibitors on cytoplasmic viscosity, surface rigidity and cleavage of the egg was investigated by Butros (1956). The egg of *Ilyanassa* continues to cleave in the presence of cyanide (Clement 1940).

7.5.2. Biochemical studies

Since most of the biochemical work on *Ilyanassa* has been covered in reviews by Collier (1965a, 1966), only the main findings will be mentioned here.

An electrophoretic analysis of hydrolytic enzymes occurring at successive stages of development in *Ilyanassa* was made by Morrill and Norris (1965). Twenty-six electrophoretically mobile enzymic bands were found. Seven bands occurred throughout development, 6 were detected only in extracts of embryos up to 4 days of age, and the remaining 13 appeared on and after the 4th day of development. During the period of differentiation of the organ primordia, between the 3- and 5-day stages, 5 bands disappeared and 6 others appeared.

Much of the biochemical work on *Ilyanassa* has been concerned with the composition or activity of the polar lobe and the role of the lobe in development.

In a study of alanylglycine dipeptidase activity during the development of *Ilyanassa*, Collier (1957) compared the activity of this enzyme in the whole egg, isolated AB, CD, C, and D blastomeres, and in the polar lobe isolated at first cleavage. When the volume of hyaline protoplasm in each of these entities was determined, it was found that the activity of the enzyme was proportional to the volume of the hyaline protoplasm. Hence no special relation was found between the distribution of the enzyme and the segregation of the lobe material at the first and second cleavages. The volume of hyaline protoplasm determined for each of the entities mentioned, expressed as percentage of the amount of hyaline protoplasm in the whole egg, was as follows: polar lobe, 16.5; lobeless egg, 84.2; AB blastomere, 42.1; CD blastomere, 57.8; C blastomere, 20.3; D blastomere, 39.8.

As noted earlier the polar lobe is rich in yolk particles, and has relatively few lipid droplets and mitochondria. In these respects it resembles the lower hemisphere of the whole egg. The principal biochemical distinction of the polar lobe so far indicated is a relatively higher content of nucleic acid precursors than is found in other regions of the egg. Berg and Kato (1959) found isolated polar lobes to contain 75% more acid soluble material with an absorption spectrum typical for purine and pyrimidine compounds than the remaining egg cytoplasm, on a per unit volume basis. At the 2-cell stage, the CD blastomere was found to have 47% more acid soluble material than the AB blastomere per unit volume, a difference which could be accounted for mainly by transference of this material by the polar lobe. On the basis of paper chromatography the acid extracts were tentatively regarded as containing mainly polynucleotides and lesser amounts of nucleotides. A parallel between the distribution of yolk and acid soluble material suggested an association of one or more polynucleotides with yolk. Collier (1960a) found a significantly greater concentration of total, acid-soluble, and phospholipid phosphorus in the polar lobe and CD blastomere than in the whole egg or AB blastomere. On the basis of these and other findings Collier (1965a) has suggested the possibility that the failure of the lobeless embryo to undergo normal differentiation is the result of a deficiency in nucleic acid precursors, and that this deficiency produces a selective repression of informational RNA synthesis. If this is the case, it appears unlikely that the significant nucleic acid precursors are associated with the yolk of the polar lobe, since nucleated vegetal fragments of the egg from which most of the yolk has been removed can form a polar lobe and differentiate lobe-dependent structures (Clement 1968).

Collier and McCann-Collier (1962) have reported the DNA content of the egg and sperm of *Ilyanassa*, and Collier (1963) has investigated the synthesis of DNA in the embryo.

Using quantitative microchemical methods, Collier (1960b) found the concentration of RNA to be higher in the AB than in the CD blastomere. The concentration of RNA in the polar lobe and the CD cell was estimated to be approximately the same. The possibility remains, however, as Collier (1965a) has pointed out, that the RNA of the lobe may be of qualitative significance.

Collier (1961a) has studied the synthesis of RNA and protein in the normal *Ilyanassa* embryo, and subsequently (Collier 1961b) the effect on protein synthesis produced by removal of the polar lobe. Protein synthesis was judged by the incorporation of C^{14}-leucine. No demonstrable synthesis of RNA or protein occurred during the first two days, but active synthesis of both set in before the appearance of morphological differentiation. Removal of the polar lobe at first cleavage was found to interfere with the embryo's capacity to incorporate C^{14}-leucine into protein, and it was concluded that the decreased rate of protein synthesis in the lobeless embryo could not be accounted for by its smaller size as compared with the normal embryo.

An RNA component with a base composition similar to that of DNA, and judged to be messenger RNA, has been separated by Collier (1965b) from *Ilyanassa* embryos undergoing active differentiation (5 days old, reared at 19 °C). According to this author, the synthesis of messenger RNA has also been observed in advance of organogenesis in the 3-day embryo.

Davidson et al. (1965) report evidence that removal of the polar lobe at first cleavage in *Ilyanassa* results in a diminished rate of RNA synthesis which is detectable by 18 hr after first cleavage. This was interpreted as suggesting that the polar lobe cytoplasm may mediate gene activation during early embryogenesis.

Clement and Tyler (1967) found that polar lobes of *Ilyanassa*, detached at first cleavage, are capable of incorporating labeled amino acid into protein, and that this ability persists for at least 24 hr after isolation. The results indicate the presence of long-lived messenger RNA in the polar lobe, and suggest that the morphogenetic influence of the lobe may be correlated with its ability to synthesize protein.

A central problem of embryology today is that of the regulation of gene activity during development. We would like to know how certain genes come to expression in one cell and not in another. It seems plain that one of the mechanisms of gene regulation is a differential distribution of cytoplasmic factors during the cleavage process. A great advantage of the *Ilyanassa* egg is that a potent cytoplasmic agent of this sort is almost completely segregated in the form of the polar lobe at the time of first cleavage. We are thus able not only to study the effects of its removal but also to examine it in isolation. Through each of these approaches we may hope to learn more of its real nature.

References

ATKINSON J. W., 1968. A comparative study of histogenesis and organogenesis in normal and lobeless embryos of the marine prosobranch gastropod, *Ilyanassa obsoleta*. Ph.D. Dissertation, Emory University.

BERG W. E. and Y. KATO, 1959. Localization of polynucleotides in the egg of *Ilyanassa*. Acta Embryol. Morphol. Exptl. *2*, 227.

BOBRETZKY N., 1877. Studien über die embryonale Entwicklung der Gastropoden. Arch. Mikr. Anat. *13*, 95.

BUTROS J. M., 1956. Simultaneous effects of metabolic inhibitors on the viscosity, surface rigidity and cleavage in *Ilyanassa* eggs. J. Cellular Comp. Physiol. *47*, 341.

CATHER J. N., 1958. Fixing and staining the chromosomes in eggs of invertebrates. Stain Technol. *33*, 146.

CATHER J. N., 1959. The effects of X-radiation on the early cleavage stages of the snail, *Ilyanassa obsoleta*. Radiation Res. *11*, 720.

CATHER J. N., 1963. A time schedule of the meiotic and early mitotic stages of *Ilyanassa*. Caryologia *16*, 663.

CATHER J. N., 1967. Cellular interactions in the development of the shell gland of the gastropod, *Ilyanassa*. J. Exptl. Zool. *166*, 205.

CLEMENT A. C., 1935. The formation of giant polar bodies in centrifuged eggs of *Ilyanassa*. Biol. Bull. *69*, 403.

CLEMENT A. C., 1940. Effects of cyanide on cleavage in eggs of *Ilyanassa* and *Crepidula*. Biol. Bull. *79*, 369.

CLEMENT A. C., 1952. Experimental studies on germinal localization in *Ilyanassa*. I. The role of the polar lobe in determination of the cleavage pattern and its influence in later development. J. Exptl. Zool. *121*, 593.

CLEMENT A. C., 1956. Experimental studies on germinal localization in *Ilyanassa*. II. The development of isolated blastomeres. J. Exptl. Zool. *132*, 427.

CLEMENT A. C., 1960. Development of the *Ilyanassa* embryo after removal of the mesentoblast cell. Biol. Bull. *119*, 310.

CLEMENT A. C., 1962. Development of *Ilyanassa* following removal of the D macromere at successive cleavage stages. J. Exptl. Zool. *149*, 193.

CLEMENT A. C., 1963. Effects of micromere deletion on development in *Ilyanassa*. Biol. Bull. *125*, 375.

CLEMENT A. C., 1967. The embryonic value of the micromeres in *Ilyanassa obsoleta*, as determined by deletion experiments. I. The first quartet cells. J. Exptl. Zool. *166*, 77.

CLEMENT A. C., 1968. Development of the vegetal half of the *Ilyanassa* egg after removal of most of the yolk by centrifugal force, compared with the development of animal halves of similar visible composition. Develop. Biol. *17*, 165.

CLEMENT A. C. and J. N. CATHER, 1957. A technic for preparing whole mounts of veliger larvae. Biol. Bull. *113*, 340.

CLEMENT A. C. and F. E. LEHMANN, 1956a. Über das Verteilungsmuster von Mitochondrien und Lipoidtropfen während der Furchung des Eies von *Ilyanassa obsoleta* (Mollusca, Prosobranchia). Naturwissenschaften *43*, 478.

CLEMENT A. C. and F. E. LEHMANN, 1956b. The distribution of mitochondria and lipid droplets during early cleavage in *Ilyanassa obsoleta*. Biol. Bull. *111*, 300.

CLEMENT A. C. and A. TYLER, 1967. Protein-synthesizing activity of the anucleate polar lobe of the mud snail, *Ilyanassa obsoleta*. Science *158*, 1457.

COLLIER J. R., 1957. A study of the alanylglycine dipeptidase activity during the development of *Ilyanassa obsoleta*. Embryologia (Nagoya) *3*, 243.

COLLIER J. R., 1960a. The localization of some phosphorous compounds in the egg of *Ilyanassa obsoleta*. Exptl. Cell Res. *21*, 548.

COLLIER J. R., 1960b. The localization of ribonucleic acid in the egg of *Ilyanassa obsoleta*. Exptl. Cell Res. *21*, 126.

COLLIER J. R., 1961a. Nucleic acid and protein metabolism of the *Ilyanassa* embryo. Exptl. Cell Res. *24*, 320.

COLLIER J. R., 1961b. The effect of removing the polar lobe on the protein synthesis of the embryo of *Ilyanassa obsoleta*. Acta Embryol. Morphol. Exptl. *4*, 70.

COLLIER J. R., 1963. The incorporation of uridine into the deoxyribonucleic acid of the *Ilyanassa* embryo. Exptl. Cell Res. *32*, 442.

COLLIER J. R., 1965a. Morphogenetic significance of biochemical patterns in mosaic embryos. *In:* R. Weber (ed.), The biochemistry of animal development, vol. 1, pp. 203–244, Academic Press, New York and London.

COLLIER J. R., 1965b. Ribonucleic acids of the *Ilyanassa* embryo. Science *147*, 150.

COLLIER J. R., 1966. The transcription of genetic information in the spiralian embryo. *In:* A. A. Moscona and A. Monroy (eds.), Current topics in developmental biology, vol. 1, pp. 39–59, Academic Press, New York and London.

COLLIER J. R. and M. MCCANN-COLLIER, 1962. The deoxyribonucleic acid content of the egg and sperm of *Ilyanassa obsoleta*. Exptl. Cell Res. *27*, 553.

COLLIER J. R. and M. MCCANN-COLLIER, 1964. Shell gland formation in the *Ilyanassa* embryo. Exptl. Cell Res. *34*, 512.

CONKLIN E. G., 1897. The embryology of *Crepidula*. J. Morphol. *13*, 1.

CONKLIN E. G., 1902. Karyokinesis and cytokinesis in the maturation, fertilization and cleavage of *Crepidula* and other Gasteropoda. J. Acad. Nat. Sci. Phila., ser. 2, *12*, 1.

CRAMPTON H. E., 1896. Experimental studies on gasteropod development. Arch. Entwicklungsmech. Organ. *3*, 1.

CROWELL J., 1964. The fine structure of the polar lobe of *Ilyanassa obsoleta*. Acta Embryol. Morphol. Exptl. *7*, 225.

DAN K. and J. C. DAN, 1942. Behavior of the cell surface during cleavage. IV. Polar lobe formation and cleavage of the eggs of *Ilyanassa obsoleta* Say. Cytologia (Tokyo) *12*, 246.

DAVIDSON E. H., G. W. HASLETT, R. J. FINNEY, V. G. ALLFREY and A. E. MIRSKY, 1965. Evidence for prelocalization of cytoplasmic factors affecting gene activation in early embryogenesis. Proc. Natl. Acad. Sci. U.S. *54*, 696.

FEIGENBAUM L. and E. GOLDBERG, 1965. Effect of actinomycin D on morphogenesis in *Ilyanassa*. Am. Zoologist *5*, 198.

MORGAN T. H., 1933. The formation of the antipolar lobe in *Ilyanassa*. J. Exptl. Zool. *64*, 433.

MORGAN T. H., 1935a. The separation of the egg of *Ilyanassa* into two parts by centrifuging. Biol. Bull. *68*, 280.

MORGAN T. H., 1935b. Centrifuging the egg of *Ilyanassa* in reverse. Biol. Bull. *68*, 268.

MORGAN T. H., 1935c. The rhythmic changes in form of the isolated antipolar lobe of *Ilyanassa*. Biol. Bull. *68*, 296.

MORGAN T. H., 1936. Further experiments on the formation of the antipolar lobe of *Ilyanassa*. J. Exptl. Zool. *74*, 381.

MORGAN T. H., 1937. The behavior of the maturation spindles in polar fragments of eggs of *Ilyanassa* obtained by centrifuging. Biol. Bull. *72*, 88.

MORRILL J. B., 1961a. Effects of cobaltous chloride on the eggs and embryos of *Ilyanassa obsoleta*. Biol. Bull. *121*, 399.

MORRILL J. B., 1961b. Effect of lithium chloride on the number of eyes in *Ilyanassa obsoleta* veligers. Biol. Bull. *121*, 399.

MORRILL J. B. and E. NORRIS, 1965. Electrophoretic analysis of hydrolytic enzymes in the *Ilyanassa* embryo. Acta Embryol. Morphol. Exptl. *8*, 232.

PELSENEER P., 1911. Recherches sur l'embryologie des gastropodes. Mém. Acad. Roy. Belg., cl. sci., sér. 2, *3*, 1.

RAVEN C. P., 1958. Morphogenesis: The analysis of molluscan development, pp. 68–69, Pergamon Press, New York, London, Paris, Los Angeles.

SCHELTEMA R. S., 1962. Pelagic larvae of New England intertidal gastropods. I. *Nassarius obsoletus* Say and *Nassarius vibex* Say. Trans. Am. Microscop. Soc. *81*, 1.

SCHELTEMA R. S., 1967. The relationship of temperature to the larval development of *Nassarius obsoletus* (Gasteropoda). Biol. Bull. *132*, 253.

STYRON C. E., 1967. Effects on development of inhibiting polar lobe formation by compression in *Ilyanassa obsoleta* Stimpson. Acta Embryol. Morphol. Exptl. *9*, 246.

G. Reverberi (ed.), Experimental embryology of marine and fresh-water invertebrates

CHAPTER 8

Fresh water gastropoda

OSWALD HESS

Institut für Allgemeine Biologie der Universität Düsseldorf

8.1. Introduction

In order to analyse the processes of embryonic development and differentiation of fresh water molluscs experiments have been carried out previously with two species, the pulmonate snail *Limnaea stagnalis*, and the prosobranch snail *Bithynia tentaculata*. This review, therefore, deals mainly with these two species. More detailed articles on the same subject have been published some time ago (Raven 1967; Hess 1962).

Unlike the majority of marine gastropods, fresh water species usually do not develop free living larvae. Fresh water species, therefore, have some very characteristic properties. They show 'direct development', that is, development proceeds directly within the egg capsule until the formation of a fully differentiated young snail. No true metamorphosis is observed. However, a more or less typical trochophora larva is usually formed. It can swim freely by rotating within the egg capsule, which is filled with a fluid. This fluid is of importance for growth because it is rich in nutritive material which is taken up by the embryo during the course of development. Whereas the trochophora stage is quite normally formed, thc veliger stage is in most cases not typically developed by the embryos of fresh water gastropoda. In particular, the vela which are so prominent in larvae of many marine gastropoda are not formed.

8.2. Ovoposition

8.2.1. Limnaea stagnalis

The eggs of fresh water pulmonates, including those of *L. stagnalis*, are typically laid in the form of egg batches which contain from 20 to 80 or even more egg capsules. Pulmonate snails are a hermaphroditic species. However, they are usually sterile themselves. During copulation spermatozoa are exchanged between both partners. Insemination occurs within the oviduct. In addition, during the formation of an egg batch insemination of the individual eggs occurs one after the other in the sequence in which they are included in the egg mass. Later, meiotic and first cleavage divisions occur in the same sequence. Therefore, one observes the first phases of development moving as a wave along the eggs of an egg batch.

At the metaphase stage of the first meiotic division an inhibition of oogenesis regularly occurs. This repression is released by the insemination. As this takes place directly before egg deposition, protrusions of the first and second polar bodies can be observed within freshly laid egg capsules some 30 to 60 min after deposition. Both meiotic divisions are accompanied by phases of high surface motility during which the eggs form lobe-like or amoeboid protrusions. In contrast to this, the eggs, which in *L. stagnalis* have an average diameter of 120 μ, are spherical at the stages before the anaphase of the first meiotic division, between the first and the second meiotic division, and between the second meiotic division and the first cleavage.

Three different egg membranes can be distinguished (fig. 1a):

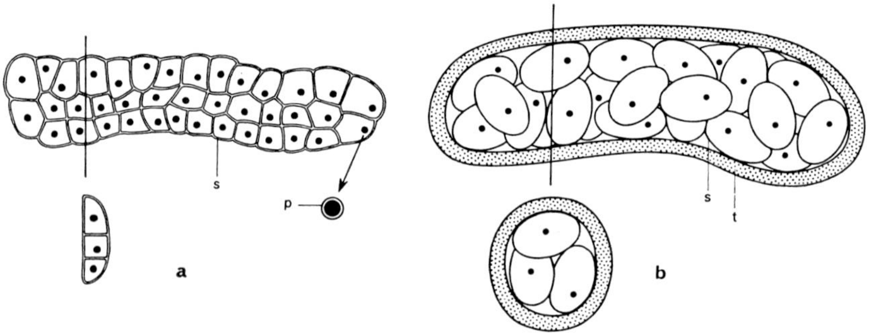

Fig. 1. Egg batches of (a) *B. tentaculata* and (b) *L. stagnalis*. Egg cells swim freely within nutritive liquid of egg capsule. p, s, t: primary, secondary and tertiary egg membranes. Detail (a): single egg cell with primary egg membrane at high magnification.

(1) The primary or vitelline membrane is excreted by the cell cortex of the oocyte. In Limnaea, as is the case in most species, it is a very thin and delicate membrane, nearly invisible in the light microscope but definitely seen with the aid of the electron microscope.

(2) The secondary membrane or chorion is formed by follicle cells. It envelops a rather large capsular space which is filled with a nutritive liquid. The egg cell (and during later stages the embryo) swims freely within the capsule liquid.

(3) The tertiary membrane or jelly coat is formed by the cells of the oviduct. It is a jelly skin which enwraps the free egg capsules to form an united egg batch.

8.2.2. Bithynia tentaculata

Bithynia, as are most of the prosobranch snails, is bisexual. In this species oogenesis is also stopped at the metaphase of the first maturation division. Insemination occurs in the oviduct of the female just before deposition of the eggs. Again, the arrest of development is released after the entrance of the sperm. Both maturation divisions occur soon after deposition. The extrusion of the polar bodies is, also in this species, accompanied by a phase of motility in the egg cortex (fig. 3a). In correct order, one egg after the other within one egg mass passes through the consecutive divisional stages.

Unlike Limnaea, no tertiary egg membrane is formed in Bithynia. The egg capsules are laid one after the other and are pasted together by the female on a substrate such as the leaves of a water plant. The capsules are arranged in two or three rows and the egg masses contain between 15 and 80 single egg capsules (fig. 1a). In Bithynia, the diameter of the uncleaved eggs averages 170 μ.

8.3. Oogenesis

During the 'classical' period of experimental analysis of embryonic development two main types of eggs or modes of differentiation have been distinguished: the so-called regulation eggs and the mosaic eggs. The principal differences between these types is the stage at which the differentiation capacities of the egg or of embryonic areas are irreversibly determined. In eggs of the regulative type this determination occurs at a comparatively late stage. Therefore, during embryonic development adaptations are possible, if changes of the topographical relationships between embryonic areas of different morphogenetic value have occurred. In eggs of the mosaic type determinations occur at rather early embryonic stages, often even before the beginning of the first cleavage division, and adaptations to changed situations are not possible. In addition, in embryos of the regulation type the differentiation of a morphogenetic area very often depends upon its situation within the whole embryo and is influenced or even directed by surrounding tissues. In contrast to this, in embryos of the mosaic type differentiation not only occurs rather early but very often is also irreversible. Therefore, determined morphogenetic areas are able to differentiate only according to their origin and their 'imprinted determination'. After experimentally induced changes of the topographical relationships between different parts of the embryo no regulation is possible. The area differentiates 'autonomously' or 'independently' according to its origin.

For a long time the embryos of the gastropoda were considered to be typical examples of the mosaic type of differentiation. However, today it is generally accepted that there are no sharp but only gradual differences between the two types of differentiation. During recent years also some processes of 'dependent' or 'regulative' differentiation have been found in gastropoda. However, it is true that in these species the majority of the differentiation processes occur in areas which have been irreversibly determined already during very early stages. A good deal of decisive determination can, therefore, be expected to occur at very early stages, mostly before the first cleavage. Indeed, important phases in embryonic development of gastropoda take place as early as in oogenesis. Thus, the processes of oogenesis turn out to be of considerable importance for later embryonic development.

Oogenesis has been studied in detail in the pulmonate *L. stagnalis* (Raven 1961). The formation of the female germ cells occurs within a so-called germinal epithelium. Only part of the cells of this epithelium become germ line cells. Most of the cells, however, are determined to differentiate into nurse cells which form a type of follicle epithelium around the true germ line cells. During oogenesis the oocytes grow considerably and protrude into the central cavity of the ovary. Yolk formation takes place by a cooperation of the follicle and nutritive cells with the cells of the germinal epithelium to which the oocyte is still connected by a slender stalk.

In the stage of the unmaturated egg the cells are clearly radially symmetrically organized. The egg has an animal–vegetative axis which is indicated by the eccentrical location of the egg nucleus with its special clear perinuclear cytoplasm. There is some evidence that the vegetal pole of the egg coincides with the point at which the growing oocyte had been attached to the germinal epithelium.

Before the beginning of the first cleavage division a drastic change in the distribution of the components of the egg cytoplasm occurs. Before the egg enters into the first cleavage division, the more or less homogeneously distributed egg components are dislocated and a specific pattern is developed. This process has been called 'ooplasmic segregation'. The most conspicuous result of the ooplasmic segregation is the formation of polar plasms which in different species are found either at the animal or at the vegetal pole or at both (fig. 2a).

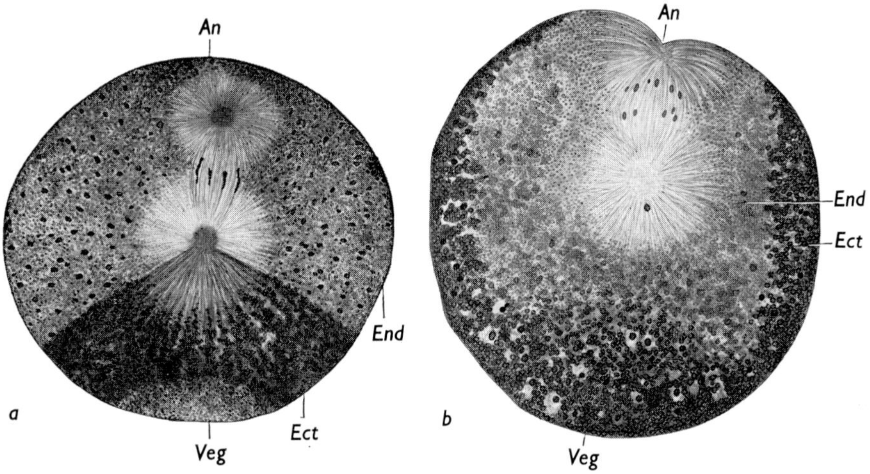

Fig. 2. Ooplasmic segregation in the egg cell of *L. stagnalis*. (a) Immediately after deposition. Early anaphase of first maturation division. Vegetative polar plasm (Ect) in the form of a sector at the vegetative poles (Veg). (b) Late anaphase of first maturation division. Depression of the egg surface at the animal pole (An) where the distal aster of the spindle has come into contact with the surface. Vegetative polar plasm (Ect) has spread beneath the egg cortex far into the animal half of the egg and now surrounds the inner plasm (End) except for the region directly at the animal pole. (Drawings from histological sections through eggs, by courtesy of Prof. C. P. Raven, Utrecht, The Netherlands.)

There is good reason to assume that the pattern in which the components of the egg cytoplasm are rearranged is realized by interactions between these components and the cortex of the egg. The architecture of the egg cytoplasm can most easily be changed experimentally by centrifugation. Immediately after centrifugation an active shift of the displaced egg inclusions starts. If the process of redistribution is not interrupted by the beginning of the first cleavage division, the original pattern in the egg is completely re-established. The only component of the egg which is not displaced by strong centrifugation is a thin cytoplasmic layer in the cortex of the egg. This membrane is assumed to be the carrier of a pattern, which can not be demonstrated with the available techniques, and in which specific local affinities to certain components of the egg cytoplasm exist. It is by this (hypothetical) affinity that the characteristic architecture of the egg cell is produced in the same way that it is reconstituted after centrifugation.

One has to conclude that this pattern is imprinted in the egg cortex during oogenesis. It seems likely that the orientation of the growing oocyte in the gonad and the arrangement of follicle and nutritive cells around the egg cell play a decisive role in the imprinting process. However, there is no proof for this at the moment. One important finding has been reported for the egg of *L. stagnalis* (Raven 1963). The egg of this species is in fact not radially symmetric at ovoposition, but is organized in a bilateral symmetry. In the equatorial zone of the egg cell several areas can be found which consist of a special differentiated cytoplasm. In histological preparations of sections through uncleaved eggs the staining of these six 'subcortical patches' is different from that of the surrounding cytoplasm. These six subcortical special areas of the egg cytoplasm are arranged in an asymmetrical pattern around the equator of the egg. In addition, a vegetative polar plasm is formed during the ooplasmic segregation (fig. 2). It is composed of a dense mass of special types of protein yolk granules. The pole plasm has the shape of a sector with about 110° the apex of which points towards the centre of the egg. However, the polar plasm is situated slightly oblique with respect to the animal–vegetative axis. Four or five of the subcortical cytoplasmic patches are located on that side of the egg where the vegetative polar plasm is highest, the remaining one or two patches are found on the opposite side of the egg. After centrifugation, the subcortical patches have disappeared. However, they are re-formed within a few hours and they reappear at their typical locations. This even happens if cleavage has started before complete reformation. Thus, the egg of *L. stagnalis* receives by both the pattern of the arrangement of the six cytoplasmic patches and by the oblique orientation of the vegetative polar plasm a dorsoventral organization.

There is some evidence that the rearrangement of the described six subcortical patches reflects the position of the six inner follicle cells which surround the oocyte during the last phase of its growth. These cytoplasmic structures arise during the passage of the egg down the genital tract, presumably by the selective accumulation of particular egg components beneath certain areas of the cortex. It was, therefore, concluded that the cytoplasmic architecture reflects a pre-existent pattern in the egg cortex which in turn arises during oogenesis by interactions between the oocyte and the surrounding structures in the gonad.

8.4. Cleavage

The eggs of the fresh water gastropoda are cleaved according to the well-known spiral pattern which is generally observed in molluscs. The first two cleavages divide the egg in a meridional direction into four blastomeres which form the quadrants A, B, C and D (figs. 4a–e, 5a). This is followed by four cleavage divisions in which the cleavage furrows are oriented in the equatorial plain. Moreover, the divisions are strongly inequal (fig. 4h). The smaller blastomeres are oriented towards the animal pole. Thus, four quartets of micromeres are formed. In addition, the spindles of these cleavage divisions are not oriented exactly perpendicular to those of the first two divisions but are shifted in a lateral direction. Therefore, if viewed from the animal pole, the micromeres are not located exactly above their corresponding macromeres, but are displaced laterally (figs. 4f–h, 5b, c). In consecutive cleavage divisions, spindles become oriented alternately in a clock-wise and an anti-clock-wise fashion. Thus, dexiotropic divisions are followed by laeotropic divisions (figs. 5b, c). In this way, all blastomeres become arranged in a strongly regular pattern (figs. 4i, 5c).

The cytoplasmic components of the uncleaved egg exhibit a characteristic pattern, as was previously described. During cleavage the same egg becomes subdivided into blastomeres which are again arranged in a very regular pattern. Thus, the blastomeres of the gastropod embryo must be highly differentiated with respect to the composition of those areas of the egg cytoplasm which have been transferred to them. Raven (1967) was able to trace the pattern of the six subcortical cytoplasmic patches which are found in the equatorial region of the uncleaved egg of *L. stagnalis* (see above) through the cleavage divisions. Just before the beginning of the first cleavage division the six patches extend latitudinally and fuse. During the first and the second cleavage division their material is distributed in a regular way among the four blastomeres so that the four quadrants of the embryo exhibit characteristic differences which coincide with the median plane of the future embryo.

There is good reason to assume that quantitative differences in the cytoplasm are rather quickly transformed also into qualitative differences. Differences in yolk content, in the available number of mitochondria, and so on, may influence the metabolism in the cell in such a way that qualitative differences shortly occur which again react upon the activity of the genetic material in the nucleus which in turn causes additional differentiation. In this way differentiation starts with minor quantitative differences in the composition of the cytoplasm and leads rather quickly to considerable qualitative differences.

From this follows that cleavage continues the processes which have started with pattern formation in the uncleaved egg cell by the ooplasmic segregation. The ooplasmic architecture is stabilized by a transformation of segregated differentiated areas of the egg cell into separated blastomeres. It seems to be a general rule that differentiation in animals occurs in a sequence of events in which quantitative differences are first created, these differences are then stabilized by transformation into cells with quantitative differences, and this in turn causes the occurrence of qualitative differences. Moreover, it has been shown for several different animal species, namely Amphibia, sea urchins, and insects (see Brown 1966; Harris and Forrest 1967a, b; Monroy and Gross 1967; Nemer 1967; Nemer and Infante 1967; Spirin 1966) and also in one gastropod (Clement and Tyler 1967) that the early processes of embryonic development are guided by gene products which have been stored in the egg cell during oogenesis. Thus, one has to take into consideration the possibility that during ooplasmic segregation not only coarse egg cytoplasmic components, such as yolk globules, mitochondria, and the elements of the endoplasmic reticulum are unevenly distributed to different regions of the egg cell, but also that primary gene products (perhaps messenger RNA in special stabilized and programmed storage forms) are differentially distributed. There are, for instance, also observations of differential distribution of ribosomes. On the other hand, it has been shown that in the unmaturated egg ribosomes can be loaded with messenger RNA molecules and that this complex is reversibly inactivated by some kind of coat, masking, or repressor protein (Monroy and Tyler 1967). In addition, specific morphogenetic substances have been postulated by many of the earlier investigators. In the view of modern molecular biology one would suppose that such substances might in fact occur in the cytoplasm of eggs and young embryos and might be identical with specific preformed species of messenger RNA molecules (Davidson et al. 1965).

In any case it seems quite clear that factors controlling the cleavage pattern under these circumstances must play a decisive role during embryonic differentiation at least in cases in which, like in gastropoda, a very regular cleavage pattern is observed. This is also clearly demonstrated by the well-known fact that disturbances of the cleavage pattern, which can easily be induced experimentally, regularly cause severe defects of the embryos.

In fresh water gastropoda, as a rule the first two cleavage furrows occur meridionally. The blastomeres of the 4-cell stage are often of equal size.

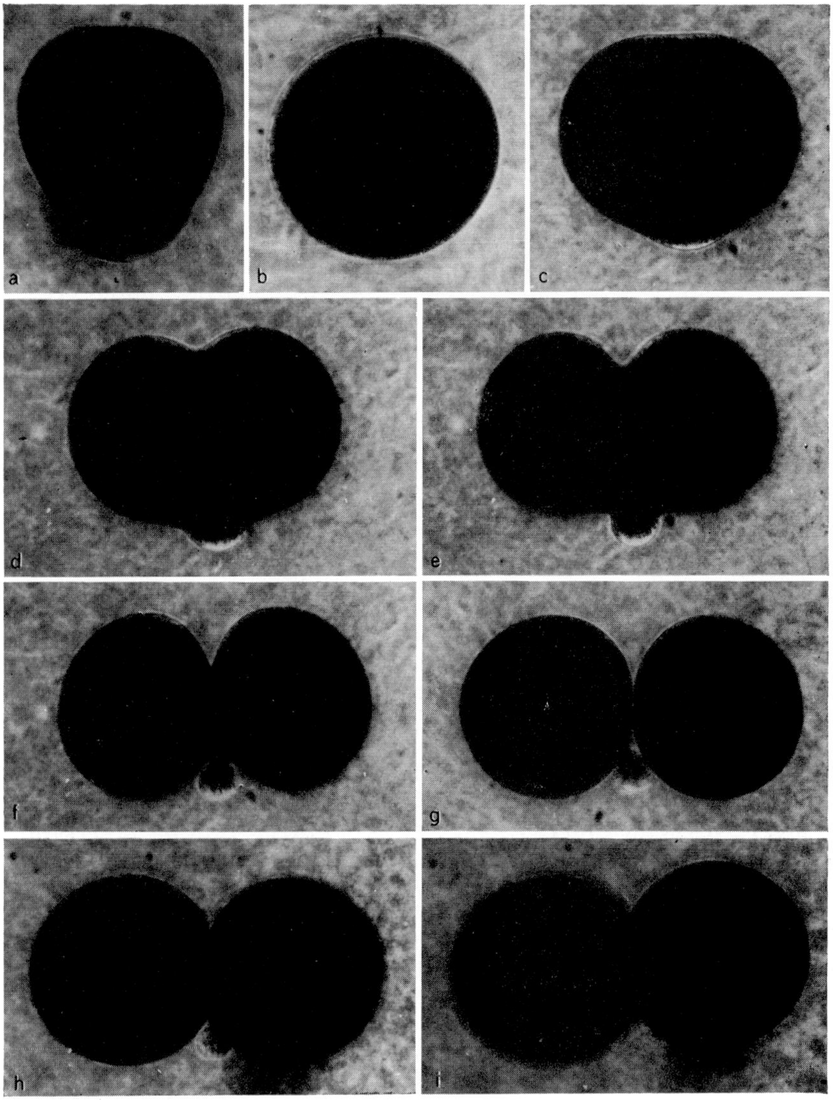

Fig. 3. Meiosis and first cleavage division in *B. tentaculata.* (a) First maturation division. Pear-shaped deformation of the egg. Extrusion of the first polar body. (b) Mature egg. (c) Beginning of first cleavage division. (d) 6 min later: protrusion of the polar lobe at the vegetative pole, indentation of the first furrow at the animal pole. (e) 5 min later. (f) 3 min later. (g) 6 min later: maximum separation of half blastomeres, polar lobe connected with the rest of the egg by a thin stalk only. (h) 48 min later: polar lobe fuses with the (right) CD blastomere. (i) 6 min later: fusion of polar lobe with the CD blastomere nearly complete. (Phase contrast photographs, approximately 140 ×.)

In spite of this, the differentiation capacities of the four quadrants are not the same: the D-quadrant has specific capacities to differentiate into mesoderm and its descendants which the other quadrants do not possess. In a number of marine species this uniqueness of the D-quadrant is also directly seen, because in these species the D-quadrant is considerably larger than the three others. This difference in size is attained in one of two ways: firstly, the two cleavage divisions can be inequal resulting in one large and three smaller blastomeres. Secondly, a special mechanism may be used with which specific regions of the mature egg are transformed undivided into only one blastomere. In these cases a so-called polar lobe is formed at the vegetative pole just prior to the first cleavage. This polar lobe consists of the vegetative polar plasm and is in some cases quite large. The rest of the egg cytoplasm is equally divided during the cleavage. After completion of the cleavage the polar lobe is connected only by a narrow stalk with the egg which, therefore, has the appearance of a 3-cell stage (trefoil stage). Then the polar lobe fuses with one of the blastomeres which is then designated as the CD-cell. During the second cleavage the same event occurs in the CD-blastomere and the lobe fuses with the D-cell. Thus, the D-quadrant receives the whole vegetative polar plasm and its size may be considerably larger than that of the other three blastomeres. It is known from many experiments with marine species that the polar lobe contains morphogenetic substances which are needed for the differentiation of organs with mesodermal origin.

In *L. stagnalis* no polar lobe is observed and the four quadrants appear to be of equal size. In contrast, a polar lobe is formed in *B. tentaculata* during the first cleavage division (fig. 3). However, as opposed to the observations with marine gastropoda, in this species the polar lobe is so small that it is hardly observable. It fuses with the D-quadrant, but because of the smallness of the lobe the D-cell can not be distinguished by its size from the three other blastomeres. So far, no experiments have been executed with *B. tentaculata* to elucidate the morphogenetic role of this small polar lobe. However, it is well known that mesoderm originates also in this species from descendants of the D-quadrant.

If components of the egg cytoplasm are displaced by centrifugation, the cleavage pattern is not severely changed. Thus, it seems that the pattern of the components in the egg cytoplasm is not important for the orientation of the cleavage furrows. However, one might assume that also the cleavage pattern is determined by specific properties of the egg cortex which seems to be the only part of the egg which is not changed during centrifugation. The cleavage pattern seems to be regulated by interactions between certain

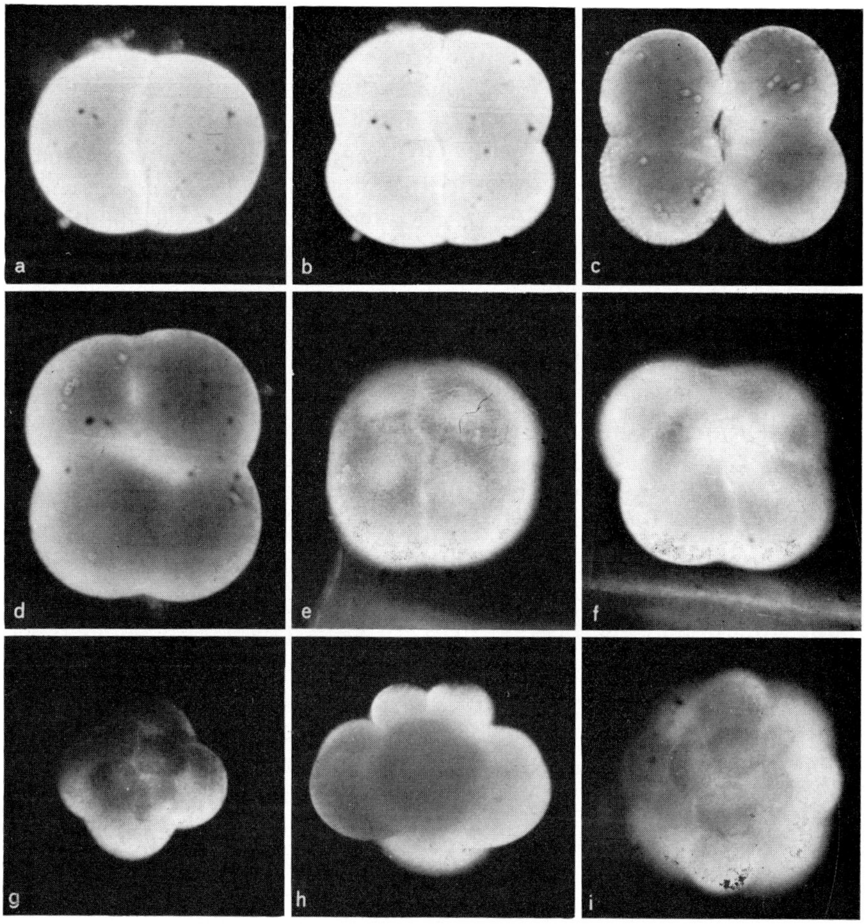

Fig. 4. Second to fourth cleavage division in *B. tentaculata*. (a) Interphase between first and second cleavage. (b) Beginning of second cleavage division, elongation of blastomeres in the direction of the first furrow. (c) End of second cleavage division. Note separation of blastomeres in the second furrow and their firm connection in the first furrow. (d) Four-cell stage. (e) End of interphase between second and third cleavage division, viewed from the animal pole: nuclei are very near to the animal pole and therefore visible through the cytoplasm as light areas. (f) Beginning of the third cleavage division, viewed from the animal pole, dexiotropic orientation of spindles. (g) Eight-cell stage, from the animal pole. Quartet of primary micromeres dexiotropically dislocated. (h) Same stage viewed from the side, to demonstrate the size difference between micromeres and macromeres. (i) Sixteen-cell stage, viewed from the animal pole, quartet of animal descendants of primary micromeres (cells $1a_1$–$1d_1$) clearly visible, rest of the embryo out of focus; however, the four secondary micromeres (2a–2d) and the four secondary macromeres (2A–2D) can be seen. (Dark field photographs, approximately 132 × (except fig. 4g).)

regions of the cortex and the centrioles. The orientation of the cleavage spindles and also of cleavage furrows can be determined by the localization and orientation of the centrioles. The presumed pattern in the egg cortex which is responsible for the orientation of the cleavage divisions must be imprinted during oogenesis. If one could find genes which are involved in the process of the imprinting of the pattern for cleavage, a remarkable consequence would be expected: since the cleavage of the embryo would then be determined during oogenesis, cleavage would follow the genotype of the mother and this could be different from that of the developing embryo itself. In Limnaea this has indeed been demonstrated experimentally. Limnaea, like most of the fresh water gastropoda, possesses a shell which is dextrally coiled. Sometimes, however, individuals with a sinistrally coiled shell are found. It has turned out that the property 'Possession of a dextrally or sinistrally coiled shell' is inherited. In addition, an interesting and in the context discussed here very important correlation between the directions of shell coiling and the orientation of the cleavage furrows has been found. According to the normal spiral type of cleavage dexiotropic cleavages are followed by laeotropic ones and vice versa. In eggs of normal specimen of Limnaea, the third, fifth, seventh, etc. cleavage steps are dexiotropic, correspondingly the fourth, sixth, etc. divisions are laeotropic. However, in individuals with sinistrally coiled shells inversed cleavage was found (fig. 5). Thus, genes controlling the coiling of the shell also control in some way the orientation of the cleavage furrows and, therefore, their phenotypic expression becomes

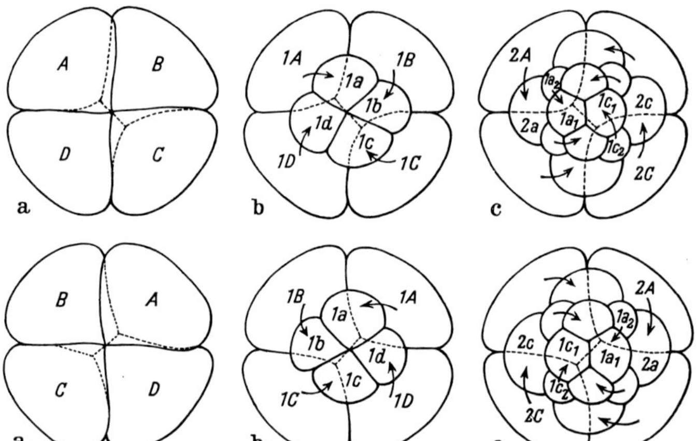

Fig. 5. Inverted spiral cleavage. Orientation of third and fourth cleavage divisions in animals with (a–c) dextrally and (a_1–c_1) sinistrally coiled shells.

manifest already during the very early stages of embryonic development. Crosses have been performed between animals with dextral and sinistral shells (fig. 6). It was found that the direction of the coiling of the shell is

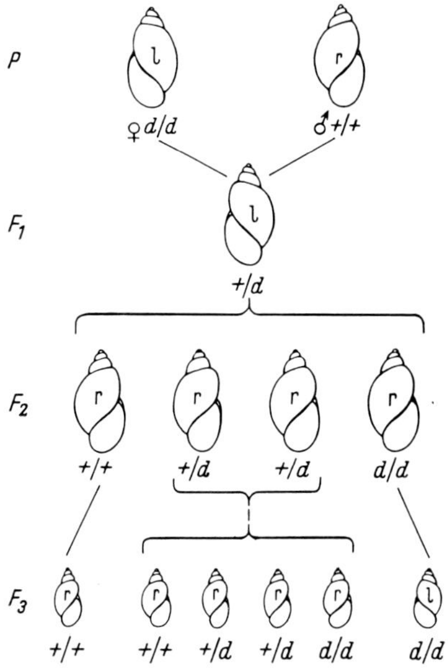

Fig. 6. Inheritance of the direction of coiling in Limnaea. Explanations. l, sinistrally, r, dextrally coiled shell; +, wild type (dextrally), d, mutant (sinistrally) allele of the gene which is responsible for the orientation of shell coiling. For further explanations, see text. (After Kühn, changed.)

determined by one single pair of Mendelian alleles (Diver 1925; Sturtevant 1923). The normal allele causing a dextral shell (and dexiotropic cleavage during the uneven division steps at the same time) turned out to be dominant over the mutant 'sinistral shell'. However, directions of cleavage and coiling always follow the genotype of the mother. Thus, snails with a sinistral shell which are homozygous for the recessive mutant allele produce only progeny which possess without any exception sinistral shells. The genotype of the mating partner is of no influence on the direction of shell coiling of the progeny. For instance, in cases where homozygous mutant females are mated to a wild type partner so that all progeny becomes heterozygous, all F_1 individuals have a sinistral shell. Thus, here the direction of the shell and also of cleavage does not correspond to the genotype of the animals, as the wild type allele is dominant over the mutant and, therefore, heterozygotes should show the wild phenotype. In addition, if a heterozygous animal is mated to a homozygous mutant one (fig. 6), the progeny of such a cross will be 50% heterozygous, and 50% homozygous mutant. Despite this, all animals of the next generation possess a shell which is dextrally coiled. Thus, the expression of the property 'direction of the shell and cleavage divisions' is always determined by the genotype of the mother and not by the genotype of the developing individual itself. This is a very clear demonstration of the principle of 'predetermination'. As according to modern ideas regulation of early embryonic processes is generally directed by preformed messenger molecules, similar events can be expected to occur very frequently. However, so far only a few cases are known in which gene expression becomes phenotypically evident to the observer already during the early stages of development.

The pattern which is supposed to be imprinted into the egg cortex and which then regulates the cleavage processes must be highly stabile. This is nicely demonstrated by the observation that isolated blastomeres show a partial cleavage. This means isolated blastomeres cleave in nearly exactly the same way as they would have done in the complete embryo (fig. 7).

8.5. Gastrulation

Gastrulation takes place by two different processes, invagination of entodermal material into the blastocoel and epibolic growth of the marginal regions of the ectoderm above parts of the entoderm. According to the

---

Fig. 7. Partial cleavage of separated blastomeres of *B. tentaculata*. (a) Pair of half blastomeres briefly after separation. (b) 15 min later: beginning of second cleavage division in one of the blastomeres. (c) 15 min later: second cleavage division in both halves. (d) 90 min later: interphase between second and third cleavage, pair of 2/4 embryos. (e) 80 min later: beginning of third cleavage, again earlier in the upper half embryo. Note inequality of cleavage and the formation of (duets of) micromeres in each half embryo. (f) 20 min later: end of the third cleavage division. (g) Single half embryo of the same stage, viewed from the animal pole and at higher magnification. Note the dextral dislocation of the duet of micromeres. (h) 100 min later: interphase between third and fourth cleavage. Pair of 4/8 embryos. (i) Stage after the fourth cleavage division. One 8/16 partner, two large secondary macromeres, two secondary micromeres of medium size, four small descendants from the former duet of primary micromeres: all cells have cleaved as they would have done in a whole embryo. (Dark field photographs, approximately 144 ×.)

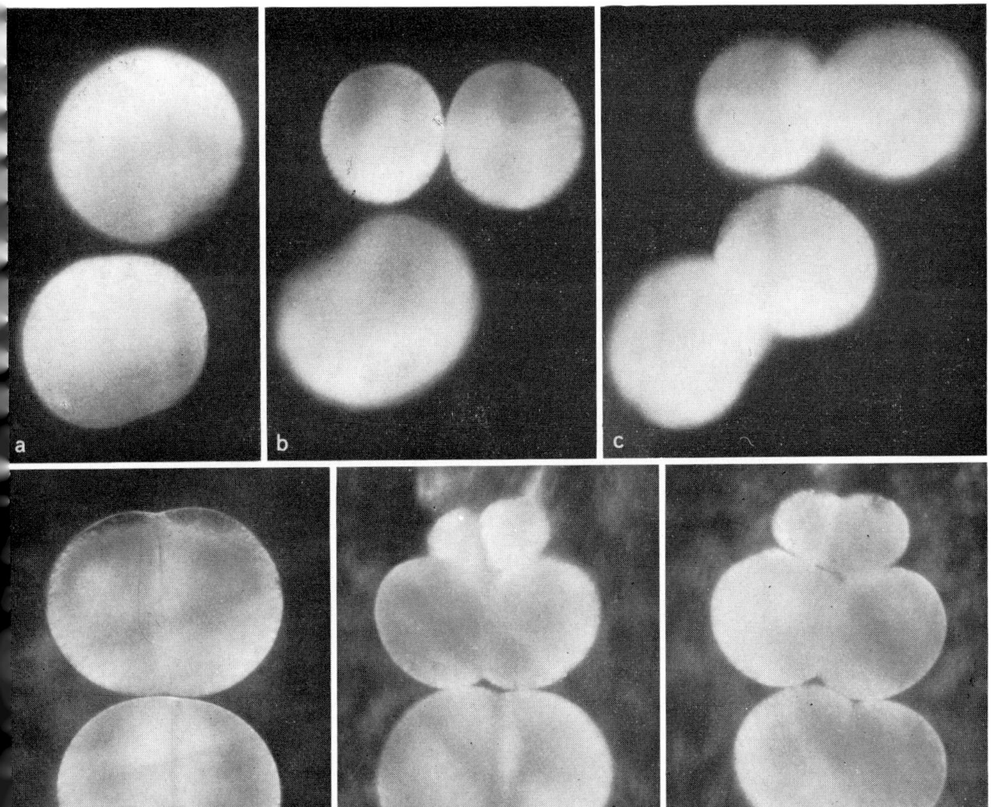
a
b
c
d
e
f
g
h
i

type of blastula which depends upon the size of the blastocoel and the amount of yolk present in the entoderm cells (fig. 8a), either the first or the second type of movement is dominant. For instance, in the embryos of *L. stagnalis*

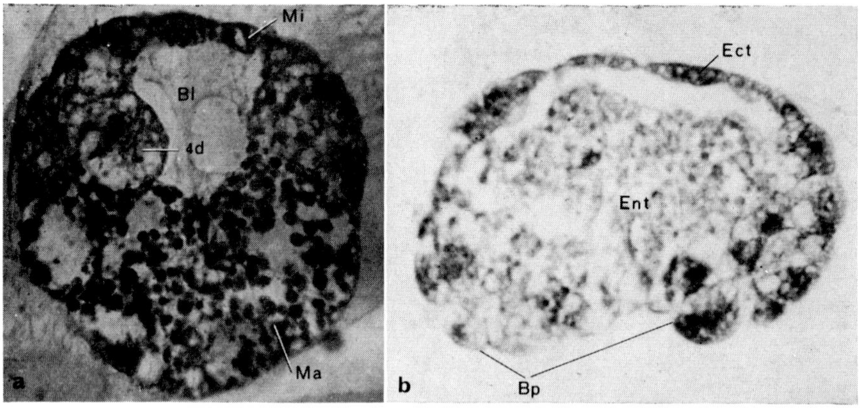

Fig. 8. *B. tentaculata.* (a) Blastula stage. (b) Gastrula stage. Explanations: Bl, blastocoel; Bp, blastopore; Ect, ectoderm; Ent, entoderm; Ma, macromeres; Mi, micromeres; 4d, mesentoblast. (Photographs of 6μ paraffine sections through embryos, approximately 175 $\times$.)

invagination of the entoderm is more pronounced. In contrast, in embryos of *B. tentaculata* both processes participate at about equal rates to gastrulation (fig. 8b).

The two processes mostly take place independently of each other. Some insight into the cooperation of these two processes has been obtained from observations of the gastrulation in half embryos of *B. tentaculata* (Hess 1956a). In this fresh water prosobranch snail gastrulation starts with the formation of a comparatively shallow pit at the vegetative pole. This arises by changes of the shape of the entoderm cells in this region. The external surface of these cells is reduced, at the same time their inner parts increase in width. The invagination is completed as the marginal ectoderm grows epibolically over the entoderm (fig. 8b). Such a growth of one cellular layer over the other requires certain properties in the cell surface of the two germinal regions. In half embryos, which develop from isolated blastomeres, gastrulation starts normally with a shallow invagination of the entoderm. During the following phase only a part of the ectoderm is capable of growing epibolically over the entoderm. In some regions of the half embryo, however, this movement appears to be inhibited. Despite this, growth of the ectoderm seems to be normal. Thus, a peculiar asymmetric gastrula is formed. Because one side of the ectoderm is not able to glide over the entoderm as would be normal, it has to advance in the direction of the animal pole. As a result a vesicle is formed. This vesicle which enlarges rather quickly, is filled with liquid. Because the liquid develops some pressure, the continuation of the invagination of the entoderm is not only inhibited, but, on the contrary, the early shallow invagination of the archenteron is finally everted. In this

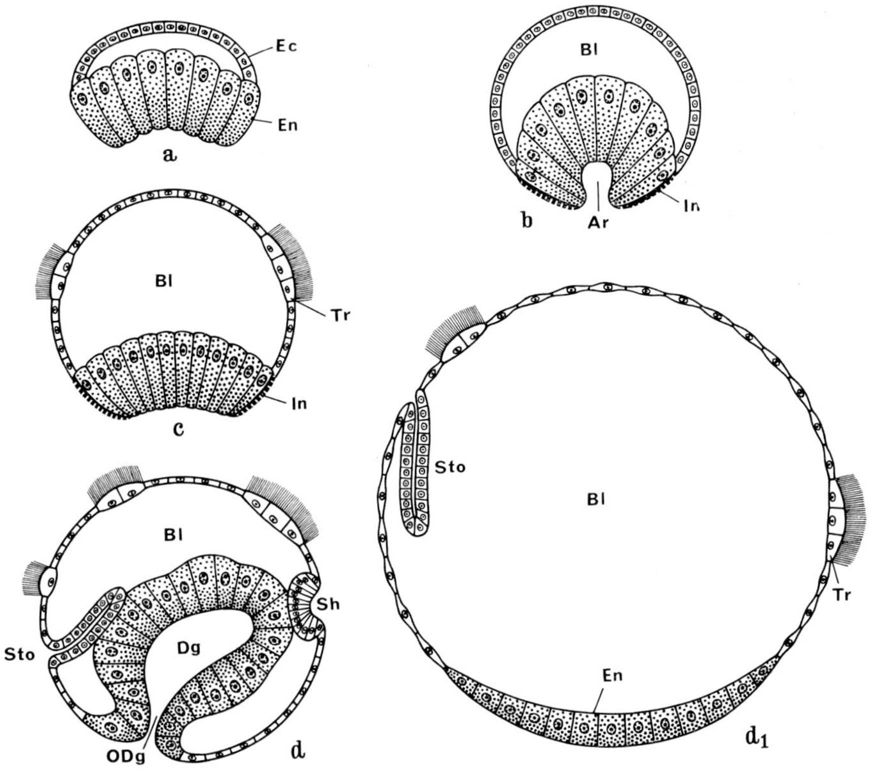

Fig. 9. Development of exogastrulated embryos in *B. tentaculata*. Diagrammatic drawings of sagittal sections. (a) Normal beginning of gastrulation with weak invagination of the entoderm. (b) Normal formation of the archenteron. The epibolic growth of the ectoderm over the entoderm is, however, inhibited. (c) Exogastrulated embryo. (d) Older exogastrulated embryo after the delayed secondary invagination of a part of the entoderm (anterior diverticulum of the digestive gland); induction of a shell gland and formation of a (closed) stomodaeum. (d_1) Exogastrulated embryo of the same age in which the anterior diverticulum of the digestive gland failed to invaginate. Formation of the stomodaeum, however, no shell gland. Explanations: Ar, archenteron; Bl, blastocoel; Dg, digestive gland; Ec, ectoderm; En, entoderm; In, inhibited areas of epibolic growth; Sh, shell gland; Sto, stomodaeum; Tr, cells of trochal ring.

way, half embryos of *B. tentaculata* develop into exogastrulated embryos.

This observation was explained by assuming that the epibolic growth of certain marginal parts of the ectoderm over parts of the entoderm needs some kind of positive affinity between these two tissues. Such an affinity might lack in those parts of the entoderm which normally lie in the middle of the embryo and only by the isolation operation reach the surface in half embryos. If this part which in normal embryos never comes into contact with the ectoderm did not possess positive affinities, the epibolic growth of the ectoderm would in consequence not be possible in such regions.

Unlike the observations just described, in embryos of *L. stagnalis* epibolic growth of the ectoderm does not seem to play a similarly important role during gastrulation. This follows not only from observations of the course of gastrulation in normal embryos, but also from the fact that half embryos of this species are mostly able to gastrulate normally.

Also in complete embryos of *B. tentaculata* exogastrulation can be induced experimentally. Treatments of different kinds may cause exogastrulation. A rather high percentage of exogastrulation is, for instance, recovered after treatment of early cleavage stages with lithium chloride. It was observed that in lithium treated embryos gastrulation seemed to start quite normally with the formation of a shallow invagination of the entoderm (fig. 9a). Then, however, gastrulation is not completed because the epibolic growth of the ectoderm seems to be disturbed (fig. 9b). Since growth in the ectoderm continues normally whereas this tissue is unable to glide over the entodermal material, a vesicle is formed which lifts off in the direction of the animal pole. Later, the invagination of the entoderm is again everted, presumably because it is pushed out by the pressure of the liquid which is accumulated in the embryonic vesicle (fig. 9c). Thus, the embryo has developed into an exogastrula which this time originated from a complete embryo. This observation again can be explained with the assumption that specific properties in the surface of either ectoderm, entoderm or both types of cells are changed by the treatment with lithium chloride in such a way that the gliding of one layer of cells over the other is at least locally inhibited. This assumption receives support from the fact that there is good evidence from other experiments that the main target of the action of lithium chloride seems to be indeed the cortex of the embryonic cells.

Besides its effects on epiboly, lithium chloride seems to be able to affect also the invagination of the entoderm to some extent. Thus, in embryos of *L. stagnalis* a treatment with lithium chloride may also cause exogastrulation. As a rule in embryos of this species which exogastrulate after lithium chloride treatment no invagination of the entoderm is observed. The blastocoel is in these cases continuously enlarged and in this way the blastulae develop directly into exogastrulated embryos. For this kind of developmental disturbance the effect of lithium chloride is highly phase specific. The sensitivity of Limnaea embryos against lithium chloride changes rhythmically and synchronously with the phases of the mitotic cycles.

Inducing exogastrulations is not a specific effect of lithium chloride. Exogastrulations occur also after various kinds of other injurious treatments. All these treatments, however, seem to have in common an influence on the redox system of the cells. Thus, exogastrulation occurs frequently after lack of oxygen during cleavage stages, or after treatment with potassium cyanide, after heat shocks, etc. Exogastrulation may also be caused by treatments with various other ions. However, usually it is necessary to use much higher concentrations of these components than lithium chloride to produce similar results. The effects can partly be compensated for by combination of different salt solutions each of which alone would cause exogastrulations. This demonstrates the complexity of the directing mechanisms of which nearly nothing is known so far (Raven 1952; Raven et al. 1956).

8.6. Embryogenesis

In gastropoda, the gastrula develops into a trochophora larva. This stage is also found among the fresh water species which do not have free living larva. Both in Limnaea and in Bithynia a typical trochophora is formed. It rotates with the aid of its trochs in the nutritive liquid of the egg capsule.

The older trochophora larvae exhibit the typical tripartite organization of molluscs in head, foot and the visceral mass with the shell gland. This typical organization of the trochophora is developed from the gastrula in the following way (fig. 10): the late gastrula stage typically shows an increased growth of the post-trochal regions of the dorsal ectoderm. In this way the blastopore is displaced to the ventral side of the embryo. Afterwards, also in the ventral region of the ectoderm growth commences and leads to the formation of an ectodermal evagination in the post-trochal region of the ventral side. This evagination differentiates into the foot anlage (fig. 10a). At the same stage a part of the ventral ectoderm sinks into the blastocoel and forms the stomodaeum. The archenteron separates into the anterior and posterior lobes of the midgut gland (= anterior and posterior diverticula of the digestive gland), and into the midgut which develops the proctodaeum. Thus, the basic plan of the future young snail has been completed.

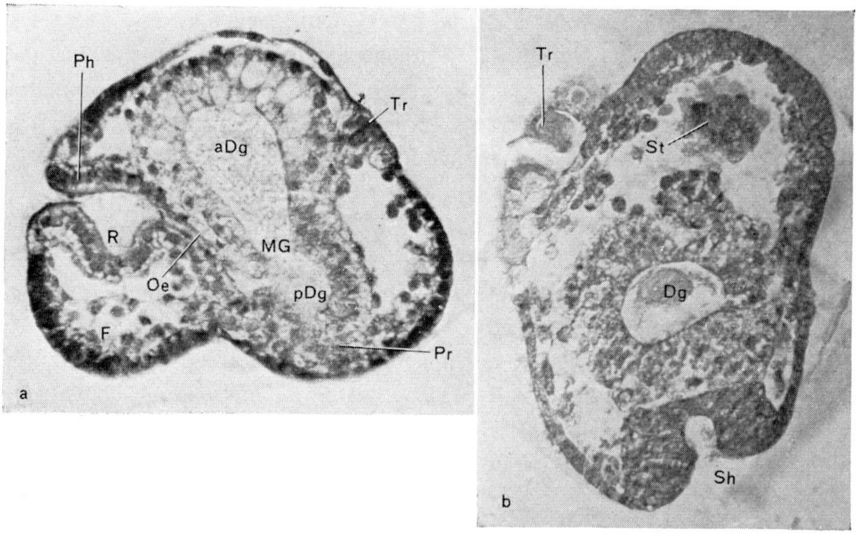

Fig. 10. Young trochophora larva of *B. tentaculata.* (a) sagittal, (b) frontal. Explanations: aDg, pDg, anterior and posterior diverticula of digestive gland; F, foot anlage; MG, midgut; Ph, pharynx; Pr, proctodaeum; Oe, oesophagus; R, radular sac; Sh, shell gland; St, stomodaeum; Tr, trochal cells. (Photographs of 6μ paraffine sections through embryos.)

Little is known about the mechanisms of the differentiation processes during this part of the development. Some information, however, is available about the processes occurring in the pretrochal region of the Limnaea embryo. Some time ago Raven observed that a typical effect of a treatment with lithium chloride is the induction of malformations in the head region. These injuries can be classified according to the degree of gradual 'cyclocephalic' malformations: synophthalmia, monophthalmia (= cyclopia), anophthalmia, and acephaly.

The differentiation in the pretrochal region of the Limnaea embryo starts with an easily recognizable and characteristic pattern of blastomeres. This whole ectodermal region is formed by cells which are derived from the first three quartets of micromeres. The cell divisions of the micromeres and their descendants are regulated in such a way that small and large cells are formed which are arranged in a characteristic pattern: directly at the animal pole of the embryo a small area of large ectodermal cells is found. They form the so-called apical plate. Just beneath the apical plate on both sides two regions composed of small ectodermal cells can be observed. These two areas are the 'cephalic plates'. They are surrounded again by big-celled ectoderm which during further development gives rise to the head vesicle. Finally, this tissue is bordered by the cells of the trochal ring which are especially large. Similar topographical relationships of ectoderm cells of different size classes are found in exogastrulated embryos. This demonstrates that the development of the characteristic cellular pattern in the ectoderm of the head region is independent from the underlying entodermal material.

It is possible to change experimentally the sequences of cell divisions in the head region, for instance, by a treatment with lithium chloride (Verdonk 1965). In this way, the pattern formed by small- and large-celled ectoderm in the head region may be altered. Typical for such types of effects is the observation that in the dorsal region of the apical plate the descendants of one particular blastomere, which normally do not divide and, therefore, form large-celled ectoderm, continue to divide in embryos treated with lithium chloride and change into small-celled ectoderm. The small-celled ectoderm formed in this way bridges the gap between the two cephalic plates. Thus, instead of two distinct small-celled areas only one large area is found. In consequence, during further development embryos of that type show the typical head malformations of the cyclocephalic series. These experiments might be judged as a good demonstration for the fact that comparatively minor primary changes of cell divisions during the cleavage stage might finally result in severe deviations in the normal topographical relationships of differentiated organs.

8.7. Organogenesis

Typical for the following period of development after the formation of the trochophora larva is the differentiation of the definitive organs of the adult snail. Our knowledge about these processes is rather poor. However, some experimental results have been received about the formation of the shell gland and the stomodaeum. Thus, this chapter will mainly deal with problems of the formation of these two prominent organs. In addition, some information about the differentiation of several head organs such as eyes, statocysts, or tentacles is also available.

8.7.1. The shell gland

The shell gland appears for the first time in young trochophore larvae at the dorsal side of the post-trochal ectoderm. At first the primordium of the shell gland consists of a thickening of ectodermal cells (fig. 11b). Later, in the centre of the thickened region an invagination of a part of the enlarged cells takes place (fig. 11c). During further development the shell gland is

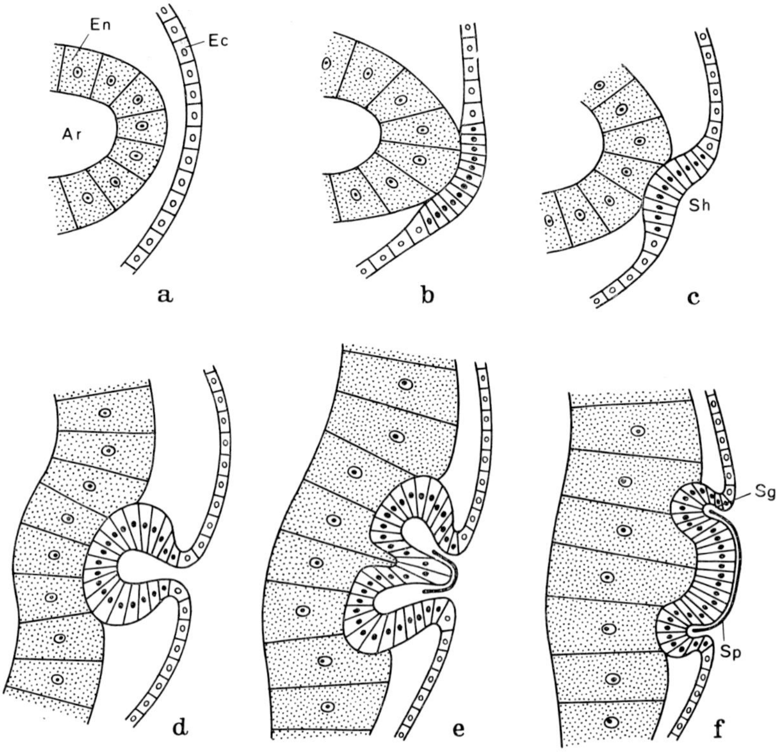

Fig. 11. Developmental stages during the formation of a shell gland. Explanations: Ar, archenteron; Ec, ectoderm; En, entoderm; Sg, shell groove; Sh, shell gland; Sp, shell primordium.

displaced to the left side of the embryo. Simultaneously, the shell gland enlarges more and more (figs. 10b, 11d). The central parts of the invagination rise again to the surface of the embryo (fig. 11e). Thus, the shell gland changes from the form of a shallow trough to a circular groove which in still later stages becomes the mantle fold. The groove surrounds an area which is called the shell field. This field secretes the shell primordium in the form of a thin larval shell composed of conchioline. During development the shell groove extends more and more (fig. 11f) and the shell (mantle) field finally covers large parts of the visceral hump.

Although in exogastrulated embryos the histological differentiation of many types of tissues appears to be quite normal (for instance in the head region), neither in Limnaea nor in Bithynia has a shell gland ever been found in such disturbed embryos.

After treatment of embryos of *L. stagnalis* with lithium chloride the invagination of the entoderm during gastrulation sometimes happens not to be completely suppressed. Then a more or less deep invagination of the archenteron may be found. In contrast to the completely exogastrulated embryos such types may sometimes develop a shell gland. It invariably appears exactly at that position where the tip of the archenteron touches the ectoderm from the inner side of the embryo. In some of these cases the direction of the invagination of the archenteron is changed and the tip of the archenteron comes into contact with the inner side of the ectoderm in the pretrochal region. In such embryos a shell gland is developed in the pretrochal region where it is derived from cells which normally develop into organs of the cerebral plates, e.g. eyes, tentacles, or cerebral ganglia (fig. 12). Raven concluded from this finding that the shell gland is induced

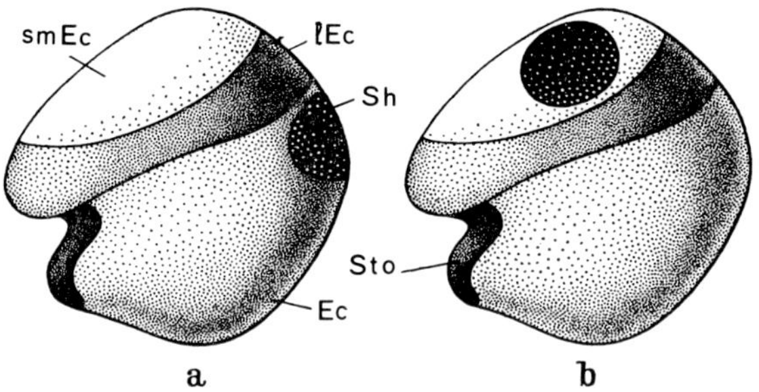

Fig. 12. Topographic relationships in *L. stagnalis* (a) of a normal embryo, (b) after disturbance of entodermal invagination and formation of a shell gland in the pretrochal region. Explanations: lEc, smEc, large-celled and small-celled ectoderm; Sh, shell gland; Sto, stomodaeum. (After Raven, changed.)

in the ectoderm by contact of the tip of the archenteron with the covering ectoderm. This was the first assumption of the occurrence of a non-autonomous, dependent process of differentiation in a molluscan embryo. During the normal development of all shell-carrying molluscs a stage can regularly be found, in which the tip of the archenteron contacts the newly developed shell gland. Therefore, Raven proposed the hypothesis that this type of differentiation of the shell gland might generally be true.

Soon after Raven's first experiment this assumption could be confirmed for the fresh water prosobranch *B. tentaculata* (Hess 1956a, b). In exogastrulated embryos of this species a very curious phenomenon has been found. In young trochophore larvae the entoderm grows considerably. At the same

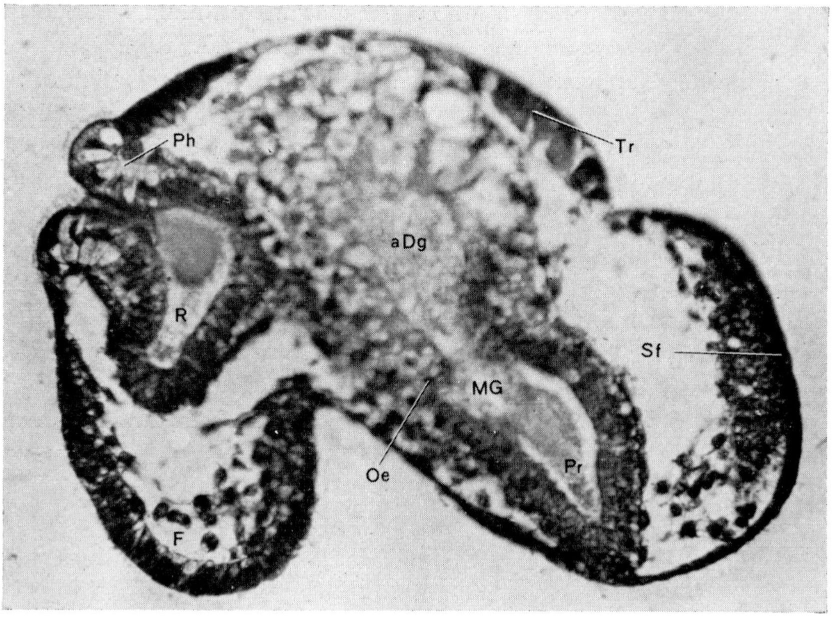

Fig. 13. Trochophore larva of *B. tentaculata*. Explanations: aDg, anterior diverticulum of the digestive gland; F, foot; MG, midgut; Oe, oesophagus; Ph, pharynx; Pr, proctodaeum; Sf, shell field (the shell grooves are not visible); Tr, trochal cells. (Photograph of a 6μ sagittal paraffine section through an embryo.)

time the archenteron is subdivided into the anterior and posterior diverticula of the digestive gland and into the midgut itself (fig. 13). Especially the formation of the large anterior digestive gland diverticulum is accompanied by a strong elongation growth of the entoderm and by a characteristic evaginational movement of the tissue. It turned out that this special movement of a particular part of the entoderm is an autonomous process: it can also be seen in corresponding stages of exogastrulated embryos. In many exogastrulated embryos of Bithynia, as a consequence of such a movement, a secondary and delayed invagination of that part of the entoderm occurs which is differentiated into the anterior diverticulum of the digestive gland (figs. 9d, 14). The invagination of the digestive gland occurs at a much later stage than the invagination of the archenteron occurs in normal embryos. At the time of the formation of the digestive gland normal embryos already possess a large and well developed shell gland. Exogastrulated embryos, no matter whether they arise from half embryos after separation of blastomeres or whether they arise after treatment of whole embryos with

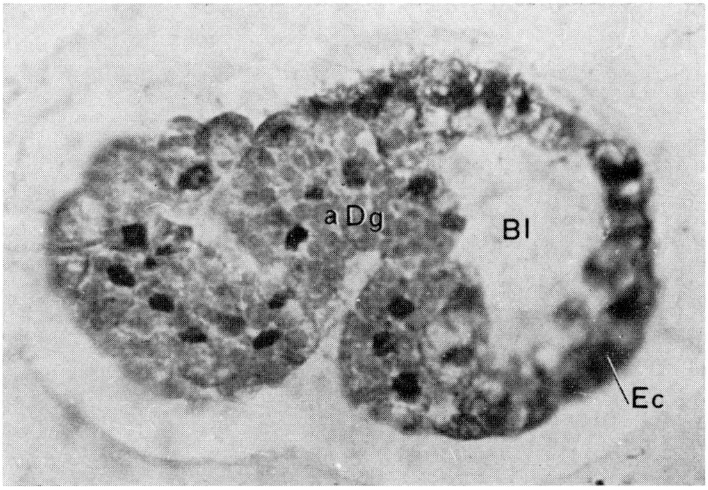

Fig. 14. Exogastrulated embryo of *B. tentaculata*. Secondary delayed invagination of a part of the entoderm during the formation of the anterior diverticulum of the digestive gland. Explanations: Bl, blastocoel; aDg, anterior diverticulum of the digestive gland; Ec, ectoderm. (Photograph of a 6μ paraffine section through an embryo.)

lithium chloride, never possess a shell gland before the invagination of the anterior digestive gland diverticulum, but always possess this organ after that event (fig. 9d). In those cases where the differentiation of the digestive gland does not result in a (secondary and delayed) invagination of a part of the entoderm into the blastocoel, a shell gland is invariably lacking (fig. 9e). This has been checked for completely exogastrulated embryos as old as 40 days (for comparison: emergence of young, normally developed snails occurs at about the twentieth day after ovoposition). The shell gland which is developed at a very late stage in these cases is always found exactly at that position where the tip of the digestive gland diverticulum comes into contact with the inner wall of the ectoderm. There is hardly any other explanation for this finding than the assumption of a contact induction according to the hypothesis of Raven.

More evidence for this hypothesis has been received from observations with half embryos: in Bithynia, half embryos can be produced by separation of blastomeres. However, such a microsurgical operation is only possible during the first two cleavage divisions. In addition, it is possible to do this operation inside the egg capsule in such a way that both halves of the egg survive. It is therefore possible to raise single-egg twins together within their egg capsule and the differentiation capacities of these pairs can be directly investigated. The first cleavage division separates the egg approximately into an anterior (AB) and a posterior (CD) half (fig. 15). If both

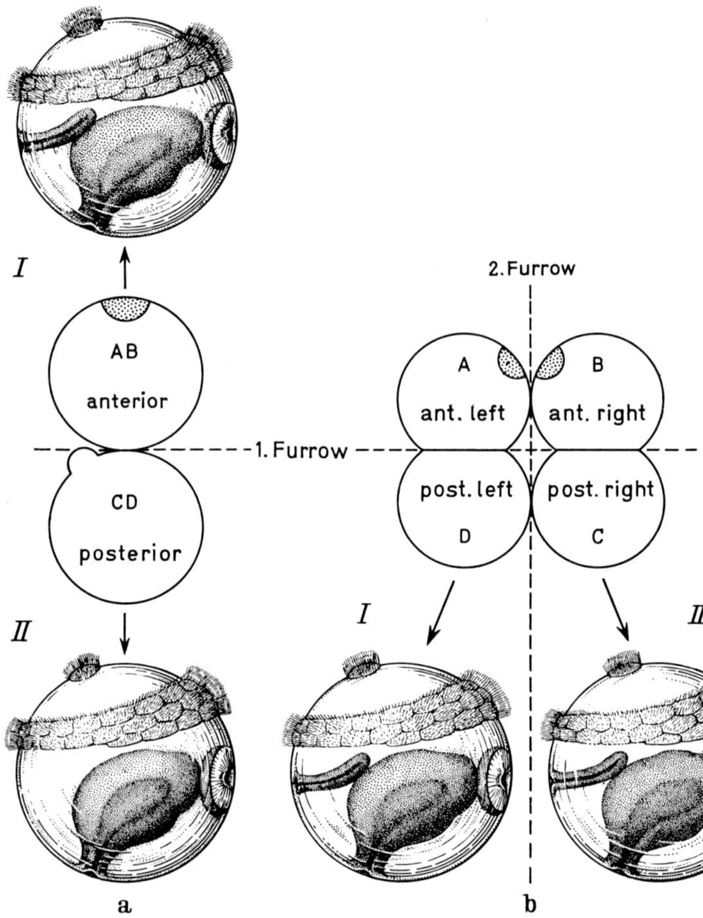

Fig. 15. Diagrammatic drawing of the microsurgical operation for separation of blastomeres during the first (a) and the second (b) cleavage division in *B. tentaculata*. Dotted: morphogenetic area which differentiates into the stomodaeum.

halves are raised together, each twin partner is able to develop a shell gland provided that in both halves the secondary invagination of the anterior digestive gland diverticulum has occurred. However, the presumptive area of that germ material which normally forms the shell gland is exclusively located within the posterior part (CD) of the embryo. If the shell gland would be developed by an autonomous differentiation of a morphogenetic area which is determined at an early stage of development, only one partner of the twins, namely that derived from the CD blastomere should be able to form a shell gland after a separation of blastomeres during the first cleavage division. As both halves of the twin do develop a shell gland, the only possible explanation which remains is the assumption of a dependent, regulative formation of that organ.

Similar results have been obtained for *L. stagnalis*. Also in this species the anterior half of an egg which is derived from the AB blastomere is able to form a shell gland (Hess 1957). The only necessary prerequisite for the formation of a shell gland in these embryos is the invagination of the entoderm, at least partly.

The experiments performed with half embryos of *B. tentaculata* provide further information about the differentiation of the shell gland: originally, Raven assumed in his hypothesis that only a comparatively small part of the archenteron, namely the small-celled presumptive material of the midgut which is located at the very tip of the archenteron, had the capacity to induce a shell gland after contact with ectoderm. However, in Bithynia this material is not invaginated during the secondary and delayed dislocation of the material of the digestive gland into the blastocoel. In older embryos of this type the small-celled midgut tissue is still found at the surface of the embryo. Thus, in these cases at least the small-celled ectodermal material can not act as an inducer of the shell gland. Therefore, one has to assume that considerably larger parts of the entoderm, in particular also parts of the large-celled entoderm and perhaps even all of the entodermal material is able to act as inducer. In addition, the results demonstrate that in Bithynia the entoderm retains the capacity for induction of a shell gland for a rather long period. On the other hand, the ectoderm also retains its competence to act upon inductive influences coming from underlying entodermal material for an equally long period.

These findings again could be confirmed for *L. stagnalis* in experiments with exogastrulated half embryos. If such embryos are cultivated for especially long periods, for instance 30 or more days, one observes that single cells or groups of cells from the exogastrulated entoderm show a tendency to move into the cavity of the exogastrula sphere. One often gets the impression that the cells immigrate actively with amoeboid deformations into the cavity. It can happen that in embryos which show such an immigration of single entodermal cells a shell gland is formed. Then the formation of the organ occurs at a very late stage of development. In all embryos of this type one or a few single entodermal cells are always found in contact with the new shell gland. Thus, the capacity to induce a shell gland is retained in the entoderm cells for a very long period. In addition, large parts of the entoderm and not only its small-celled fraction of the midgut primordium are capable of inducing a shell gland. The latter statement follows from the observation that most of the immigrated entodermal cells which obviously have induced the formation of a shell gland in these exceptional cases belong to the large-celled type which normally develops into digestive gland tissue. Finally, these findings demonstrate that an induction of a shell gland does not necessarily need a sublayer of closed entodermal tissue but can also be provoked by single cells.

A question which in both Limnaea and Bithynia has not yet been solved is whether there exists a limitation of the competent area in the ectoderm. Raven has demonstrated experimentally that shell glands can develop in pretrochal regions. Thus, the competent area is clearly much larger than the area which normally differentiates into this organ. However, so far a shell gland has never been found in the vicinity of the stomodaeum, e.g. on the ventral side of the ectoderm. In Bithynia the secondary invagination of the entodermal material during the formation of the anterior digestive gland diverticulum is always directed to the dorsal side. Thus, the formation of the shell gland on this side of the embryo is what one would expect. In addition, in older half embryos of Limnaea after complete exogastrulation the shell gland is always found opposite the stomodaeum, e.g. again in the dorsal part of the embryo. In these cases the shell gland is induced by single immigrated entodermal cells. Presumably, the cells have a tendency to move preferentially towards the dorsal parts of the embryo. These findings might indicate the existence of some kind of directional activity in the embryos which forces the archenteron or the digestive gland material and also single immigrating entodermal cells into a dorsal direction. It is also possible to assume that the competence to react to shell gland induction is restricted by some inhibiting effect which spreads from the region of the stomodaeum. The inhibited area could possibly become larger with the age of the developing embryo. This would also explain why differentiation of a shell gland, also after the immigration of single entodermal cells, occurs exclusively in the dorsal part.

As far as can be seen shell glands are in all cases of similar size. Specifically, there are no size differences in shell glands of whole embryos and embryos that have developed from separated blastomeres. Shell glands which have been developed in completely exogastrulated embryos of *L. stagnalis* after the immigration of single entodermal cells are also of normal size. Thus, it seems as if the strength of an induction does not play a role here. There might be a threshold value which has to be reached if a shell gland is to be formed, but once the value has been reached an induction will take place and the further differentiation of the induced material will occur in a 'self-regulative' fashion. It is also important to mention that in no case so far observed, more than one shell gland per embryo has been found. This is true although in completely exogastrulated embryos immigrated entodermal cells are dispersed over large parts of the embryos and one would, therefore, expect the induction of two or even more separate shell glands. The fact that this was never observed indicates the immediate inhibition of the differentiation of any further shell gland once the first gland anlage has been established and started to develop.

During recent years some experiments have been published which demonstrate that this type of induction of the shell gland might also exist in other species of molluscs. For instance, evidence for an induction of the shell gland by entodermal material has been found in the marine prosobranch snail *Ilyanassa obsoleta* (Cather 1967; see the chapter by Clement in this volume).

8.7.2. The stomodaeum

The formation of the stomodaeum is principally different from the development of the shell gland as just described. In normal development a small area of ectodermal cells in the vicinity of the blastopore invaginates into the blastocoel to form the stomodaeum during the transformation of the gastrula into a trochophora larva (figs. 9d, 10a). An ectodermal tube is formed the inner end of which remains closed in the beginning stages. Only later it opens into the cavity of the midgut which in the meantime has been developed. The next step in the differentiation of the stomodaeum is its subdivision into pharynx, the radular sac and the oesophagus (figs. 10a, 13). During the early stages in which the funnel-shaped stomodaeum is still closed its cells are already histologically different from the rest of the ectoderm and can easily be distinguished. Later, differentiation proceeds and the oesophagus, the radular sac and the pharynx are built up by cells of a specific and easily recognizable type. In exogastrulated embryos of Limnaea Raven has found a small area of ectodermal cells which was located between trochal cells and entoderm. These cells were histologically differentiated and showed the characteristic properties of stomodaeum cells. In some cases a shallow indentation has also been observed. Obviously, stomodaeum material is able to differentiate also in completely exogastrulated embryos.

In Bithynia, exogastrulated half or whole embryos regularly form a stomodaeum, the material of which is deeply invaginated into the cavity of the vesicular embryo (fig. 9d). The invagination of the stomodaeum and the differentiation of its cells are also observed in those cases where no secondary delayed invagination of the anterior ventriculum of the digestive gland had occurred (fig. 9e). Thus, the cells of the stomodaeum are capable of independent invagination. However, in the cases so far described no further differentiation into pharynx, radular sac and oesophagus has been observed. In addition, the stomodaeum remains closed, perhaps because there does not exist a midgut cavity in these embryos into which the stomodaeum normally opens.

Further results on the differentiation of the stomodaeum have been obtained by an analysis of half embryos in Bithynia. As has already been mentioned, half embryos are produced by separation of blastomeres during the first and second cleavage divisions (fig. 15). If the operation is performed during the second cleavage it is only successful if the separation follows the second cleavage furrow. In the furrow of the first cleavage the blastomeres are firmly connected at this stage and, therefore, are invariably destroyed when separation is attempted in this direction. Thus, two types of single egg twin pairs can be distinguished: after separation during the first cleavage division only approximately anterior (AB) and posterior (CD) halves are received; in contrast to this, after separation during the second cleavage division, exclusively lateral halves are produced, namely left ones (AD) and right ones (BC). One always finds that among a twin pair derived from separation during the first cleavage only one partner possesses a stomodaeum whereas the second partner does not have this organ. Both partners may, however, have a shell gland. In contrast, all half embryos that arise from a separation during the second cleavage division possess a stomodaeum.

To explain this finding it is assumed that the stomodaeum develops from a morphogenetic area which is already determined before the first cleavage division and which possesses the capacity for an autonomous and independent differentiation. This area appears to be located in the region of the AB blastomere. After separation during the first cleavage division this area remains complete within the anterior half of the embryo. In contrast to this, the area of the stomodaeum may be divided after isolation of blastomeres during the second cleavage division. Thus, each half embryo receives approximately one half of the morphogenetic area. Each partial area seems to possess the capacity for the normal formation of a stomodaeum anlage. In complete agreement with corresponding observations during the formation of shell glands it seems that there is no difference in size between a stomodaeum derived from a whole morphogenetic area and one derived from only half an area.

A detailed study of the further development and differentiation of the stomodaeum in Limnaea has been executed by Raven (1958). After treatment of young embryos with lithium chloride, heat shocks, or centrifugation, an eversion of the stomodaeum tissue is observed in some cases. The differentiation of the oesophagus takes place normally. A pair of salivary glands and the radular sac are also formed. During the development of the salivary glands some dependency upon the oesophagus seems to exist. Thus, one may presume that here again some kind of dependent differentiation takes place. The radular sac is often quite normally differentiated. It opens independently on the surface, and it may contain a small radula composed of several rows of teeth which may extend to the outer surface of the everted pharynx. Muscles are also formed. They seem to insert at the 'correct' positions. These findings demonstrate that the determination of the head organs, including the nervous system, becomes irreversibly defined at an age of 3 to $3\frac{1}{2}$ days, e.g. during the early trochophore stage. After this the anlagen of the organs seem to be irreversibly determined, and they differentiate normally also in those cases where the topographical relationships have been severely changed.

8.7.3. The organs of the head

There are some observations dealing also with the development of the head organs, such as tentacles, eyes and statocysts. As was already mentioned, it is possible to produce half embryos of *B. tentaculata* during the first two cleavage divisions by separation of blastomeres. The microsurgical operation is performed within the egg capsules and both halves of an egg may survive. As the pairs of single-egg twins are raised within their original egg capsule, a direct comparison of the differentiation capacities of the half embryos derived from the same egg cell is possible. At late developmental stages both partners of a twin pair may possess quite often antennae, eyes, or statocysts. Thus, material originally included in only one egg is capable of developing three or four of each of these organs instead of normally only a single pair. Similarly, duplications of antennae, eyes and statocysts have been obtained in whole embryos of *L. stagnalis* after various kinds of experimental influences, such as heat shocks or changes of the ionic environment. This demonstrates very clearly that during the differentiation of these organs regulations are possible. At the moment, however, no further data on these processes are available.

8.8. Concluding remarks

It is not possible, at the moment, to give a complete picture of the differentiation events taking place during the embryonic development of gastropod species. Our knowledge in this field is very incomplete. One may, perhaps, tentatively distinguish between two periods during the course of development of the gastropod eggs. During the first period a predominance of independent autonomous differentiation processes is found. This period includes development up to the formation of the trochophore larva which already possesses the general basic tripartite organization of a molluscan organism into head, foot and visceral mass. In contrast, during the second period of development more and more dependent differentiations come into action. Here, organ anlagen have to be induced in some tissue by some other tissue and regulation according to the actual topographical relationships within the individual embryo can occur. Therefore, we have no further reason today to distinguish between so-called mosaic and regulative embryos.

References

BROWN D. D., 1966. The nucleolus and the synthesis of ribosomal RNA during oogenesis and embryogenesis of *Xenopus laevis*. Natl. Cancer Inst. Monograph. *23*, 297–309.

CATHER J. N., 1967. Cellular interactions in the development of the shell gland of the gastropod, *Ilyanassa*. J. Exptl. Zool. *166*, 205–224.

CLEMENT A. C. and A. TYLER, 1967. Protein-synthesizing activity of the anucleate polar lobe of the mud snail, *Ilyanassa obsoleta*. Science *158*, 1457–1458.

DAVIDSON E. H., G. W. HASLETT, R. J. FINNEY, V. G. ALLFREY and A. E. MIRSKY, 1965. Evidence for the prelocalization of cytoplasmic factors affecting gene activation in early embryogenesis. Proc. Natl. Acad. Sci. US *54*, 696–704.

DIVER C., 1925. The inheritance of inverse symmetry in *Limnaea peregra*. J. Genet. *15*, 113.

HARRIS S. E. and H. S. FORREST, 1967a. RNA and DNA synthesis in developing eggs of the milkweed bug, *Oncopeltus fasciatus*. Science *156*, 1613–1615.

HARRIS S. E. AND H. S. FORREST, 1967b. Inhibition by certain pteridines of ribosomal RNA and DNA synthesis in developing *Oncopeltus* eggs. Proc. Natl. Acad. Sci. US *58*, 89–94.

HESS O., 1956a. Die Entwicklung von Halbkeimen bei dem Süßwasser-Prosobranchier *Bithynia tentaculata*. Arch. Entwicklungsmech. Organ. *148*, 336–361.

HESS O., 1956b. Die Entwicklung von Exogastrula-Keimen bei dem Süßwasser-Prosobranchier *Bithynia tentaculata*. Arch. Entwicklungsmech. Organ. *148*, 474–488.

HESS O., 1957. Die Entwicklung von Halbkeimen bei dem Süßwasser-Pulmonaten *Limnaea stagnalis*. Arch. Entwicklungsmech. Organ. *150*, 124–145.

HESS O., 1962. Entwicklungsphysiologie der Mollusken. Fortschr. Zool. *14*, 130–163.

MONROY A. and P. R. GROSS, 1967. The control of gene action during echinoderm embryogenesis, in 'Morphological and biochemical aspects of cytodifferentiation'. S. Karger, Basel, pp. 37–51.

MONROY A. and A. TYLER, 1967. The activation of the egg. Fertilization. Academic Press, New York, pp. 369–412.

NEMER M., 1967. Transfer of genetic information during embryogenesis. Progr. Nucl. Acid Res. Mol. Biol. *7*, 243–301.

NEMER M. and A. A. INFANTE, 1967. Early control of gene expression, in 'The control of nuclear activity' ed. by L. Goldstein. Prentice-Hall, Englewood-Cliffs, N.J., pp. 101–127.

RAVEN C. P., 1952. Morphogenesis in *Limnaea stagnalis* and its disturbance by lithium. J. Exptl. Zool. *121*, 1–72.

RAVEN C. P., 1958. Abnormal development of the foregut in *Limnaea stagnalis*. J. Exptl. Zool. *139*, 189–246.

RAVEN C. P., 1961. The storage of developmental information. Pergamon Press, London.

RAVEN C. P., 1963. The nature and origin of the cortical morphogenetic field in *Limnaea*. Develop. Biol. *7*, 130–143.

RAVEN C. P., 1967. Morphogenesis: The analysis of molluscan development, 2nd ed. Pergamon Press, London.

RAVEN C. P., 1967. The distribution of special cytoplasmic differentiations of the egg during early cleavage in *Limnaea stagnalis*. Develop. Biol. *16*, 407–437.

RAVEN C. P., A. C. DRINKWAARD, J. HAECK, N. H. VERDONK and L. A. VERHOEVEN, 1956. Effects of monovalent cations on the eggs of *Limnaea*. Pubbl. Staz. Zool. Napoli *28*, 136–168.

SPIRIN A. S., 1966. On 'masked' forms of messenger RNA in early embryogenesis and in other differentiating systems, in 'Current topics in developmental biology', vol. 1, ed. by A. Monroy and A. A. Moscona. Academic Press, New York, pp. 1–38.

STURTEVANT A. H., 1923. Inheritance of directing of coiling in *Limnaea*. Science *58*, 269.

VERDONK N. H., 1965. Morphogenesis of the head region in *Limnaea stagnalis*. Thesis, Utrecht, pp. 1–134.

G. Reverberi (ed.), Experimental embryology of marine and fresh-water invertebrates

CHAPTER 9

Dentalium

G. REVERBERI

Zoological Institute, University of Palermo, Italy

9.1. Introduction

The Dentalium egg has long attracted the interest of researchers, and embryology has received very valuable contributions from studies of this egg. The early studies concerned normal development. This was first described by Lacaze-Duthiers (1857) and afterwards by Kowalewsky (1883). Soon the egg was experimentally investigated, and in this context the names of Délage (1899), Wilson (1904) and Schleip (1925) should be mentioned. After this it did not attract further interest and only recently have a few papers been published (Arvy 1950; Reverberi 1958; Verdonk 1968a, b).

9.2. The material

Dentalium is a marine mollusc which belongs to the small group of Scaphopods. It lives in the mud, from which it is easily collected. The animals have separate sexes which, however, cannot be distinguished exteriorly. Hermaphroditic animals are occasionally found; their eggs, if autofertilized, develop normally (Reverberi unpubl.). The mature animals lay eggs or sperm spontaneously; the period of maturity extends from June to October in Naples. The egg is relatively large and is highly suited for embryological studies.

9.3. The ovarian egg

9.3.1. The ovarian egg has been investigated very little. The most recent paper on this subject is that by Arvy (1950). According to this author the very young oocyte (15 μ) has a large germinal vesicle, surrounded by a highly basophilic and pyroninophilic cytoplasm. Inside the germinal vesicle a small pyroninophilic nucleolus can be seen; the pyronine staining in both cytoplasm and nucleolus disappears after treatment with ribonuclease. At a later stage of development the nucleolus is larger and in budding activity and the buds are also pyroninophilic. The cytoplasm is homogeneous and contains small mitochondria mostly localized around the nucleus or in the region of the attachment of the oocytes to the wall of the ovary. At a later stage of development the nucleolus is still budding but is heterogeneous, with pyroninophilic and pyroninophobic zones. The cytoplasm is less basophilic and the mitochondria have the same distribution as noted earlier: in the proximity of the mitochondria, chromophobic inclusions are observed which are related to vitellogenesis. In the vitellogenetic oocytes the cytoplasm is no longer basophilic, with the exception of a region which extends from the point of attachment of the oocyte to the ovary up to the germinal vesicle; this region has a conical form.

9.3.2. Personal observations (unpubl.) add more data to the observations of Arvy. In previtellogenetic oocytes the nucleolus is large and compact; it stains intensely with azan. Besides the nucleolus one notices in the nucleus some small irregular bodies which also stain intensely with azan: they have probably originated from the nucleolus by budding (figs. 1a, b). The nucleolus is often heterogeneous, being composed of two parts which stain differently. In vitellogenetic oocytes the nucleolus no longer has a compact appearance, and stains very poorly. The nucleolus is lacking in mature oocytes. At the end of its growth the oocyte contains a large germinal vesicle surrounded by a distinct membrane, and the cytoplasm is filled with granules of different constitution and size. Some are large (they are probably pigment-granules) and stain intense blue with azan: others (yolk-granules?) stain intense red. Their distribution is characteristic, the blue granules being situated centrally, the red granules peripherally. The cortex shows large vacuoles which stain blue (fig. 2). A remarkable localization is that of some very small granules, which are situated at one pole, probably the antipole.

9.3.3. These structures and their localizations can be further visualized by means of the electron microscope. In previtellogenetic oocytes (fig. 3) a very large germinal vesicle with a double membrane pierced by numerous pores which are crowded with small granules can be seen. The cytoplasm contains the endoplasmic reticulum, which is not very organized however; a few mitochondria and some pigment-granules are also present. The

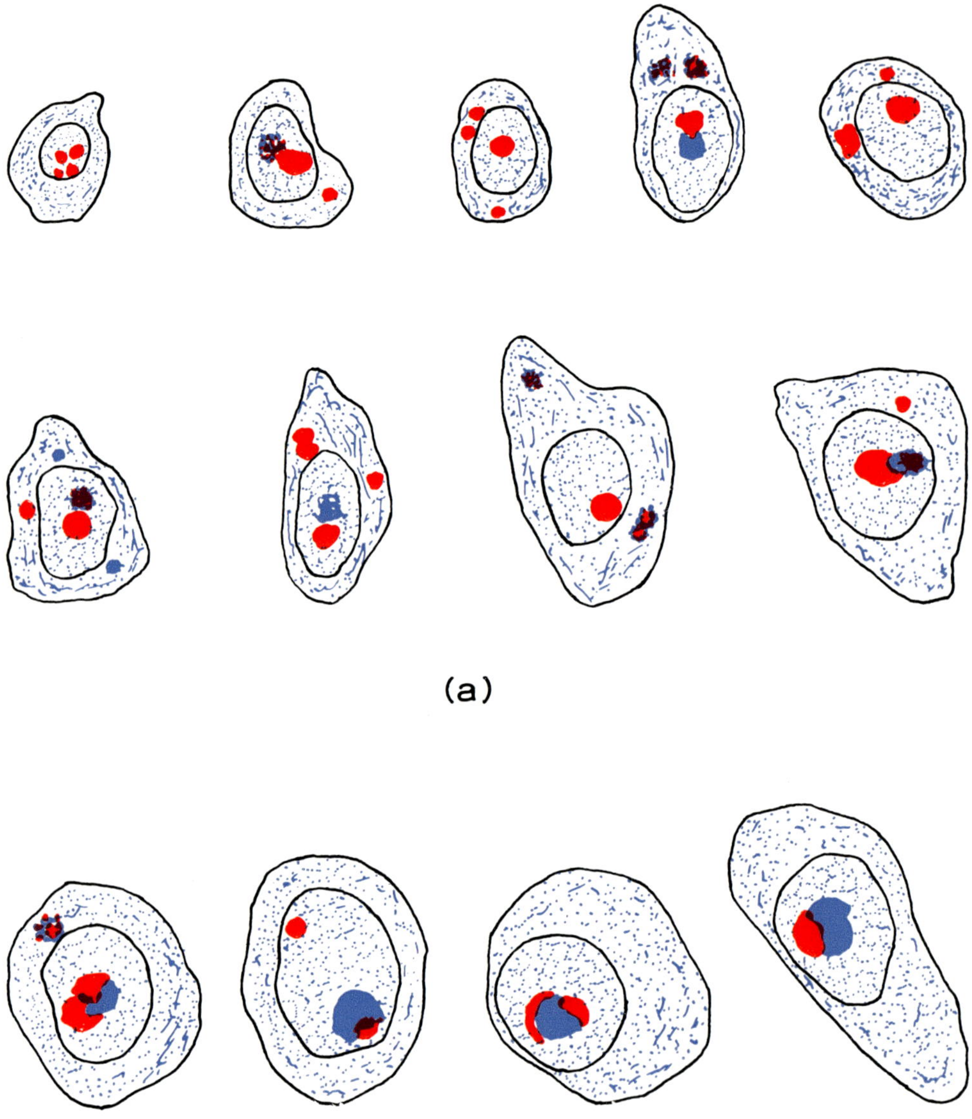

Fig. 1. Previtellogenetic oocytes of Dentalium stained with azan (a) and Rhodanil blue (b). Inside the large germinal vesicle one or more nucleoli can be seen; sometimes the nucleolus shows two regions differently stained. Small bodies which stain as the nucleoli are also frequently observed in the cytoplasm (original).

perinuclear region contains numerous small vesicles. The vitellogenetic oocytes have (fig. 4) a cytoplasm crowded with large yolk and pigment-granules; the region which probably corresponds to the point of attachment of the egg to the ovary (the vegetal pole) is, however, extremely rich in mitochondria and very small yolk-granules. The periphery is occupied by many large cortical granules and other still unidentified granules; numerous microvilli project from the plasma membrane (fig. 5).

9.4. *The ripe free egg*

The mature egg has been accurately described by Wilson. If obtained directly from the ovary by laceration, it appears as a yellow or red pigmented disc. The center is occupied by a large germinal vesicle. Some minutes after shedding this collapses and the vitelline membrane also breaks down. The egg rounds up, a jelly-like envelope is formed and the egg floats up from the bottom. The rupture of the vitelline membrane is caused by the jelly. The

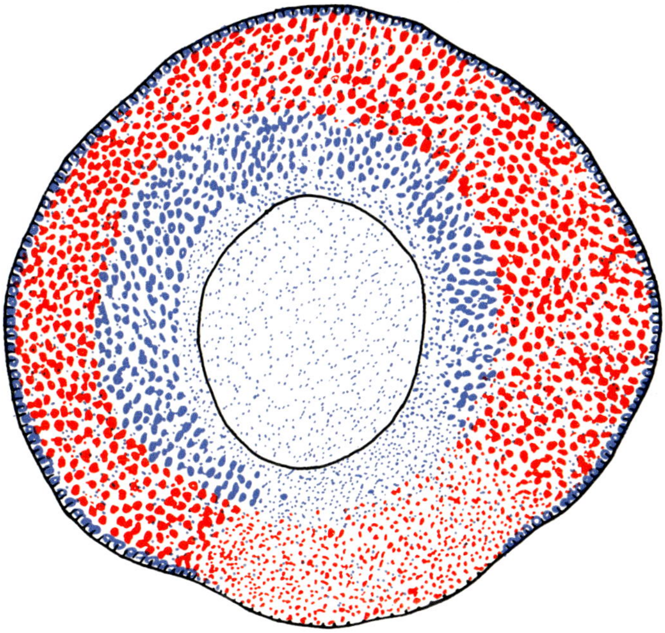

Fig. 2. Mature oocytes stained with azan: note the large germinal vesicle without the nucleolus, the cytoplasm filled with granules of different dimensions and colours, the peripheral nuclear region, the fine granular region at the vegetal pole and the cortical granules (original).

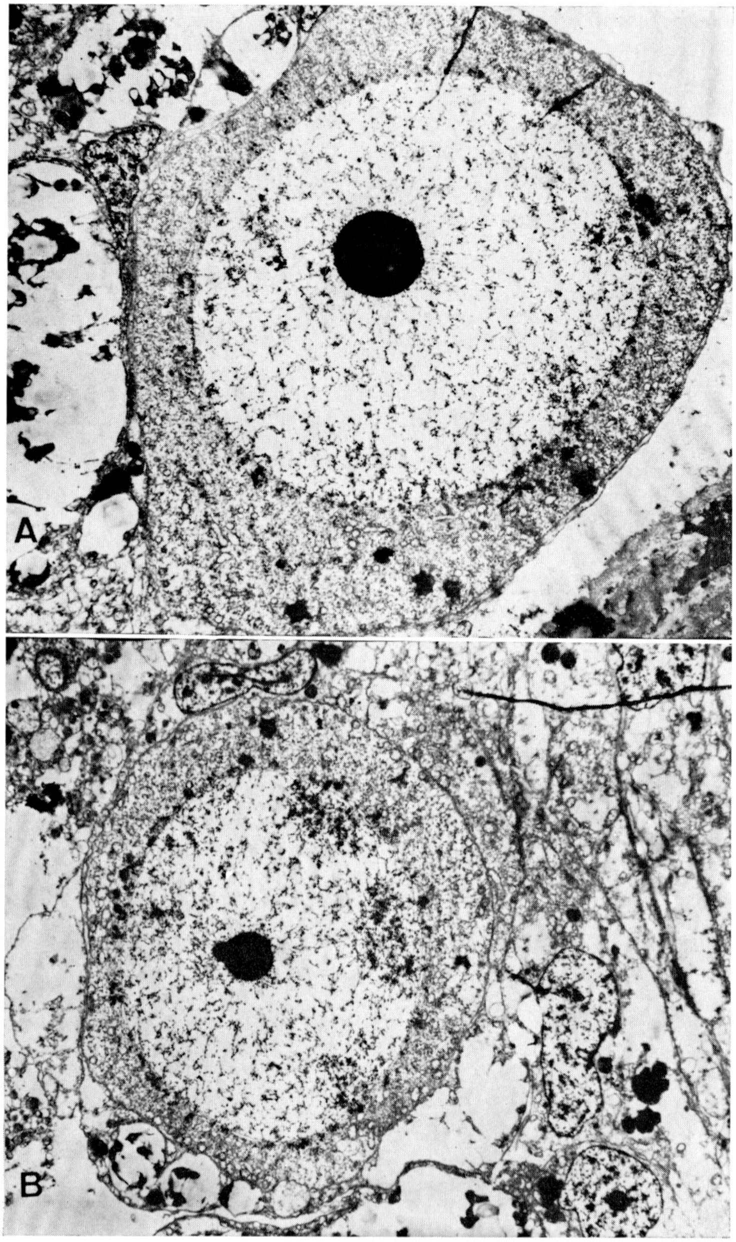

Fig. 3. Previtellogenetic oocytes (original electron micrograph).

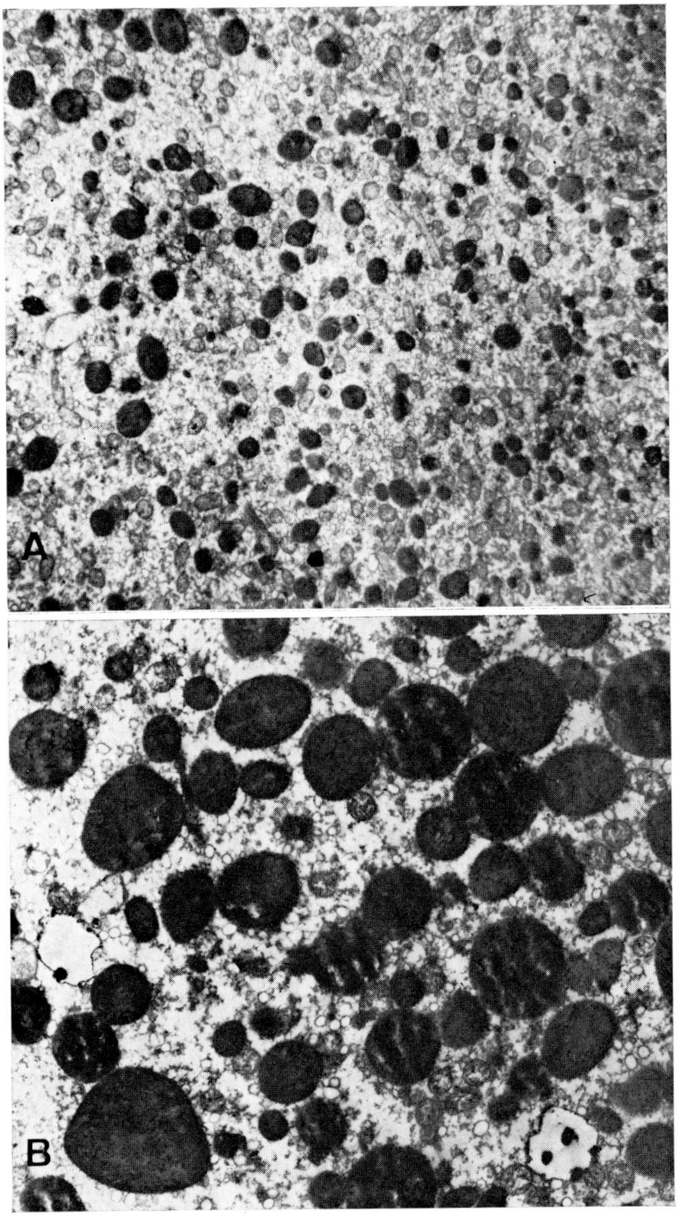

Fig. 4. Vitellogenetic oocytes: the vegetal region of the egg is made up of numerous mitochondria and small yolk-granules; B is a lateral region of the egg with large yolk-granules and very few mitochondria (original electron micrograph).

egg now escapes, the shrunken vitelline membrane remaining around the egg as a halo which is subsequently lost. The egg, now surrounded only by the jelly, shows two unpigmented areas at opposite poles, one of which is larger than the other. In the animal area is a central spot which probably corresponds to the meiotic spindle; the vegetal area, which bulges slightly outwards, corresponds to the point of attachment of the egg to the ovary. With Janus green the animal and the vegetal area stain green; these areas also stain blue after treatment for the Nadi reaction (cf. fig. 8). The coloration seems due to the presence of mitochondria (Reverberi 1958). The egg has been studied cytologically by Wilson. Of particular importance is the vegetal region which, unlike the other regions, does not contain any yolk- or pigment-granules and is made up only of a granular plasm which stains intensely with Congo red (fig. 6). Electron microscope studies are greatly needed: we can only show a figure in which the periphery is occupied by large cortical granules and the plasma membrane raised in long, thick microvilli (fig. 5b).

9.5. *Fertilization, segmentation and larval stage*

9.5.1. The rupture of the germinal vesicle is a necessary condition for the development of the egg. This has been shown by Délage (1899) who remarked that anucleated fragments derived from eggs with the germinal vesicle still intact are unfertilizable. According to Délage the germinal vesicle contains a substance which is responsible for ripeness of the egg. The spermatozoon enters at the vegetal pole. Before entering it has to overcome the imposing barrier of jelly which surrounds the egg: how this is done is not known. After fertilization the egg extrudes the two polar globules; at each extrusion the egg changes its shape, becoming slightly pyriform due to a small hernia which appears at the vegetal pole.

9.5.2. The pattern of segmentation (fig. 7) recalls in nearly every detail that of Ilyanassa. The first segmentation is preceded by the formation of a large lobe at the vegetal pole (L_1) which contains all the antipolar material. However, since the surface of the lobe is much larger than that of the original lower polar area, it is probable that part of the material situated in the anterior region of the egg also flows into the lobe. After formation of the first polar lobe the egg cleaves into two equal blastomeres, AB and CD: CD is connected with the polar lobe. This stage is termed the trefoil stage and does not persist for long, because the polar lobe is soon absorbed into the blastomere CD, which, consequently, becomes much larger than AB.

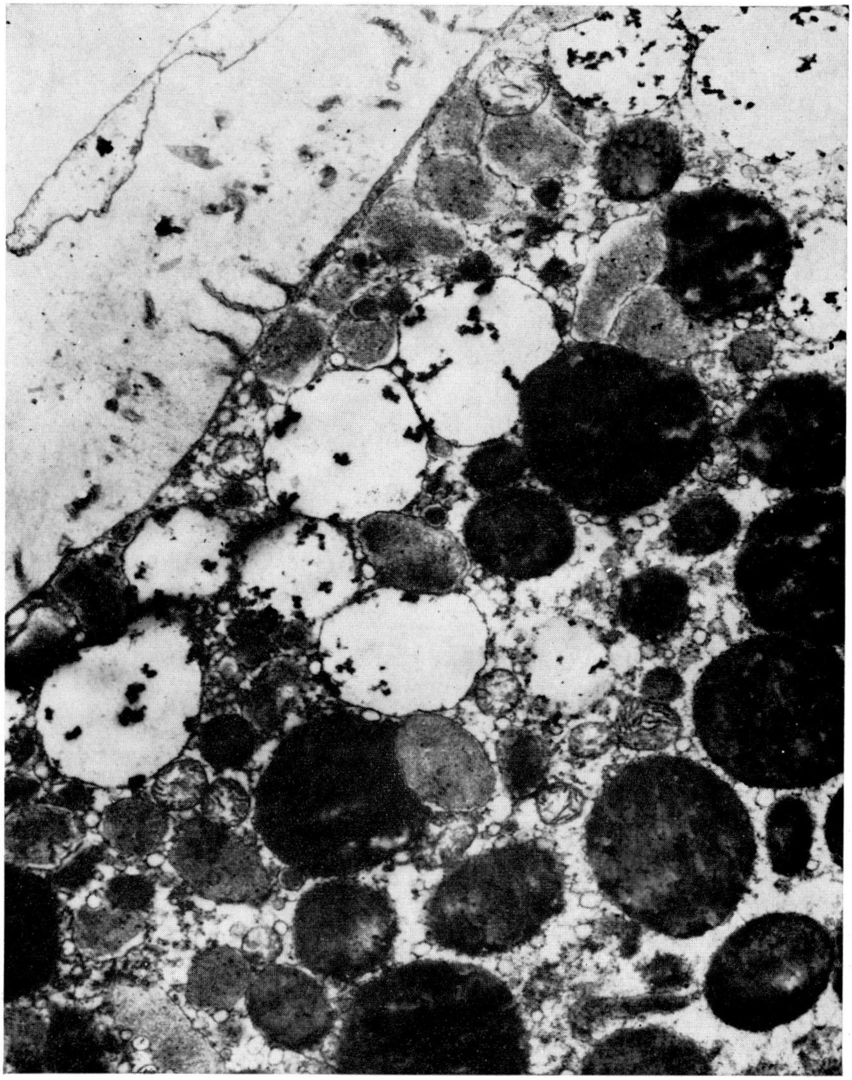

Fig. 5a. The cortical region of a mature oocyte: note the large cortical granules and other compact granules and microvilli (original).

The 2nd segmentation is ushered in by the reappearance of the lobe (L_2). This will be absorbed into the D-cell, which finally becomes larger than A, B and C. Before the egg enters into the 3rd segmentation the polar lobe reappears (L_3); the third cleavage follows, giving rise to the first quartet

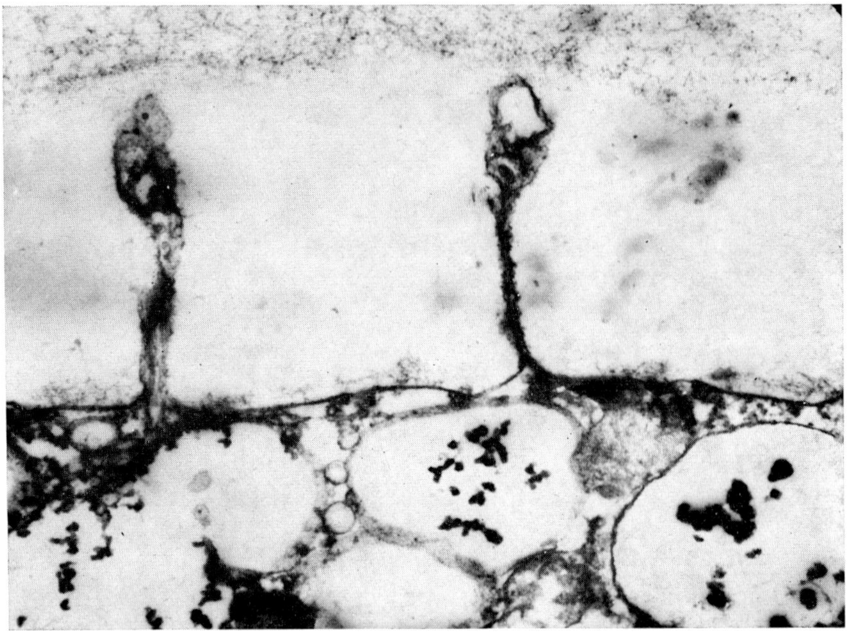

Fig. 5b. The cortical region of a ripe egg: note the large cortical granules, the microvilli which extend into the perivitelline space and the external layer of jelly (original).

of micromeres (1a, 1b, 1c, 1d). A second quartet of micromeres (2a, 2b, 2c, 2d) is formed by a new (4th) cleavage. The 2d cell is larger than the others and is the first somatoblast. Part of the material of the polar lobe is passed on to it. The fifth segmentation produces a third quartet of micromeres (3a, 3b, 3c, 3d); a fourth quartet is formed by the sixth cleavage (4a, 4b, 4c, 4d). The 4d cell is the second somatoblast, and also receives part of the vegetal plasm.

The animal plasm, on the other hand, is at first segregated in the animal region of the macromeres (A, B, C, D) and subsequently in the quartets of micromeres which will give rise to the trochoblasts. The segregation of the animal and vegetal plasms can be followed by the Nadi treatment or Janus green staining (fig. 8).

9.5.3. The larva is a trochophore (fig. 9) bearing a long tuft of sensory cilia at the apex, and three rows of cilia at the equator (the 'trochus'). The apical tuft arises from certain cells which together constitute the 'apical plate'; the equatorial cilia, on the other hand, arise from the trochoblasts.

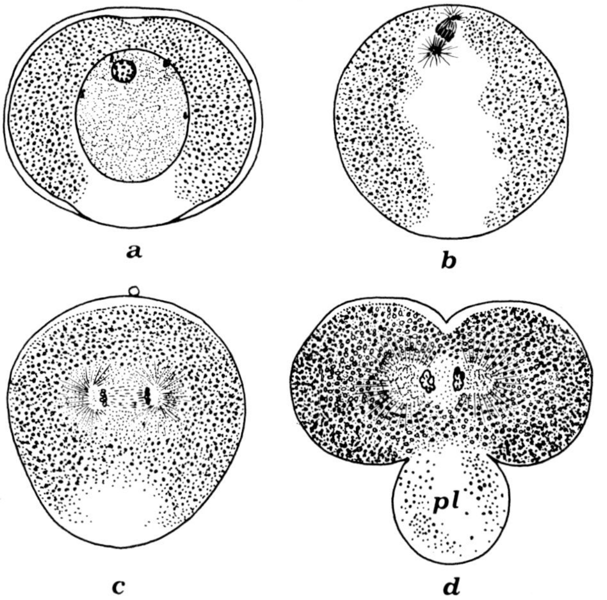

Fig. 6. Eggs at different stages: a, b, before fertilization; c, beginning of the first cleavage; d, trefoil stage; pl, polar lobe. (Redrawn from Wilson 1904.)

A very short tuft of hairs is present at the antipole and the anterior region of the trochophore is also covered with short cilia. In the post-trochal region there are the rudiments of the mantle fold and shell gland. At a more advanced stage the larva has a well-developed post-trochal region, while the pretrochal region becomes reduced. A shell and a trilobate extensible foot will finally be formed and the animal then sinks to the bottom and begins to excavate the mud for food.

9.6. The development of the fragments of the egg

9.6.1. According to Délage (1899) the fragments obtained from the unfertilized egg after the collapse of the germinal vesicle develop (if fertilized) into dwarf but otherwise normal trochophores. Wilson (1904) however, has remarked that only the fragments containing the vegetal area are able to develop into normal trochophores (fig. 10): before cleaving they form the polar lobe (of correct proportional volume), divide like an entire egg of diminished size, and, finally, give rise to normal dwarf trochophores. The upper, nucleated fragments, on the other hand, do not form the polar lobe,

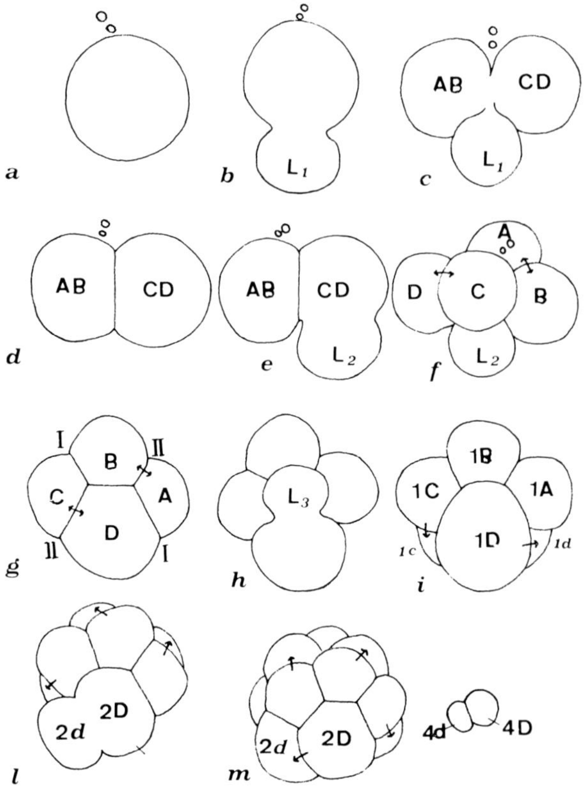

Fig. 7. Early cleavage of the egg; l, l_2 and l_3, the first, second and third polar lobes. (Redrawn from Wilson 1904.)

divide into two equal blastomeres, form the quartets of micromeres and finally give rise to trochophores without the apical tuft and post-trochal region. From the fragments obtained by a vertical cut, and which, therefore, contain a part of the vegetal plasm, a small complete trochophore always develops. These results lead to the conclusion that the vegetal plasm is essential for the genesis of certain 'structures'.

9.6.2. No less interesting are the results obtained with the fragments of the fertilized egg. If this is cut equatorially, an animal and a vegetal fragment are obtained. The animal fragment, which has both nuclei, cleaves without forming the polar lobe and finally gives rise to a larva which lacks the apical tuft and the mesodermic region. The vegetal fragment has no nucleus and

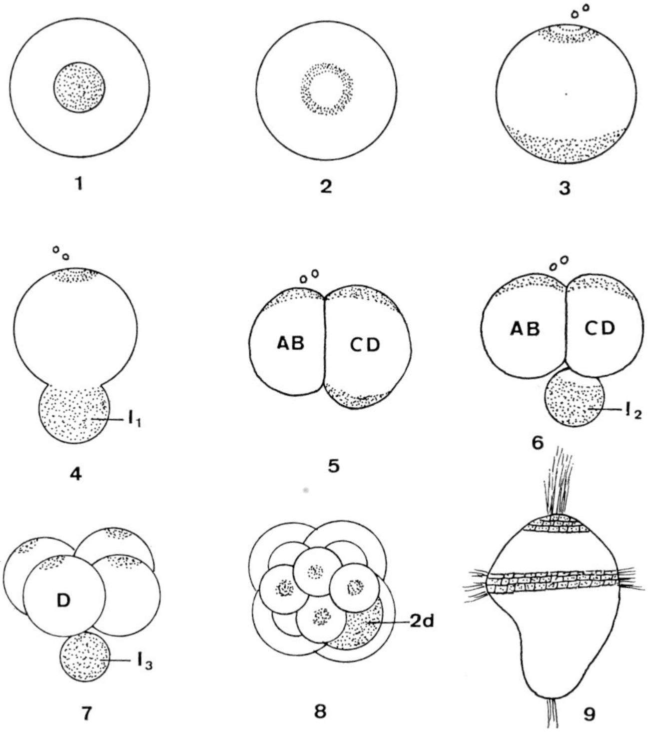

Fig. 8. An egg at different stages of development: the egg was stained with Janus green; the intensely coloured regions are indicative of the presence of large quantities of mitochondria (Reverberi 1958).

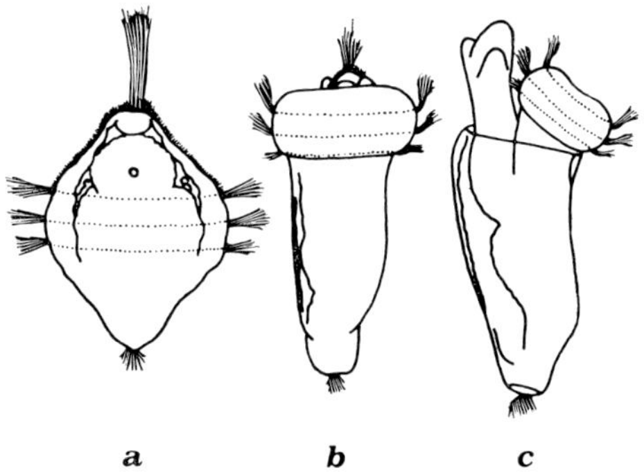

Fig. 9. Normal trochophores of 24 hr (a) and 72 hr (b) and (c). (Redrawn from Wilson 1904.)

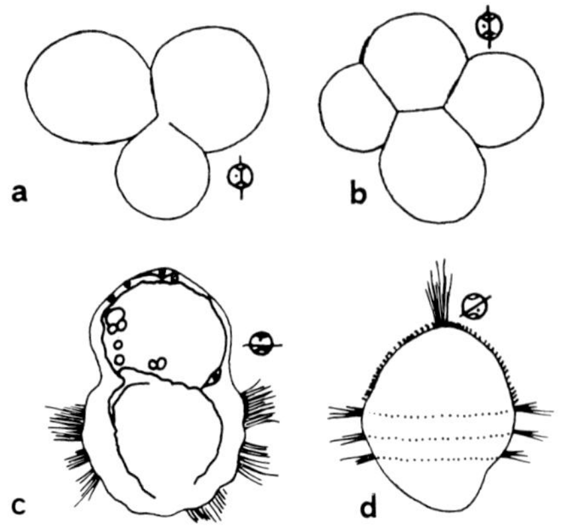

Fig. 10. Development of egg fragments: (a) and (b) segmentation of fragments containing the lower polar area; (c) development of a trochophore from a fragment without the lower area; (d) dwarf trochophore from a fragment with the lower polar area. (Redrawn from Wilson 1904.)

does not cleave; it only forms polar lobes synchronously with the segmentations of the corresponding animal fragment; the polar lobes are always of the same size as in normal eggs, in other words there is no regulation of size.

9.7. Removal of the polar lobe (fig. 11)

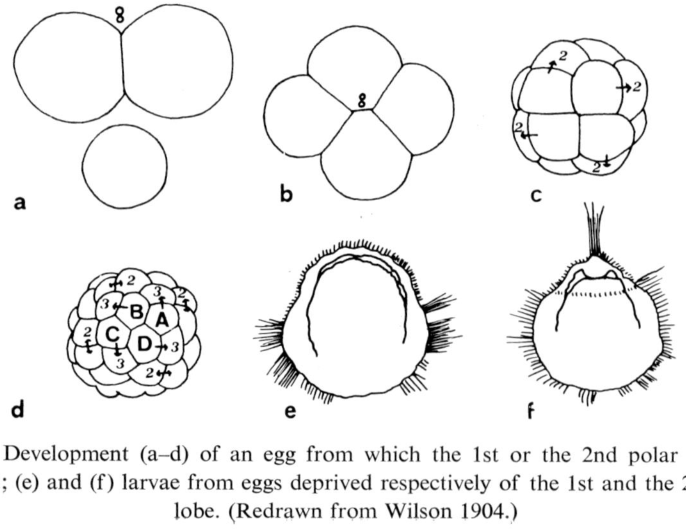

Fig. 11. Development (a–d) of an egg from which the 1st or the 2nd polar lobe was removed; (e) and (f) larvae from eggs deprived respectively of the 1st and the 2nd polar lobe. (Redrawn from Wilson 1904.)

The polar lobe can easily be removed at different times (L_1, L_2, or L_3). After isolation the polar lobe shows a periodical amoeboid activity (Wilson 1904) which is synchronous with the segmentation of the egg. It would thus seem that it has a memory; it should, however, be noted that this memory is exclusively cytoplasmic, since the lobe does not contain a nucleus. The results of removal of the polar lobe are the following:

9.7.1. Removal of the L_1. The egg cleaves without L_2 and L_3; AB and CD are the same size, and cannot be distinguished: also A, B, C and D, which are formed at the 2nd segmentation, are the same size.

Gastrulation is normal but the trochophore lacks all the post-trochal region; the three ciliary girdles are, however, present, although irregularly arranged. The trochophore has no shell gland or mantle fold; the apical tuft and the apical plate are also wanting.

The more developed larva does not reduce the pretrochal region and has neither foot nor shell. From these results it has been deduced that the vegetal plasm is necessary for formation of the apical plate and the post-trochal structures.

9.7.2. Removal of the L_2. The situation is not changed with regard to the post-trochal structures: however, the apical tuft is present (cf. fig. 11). This means that L_2 no longer has the specific substance necessary for the formation of the polar tuft: evidently during the period between the first and the second segmentation it has moved up and been segregated at the animal pole of the D-cell from which it is afterwards segregated in the 1d-cell.

9.8. Isolation of AB and CD (fig. 12)

At the 2-cell stage the two blastomeres can be isolated very easily. As mentioned, after the reabsorption of the polar lobe, the blastomere CD is larger than AB. After isolation, the blastomeres round up and cleave as if still forming part of the entire embryo. AB does not form any polar lobes, and develops into a larva which has neither post-trochal structures (mesoderm, foot, shell and mantle rudiments) nor apical tuft. CD, on the other hand, forms the polar lobes, cleaves normally, forms the first and the second somatoblasts, and finally gives rise to a larva which differs from normal only in having the post-trochal region too large and the pretrochal region too narrow.

Fig. 12. Development of isolated blastomeres AB and CD. (Redrawn from Wilson 1904.)

9.9. *Isolation of A, B, C and D (fig. 13)*

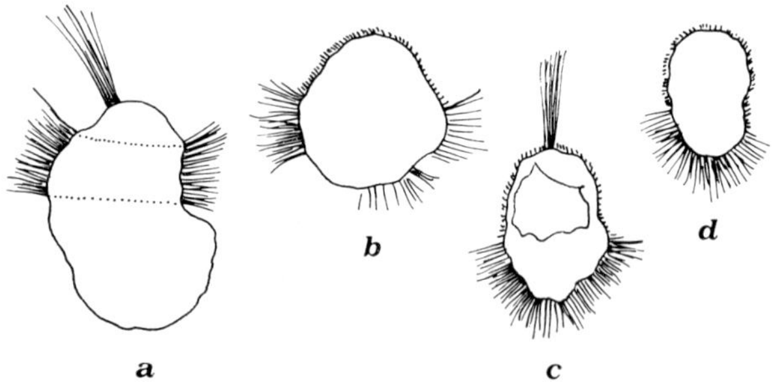

Fig. 13. Larvae derived respectively from isolated blastomeres D(a), A(b), 1d(c), 1c(d). (Redrawn from Wilson 1904.)

The isolated A, B and C blastomeres develop as parts of the entire embryo, and give rise to very defective larvae without post-trochal structures and apical tuft. D, on the contrary, gives rise to a more or less normal larva.

9.10. Isolation of the first quartet of micromeres

The quartet of micromeres, if isolated, gives rise to an ectoblastic non-gastrulating embryo. It has a pyriform shape and swims actively.

It has trochoblastic cilia only in the lower region, the upper region having a fine, short ciliature. It also possesses an apical tuft which derives from the cell 1d (fig. 13).

9.11. Conclusion

From the data reported above one cannot escape the conclusion of Weiss: 'These results stress the formative role of the polar plasm and show, as it seems, incontrovertibly that tangible forerunners of certain definite embryonic formations can be detected in the cytoplasm of these eggs as early as at the beginning of segmentation' (Weiss 1939). What are these forerunners? The early embryologists laid great emphasis on the cytoplasmic substances. However, 'organ-forming' plasms have never been detected; on the contrary certain cytoplasmic plasms can be displaced or removed without any consequence for the forming larva. The modern embryologists have abandoned the idea of plastic 'organ-forming' substances. Some of them localize the developmental potentials in the cortex (Verdonk 1968a); others still acknowledge the importance to the cytoplasmic materials, although not as plastic substances but as activators of the genes.

This seems the right way to go on in the future.

References

ARVY L., 1950. Données histologiques sur l'ovogénèse chez *Dentalium entale* Deshayes. Arch. Biol. (Liège) *61*, 187–195.

BOISSERVAIN M., 1904. Beiträge zur Anatomie und Histologie von Dentalium Jena. Naturwiss. *38*, 553–570.

DELAGE Y., 1899. Études sur la mérogonie. Arch. Zool. Exptl. Gen. *7*, 383.

FOL H., 1889. Sur l'anatomie microscopique du Dentale. Arch. Zool. Exptl. Gen. *7*, 91–148.

KOWALEWSKY A., 1883. Étude sur l'embryogénie du Dentale. Ann. Musée Hist. Natl. Marseille *1*, n. 7, 1–54.

LACAZE-DUTHIERS H., 1857. Histoire de l'organisation et du développement du Dentale. Ann. Sci. Natl. Zool. *4*, 171–255.

REVERBERI G., 1958. Selective distribution of mitochondria during the development of the egg of Dentalium. Acta Embryol. Morphol. Exptl. *2*, 79–87.

SCHLEIP W., 1925. Die Furchung dispermer Dentalium-Eier. Arch. Entwicklungsmech. Organ. *106*, 86–123.

VERDONK N. H., 1968a. The effect of removing the polar lobe in centrifuged eggs of Dentalium. J. Embryol. Exptl. Morphol. *19*, 33–42.

VERDONK N. H., 1968b. The relation of the two blastomeres to the polar lobe in Dentalium. J. Embryol. Exptl. Morphol. *20*, 101–105.

WEISS P., 1939. Principles of development. Holt & Co., New York.

WILSON E. B., 1904. Experimental studies in germinal localization. I. The germ regions in the egg of Dentalium. II. Experiments on the cleavage mosaic in Patella and Dentalium. J. Exptl. Zool. *1*, 1–268.

G. Reverberi (ed.), Experimental embryology of marine and fresh-water invertebrates

CHAPTER 10

Cephalopods

JOHN M. ARNOLD

Kewalo Laboratory, The Pacific Biomedical Research Center of the University of Hawaii and The Marine Biological Laboratory of Woods Hole, Mass., U.S.A.

10.1. Introduction

The Mollusca comprise the second largest animal phylum and as such form an extremely diverse group. The Cephalopoda differ greatly from the rest of this phylum, being different not only in their general body orientation, their greatly modified foot, the high intelligence of the Octopoda, but also in their pattern of development and differentiation. Spiral cleavage, thought to be so characteristic of the annelids and molluscs, is lacking in the highly telolecithal eggs of the cephalopods. Development is direct to a miniature adult with none of the larval stages considered to be characteristic of the Spiralia. Apparently differentiation is not of the classical mosaic type. Therefore, this chapter will consider a group that differs in many respects from the rest of the Mollusca; a point that will become increasingly clear.

There has been a surprising lack of experimental work done on cephalopod embryos when they are compared to the rest of the Mollusca. Clarke (1966) claims the Cephalopoda are the least known class of animals and this also applies to many aspects of their development. In part, this is related to the sporadic availability of the embryos but recent techniques for stimulating egg laying in Octopus (Wells and Wells 1959) and Loligo (Arnold 1962) have been developed so more work can be hoped for in the near future. Most of this discussion will emphasize the embryo of Loligo since this genus is better known developmentally than most of the other genera, and seems to be fairly generalized and intermediate in its developmental pattern. More specifically, most reference will be made to the development of *Loligo pealii* as the author is most familiar with this animal and therefore is biased. Unless otherwise stated many of the observations on *L. pealii* are original and as yet, unpublished. Unless reference is made to the contrary this 'type specimen' approach will be used and the author assumes the responsibility for the information presented.

10.2. Gametes and fertilization; structure and function of the spermatophore

The cephalopod spermatophore represents one of the most highly evolved and beautifully organized structures involved in invertebrate reproduction. No discussion of cephalopod development would be complete without mention of the spermatophore. Although frequently varying in fine details of structure still the spermatophores are rigidly organized and uniform enough to be of taxonomic significance (Marchand 1913). The discussion presented here is primarily derived from *L. pealii* although it is general enough to apply to most species described in any careful detail. Drew (1911) has described the structure and ejaculation of the spermatophore of *L. pealii*. Essentially the spermatophore is made up of three major units: the sperm mass; the ejaculatory apparatus and the cement body; and the tunics, membranes and fluid-filled spaces which surround the whole 'organ'. The sperm mass is composed of densely packed, highly oriented spermatozoa covered by the inner tunic (fig. 1). Estimates of the number of spermatozoa vary between 7,200,000 and 9,600,000 (Austin et al. 1964) per spermatophore. The sperm heads are so highly oriented that whole spermatophores have been used for X-ray defraction studies of 'living' DNA (Wilkins 1956). The ejaculatory apparatus and the cement body are anterior (toward the future open end) to the sperm mass and make up a very complex structure shown in figure 2. The most striking part of the spermatophore (fig. 2) is the coiled spiral filament which functions in keeping the lumen of the ejaculatory apparatus open rather than acting as a coil spring (Drew 1919). The cement body contains a substance which attaches the sperm mass and its coverings to the inside of the mantle or seminal receptacle of the female when ejaculation takes place. The tunics and fluid-filled spaces function not only in covering and containing the sperm mass and ejaculatory apparatus but also as the motive force behind the 'explosion' of the spermatophore.

According to the work of Drew (1919) and Austin et al. (1964) the steps in ejaculation of the spermatophore are as follows. The male takes a few to several spermatophores from his genital opening with his hectocotylized arm; the cap thread is pulled, and this in turn initiates the rupture of the spermatophore. The elastic nature of the outer tunic forces the sperm mass

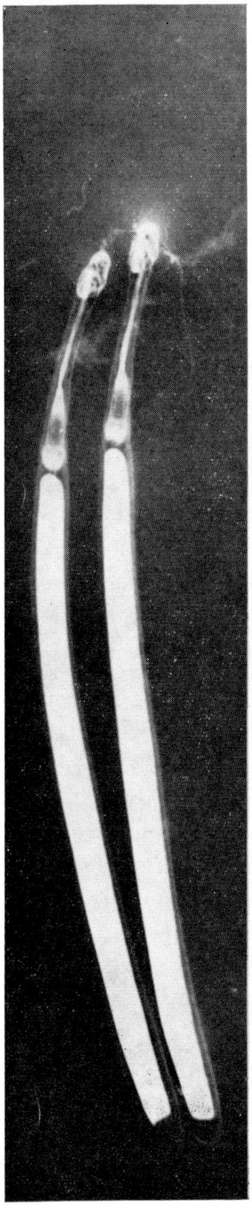

Fig. 1

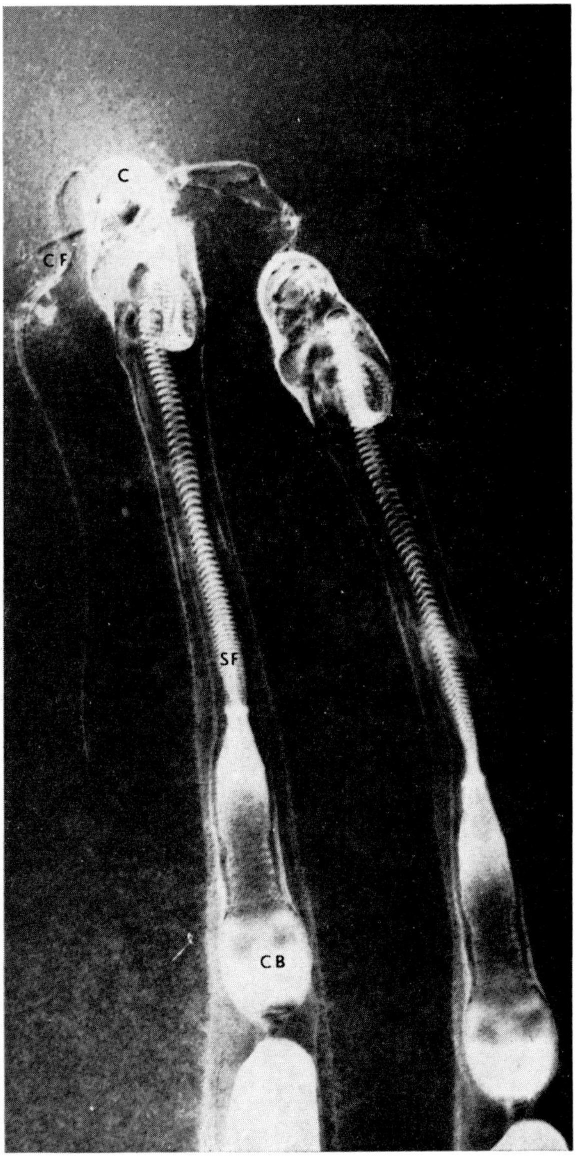

Fig. 2

Fig. 1. The spermatophore of *L. pealii.* The large white mass is composed of highly oriented sperm. The ejaculatory apparatus and the cap thread are also evident. ca. 12 ×

Fig. 2. The ejaculatory apparatus at higher magnification showing the cement body (CB), the spiral filament (SF), the cap (C), cap thread (CF) and tunic and membranes. ca. 34 ×

and the cement body to travel through the lumen of the ejaculatory apparatus and the sperm mass becomes incased in the membranes which formerly surrounded the ejaculatory apparatus. The cement body is broken and its contents are smeared on the end of the sperm reservoir formed by the membranes of the ejaculatory apparatus. The spiral filament frequently breaks up or remains attached to the tunics when they detach from the newly formed sperm reservoir. The whole process takes a few to several seconds and can be adequately observed only by high speed cinematography or by slowing the process down with a magnesium chloride solution in sea water.

Austin et al. (1964), and Hamon (1939) have investigated the chemical composition of the various structures of the spermatophore which are, it must be remembered, all acellular. By staining reaction Austin et al. (1964) demonstrated the major components to be a complex of proteins and polysaccharides with variations occurring between the different tunics and the membranes. The outer layers are probably mucopolysaccharides and the inner mucoproteins. The cement is a highly alkaline substance. Austin et al. (1964) also made a number of metabolic studies which demonstrated the spermatophores could convert glucose and fructose to lactic acid. This is highly unusual for invertebrate spermatozoa. The fluid of the spermatophore contains a factor which depresses respiration and motility when added to a suspension of the spermatozoa. Although many papers describe the structure and parts of the ejaculation of the cephalopod spermatophore the two referred to above are most significant.

The development and formation of the spermatophore is of considerable interest. Drew made cursory observations of the process in *L. pealii* but Blanquaert (1925) gives considerable detail and micrographs of the spermatophore development in several decapods. Essentially the spermatozoa leave the gonad and travel down a duct through a number of vesicles surrounded by glands which secrete the membranes and tunics which come to enclose the sperm mass. In the first vesicle the sperm mass is twisted into a dense helix and covered with membranous coverings and is embedded in a viscous material. Exactly how each spermatozoon comes to be so highly oriented with regard to the others is unknown. In subsequent vesicles the axis of the ejaculatory apparatus is formed (in *L. pealii* this later disappears but in some species it persists) as are the cement gland, membranes, and tunics. The developing spermatophores continue to spiral and twist about their axes as they pass down the length of the seminal duct and the cap is finally formed by a spiralling of the cap thread at the future opening end. The fully formed spermatophores are stored in 'Needham's sac' until copulation takes place. The amazing fact is that other than the spermatozoa the spermatophore is completely acellular and a highly organized product of secretion.

Fertilization in the cephalopods is always internal and copulation is accomplished by means of a specialized arm or arms of the male. A detailed description of the variations in copulation and fertilization in the cephalopods can be found in Lane (1960) but a few comments will be included here for the sake of completeness. In the Octopus the third right arm is modified to form the hectocotylus which is inserted into the end of the female's oviduct. A temporary fold runs the length of the arms and the spermatophores are transferred by an unknown mechanism (Von Orelli 1962). Ejaculation of the spermatophores occurs at some unknown spot and the sperm are carried up the oviduct to the ovary by peristalsis. Copulation usually occurs 'at a distance' with the male contacting the female only with hectocotylus. Copulation may last several minutes to several hours (Wirz in Lane 1960; Wells 1962) and once begun is quite peaceful. The Octopoda have an undeserved reputation in this respect. Fertilization presumably takes place in the proximal portions of the oviduct through the micropyle in the distal end of the chorion.

Copulation and transfer of the spermatophores in the Decapoda is somewhat different. There have been many observations made on copulation in Sepia (Tinbergen 1939; Bott 1938), Loligo (Drew 1911; Arnold 1962), Sepioteuthis (Arnold 1965b), and other genera and for details reference should be made to those papers. Essentially, the process in *L. pealii* is as follows: the male reaches into his mantle and withdraws several spermatophores from his genital opening. At this time the cap thread is pulled and the spermatophores burst. Copulation can occur in one of two positions; head to head or with the animals paralleling each other. In the head to head position the spermatophores are transferred to the seminal receptacle below the mouth of the female. In the second, far more common position, the spermatophores are deposited inside the mantle, frequently on the distal end of the oviduct. Fertilization apparently occurs in the oviduct, despite the observation of Drew (1911) that sperm were not found there. Jelly covered eggs isolated from the lower part of the oviduct have been fertilized. Several layers of the egg jelly are deposited around the eggs as they pass down the oviduct and this represents several millimeters of viscous material for the spermatozoa to penetrate. Immediately upon laying each egg is seen to have a small cloud of several hundred spermatozoa surrounding the micropyle and frequently one or two can be seen swimming in the space between the egg surface and the inside of the chorion. Drew (1911) states that the female *L. pealii* may hold the eggs between her arms so that the spermatozoa from the buccal seminal receptacle may penetrate the egg jelly and fertilize the eggs. In cases where the egg capsule was dropped significantly fewer eggs were fertilized. Artificial fertilization of eggs isolated directly from the ovary is possible, but since the chorions do not swell away from the egg surface in the absence of the egg jelly, development stops in the mid blastoderm stages (Arnold unpublished).

10.3. Development and structure of the egg and its accessory coats

No doubt the most influential factor in cephalopod development which sets this group developmentally apart from the rest of the Mollusca is the large telolecithal egg. Because of the large yolk mass, cleavage is meroblastic and spiral cleavage is not present. Development is direct and the characteristic spiralian larvae are absent. There is some variation in the size and shape of the cephalopod eggs which correlate with neither the taxonomic position nor the size of the adult animal. For example: Eledone has an elongate egg 13 × 4.5 mm, Argonauta has an ovate egg 0.8 × 0.6 mm, while the eggs of several species of Octopus are all elongate and vary between 3.5 mm by 1 mm to 1.5 mm by 0.75 mm. Despite this all of these animals are in the same suborder. Further examples can be found in the decapods where *Sepioteuthis sepioidea* has eggs 6 mm by 3 mm and *L. pealii* has eggs 1.6 mm by 1 mm. These two genera are in the same family. The size of the adult has little to do with the size of the eggs as exemplified by the very large *Octopus dofleini* having eggs only about 4 mm by 2 mm while *O. macropus* having about 1/30th the mass has roughly similar size eggs. The shape of the eggs seems to be determined in large part by the rigidity of the surrounding chorion. In *O. vulgaris*, for example, removal of the chorion allows the eggs to change from an elongate shape roughly three times longer than wide to an ovate or subspherical shape. Dechorionation of the *S. sepioidea* egg results in breakage since apparently the egg cannot support its own weight by the strength of the cortex alone. The yolk mass is quite fluid and the rigidity of the egg is externally derived until the later stages of development; possibly even until cellulation is completed. The eggs of Nautilus have only been observed once (Willey 1897) and were laid singly in a double casing 45 × 16 mm. The egg itself was quite large, 17 mm in length.

In *L. pealii* the eggs at laying or fully matured in the ovary are ovate,

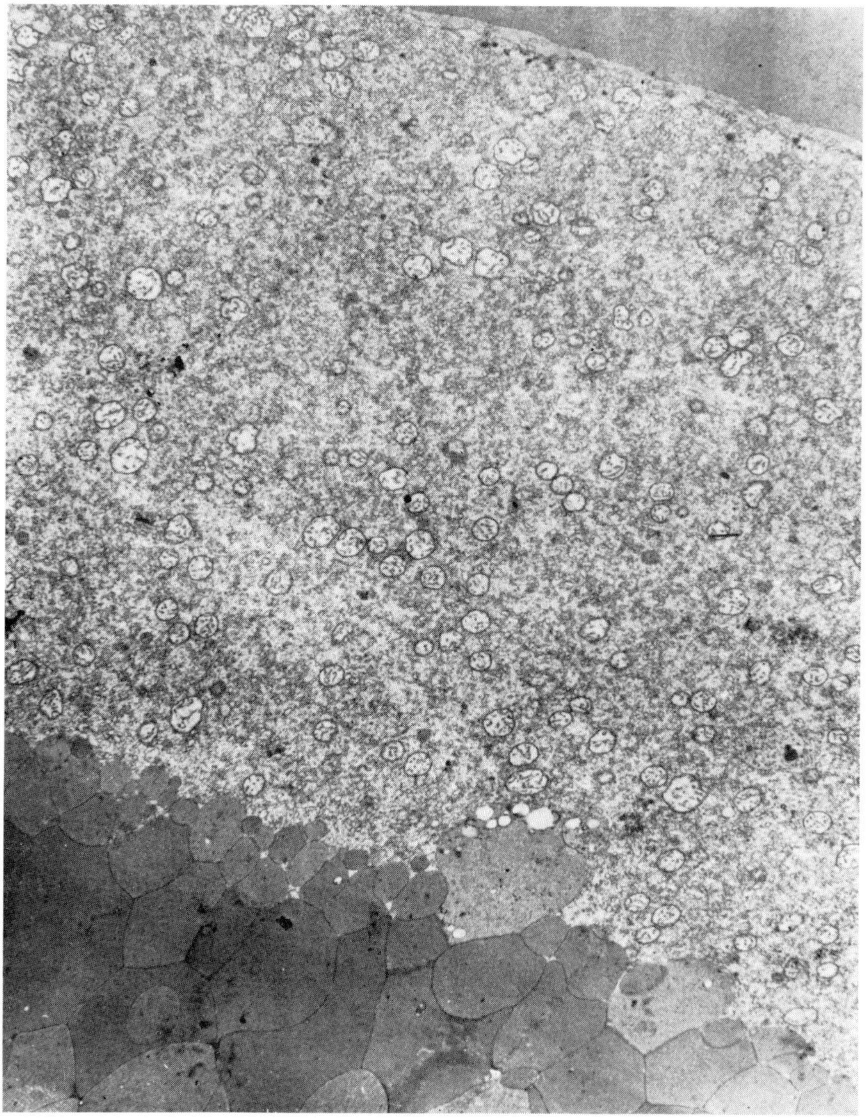

Fig. 3. Section of an uncleaved egg of Loligo showing the general appearance of the cytoplasm, the chorion and yolk platelets. ca. 5,700 ×

translucent and homogeneous in appearance. Membrane bound yolk platelets a few micra in diameter are tightly concentrated in the center and are surrounded by a very thin layer of cytoplasm (fig. 3). This cytoplasmic layer contains many small mitochondria, Golgi figures, and small vesicles. The blastodisc does not appear until after fertilization. There seems to be no vitelline membrane (Korschelt 1892) although in some electron micrographs a thin covering on the cells of the blastoderm is present. At the animal pole there is a clear area in an otherwise granular cytoplasm which contains the nucleus. Meiosis does not occur until fertilization is accomplished.

The eggs of the Octopoda are without any accessory jellies and are covered only by the chorion. At the end opposite from the micropyle the chorion is drawn out into a long filament and this may either be woven into an egg mass or individually cemented onto the substrate depending upon the species. There seems to be some correlation between the size of the egg and whether or not it is incorporated into an egg mass; larger eggs are laid singly. In *O. vulgaris* the eggs are incorporated into masses with the chorion filaments tightly woven and cemented into a common central strand. The base of this central strand is cemented to the overhead substrate of the animal's lair and the eggs are tended by the female who constantly flushes them with water from her siphon and runs the tips of her arms over them, removing small particles and organisms. The brooding female, after depositing as many as 100,000 eggs over a period of a few weeks, undergoes behavior changes, refuses to feed, and eventually dies. The synchrony and rate of laying make this animal attractive from an experimental standpoint.

In the decapods (Sepia, Loligo, Sepioteuthis, Ommastrephes) there is usually some type of accessory coat laid down upon the eggs as they pass down the oviduct. In Sepia the large eggs are laid individually in small groups (Tinbergen 1939) and the female frequently squirts ink on the eggs just after attachment, so the thin layer of jelly is stained black. This does not seem to be necessary for development however. There is some brooding of the eggs by the female. In Sepioteuthis three or four eggs are contained in a common layer of jelly (Arnold 1965b). At least three layers of jelly are present: a layer directly folded around the eggs and stretched to the attachment end; a thicker layer covering this and giving the egg capsule a smooth outline; and finally an outer layer of tougher jelly, possibly formed by the hardening contact with sea water. In Loligo there may be up to 200 eggs in an egg string (Arnold 1963; Fields 1965) and the structure of the egg string is fairly complex (fig. 4). These egg strings are attached to the substrate or more commonly to other egg strings until a large common egg mass is built-up by several females over a period of a few days. McGowan reported an egg mass of *L. opalescens* that was forty feet in diameter. Arnold

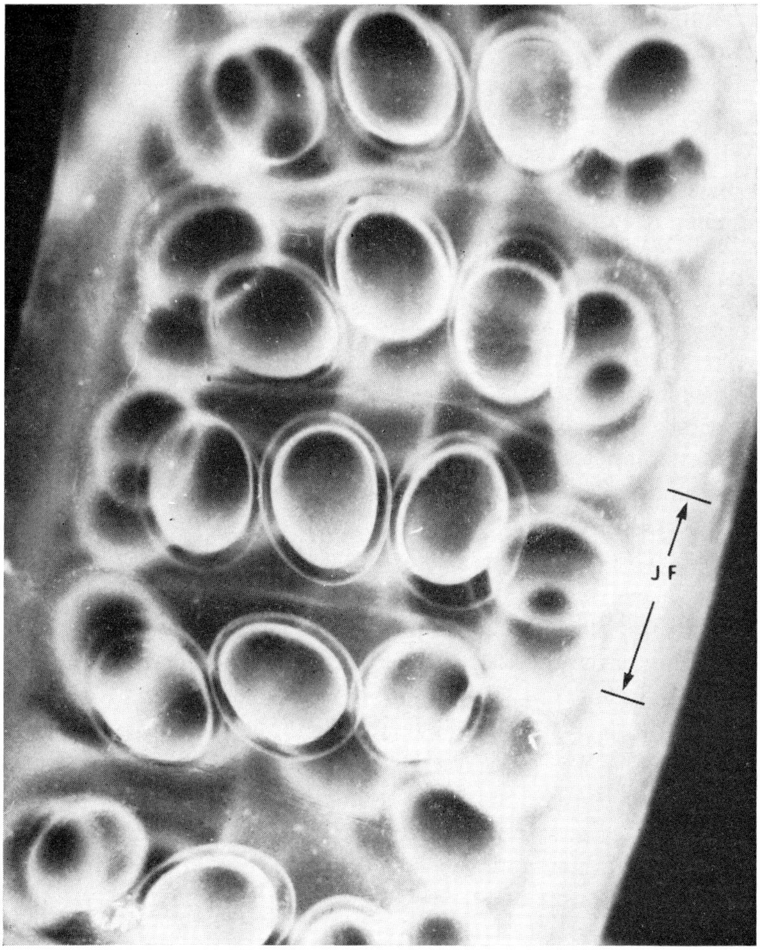

Fig. 4. Part of the *L. pealii* egg string by dark field illumination. Each embryo is surrounded by a chorion and all are contained in a fold of jelly (JF). The jelly folds spiral about and extend to the end of the egg string, forming a central core. The whole structure is covered with additional jelly and at least two tunics. ca. 12 ×

(1962) has used the visual stimulus of the egg mass to obtain egg laying at convenient times for his work on *L. pealii*.

Hamabe (1961) has reported another type of decapod egg mass. He confined *Ommastrephes sloanii pacificus* in sunken barrels and had egg laying occur under these conditions. In this case a large mass of jelly was secreted by the oviductal glands and the eggs and secretion of the nidamental gland containing the egg mass was injected into the center of this mass. This egg mass filled from one third to two thirds of the barrel and contained from 300 to 4,000 eggs. Hamabe assumed this process took about two hours. These egg masses, in nature, would have an indefinite shape and probably be deposited in natural cavities in rocks, etc. This type of egg mass and the subsequent development of the embryo closely resemble Grenacher's unidentified embryo suggesting his embryo may have been a related genus or species.

Development of the cephalopod egg and other telolecithal eggs has been of interest to several investigators but only the work of Yung Ko Ching (1930), Loyez (1905), Bergmann (1903), and Konopacki (1933) will be mentioned here. Yung Ko Ching found the early stages of prophase to occur in relatively young animals and Callan (1957) has reported that Loligo has lampbrush chromosomes. Apparently meiosis is arrested in prophase until stimulated to completion by fertilization. Loyez and Yung Ko Ching divided oogenesis into three phases: (1) a period of multiplication of the follicular epithelium, during which folds of follicular epithelium appear and penetrate into the growing oocyte; (2) a period of active secretion of the follicular folds to cause growth of the yolk in the developing oocyte and formation of the chorion; and (3) a regression and disappearance of the folded epithelium. Yung Ko Ching showed figures with material being transported through holes passing through the developing chorion. Loyez and Bergmann showed blood vessels present deep in the folds and claimed transfer occurred from blood vessels to the epithelium and into the growing oocyte. By far the most elegant work has been done by Konopacki (1933) using histochemical methods. He demonstrated transfer of hemocyanin and other proteins from the maternal blood to the developing oocytes by cells of the follicular epithelium. Lipoids from the sinus spaces also penetrate this epithelium. He further claimed that nucleoproteins were broken down in the follicular cells and the precursors were incorporated into the developing oocytes for a possible later utilization during embryogenesis. The chorion according to Konopacki is formed by the follicular cells from glucoproteins and mucoids and laid down upon the surface of the growing oocyte. This is very reminiscent of the modern work of Roth and Porter (1964) on insect eggs and Anderson (1967) on rabbit eggs.

10.4. Fertilization, polar body formation and formation of the blastodisc

Hoadley (1930) described the events of fertilization in *L. pealii* following penetration of the micropyle by the sperm. The large spermatozoa can easily be seen inside the chorion and the author on several occasions has seen more than one spermatozoa in the intrachorionic space. What the nature of the block to polyspermy can be is unknown but the literature contains no reference to diaspermic development. Once the sperm has penetrated the egg surface two events occur: meiosis resumes and the first polar body is given off, and cytoplasmic streaming establishes an enlarged blastodisc in the region directly below the micropyle (Hoadley 1930; Arnold 1965a). In some cases there seem to be regular channels which are followed, although a generalized shift occurs as well. Spek (1934) noticed a shift toward a higher pH at the animal pole concurrent with this cytoplasmic streaming; the vegetal pole also became more acidic.

Between twenty and thirty minutes after the sperm has entered the egg the first polar body is given off (Arnold, stage 2, fig. 5) and the second polar body follows after about one hour post-fertilization. Hoadley (1930) made an extensive study on the fate of the polar bodies in *L. pealii* and found that the polar bodies could divide a number of times and five polar bodies were not rare. Hoadley concluded '. . . the development of the polar bodies is dependent upon the presence of cortical ectoplasmic material in that organ or at least, on the ratio between the amount of ectoplasmic substance and the remaining cytoplasm and nuclear material.' In *L. pealii* (18 °C) the polar bodies lie slightly below the apex of the animal pole and the first cleavage furrow seems to run to one side or the other of the polar body. In *O. vulgaris* the cytoplasmic streaming seems to occupy a much longer time than the three hours required for first cleavage in *L. pealii*. Naef (1928) puts first cleavage time at nine hours but the author has experienced twelve to fourteen hours as more common time (24°–26 °C).

10.5. Early development cleavage

As has been pointed out before the large yolk mass of the cephalopod egg has greatly modified cleavage so that the typical molluscan spiral cleavage is absent. Some attempts have been made to force the comparison of the spiral pattern with the meroblastic cleavage found in the cephalopod but this is merely a pedantic exercise and serves little purpose in our discussion (see

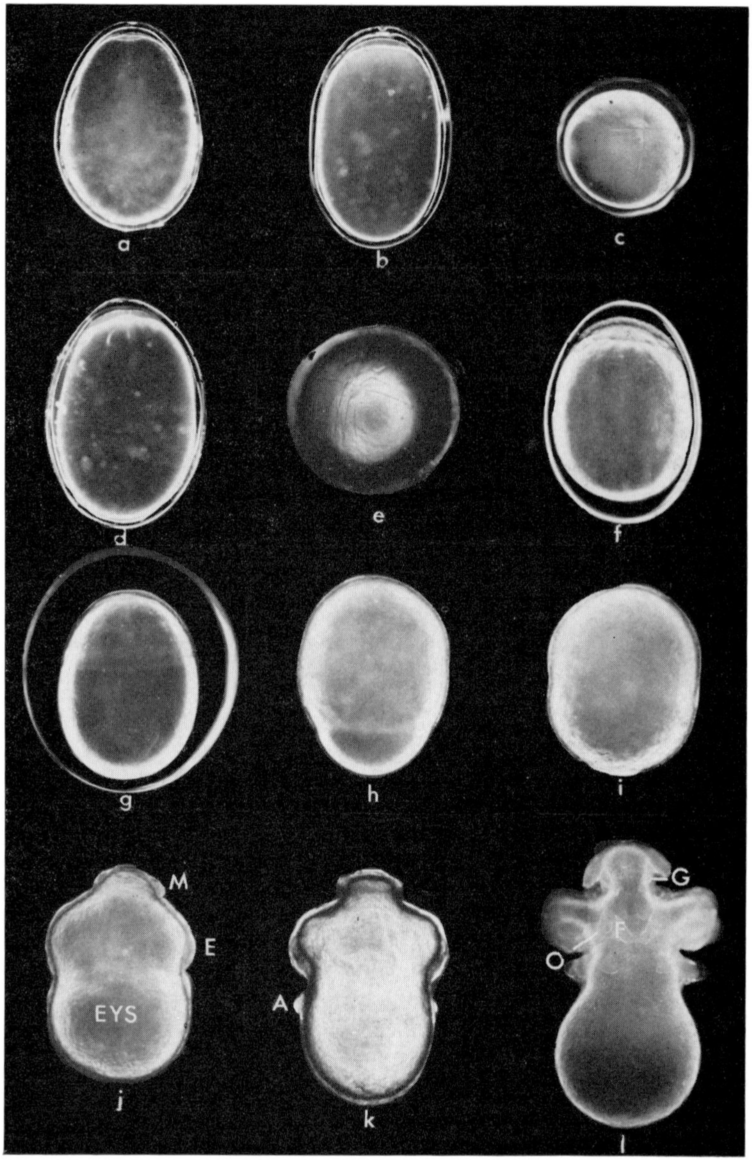

Fig. 5. Representative normal stages of development of *L. pealii:* (a) stage 1 viewed from one side. On the left is the future embryonic anterior. (b) stage 2; front view. (c) stage 5; second cleavage. (d) stage 7; side view. Note the slight asymmetry of the blastoderm. (e) stage 7 from above. The bilateral cleavage pattern is evident. (f) stage 10; early blastoderm. (g) stage 14; spreading blastoderm. (h) stage 17. (i) stage 18; organ primordia are beginning to appear. (j) stage 19; the lower half is becoming the external yolk sac (EYS).

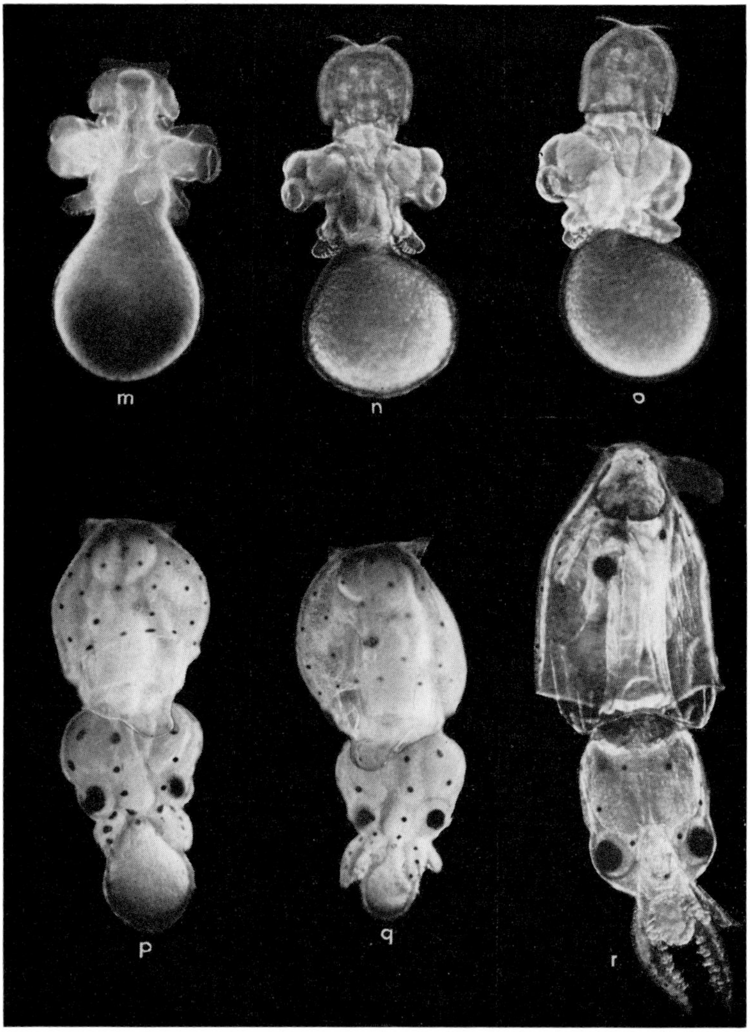

Fig. 5 *(continued)*.

The eye primordia (E) are evident laterally above the 'Waist' and the mantle (M) and shell gland are evident at the apex. (k) stage 20; the arm primordia (A) are prominent. (l) stage 23; the gill primordia (G), funnel folds (F), and otocysts (O) are evident. (m) Stage 24; further downward growth of the mantle. (n) Stage 25+; the lens is evident in the optic vesicle. (o) stage 27. (p) stage 28. (q) stage 29; pigmentation beginning in the ink sac. (r) stage 30; newly hatched larvae with the rudiment of the yolk sac. ca. 20 ×

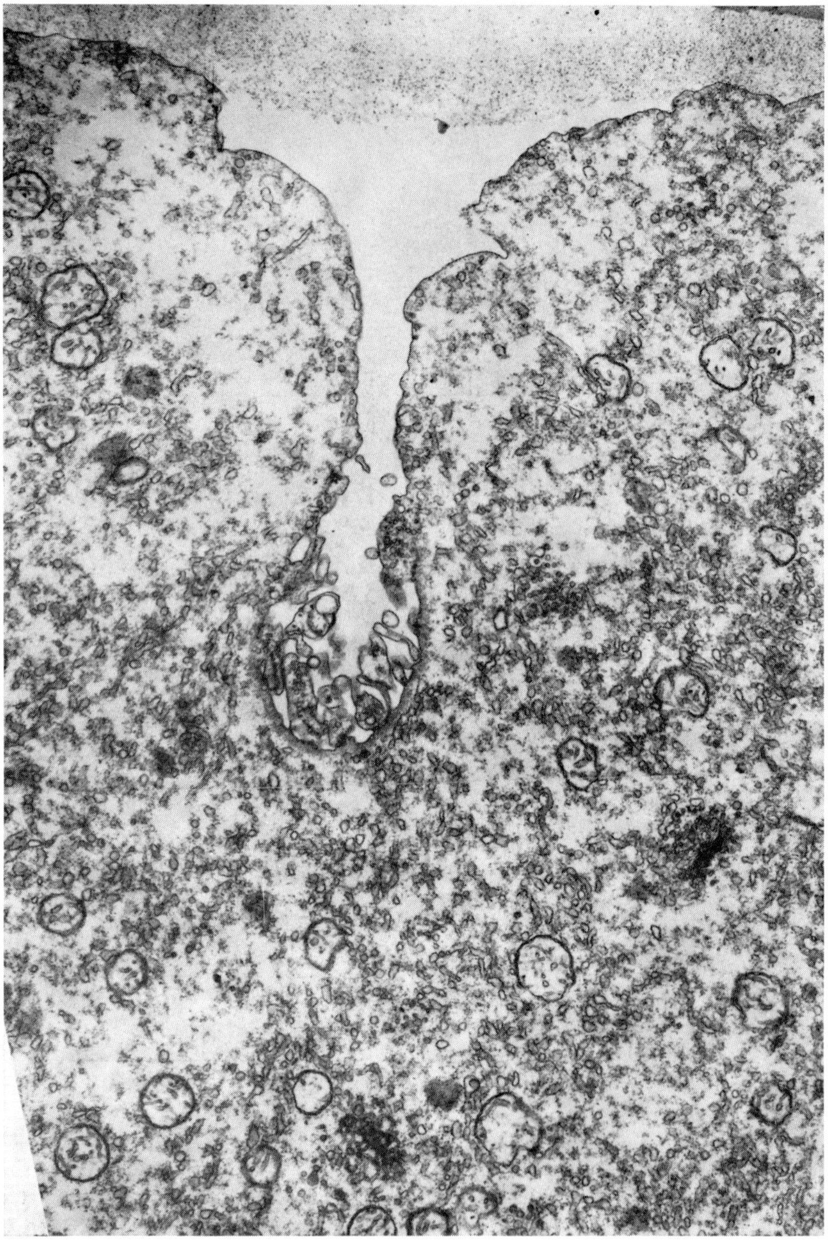

Fig. 6. Beginning first cleavage in the Loligo egg. Note the fibrous material at the base of the cleavage furrow. See text for details. ca. 11,000 ×

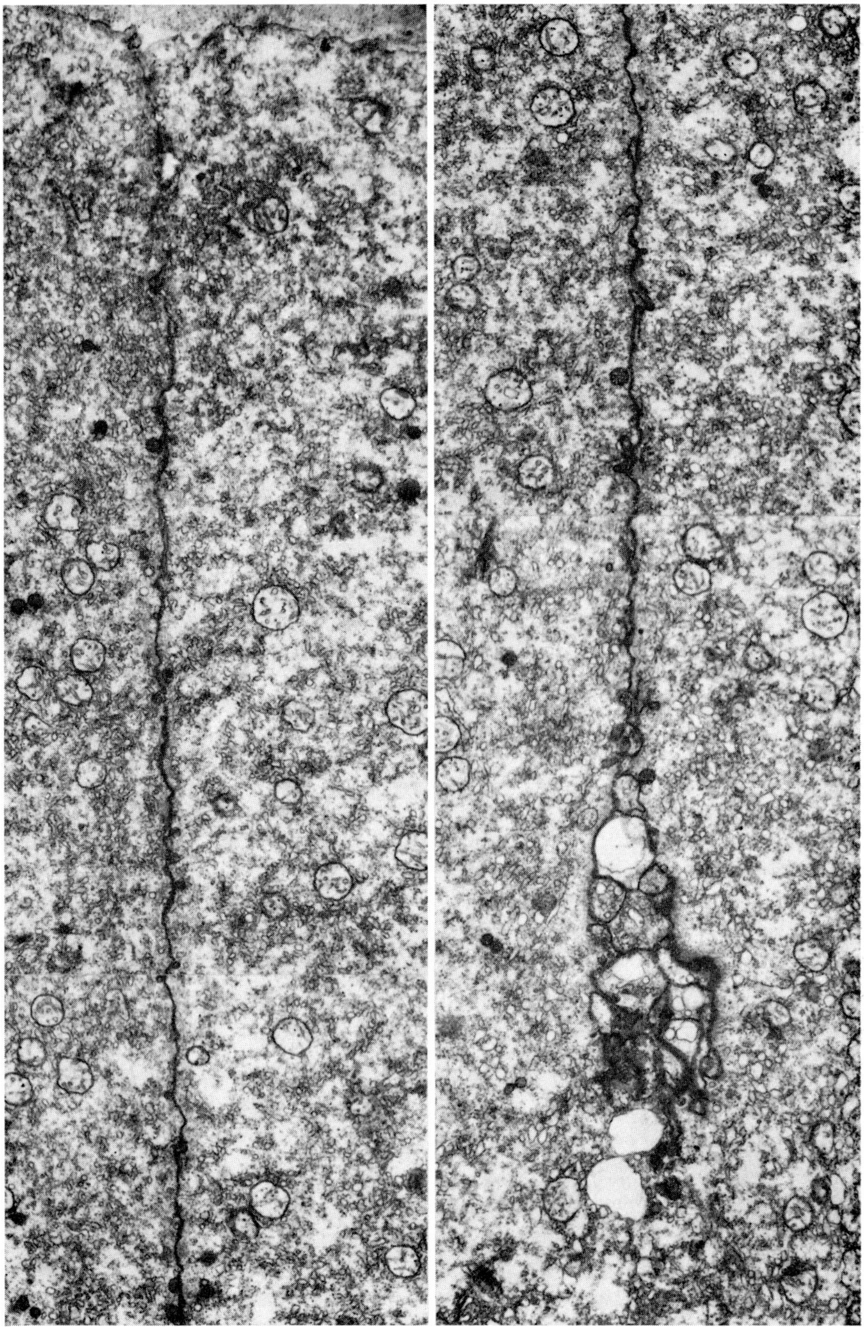

Fig. 7. Later stages of first cleavage. ca. 7,500 ×

Naef 1928). For purposes of this discussion the cleavage period will be considered to last from the onset of karyokinesis until the formation of the multilayered blastoderm. In *L. pealii*, our 'type species', this period lasts approximately 12 hours at 18 °C. Watase (1888) and Arnold (1965a) have described cleavage in this species and their work will be summarized here.

About three hours after fertilization a fairly large blastodisc has become established (see fig. 5) and the polar bodies lie approximately above the future cleavage plane. It is possible at this time to distinguish the future anterior and posterior of the embryo and the adult because the egg is fairly asymmetrical (Watase 1888). Viewed from the side the future anterior surface slopes at an angle from the animal pole to the equator while the future posterior surface is near to the vertical (fig. 5). In some larger decapod eggs such as *S. sepioidea* this is quite striking. This shape asymmetry is not evident in the eggs of three species of Octopus seen by this author but the nucleus is displaced to a position slightly below the apex of the animal pole as it is in Loligo. It must be remembered that the animal pole corresponds to the future functional posterior end of the animal while the vegetal pole corresponds to the location of the foot or functional anterior of the swimming decapod. Vialleton (1888) makes reference to a similar situation in the eggs of Sepia he observed. It is therefore possible to orient the egg in terms of future planes of symmetry before cleavage has begun, a convenient property for experimentation. The first indication of a cleavage furrow in the living egg also appears slightly below the exact apex of the blastodisc. It quickly extends to either margin of the blastodisc and extends all the way to the yolk below in clear cytoplasm. Electron micrographs at this stage (fig. 6) show the base of the furrow to end in many small cytoplasmic blebs with an area of granular to fibrous uniform material below them. This material may possibly be associated with the formation of the furrow (cf. Cloney 1966). The plasma membrane of the furrow may possibly arise by fusion of many vesicles found in the cytoplasm near the furrow as they seem to be oriented toward it. Fig. 7 is a micrograph of a somewhat later cleavage furrow where membrane opposition has taken place and the base of the furrow is isolated from the surface of the egg. The cytoplasmic blebs are still distinguishable and the fibrous material is evident against the membrane at the base of the furrow (Arnold 1968b, c, 1969).

The second cleavage furrow occurs slightly below the exact apex of the egg and is at right angles to the first (fig. 5). In *L. pealii* this takes place about fifty minutes to one hour after formation of the first cleavage furrow. A more detailed chronology will be found in Arnold (1965b). Third cleavage is asymmetrical and divides the two anterior blastomeres at a slightly faster rate than the posterior blastomeres. The rigidly bilateral pattern of cleavage is maintained; however, cleavage has now become unequal, a fact which has delighted workers in comparative embryology. This has lead Naef (1928) to use the unfortunate terms micromeres and macromeres, which, because their fates are not analogous to the gastropods, should be dropped. Fourth cleavage is even more asynchronous and unequal with anterior cells dividing first and being considerably smaller than the posterior cells. Cleavage continues through a number of intermediate stages and a 26-cell stage is present briefly. Despite the anterior to posterior asynchrony the two sides remain quite synchronous and remain mirror images of one another (fig. 5). In Octopus embryos the cleavage is somewhat less regular and the cells do not meet so neatly at their margins. The author has not observed the living materials to have the intercellular gaps so prominently figured in Naef (1928, plate 24). At about the 16-cell stage (stage 7, Arnold 1964a) the lower surfaces of the central blastomeres begin to cut free from the underlying yolk mass. Apparently this takes place by fusion of anastomosing channels which spread out from the original cleavage furrow and by contraction of filaments similar to those described above (fig. 8). In this way cells in the center of the blastoderm become separated from the underlying yolk by a plasma membrane. A thin layer of cytoplasm lies directly upon the yolk and yolk digestion is quite evident (fig. 9). Later this region will become incorporated into the yolk epithelium.

As cleavage continues, the cells in the center of the developing blastoderm become increasingly smaller and two populations of cells are established: the inner blastomeres and the outer blastomeres. Vialleton has used the term 'blastocones' to describe these outer blastomeres, which, in fact are a periblast similar to that found in fish embryos, except in that case the cell furrow does extend downward between the nuclei. The lower end of each of these blastocone cells is in communication through the common surrounding cytoplasm or cortex of the undivided portion of the egg. These blastocone cells are further distinguished because they are in direct contact with the underlying yolk. New cells are continually divided off the upper margin of these blastocone cells and are added to the population of small cells that make up the expanding blastoderm. The blastoderm also undergoes continued division of the original central population. These processes continue until much of the surface of the egg has been covered with cells (fig. 5).

In a relatively small egg like that of *L. pealii* the blastocone cells are not

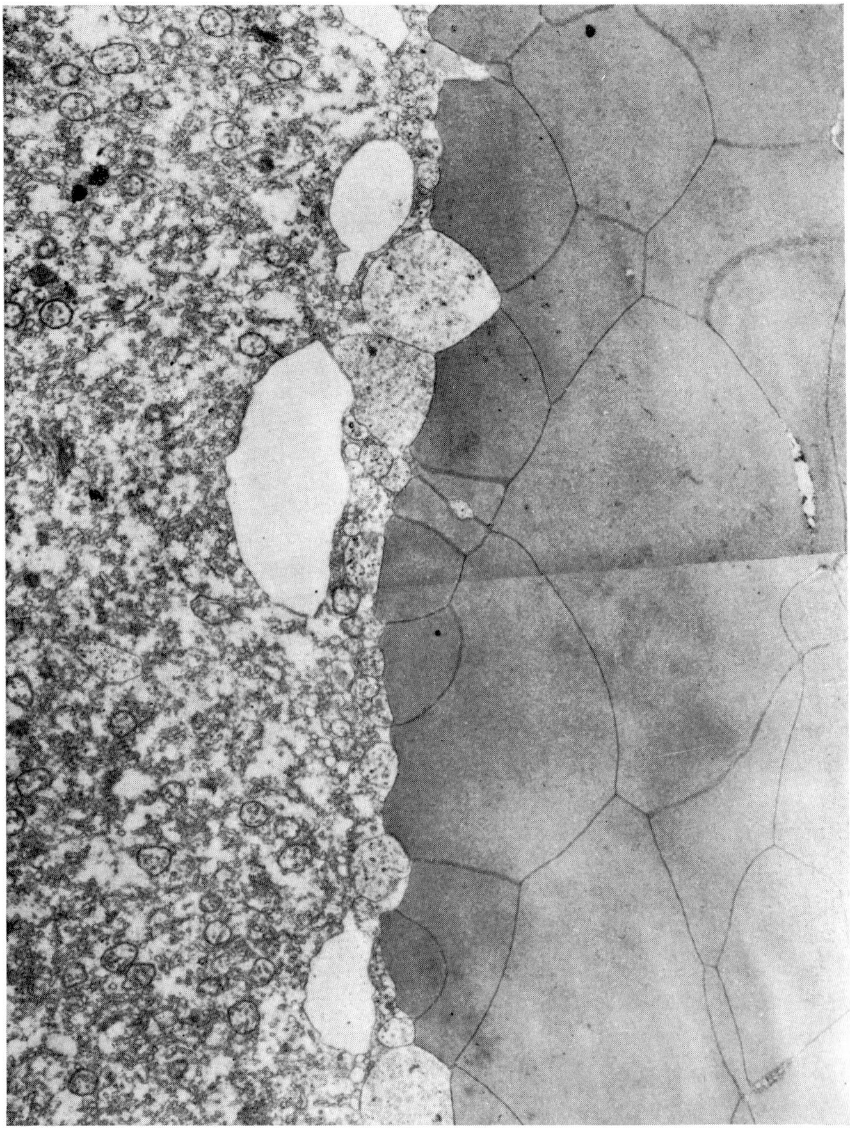

Fig. 8. Furrow formation between the future blastoderm cells and yolk epithelium. The large vesicles anastomose and fuse to form the furrow. ca. 14,300 ×

very evident but larger eggs such as *S. sepioidea*, or *Sepia officinalis* show them quite prominently. Their precise developmental fate is unknown as is the fate of any of the cells of the young blastoderm. Arnold (unpublished)

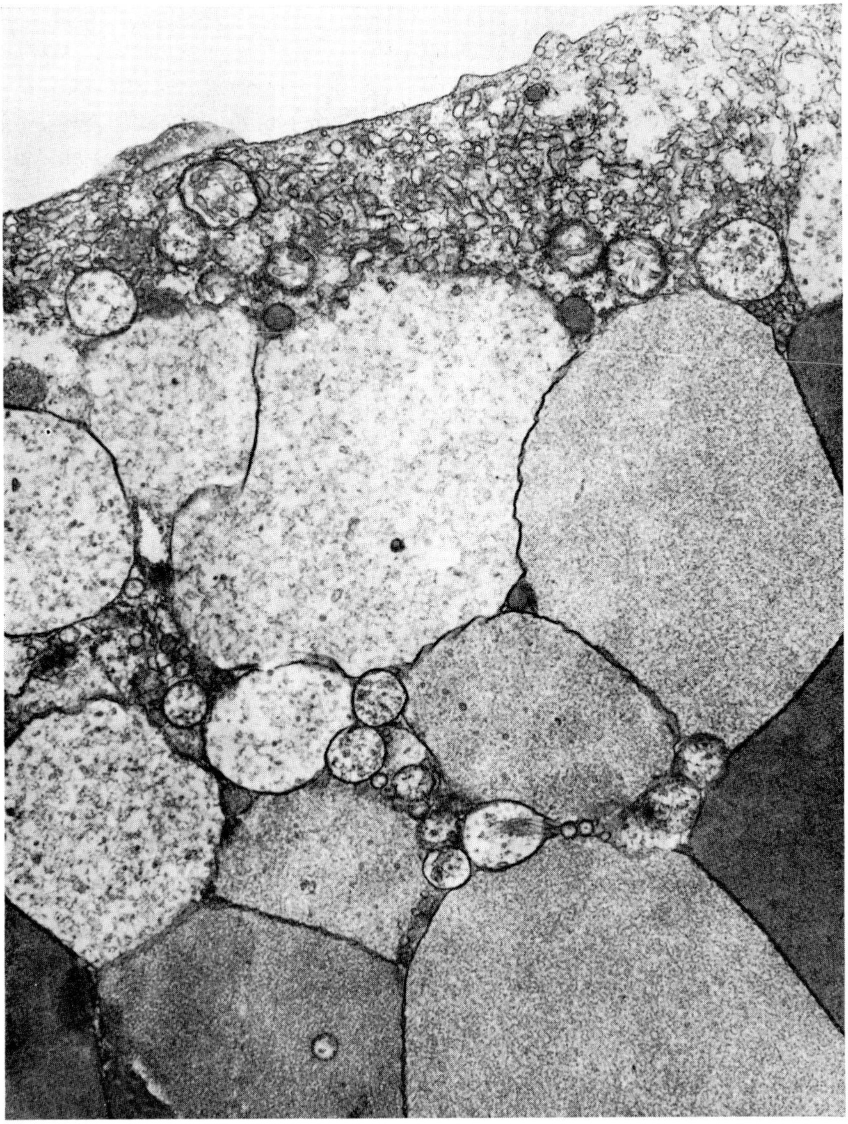

Fig. 9. Digestion of yolk platelet in the yolk epithelium directly below the blastoderm at stage 11. The nuclei are peripheral to this region. The membrane in the individual platelets persists until most of the yolk is gone. ca. 17,000 ×

has irradiated with an ultraviolet microbeam a few cells of the four- and eight-cell embryo of Octopus and Loligo embryos; these embryos continue to live although the irradiated cells do not. Division, however, becomes quite irregular and development eventually ceases. This type of experiment needs to be continued and greatly expanded. To the knowledge of the author no cell lineage studies have been made on the cephalopods and because of the lability of some of the embryonic cells the classical approaches are unfeasible.

10.6. Development of the germ layers from the blastoderm

Unfortunately the literature on the development, relative role, and final fate of the germ layers in the cephalopod is quite confused. This has been the result of many ill-conceived attempts to force cephalopod development to fit the classical lines of molluscan thought and to stretch the application of the 'Biogenetic law'. Too much energy has been devoted to speculation and too little to careful analysis of specific details of development. Sacarrao (1953) has described this state of affairs and this author has similar feelings.

The blastoderm is a single layer of cells made up of blastomeres in the center and surrounded by a ring of blastocones. This would be roughly stage 9 (Arnold 1965b) (figs. 10, 11, 12, 13). The blastomeres are separated from the underlying yolk by a plasma membrane and an irregular space with many areas of close junction existing between the cell and the yolk plasma membrane. The blastocones, or peripheral ring of cells, are continuous

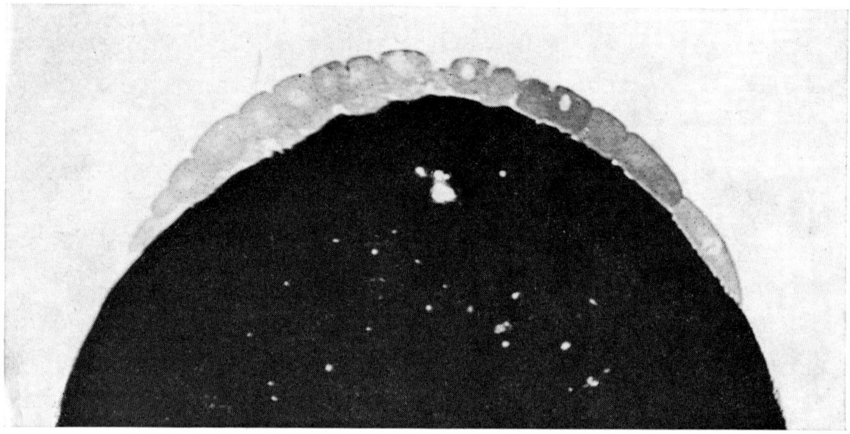

Fig. 10. Section of the stage 9 blastoderm. In this and figs. 12, 13, and 14 note the position of the mitotic figures. See text for details. ca. 143 ×

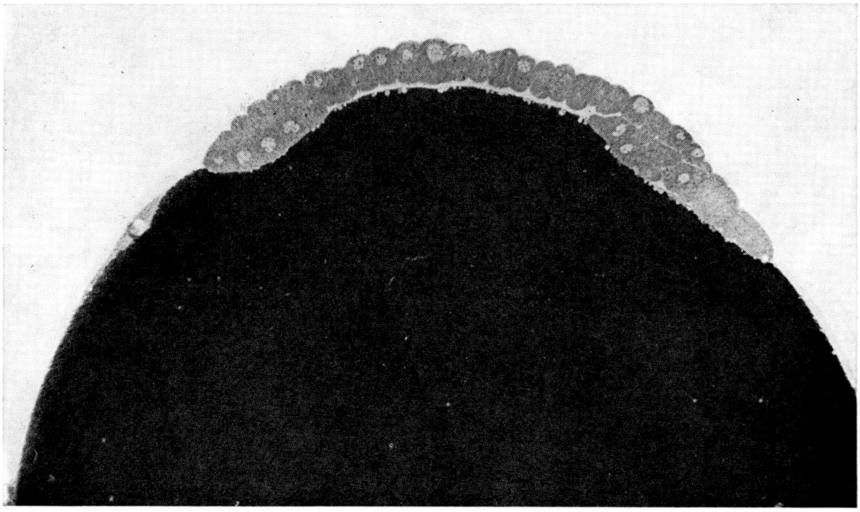

Fig. 11. Stage 9+ blastoderm in section. Note the one blastocone cell beyond the blastoderm proper. ca. 143 ×

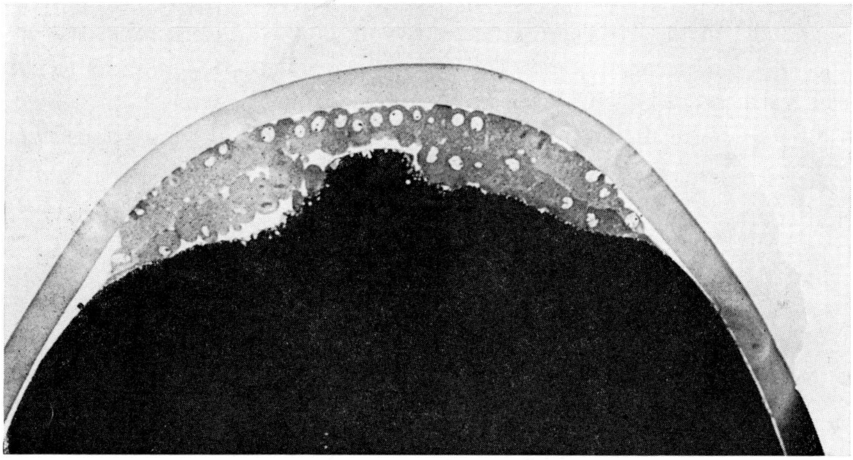

Fig. 12. Stage 10 showing formation of the lower blastoderm layer. ca. 143 ×

with the yolk below and with each other at their vegetal pole end. The *L. pealii* embryo at this time is composed of about 60 to 100 cells.

In the cephalopod embryo the yolk becomes covered by a special syncytial layer of cells called the yolk epithelium (Lankester 1875) which is digestive

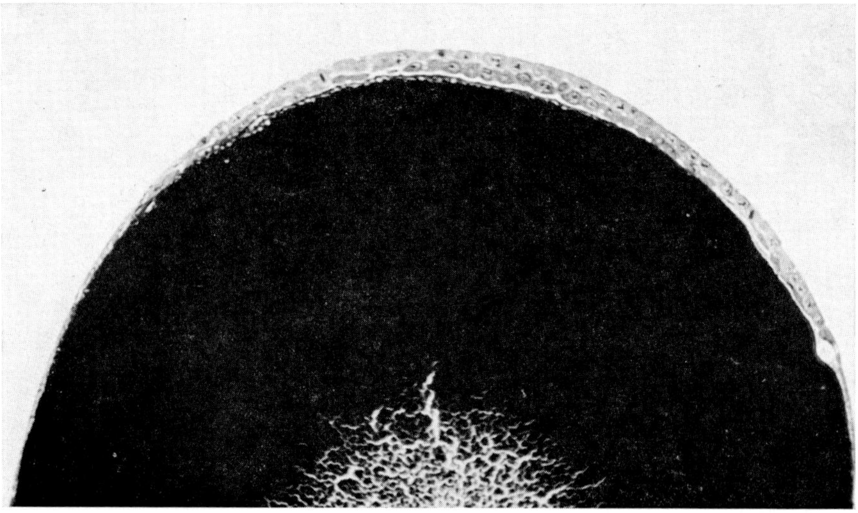

Fig. 13. Stage 13 after further cell division and reduction of the lower blastoderm layer to a single cell layer. ca. 143 ×

in function and morphogenetic in significance (see Arnold 1965c). Unfortunately, in the earlier literature this yolk epithelium has been referred to as the vitelline membrane. There has been much speculation as to the origin of the nuclei of this layer and a number of ideas have been presented. In the opinion of the author the view presented here is the most consistent

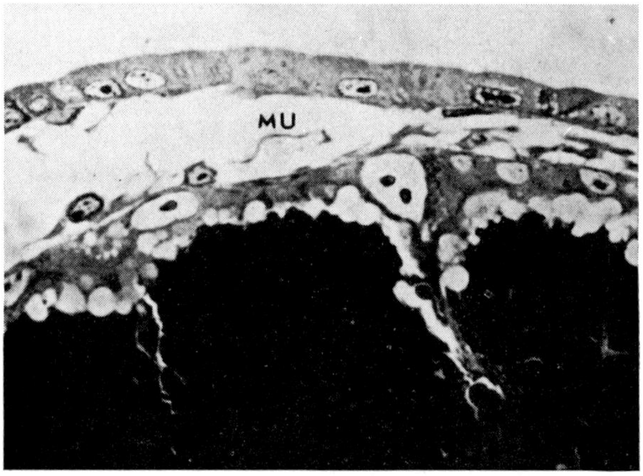

Fig. 14. Muscle bands (MU) in the developing hemal spaces of the stage 20 embryo. The yolk epithelium nuclei are prominent. ca. 400 ×

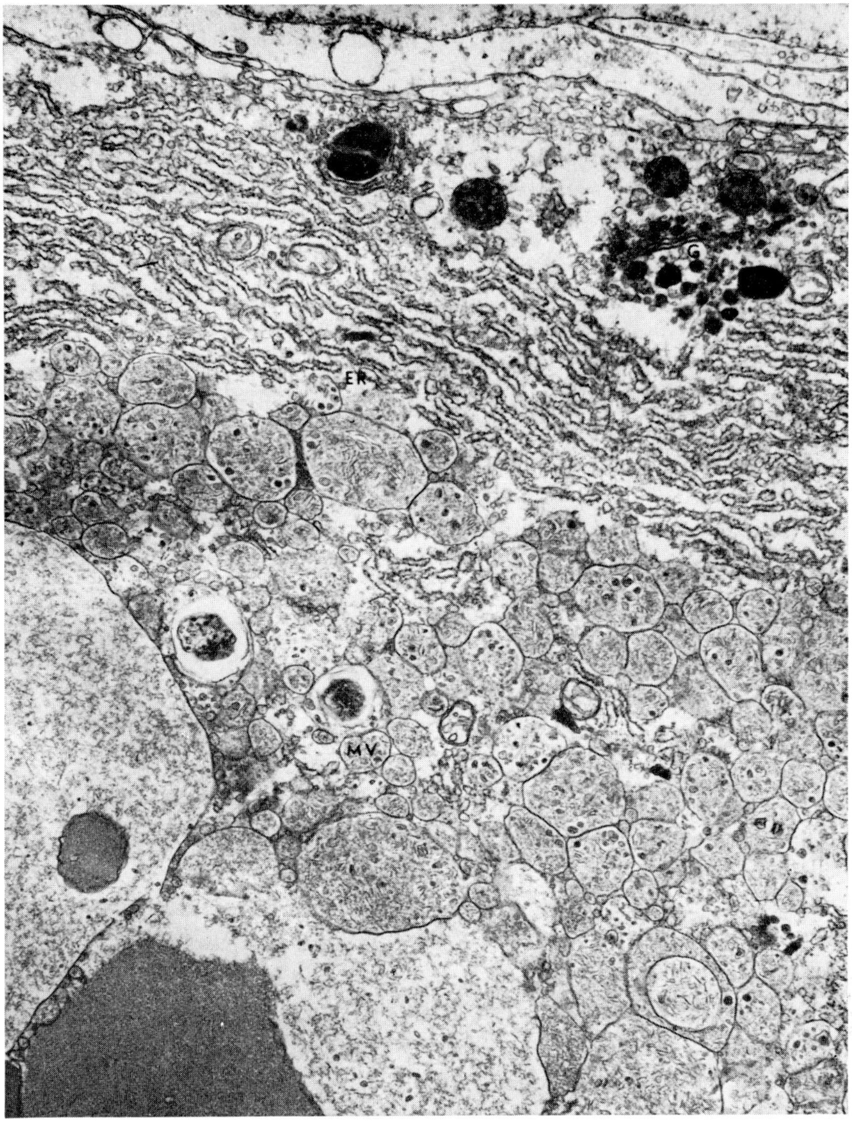

Fig. 15. Yolk digestion in the external yolk sac. Breakdown occurs within the platelet and nutrients are transferred through the cytoplasm to the hemal sinus. This apparently involves the multivesiculate bodies (MV), endoplasmic reticulum (ER), and the Golgi apparatus (G). ca. 12,000 ×

with the literature and fits the data he has available to him from personal experience with *L. pealii*, three species of Octopus, and *S. sepioidea*. The blastocones continue to divide as do the blastomeres at stage 9 but apparently some of the nuclei from the blastocones become incorporated into the layer beneath the blastoderm itself. Exactly how this occurs is not clear, but a nucleated ring of syncytial cytoplasm is established just inside the margin of the spreading blastoderm by about stage 10+.

Some blastocones continue to divide off blastomeres at their apical ends. The nuclei of this syncytial layer are diffuse, lightly staining, and have a very characteristic appearance in the electron microscope. A very dense endoplasmic reticulum develops around each of these nuclei and the whole layer becomes quite basophilic (fig. 14). These characteristics are retained by the yolk epithelium throughout all of the embryonic development. The yolk epithelium has an important digestive function in relation to the breakdown of the yolk platelet (see fig. 15). The yolk epithelium continues to expand beneath the margin of the blastoderm and eventually the nuclei of the yolk epithelium are found directly beneath the apex of the egg. These specialized nuclei are apparently continually supplied from the marginal surface at the edge of the blastoderm. In stage 9+ and stage 10 a third layer of cells appears. This third layer is derived by marginal mitosis of the blastoderm cells themselves. Fig. 12 shows a section of a stage 9+ embryo in which mitotic figures can be seen at the margin of the blastoderm. This orientation of the mitotic figures would give rise to daughter cells that would tend to expand the blastoderm. At the very margin of the blastoderm this would give rise to a second layer of cells because the innermost daughter cells would tend to go under the existing single layer of blastoderm cells. In this way an intermediate layer of cells is established without the necessity of invoking cellular movements or 'gastrulation'. This leads to the establishment of a ring-shaped third layer of cells at the margin of the blastoderm (Arnold, stage 10 (1965a); figs. 12, 13). There have been many varying opinions as to the origin of these cells. Naef (1928) favors and illustrates an 'invagination' of cells at the margin of the blastoderm and Lankester (1875) used the term 'primitive streak' to designate an area of supposed invagination. There is no evidence to support these views and until some very sophisticated marking experiments can be done they cannot be supported. The weight of evidence supports the views of Saccarao (1954), Watase (1888), Ussow (1875), and others; this layer is derived by delamination of the upper layer. The micrographs presented here corroborate this.

The intermediate cell layer is at first horseshoe-shaped in Loligo and circular in Tremoctopus (Saccarrao 1954) but later becomes a broad ring which underlies a large portion of the blastoderm (stage 10, fig. 13). At stage 10 in Loligo there are many mitotic figures which give rise to daughter cells in such orientation that a multilayered intermediate layer is established (fig. 12). This is of a temporary nature because at stage 13 (fig. 13) the blastoderm is two layers thick with a continuous yolk epithelium below it. At stage 10 (Arnold 1965b) a central papilla of yolk is temporarily present. As the intermediate layer expands and spreads under the apex of the egg this yolk papilla disappears (stage 11).

The multilayered blastoderm continues to expand primarily by marginal division until the equator of the egg is reached (Loligo). This delineates the future organogenic region of the embryo. In larger, more yolky eggs, the embryonic shield may cover only a small fraction of the total surface of the egg. This causes an associated gross shape change in development of the embryo but not in the course of events of organogenesis. The remaining yolk mass becomes cellulated with a double layer of cells: the outer single layer and the inner syncytial yolk epithelium. This vegetal pole region will become the external yolk sac in later development.

There has been much discussion of the relative fate of the three cell layers; however, this seems pointless. Watase (1888) makes a strong case that the yolk epithelium is homologous to the endoderm. The outer layer of cells derived by spreading of the surface of the blastoderm is then ectodermal. The middle layer derived by delamination is then mesodermal or mesendodermal. Several problems immediately arise with this and all other such interpretations. If the yolk epithelium is the true endoderm then at hatching, when the embryo discards the external yolk sac and the last vestiges of the internal yolk sac are depleted and broken down, the embryo would lack endoderm. Saccarrao (1953) has postulated a double origin of the endoderm; in part from the yolk epithelium which is a transitory structure, and in part from the middle layer of cells of the late blastoderm which he terms the mesendoblast. According to Sacarrao there is a sorting out of mesoderm and endoderm immediately following the formation of this layer. Other authors have different interpretations but a lengthy discussion of their various views is not profitable. Despite the name given to it, the yolk epithelium is derived from the marginal cells surrounding the blastoderm and an intermediate layer forms by delamination. This intermediate layer together with the outer layer form all the organs of the future adult while the yolk epithelium plays a digestive and morphogenetic role.

The mechanism of spreading of the blastoderm and cellulation of the egg surface has been investigated by Arnold (1961). Apparently, the cells spread by marginal mitosis and there is no gross movement of the outermost layer over the egg surface. This was demonstrated by marking experiments, colchicine inhibition of mitosis, and time lapse cinematography.

10.7. Organogenesis in the cephalopoda

There are several possible approaches to a description of organogenesis in the cephalopods. The one used here will be primarily analytical rather than comparative. It will be necessary to describe, in a cursory fashion, the gross features of organ formation for purposes of orientation and completeness. The references included will supply the details of development. Another author is preparing a comparative account of cephalopod development to be published elsewhere. A general account of organ formation will be found in Grenacher (1874), Lankester (1875), Korschelt (1936), Bobretzky (1877), Brooks (1880), Faussek (1896), Ussow (1875), MacBride (1914), Naef (1928) and Arnold (1965a). Figs. 5 and 6 should be consulted in this discussion.

10.8. A brief survey of organ formation

At the stages when the embryonic organs first make their appearance (stages 16, 17 and 18) the embryo is composed of three layers of cells derived from the original blastoderm: the inner yolk epithelium, the middle layer which extends only to the future margin of the yolk sac, and the outer layer which covers the whole cellulated part of the embryo. In early stages there remains a vegetal region of the original egg surface to be covered by cellulation. This cellulation is completed at stage 18.

The first organ to make its appearance is the external yolk sac which includes all the area below the future arm region. It is composed of two layers of cells, the outer 'ectoderm' and the yolk epithelium. A more detailed description is given below. At the apical end of the embryo an invagination occurs which gives rise to the shell gland while the area surrounding it becomes the primitive mantle (Sacarrao 1951; Ussow 1875; Korschelt 1936). This invagination soon disappears and the mantle grows outward to form a ring which surrounds the future distal region of the embryo (stages 20 and 21). This 'ring' is attached at its inner edge and the future dorsal margin of the embryonic body. The mantle expands by active mitosis and eventually covers all but the future head region of the embryo and this establishes the mantle cavity (stages 25 and 27).

Soon after the mantle and the shell sac become evident the eyes, arms, and posterior funnel folds are observable (stage 17) on the anterior surface of the embryo (adult ventral). The eye first appears as a thickening on either side of the embryo. In the center of this placode a depression, the future retinal region, appears and gives the eye a ring-like appearance (stages 17 and 18). At stages 19 and 20 center edges of this ring invaginate and eventually fuse to form the optic vesicle. A further circular invagination on the outer surface of the optic vesicle forms the iris. The cornea is derived from still another invagination which arises behind the optic vesicle itself (stage 17). In the cephalopods with lidded eyes a fourth invagination occurs to form the lid. A detailed account of some further aspects of eye development is given below.

The arms also first appear as thickened regions on the surface of the embryo. The location of the arms marks the limit between the external yolk sac and the embryo proper. In *L. pealii* this is approximately at the equator of the egg. In *Ommastrephes pacificus* (Hamabe 1962), Grenacher's embryo, and most of the embryos known from the ommastrephids (Clarke 1966) a very small external yolk sac is formed. All of these eggs are comparatively small and measure about 1 mm or less in length. In larger eggs of Sepia and Sepioteuthis the embryonic regions develop as a germinal disc at the animal pole and the external yolk sac is many times the volume of the embryo. Development of these large-yolked forms is correspondingly slow. Individual arm primordia separate from the two bilateral primordia and increase in size to project out from the surface of the embryo. The sucker primordia appear as small clumps of cells on the lower arm surface about stage 22 in *L. pealii* and later hollow out. The second arm becomes quite prominent in the decapods at an very early stage. In some of the oceanic squid the arms develop rather slowly and the tentacles are fused together (Clarke 1966; Hamabe 1962).

The folds of tissue which give rise to the funnel, the associated musculature and the nuchal cartilage first appear on the anterior surface above and median to the eyes (stage 17). These are the posterior funnel folds. In a second region of primordia, the anterior funnel folds become obvious about stage 18 and by stage 19 they have become quite prominent. The posterior funnel folds divide into two bands longitudinally, the inner portion of which becomes the funnel retractor muscle (stage 20) and the outer part of the funnel proper. The anterior and posterior funnel folds fuse and increase in size so they stand out from the surface of the embryo. Growth occurs at the anterior end so that the median margins of the two halves eventually fuse and form a continuous tube which separates distally from the underlying layer of cells. The growing mantle eventually covers the posterior end of the funnel folds and only the distal end can be seen when the mantle is contracted.

On the future dorsal surface between the eyes and centered above the two bands of the arm primordia the stomodeal invagination appears (Grenacher 1874). Apparently all the foregut is derived from this invagination. The salivary pit is evident in a secondary invagination within the stomodeum (stage 22), although the mouth has not yet completely closed in the region nearest to the yolk sac. With further growth the foregut becomes quite small and the mouth becomes very hard to distinguish as the arms grow around it. The anal papilla appears on the midline between the funnel fold at about stage 19. The hindgut arises within the anal papilla as a single layer lying directly upon the yolk epithelium. It is apparently isolated from the ectoderm of the surface and does not arise by invagination as was described by Watase (1888). Korschelt (1892), Bobretzky (1877), and Boletzky (1967) have described the further development of the mid- and hindgut rudiment.

Boletzky (1967) has described the development of the circulatory system in Octopus. At about stage 20 a large hemal space appears between the yolk epithelium and the outer layer of cells of the external yolk sac. Similar sinuses arise in the head region (the cephalic sinus) and internal to the funnel folds and gill primordia (the posterior sinus). These two sinuses interconnect near the mouth and the anal papilla and broadly connect with the hemal space of the external yolk sac. As the statocysts, ganglion rudiments, stomadeum, and other organs develop the hemal sinus within the embryonic body becomes progressively restricted until the adult venous pattern is established. The branchial hearts arise as large blood filled spaces at the base of each gill and connect to the branchial vein before actual beating of the heart begins. Somewhat later the cephalic aorta and the posterior aorta appear and make junction with the venous system. The systemic heart arises independently as a pair of vesicles between the future pericardial regions of the coelom. These paired vesicles unite posteriorly to form the ventricle and eventually connect to the developing branchial veins. During this time blood has been circulated by pulsation of the external yolk sac but as the definitive hearts arise the development of the circumoral musculature restricts and finally stops the circulation from the external yolk sac (see below).

The circulatory system and coelom arise concurrently, but the cavities formed can be distinguished by differences in their linings. The forming coelom is lined with a more or less regular epithelium while the circulatory sinuses have a very irregular lining which in some places is formed only by the yolk epithelium (Faussek 1900). Marthy (1968) studied the organogenesis of the coelom system of Octopus. The coelom arises as paired rudiments of the kidney and gonad. The gonadal rudiments fuse to form a single structure which then connects to the kidney rudiments making one complex. This complex contains solid rudiments of the pericardium, pericardial glands, kidneys and gonad. The pericardial cavity at first is paired but increases posteriorly and medially on either side of the ventricle and eventually fuses to form the unpaired pericardial cavity behind the heart. The kidneys develop as paired structures which are in close association with the left and right vena cava and pericardial cavity. The renal-pericardial canal arises from the former connection between the kidney rudiments and the pericardial cavity and the ureter eventually connects into the pericardial funnel. The gonad primordium develops fairly late on the dorsal side of the ventricle and two groups of cells, the germinal cells with large nuclei and the smaller stromal cells appear. Ectoderm grows inward, unites with the coelomic epithelium in the pericardial region, and forms the lumen of the gonoducts. Thus the gonocoele remains incompletely divided from the pericardium. The pericardial glands develop on the dorsal side of the branchial hearts and project into the pericardium but the epithelium, at least until after hatching, remains almost undifferentiated. The original 'coelom-mesoderm' differentiates into varying types of epithelium depending on eventual location in the coelomic system (Marthy 1968).

The gill primordia become prominent on either side of the anal papilla at stage 19 and increase in size to become two small 'wings' of tissue which begin to undergo division into separate filaments about stage 25. Joubin (1885) described the process of folding and separation of individual gill filaments in Sepia and the establishment of blood circulation through these gills. The base of each gill is provided with a heart which pumps blood into the gill filaments. At hatching the gills are quite well formed but have not completed all of their ramifications. Ranzi (1930b) operated on Sepia embryos by pressing on the chorion with needles and demonstrated the gills begin their development independently but later development is correlated with the blood supply.

At stage 19 two small depressions appear on the surface just median to the eyes on the future ventral of the animal. These differentiate into the otocysts by closure of the outer surface to form a prominent vesicle which separates from the surface. This vesicle remains in communication with the surface by a small canal but eventually completely closes off from it. On one of the opposite walls the epithelium forms the cristae which are attached to the spherical or kidney-shaped otocyst by fine hairs. A sensory epithelium develops and communication is established with the central nervous system. The otocyst increases greatly in size and is quite prominent at hatching. The whole structure has gradually become incorporated into the ever expanding funnel region and completely sinks below the surface.

During the above stages there is a general shape change occurring which changes the ovate embryo into a 'larval' cephalopod. More striking is the constriction of the 'waist' which separates the yolk sac from the organogenic region of the embryo. In this 'waist' region muscular cells develop which are quite prominent and contract slowly. The head region increases in relative size as the eyes and optic ganglia become prominent. By the downward growth of the mantle the body region seems to undergo a significant increase in size, due in large part to the formation of the mantle cavity, functionally the exterior of the animal. No extra-embryonic membranes other than the yolk sac are involved with this developmental sequence.

10.9. Development of organs of special interest

In the embryonic development of the cephalopods there are several unique and unusual organs which deserve special attention and will be treated separately here emphasizing only their unique features. Details of their developmental anatomy can be found in the papers referred to in the bibliography. The organs covered here are: the yolk sac, the eyes, the nervous system and the skin and epithelial organs. To some extent the choice of these organ systems has been arbitrary but it is hoped that they will be interesting enough to provoke the reader to further study.

10.10 Development of the yolk sac

Probably the most influential organ in the development of the cephalopod embryo is the yolk sac. Sacarrao (1952, 1955) emphasized the importance of the inter-relationship between the embryo and the 'yolk organ'. Portmann (1926) and Portmann and Bidder (1928) emphasized the relationship of the development of the embryonic circulatory system and the utilization of the stored nutrients. In the course of development the yolk sac becomes divided into three regions: two internal regions (within the embryonic body proper) and one external region (attached to the embryo in the region of the mouth). The earliest development of the external yolk sac is concurrent with cellulation of the embryo but the yolk epithelium of the internal yolk sac has been evident and functional since early stages of development. The external yolk sac in Loligo comprises the entire area below the equator and does not have an extensive development of a middle layer; rather a single layer of flattened cells lie directly upon the syncytial yolk epithelium. As development continues spaces appear between these two lamina and some of the cells in the intervening area appear to differentiate into blood cells. The space continues to enlarge and becomes an extensive blood sinus which eventually connects to the developing blood vessels of the internal yolk sac. Some of the cells from the ill-defined intermediate layer apparently differentiate into single-celled muscle bands which traverse the blood sinus and connect the yolk epithelium to the outer layer of the yolk sac (fig. 14). These muscle cells cause the rhythmic contraction of the external yolk sac and circulate the blood in the hemal space. In the yolk epithelium yolk platelets are broken down and the nutrients transferred to the hemal spaces in a process that appears to involve multivesiculate bodies and possibly Golgi-derived vesicles (fig. 15). Elaborate regions of rough endoplasmic reticulum dominate the cells and cause the well known basophilia (Portmann and Bidder 1928).

As development of the embryonic organs continues the yolk sac becomes regionalized into an external yolk sac demarcated by the arms and a girdle of specialized epithelio-muscular cells, and an internal yolk sac divided into an anterior and posterior portion (fig. 16). In the early stages of organ

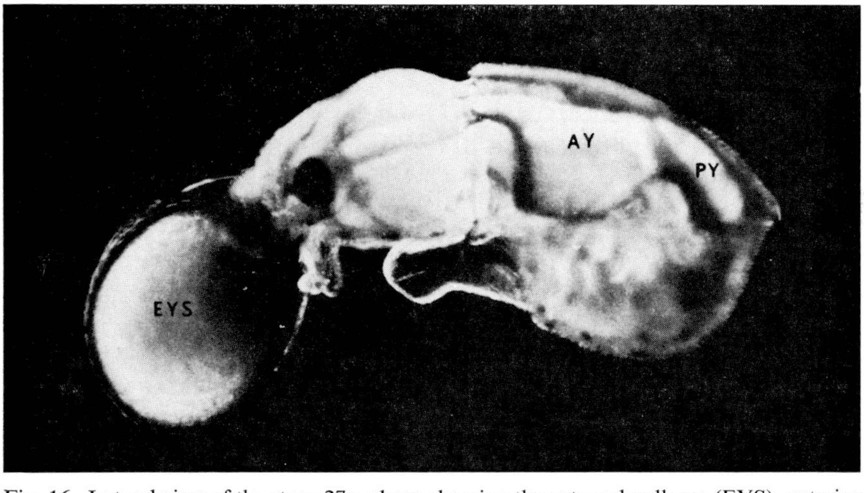

Fig. 16. Lateral view of the stage 27 embryo showing the external yolk sac (EYS), anterior internal yolk sac (AY) and posterior internal yolk (PY). ca. 37 ×

formation there are broad connections between the hemal spaces of the external yolk sac and the blood sinuses of the internal yolk sac, but as development proceeds the anterior portion is compressed into a network of small vessels which eventually disappear and the posterior blood sinuses persist only as the abdominal vein (stage 28). This limits the hemal mechanism of nutrient supply and the liver takes over the function of yolk digestion. Portmann and Bidder (1928) divided yolk absorption into four periods. The first is from cleavage to the establishment of the embryonic circulation. In this period most of the nutrient substance digested by the yolk epithelium can be directly transferred to the cells of the embryo. The second period lasts from the establishment of the circulatory system to the breakdown of the circulatory pattern by the contraction of the circumoral musculature. Period three ends when the external yolk sac is depleted although in *L. pealii* the larvae frequently hatch with a small external yolk sac which they later discard. The final period of yolk digestion is mostly post-hatching and starts with the depletion (or casting off) of the yolk sac and includes the utilization of the internal yolk material. This is accomplished when the animal is a free swimming larva. The first rudiments of the hepatopancreas arise during period two and ramify into the blood sinuses. In period three the liver becomes active in secretion while the internal yolk sac increases in size at the expense of the external yolk sac. Active digestion of yolk can be seen at this time in the yolk epithelium and in the liver. In the fourth period of yolk digestion the yolk is no longer the sole source of nutrient and the yolk sac diminishes as the liver grows to engulf it. Eventually, the last remnants of the yolk disappear and the yolk epithelium degenerates. Thus there are two functional mechanisms of yolk utilization in the cephalopod: the yolk epithelium-hemal system and the hepatic system. In addition to this the yolk epithelium also functions as an inductive morphogenetic map; a point which will be discussed below.

10.11. Development of the eye

Of all the organs of the cephalopod, the eye has probably received most attention because of its remarkable resemblance to the vertebrate eye. This similarity is a result of convergent evolution as amply illustrated by its ontogeny. The position of the retina, lens, iris, and 'cornea', in both eyes is the same so they function alike, but the comparison ends there. The vertebrate lens is cellular and focuses by changing its shape while the cephalopod lens is acellular and inflexible. The nerve processes can multiply off the outside of the retina in the cephalopod instead of running between the light and the photoreceptor to a common single nerve as is found in the vertebrate. The cornea in the vertebrate eye plays an important role in image formation while in the cephalopod it is merely a window unattached to the eyeball proper. Only three aspects of development of the eye will be covered here: the organogenesis of the whole eye, the histogenesis of the retina, and the histogenesis of the lens.

The organogenesis of the eye has been described by a number of authors (Lankester 1875; Grenacher 1895, 1898; Faussek 1900; Sacarrao 1954; etc.) and the description that follows is based on their work as well as the observations of the author. The eyes first appear as slightly thickened bilateral oval placodes just above the midline of the embryo (stage 17). The margins of these placodes begin to invaginate (fig. 17) and eventually an optic vesicle

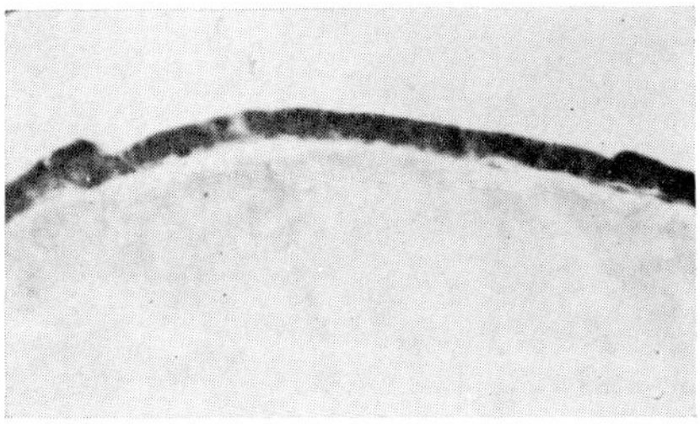

Fig. 17. Section of the developing eye at stage 18. For details of this figure and figs. 17–23 see the text. ca. 137 ×

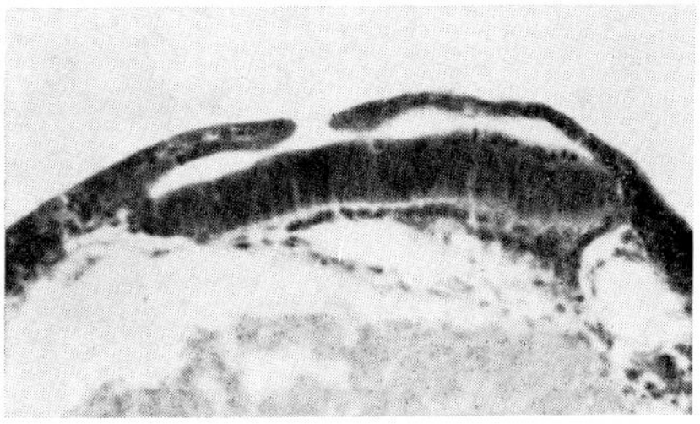

Fig. 18. Stage 20 eye during invagination. ca. 137 ×

is formed which has two easily distinguishable regions: the retinal area in the proximal part and the lentigenic area in the distal portion (figs. 18 and 19). A second invagination arises on the outside of the lentigenic area and gives rise to the iris (fig. 20). The iris becomes deeply pigmented with

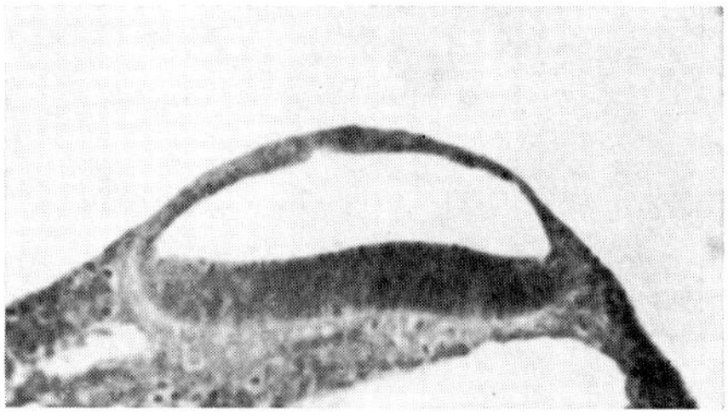

Fig. 19. Stage 22 eye during beginning lens development. The notch in the outer wall of the optic vesicle is the site of the lens primordium. ca. 137 ×

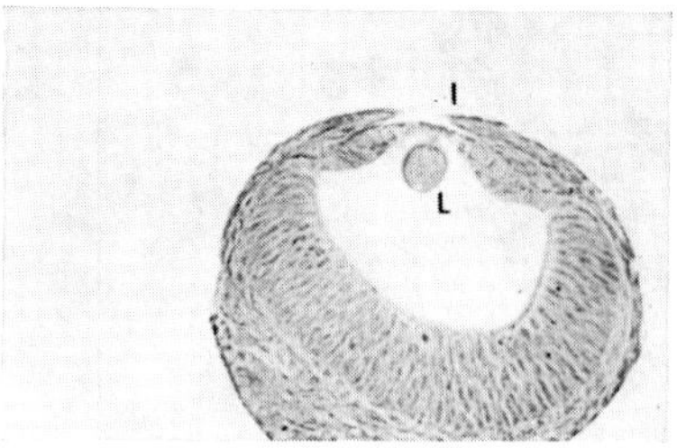

Fig. 20. Stage 26 eye. The lens (L) and the iris (I) are evident. ca. 160 ×

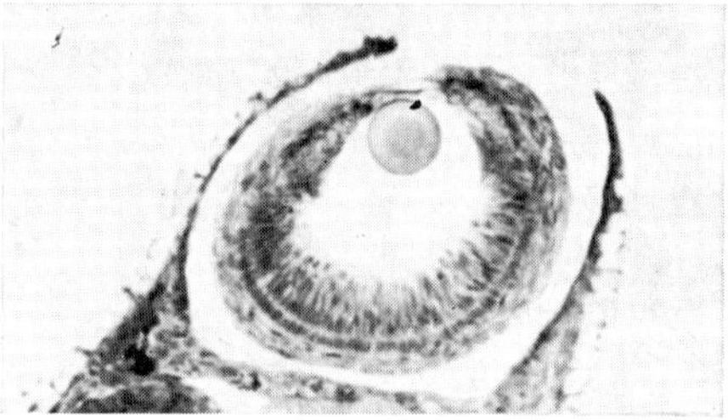

Fig. 21. Stage 27+ eye. The cornea is forming. ca. 137 ×

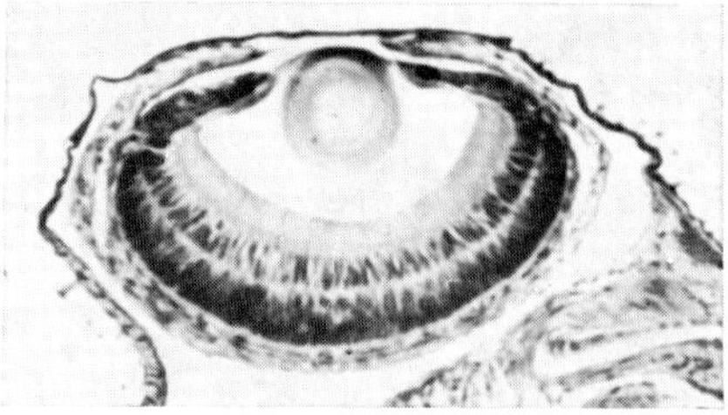

Fig. 22. Stage 30 eye (hatching). The nuclei of the retina have become restricted to two regions. ca. 137 ×

iridophores and ommatochrome granules. A third invagination occurs which forms the functional cornea (stages 27–28 in *L. pealii*) (fig. 21). In some cephalopods this remains open and in communication with the sea water, hence it is not a true anterior chamber (Tompsett 1939). In a species where a lid is present (e.g. Sepia) it arises as a fourth invagination of the surrounding skin and it is under muscular control. The distal portion of the original optic vesicle differentiates into a number of structures including the musculature that moves the lens, the lentigenic body, and the cartilage that surrounds the eyeball (fig. 22).

The development of the lens was incorrectly described by Faussek (1900) and unfortunately this error has been perpetuated in the literature despite the correct interpretation of Williams (1909). When the optic vesicle is formed a group of cells characterized by their large nuclei become evident on the distal surface of these cells (lentigenic cells, Arnold 1967a) and eventually send out cytoplasmic processes which fuse and form a lens primordium. The lentigenic body itself increases in size and cell number by addition of cells at its periphery. As the lens primordium grows to a subspherical shape the definitive lens substance is elaborated. The Golgi apparatus produces many vesicles which are transported to the lens primordium and fuse there to make an electron dense conglomerate which eventually replaces the cytoplasm of the lentigenic process. Growth of the lens continues by the application of more lentigenic processes and eventually a secondary lentigenic body forms the anterior portion of the lens. As the lens grows it displaces the lentigenic cells so that the lentigenic body changes from a disc to a ring surrounding the lens and the lens is suspended in front of the retina by the lentigenic processes. The lens continues to grow by application of new processes to its surface, elaboration of the dense lens substance, and eventual obliteration of the cytoplasm by the dense lens substance. Thus the cephalopod lens is virtually acellular and grows by accretion to its surface. Because of the number of microtubules in the lentigenic processes and the obvious necessity of transport along these processes, it was postulated that these microtubules are in some way associated with cytoplasmic transport (Arnold 1966).

The histogenesis of the retina occurs quite late in embryonic development (stage 28). Faussek (1900), Hesse (1900), Bobretzky (1877), and Grenacher (1895, 1886) among others have described various aspects of the development of the retina but Faussek's work will be followed here. While the lens is differentiating and developing to a spherical shape the future retinal tissue undergoes a mitotic phase and builds up to a thick cellular layer with many ovate nuclei evenly distributed throughout. On the inner surface a 'structureless' limiting membrane develops. This is followed by the development of the rod from the inner surface downward and a concurrent sorting out of the nuclei into two distinct layers; those of the limiting membrane cells and those of the sensory cells. Each rod cell (rhabdomere) has a single neurofibril running down its length. A particulate pigment zone develops inside the innermost layer of the nuclei and these particles eventually migrate to accommodate for changes in light intensity. Unlike the vertebrate eye the nerve fibers from each of the rod cells grow downward away from the optic vesicle and group together to form many nerve bundles, each of which makes a separate connection with the optic ganglion. Ranzi (1928, 1930a) isolated the developing eye and also used chemical treatments of the egg surface to demonstrate that the retina induces the surrounding tissue to form an optic ganglion. Therefore, three distinct layers appear in the retina: a thin limiting membrane, a layer of rod cells with the associated pigment, and an external layer of nuclei and cell bodies of the sensory and supporting cells, with its associated blood supply, and the nerve fibers. Reference to the figures will assist the reader in following this process.

10.12. Development of the nervous system

The development of the cephalopod nervous system is so complex that more than a cursory coverage is impossible here. For a broader account the reader is referred to Korschelt (1936), Ussow (1875), and Faussek (1900). The ontogeny of the nervous system is closely linked to its phylogenetic history. The ganglia that comprise the molluscan nervous system are organized into a large and complex brain in the cephalopods. These ganglia all arise as ectodermal thickenings and grow together as the embryo develops and changes shape. A brain is formed by fusion of the optic, pedal, cerebral, visceral, and brachial ganglia around the esophagus. Recently, Martin (1965) studied the development of the giant fiber system and concluded the pattern of development shown there also reflected the phylogenetic history of the Cephalopoda.

10.13. Development of the skin and epithelial organs

The skin of the cephalopods is somewhat simpler than other molluscan skin but it is elaborate enough to prevent a complete coverage of all of its organs here. Only the unique features will be covered. The recent papers of Fioroni (1963, 1962a, 1962b) are an excellent introduction to this literature.

The skin is composed of a single cell layer of epidermis and the cutis with a muscular layer below. The epidermis is derived from ectoderm and is composed of ciliated and glandular cells. These gland cells not only produce mucus but also appear to have a hatching function because of the decrease in activity and number in the post-hatching decapod larvae. The cutis is derived from mesoderm as is the muscular layer beneath it. There are a number of epidermal organs derived by modification of the epidermis and the cutis. These are the chromatophores, iridophores, Hoyle's organ and Kölliker's organ. Fioroni (1963) and Sacarrao (1954) have described the development of the cephalopod chromatophore and their description, rather than the conflicting one of Chun (1902), will be followed here. The chromatophore is composed of two major mesodermal components: the inner pigment cell and the outer radial musculature. In development the chromatophore first appears as a larger cell with an acentrically-placed nucleus which is attached to mesenchyme cells by fine cytoplasmic processes. As the future pigment cell begins to grow the mesenchyme cells arrange themselves around it and develop into the radial muscle cells. The pigment develops as a mass of granules contained in a fibrous sac rather than as a fluid (Cloney unpublished). Contraction of the radial muscles expands the sac from a subspherical shape to a disc and this increase in surface area causes the color change of the skin. The iridophores are cells specialized for reflecting light and are located around the eye, the ink sac of the decapod, and found throughout the skin in varying concentrations. These cells are unable to change color but because they are frequently overlaid by chromatophores their shiny iridescence can be modified or masked. This reflective organelle is composed of 0.1 μ thick platelets of dense material arranged in stacks (Fioroni 1965; Arnold 1967b). These stacks are membrane bound and this membrane is continuous with the plasma membrane although it appears to be derived from Golgi vesicles. The dense material develops by fusion of small granules in the area between the developing platelet membranes. Apparently the vesicles which form the membranes are aligned by unidirectional microtubules which run into the end of the developing platelets (fig. 23).

Hoyle's organ is composed of specialized gland cells on the dorsal posterior mantle of the stage 28 to stage 30 embryo. Von Orelli (1959) has described the function and gross morphology of this hatching organ and has shown that once the larvae have hatched it disappears. Fioroni (1962a) has described the development of these cells to become rather typical goblet cells. The hatching enzymes accumulate in a large vacuole in these cells. At hatching

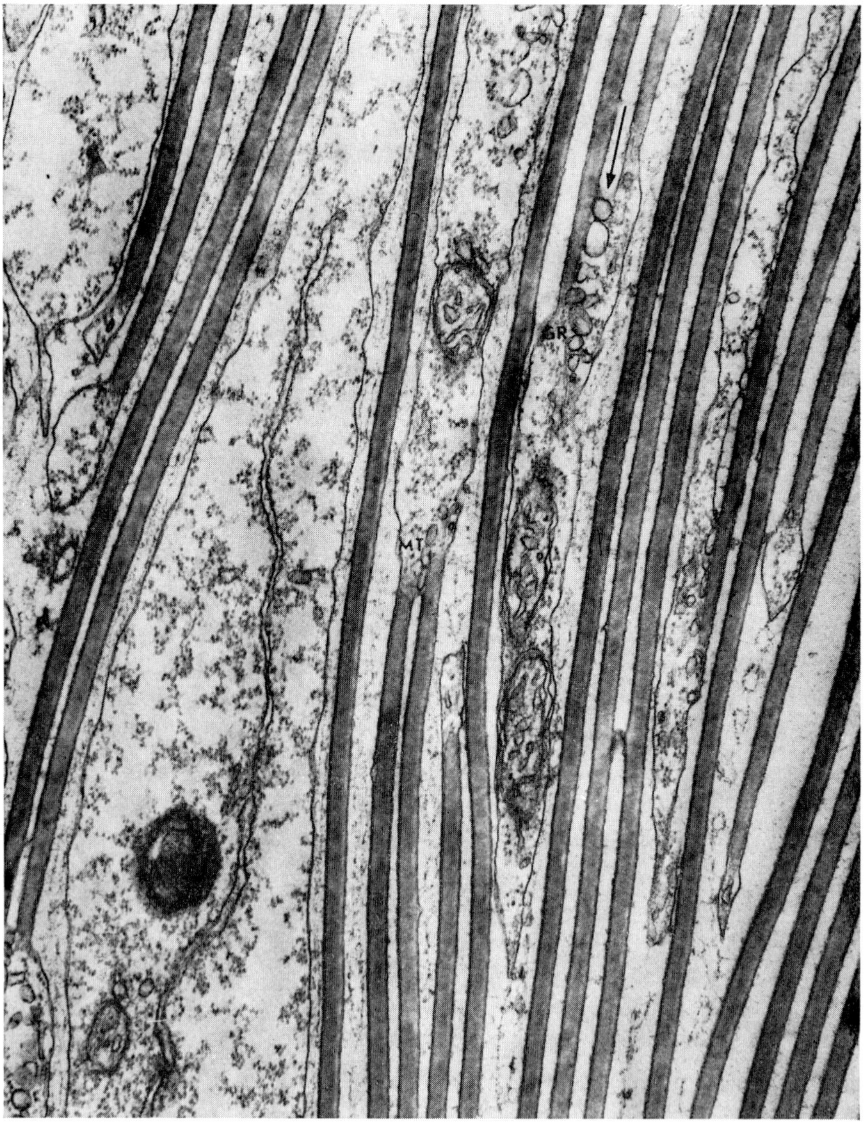

Fig. 23. Developing iridophores in the iris of *L. pealii*. The arrow indicates a region of vesicles fusing to form a membrane. Microtubules (MT) run into the end of the platelet and granules (GR) appear to be fusing to from the platelets. ca. 43,000 ×

the Hoyle organ is applied to the inside of the chorion and an opening the shape of the organ is digested through the chorion. The pressure of the intrachorionic fluid and the contractions of the mantle force the larvae out of the egg capsule. In the decapods there is a slimy degeneration of most of the skin at hatching. This may indicate that many of the gland cells of the general skin also participate in softening of the chorion (Ranzi 1931b).

In some members of the Octopoda unique epidermal organs of unknown function appear in the skin (Fioroni 1962b). These are Kölliker's organs which individually are composed of a bundle of chitinous rods contained in a cellular cup embedded in the skin. Both ectoderm and mesoderm participate in the formation of these organs. In development a basal cell sinks below the rest of the epidermis and a bundle of chitinous material develops within it. By fusion of epidermal cells to the surrounding mesodermal cells the cup is formed. Eventually the bundles of chitin become free within the cup.

10.14. An analysis of cephalopod differentiation

As in any embryonic system the cephalopod embryo presents certain basic questions concerning the mechanisms of differentiation. A relatively small amount of experimental work has been done on cephalopod embryos, probably because of the lack of a reliable source of material and the inability of the young embryos to live outside of their chorions. These technical problems have been surmounted. What follows here is an attempt to assemble the fragmentary data and to make a coherent story of it. The conclusions drawn are speculative and therefore the author assumes complete responsibility for them.

L. pealii can be caused to lay eggs by the visual stimulus of an egg mass (Arnold 1962) and Octopus egg laying is initiated after the nerve supplying the optic gland is severed (Wells 1962). Thus it is possible to obtain cephalopod embryos almost on demand. The problem of survival outside the chorion has been attacked by tissue culture techniques. The author has used a mixture of sea water, adult squid blood, and antibiotics. Recently he has found 10% horse serum to be equally successful as a substitute for the squid blood. Fedorow (1933) was able to culture pieces of nervous tissue in a mixture of 'homologous amniotic fluid', cephalopod embryo extract and Van 't Hoff's solution. Szabo and Arnold (1963) were able to culture isolated tissues and cells of the ink sac and had this material survive for as long as ten months. They used sea water and 10% horse serum, sea water and 10% squid blood, and medium 199 made up in sea water. Survival was best in the squid blood medium. Therefore these embryos appear to be fairly easy to deal with in experimental situations.

The classical question of mosaic versus regulative development was approached by Ranzi (1928, 1930a, b, 1931a, 1932) who operated on embryos by pressing on the chorion with needles. He isolated large pieces of fairly old embryos and concluded since there was no replacement of loss parts that development was strictly mosaic. However, this question is somewhat more complex when younger embryos are used.

The role of the yolk epithelium in digestion of the nutrient media was discussed above but apparently it has a greater role in embryogenesis (Arnold 1965b). At stages 16–18 large areas of the outer layers of the blastoderm were teased free from the yolk epithelium and the embryos were grown in culture. The wound healed quickly by migration of the surrounding cells and development proceeded normally. If the very early eye primordium was removed the resultant embryos did develop with both eyes despite the fact that cells used to form the new eye normally were fated to form other organs. As long as the yolk epithelium remained intact this appeared to be the case for all of the organ primordia. However, if the yolk epithelium was removed with the overlying cells the donor embryos lacked those organs from the operated area and the isolated tissues developed into complete organs. Grafting dissociated-reaggregated cell clumps onto the freshly stripped yolk epithelium confirmed the conclusion that the yolk epithelium functions as a morphogenetic inductive map with circumscribed areas of

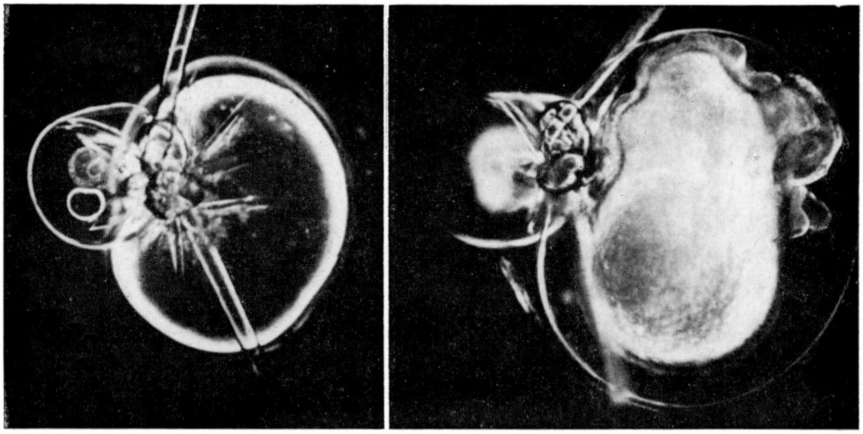

Fig. 24. Removal of part of the cortex by ligation. The arms and eye are missing on the operated side of the embryo. ca. 31 ×

the future embryo topographically laid out in its inductive machinery. The specific nature of this developmental information and how it is transferred is unknown but this seems to be a classical case of embryonic induction. Potter et al. (1965) have demonstrated electrophysiological connection between many of these cells.

The origin of this inductive morphogenetic pattern in the embryo is of special interest. Arnold (1968a) demonstrated a similar pattern exists in the egg cortex. Essentially three different experimental techniques were used. At stage 10 uncellulated areas of the egg cortex beyond the margin of the blastoderm were removed by ligation. This resulted in missing organs or defects that could be topographically related to the missing cortical region (fig. 24). A similar type of experiment using an ultraviolet microbeam irradiating apparatus was performed with much the same results (fig. 25),

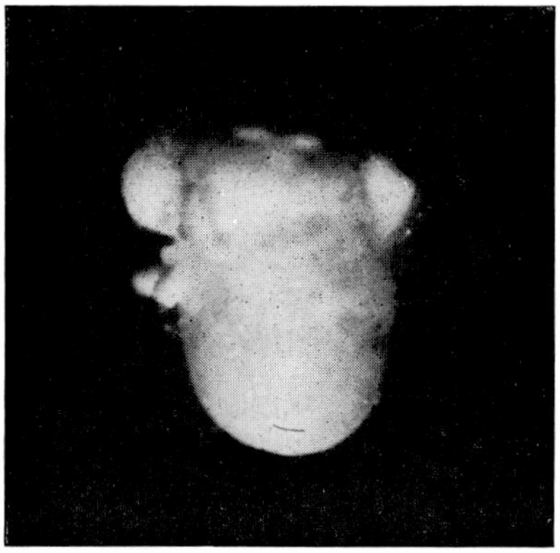

Fig. 25. Ultraviolet microbeam irradiation of the future arm region of the cortex at stage 4. Note the lack of organs in the irradiated area. ca. 25 ×

despite the fact that the irradiation was done for a few seconds at the first cleavage. Irradiation was tangential to the surface and at a distance of at least one half millimeter from the nuclei so damage to nuclear genes can be ruled out. Cellulation of the irradiated area proceeded normally but instead of an arm forming in the normal arm region a layer of undifferentiated cells resulted. A third set of experiments were performed in which the blastodisc was displaced by centrifugation to see what effect this had on organ placement. The blastodisc and second polar body formation could be displaced to the centripetal pole and cleavage would occur there. Although the organs formed on the centripetal pole were larger, the placement of the organs

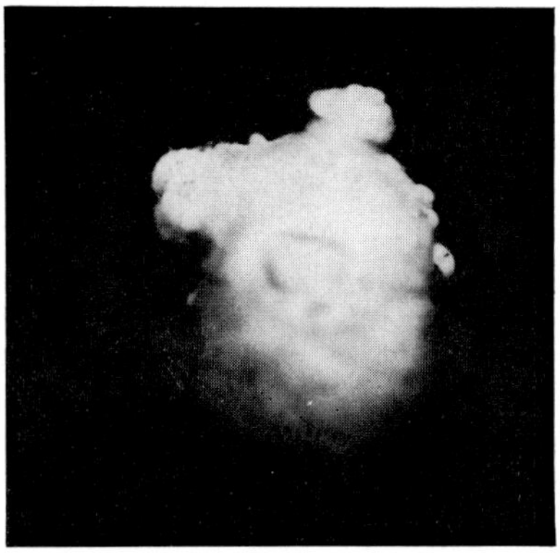

Fig. 26. Centrifugation displaced blastodisc. The organ placement is normal but the organs on the centripetal (left) side are larger due to the greater amount of cytoplasm they received. ca. 25 ×

was essentially normal (fig. 26). Together, these three experiments implicate the egg cortex as being a site of developmental information storage. These results would be in close agreement of those obtained by Curtis (1960) in the amphibian embryo, Morgan and Tyler (1936) on the Chaetopterus egg, and those of Raven (1958, 1963) with a gastropod egg. Ranzi (1928) was able to produce a cycloptic series by treatment of *L. vulgaris* with LiCl and other salts at about stage 10, demonstrating that there must be some ability of this cortical pattern to compensate for losses.

Combining the two major conclusions stated above it is possible to construct a tentative model of information transferred in the cephalopod embryo. The informational pattern in the egg cortex must logically be derived from the follicle cells of the gonad. Raven (1958) has speculated this is the case in Limnea eggs. This informational pattern is transferred to the yolk epithelium in cleavage and cellulation of the embryo, and eventually becomes an inductive morphogenetic map which regulates the placement of the organ primordia and the entire process of embryogenesis. Modern theory would lead us to believe that this would involve masked messenger RNA storage in the egg cortex and repression and derepression in induction.

10.15. Conclusions

The conclusions that can be drawn from all of the data presented here are quite diverse. It is immediately obvious that the cephalopods are a strikingly different group of molluscs because they lack spiral cleavage, have large eggs, and exhibit none of the typical larval stages of the spiralia. It is a moot question if development is mosaic or regulative. Although the pattern of embryogenesis is rigid, apparently the fate of specific cells is not. Therefore, cephalopod development seems worthy of more attention than it has received. Since the cephalopods represent an end point in one branch of animal evolution it would seem that comparison with vertebrates and other animals would be of particular importance.

I would like to thank Mrs. Joan Hahn Henrickson for reading this manuscript and helping prepare the bibliography. Drs. W. C. Summers, S. V. Boletzky, and M. C. Mercer all offered helpful advice and criticism. Much of the work reported here was performed with the aid of grants from the National Science Foundation and National Institutes of Health of the U.S.A.

References

ANDERSON E., 1967. Observations on the uptake of horseradish peroxidase by developing oocytes of the rabbit. J. Cell Biol. *35*, 160A.

ARNOLD J. M., 1961. Observations on the mechanism of cellulation of the egg of *Loligo pealii*. Biol. Bull. *121*, 380–381.

ARNOLD J. M., 1962. Mating behavior and social structure in *Loligo pealii*. Biol. Bull. *123*, 53–57.

ARNOLD J. M., 1963. Techniques for *in vitro* culture of Cephalopod embryos. Assoc. Island Marine Lab. Carib. *5*, 19.

ARNOLD J. M., 1965a. Normal embryonic stages of the squid *Loligo pealii* Leseur. Biol. Bull. *128*, 24–32.

ARNOLD J. M., 1965b. Observations on the mating behaviour of the squid *Sepioteuthis sepioidea*. Bull. Marine Sci. *15*, 216–222.

ARNOLD J. M., 1965c. The inductive role of the yolk epithelium in the development of the squid, *Loligo pealii* L. Biol. Bull. *129*, 72–78.

ARNOLD J. M., 1966. On the occurrence of microtubules in the developing lens of the squid *Loligo pealii*. J. Ultrastruct. Res. *14*, 537–539.

ARNOLD J. M., 1967a. Organellogenesis of the Cephalopod Iridophore: Cytomembranes in development. J. Ultrastruct. Res. *20*, 410–421.

ARNOLD J. M., 1967b. Fine structure of the development of the Cephalopod lens. J. Ultrastruct. Res. *17*, 527–543.

ARNOLD J. M., 1968a. The role of the egg cortex in Cephalopod development. Devel. Biol. *18*, 180–197.

ARNOLD J. M., 1968b. Formation of the first cleavage furrow in a telolecithal egg *(Loligo pealii)*. Biol. Bull. *135*, 408–409.

ARNOLD J. M., 1968c. Analysis of cleavage furrow formation in the egg of *Loligo pealii*. Biol. Bull. *135*, 413.

ARNOLD J. M., 1969. Cleavage furrow formation in a telolecithal egg *(Loligo pealii)*. I. Filaments in early furrow formation. J. Cell Biol. *41*, 894–904.

AUSTIN C. R., C. L. MANN and T. MANN, 1964. Spermatophores and spermatozoa of the squid *Loligo pealii*. Proc. Roy. Soc. (London) Ser. B *161*, 143–152.

BERGMANN W., 1903. Ueber den Bau des Ovariums bei Cephalopoden und einige Nachträge zu Eibildung der selben. Arch. Naturges. Bd. *69*, 227–236.

BOBRETZKY N., 1877. Untersuchungen über die Entwicklung der Cephalopoden. Nachr. Ges. Freunde Naturwiss. *24*.

BOLETZKY S. V., 1967. Die embryonale Ausgestaltung der frühen Mitteldarmanlage von *Octopus vulgaris* Lam. Rev. Suisse Zool. *74*, 555–562.

BOTT R., 1938. Kopula und Eiablage von *Sepia officinalis* L. Z. Morphol. Ökol. Tiere *34*, 150–160.

BROOKS W. K., 1880. The development of the squid *Loligo pealii*. Anniv. Mem. Boston Soc. N.H.

BLANCQUAERT T., 1925. L'origine et la formation des spermatophores chez les céphalopodes décapodes. Cellule Rec. Cytol. Histol. *36*, 315–356.

CALLAN H. G., 1957. The lampbrush chromosomes of *Sepia officinalis* L., *Anilocra physodes* L., and *Scylliam catulus* Cuv. and their structural relationship to the lampbrush chromosomes of amphibians. Pubbl. Staz. Zool. Napoli *29*, 328–345.

CHING M. Y. K., 1930. Contribution à l'étude cytologique de l'ovogénèse du développement et de quelques organes chez les céphalopodes. Ann. Inst. Oceanog. (Paris) *7*, 301–364.

CHUN C., 1902. Über die Natur und die Entwicklung der Chromatophoren bei den Cephalopoden. Verhandl. Zool. Botan. Ges. Wien *12*, 162.

CLARKE M. R., 1966. A review of systematics and ecology of oceanic squids. Advan. Marine Biol. *4*, 93–325.

CLONEY R. A., 1966. Cytoplasmic filaments and cell movements: epidermal cells during ascidian metamorphosis. J. Ultrastruct. Res. *14*, 300–328.

CURTIS A. S. C., 1960. Transplantation of cortical material in the African clawed toad *Xenopus laevis*. J. Embryol. Exptl. Morphol. *8*, 163.

DREW G. A., 1911. Sexual activities of the squid *Loligo pealii*. J. Morphol. *22*, 327–360.

DREW G. A., 1919. Sexual activities of squid. II. Spermatophore structure. J. Morphol. *32*, 379–436.

FAUSSEK V., 1896. Zur Cephalopoden Entwicklung. Zool. Anz. *19*, 496–500.

FAUSSEK V., 1900. Untersuchungen über die Entwicklung der Cephalopoden. Pubbl. Staz. Zool. Napoli *14*, 83–237.

FEDOROW B. G., 1933. Über die *in-vitro*-Kultur des Nervengewebes der Cephalopoden. Biol. Zentr. *53*, 41–49.

FIELDS G., 1965. The structure, development, food relations, reproduction, and life history of the squid *Loligo opalescens* Berry. Calif. Fish Bull. *131*, 1–108.

FIORONI P., 1962a. Die embryonale Entwicklung der Hautdrüsen und des Trichterorganes von *Octopus vulgaris* Lam. Acta. Anat. *50*, 264–295.

FIORONI P., 1962b. Die embryonale Entwicklung der Kölliker'schen Organe von *Octopus vulgaris* Lam. Rev. Suisse de Zool. *69*, 497–511.

FIORONI P., 1963. Zur embryonalen und postembryonalen Entwicklung der Epidermis bei zehnarmigen Tintenfischen. Verhandl. Naturforsch. Ges. Basel *74*, 149–160.

FIORONI P., 1965. Die embryonale Musterentwicklung bei einigen Mediterranean Tintenfischarten. Vie et Milieu *16*, 655–756.

GRENACHER H., 1886. Die Retina der Cephalopoden. Abhandl. Naturforsch. Ges. Halle 16, 209–657.

GRENACHER H., 1874. Zur Entwicklungsgeschichte der Cephalopoden. Z. Wiss. Zool. *24*, 369–498.

GRENACHER H., 1895. Über die Retina der Cephalopoden. Zool. Anz. *18*, 280–281.

HAMABE M., 1961. Experimental studies on breeding habits and development of the squid *Ommastrephes sloani pacificus* Steenstrup. I. Copulation. Dobutsugaku Zasshi *36*, 25–34.

HAMABE M., 1962. Studies on breeding habit and larval development of the squid *Ommastrephes soloani pacificus* Steenstrup. V. Formation of the fourth arm and the tentacle in the rhynchoteuthis larvae. Dobutsugaku Zasshi *71*, 65–70.

HAMON M., 1939. Les constituents chimiques des enveloppes des spermatophores de Cephalopodes. Compt. Rend. *208*, 387–389.

HESSE R., 1900. Untersuchungen über die Organe der Lichtempfindung bei niedern Tieren. VI. Die Augen einiger Mollusken. Z. Wiss. Zool. *68*, 379–477.

HOADLEY L., 1930. Polocyte formation and cleavage of the polar body in *Loligo* and *Chaetopterus*. Biol. Bull. *58*, 256–264.

JOUBIN L., 1885. Sur la structure et le développement de la branchie de quelques Céphalopodes de côtes de France. Arch. Zool. Exptl. Gen. *13*, 75–150.

KONOPACKI M., 1933. Mikrometabolizm podczas owagenezy u *Loligo vulgaris*. (Micrométabolisme de l'oogénèse chez *Loligo vulgaris*) 'Kosmos'. J. Soc. Pul. Natural. Kopen. *58*, 133–156.

KORSCHELT E., 1892. Beiträge zur Entwicklungsgeschichte der Cephalopoden. Festschrift. f. Leuckant, Leipzig.

KORSCHELT E., 1936. Vergleichende Entwicklungsgeschichte der Tiere. Fischer, Jena.

LANE F. W., 1960. Kingdom of the octopus. Sheridan House, New York.

LANKESTER E. R., 1875. Observations on the development of the Cephalopoda. Quart. J. Microscop. Sci. *23*, 37–47.

LOYEZ M., 1905. Recherches sur le développement ovarien des œufs méroblastiques à vitellus nutritif abondant. Arch. Anat. Microscop. Morphol. Exptl. *8*, 239–397.

MACBRIDE E. W., 1914. Textbook of embryology, vol. I. Macmillan and Co. London.

MARCHAND W., 1913. Studien über Cephalopoden II: Über die Spermatopheren. Zool. Stutt. *26*, 171–200.

MARTHY J., 1968. Die Organogenese des Coelomsystems von *Octopus vulgaris* Lam. Rev. Suisse Zool. *75*, 723–763.

MARTIN R., 1965. On the structure and embryonic development of the giant fiber system of the squid *Loligo vulgaris*. Z. Zellforsch. *67*, 77.

MORGAN T. H. and A. TYLER, 1936. The relation between entrance point of the spermatozoan and bilaterality of the egg of *Chaetopterus*. Biol. Bull. *74*, 401.

NAEF A., 1928. Die Cephalopoden, Monographie 35. Fauna e Flora del Golfo di Napoli. Vol. I.

PORTMANN A., 1926. Der embryonale Blutkreislauf und die Dotterresorption bei *Loligo vulgaris*. Z. Morphol. Ökol. Tiere *5*, 406–423.

PORTMANN A. and A. M. BIDDER, 1928. Yolk absorption in *Loligo*. Quart. J. Microscop. Sci. *72*, 301–324.

POTTER D. D., E. J. FURSHPAN and E. S. LENNOX, 1965. Connections between cells of the developing squid as revealed by electrophysiological methods. Natl. Acad. Sci. US *55*, 328–336.

RANZI S., 1928. Correlazioni tra organi di senso e centri nervosi in via di sviluppo. Arch. Entwicklungsmech. Organ. *114*, 364–370.

RANZI S., 1930a. Ulteriori richerche sulle correlazioni tra organi di senso e centri nervosi in via di sviluppo. Biol. Zool. (Naples) *1*, 131–135.

RANZI S., 1930b. Condizioni determinanti lo sviluppo delle branchie. Atti Acad. Naz. Lincei Mem. Classe Sci. Fis. Mat. Nat. *12*, 468–472.

RANZI S., 1931a. Risultati di ricerche di embriologia sperimentale sui Cefalopodi. Arch. Zool. Ital. *16*, 403–408.

RANZI S., 1931b. Sviluppo di parti isolate di Cefalopodi (Analisi sperimentale dell' embriogenesi). Pubbl. Staz. Zool. Napoli *11*, 104–146.

RANZI S., 1932. Sull' indipendenza dell' istogenesi dell' organogenesi. Atti Pontif. Acad. Sci. Nusvi Lincei *85*, 27–28.

RAVEN C. P., 1958. Morphogenesis: The analysis of Molluscan development. Pergamon Press, New York.

RAVEN C. P., 1963. The nature and origin of the cortical morphogenetic field in Limnaceae. Develop. Biol. *7*, 132–143.

ROTH T. F. and K. R. PORTER, 1964. Yolk protein uptake in the oocyte of the mosquito *Aedes aegypte*. J. Cell Biol. *20*, 313.

SACCARRAO G., 1951. Notice on the embryonic shell sac of *Octopus* and *Eledone*. Arg. Mus. Bocage *22*, 103–106.

SACARRAO G., 1952. Remarks on gastrulation in Cephalopodes. Arg. Mus. Bocage *23*, 43–47.

SACARRAO G., 1953. Sur la formation des feuillets germinatifs des Céphalopodes et les incertitudes de leur interprétation. Arg. Mus. Bocage *24*, 21–64.

SACARRAO G., 1954. Quelques aspects sur l'origin et le développement du type d'œil des Cephalopodes. Arg. Mus. Bocage *25*, 1–29.

SACARRAO G., 1955. On the ontogenetic evolution of the relations between the embryo and the vitelline organ in Cephalopods. Arg. Mus. Bocage *26*, 1–126.

SPEK J., 1934. Die bipolare Differenzierung des Cephalopoden und des Prosobranchiereies. Arch. Entwicklungsmech. Organ. *131*, 362–372.

SZABO G. and J. M. ARNOLD, 1963. Studies of melanin biosynthesis in the ink sac of the squid *(Loligo pealii)*. III. *In vitro* culture of tissues and isolated cells from the adult ink gland. Biol. Bull. *125*, 393–394.

TINBERGEN L., 1939. Zur Fortplanzungsethologie von *Sepia officinalis* L. Arch. Neerl. Zool. *3*, 323–364.

TOMPSETT D. H., 1939. Sepia. Liverpool Marine Biology Committee Memoirs XXXII. University Press, Liverpool.

USSOW M., 1875. Zoologic embryological investigations. Ann. Mag. Natl. Hist. Ser. 4, *15*, 97–113; 209–221; 317–320.

VIALLETON L., 1888. Recherches sur les premières phases du développement de la seiche *(Sepia officinalis)*. Ann. Sci. Natl. Zool. *7*, 165.

VON ORELLI M., 1959. Über die Sclüpfen von *Octopus vulgaris*, *Sepia officinalis*, und *L. vulgaris*. Rev. Suisse Zool. *66*, 330–343.

VON ORELLI M., 1962. Die Uebertragung der Spermatophore von *Octopus vulgaris* und *Eledone*. Rev. Suisse Zool. *69*, 193–203.

WATASE S., 1888. Observations on the development of Cephalopods; homology of the germ layer. Stud. Johns Hopkins Biol. Lab. *4*, 165–183.

WELLS M. J. and J. WELLS, 1959. Hormonal control of sexual maturity in *Octopus*. J. Exp. Biol. *36*, 1–33.

WELLS M. J., 1962. Brain and behavior in Cephalopods. Stanford University Press, Stanford, Calif.

WILKINS M. H. F., 1956. Structure of deoxyribosenucleic acid and nucleoprotein and its possible bearing on protein synthesis. Biochem. J. *62*, 40.

WILLEY A., 1897. The ovaposition of *Nautilus macromphalus*. Proc. Roy. Soc. (London) Ser. B *60*, 467–471.

WILLIAMS L. W., 1909. Anatomy of *Loligo pealii*. Brill, Leiden.

WIRZ K., in Lane, *loco cit*.

G. Reverberi (ed.), Experimental embryology of marine and fresh-water invertebrates

CHAPTER 11

Crustaceans

J. GREEN

Westfield College, University of London

11.1. Introduction

There is great variation in the stage of development at which a crustacean emerges from the egg. Some hatch as miniatures of their parents, but many others emerge as larvae which have to pass through a series of moults before assuming the adult form. It follows from this that there is considerable variation in the amount of organogenesis during embryonic development, and in many forms there is extensive postembryonic development.

The least well-developed crustacean larva is the nauplius. This emerges

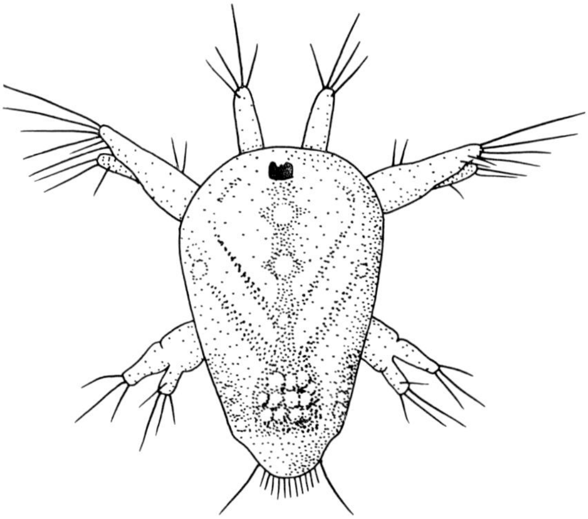

Fig. 1. Nauplius larva of Cyclops (Copepoda). Note the three pairs of appendages, single eye and unsegmented body.

from the egg with an unsegmented body and only three pairs of appendages: the antennules, antennae and mandibles (fig. 1). A single median eye is present. Sometimes the body is segmented, and the larva is then known as a metanauplius. The widespread occurrence of the nauplius larva is one of the reasons for regarding the Crustacea as being monophyletic in origin. A nauplius or metanauplius larva is found in at least one representative of each of the following groups: Anostraca, Notostraca, Conchostraca, Cladocera (Leptodora only), Copepoda, Cirripedia, Euphausiacea, and the family Penaeidae among the Decapoda. In addition the newly emerged larvae of the Cephalocarida, Mystacocarida and the Ostracoda have many features in common with the nauplius larva.

Many decapods hatch as a protozoea or a zoea larva. The protozoea has an elongated segmented abdomen, but swims by means of its antennae, while the zoea swims by means of its thoracic limbs. Well-developed lateral compound eyes and six pairs of appendages are present in the zoea (fig. 2).

A summary is given below of the state of development of the neonatae of each of the major crustacean groups, and of the way in which the eggs are carried or deposited by the females.

Subclass Branchiopoda

Order Anostraca. In this order the eggs are carried by the females in a brood pouch behind the last pair of trunk limbs. The eggs usually have hard external membranes and frequently suffer desiccation before development is completed. They are often termed resting eggs because they have the capacity to delay development while resisting freezing and desiccation. Some races of *Artemia salina* are parthenogenetic, others are bisexual. Artemia can also produce thin-shelled eggs which are retained in the brood pouch and give rise to nauplii after a few days. All Anostraca hatch as nauplii or metanauplii.

Order Notostraca. The eggs are carried by females, or hermaphrodites in some species, in pouches on the 11th pair of legs. After several days the eggs are deposited on plants or on the bottom of a temporary pool. Most species hatch as nauplii or metanauplii, but *Lepidurus arcticus* produces relatively large eggs and hatches at a more advanced stage.

Order Conchostraca. The eggs are carried between the trunk and the large bivalved carapace. Most species hatch from resting eggs as nauplii, but *Cyclestheria hislopi* has direct development and hatches as a miniature of the adult (Sars 1887).

Order Cladocera. Two types of eggs are produced by Cladocera. One does not require fertilisation and undergoes direct development in the brood pouch of the female, becoming free swimming after a few days. Some Cladocera secrete a nutritive fluid into the brood pouch: Moina and Polyphemus are examples (Gravier 1931). The embryos of these genera are incapable of development outside the brood pouch, but other genera, such as Daphnia and Chydorus are capable of normal development if removed from the brood pouch and kept in water (Rammner 1933).

The second type of egg produced by the Cladocera is a resting egg capable of withstanding freezing and desiccation. Some genera produce a single resting egg at a time, Daphnia produces two at a time, and Eurycercus produces a variable number, up to about 10. The resting eggs may be liberated to float in the water, as in Leptodora, or they may be enclosed in a modified portion of the carapace which is cast off when the female moults. This modified brood pouch is known as an ephippium. Resting eggs are normally fertilised, but in some Arctic populations males are unknown, and the resting eggs are diploid and parthenogenetic (Schrader 1926).

The resting eggs of most Cladocera give rise to young which are miniature adults, but from the resting egg of Leptodora a metanauplius emerges.

Subclass Cephalocarida

Hutchinsoniella macracantha produces a single egg in each brood. This egg is carried in an ovisac behind the last pair of trunk limbs. The larva resembles a metanauplius. In addition to the three naupliar appendages the first maxilla is present, and the second maxilla is represented by a small bilobed protuberance (Sanders 1963).

Subclass Copepoda

The eggs may be shed freely into the water (e.g. Calanus) or carried in a thin-walled sac behind the last legs (e.g. Cyclops). Some parasitic forms produce cylindrical egg strings with enormous numbers of small eggs. The Notodelphyoidea have a dorsal brood pouch projecting over the abdomen. Most copepods hatch as nauplii that undergo five moults before changing into copepodids, which resemble the adults except that their limbs are not fully developed. The Choniostomatidae, which are parasites of other Crustacea, hatch as copepodids with two pairs of swimming legs (Hansen 1897).

Subclass Mystacocarida

Derocheilocaris remanei emerges from the egg in a stage resembling a metanauplius. The antennules, antennae and mandibles are well developed and four distinct segments, including the telson are present behind the mandibles (Delamare–Deboutteville 1954).

Subclass Ostracoda

The eggs are carried between the trunk and the bivalved carapace, or else attached to water plants and stones. The larva has a large bivalved carapace, a median eye and the same three appendages as a nauplius.

Subclass Cirripedia

Many cirripedes are hermaphrodites, but dioecious species are known as well as forms with complemental males. These are dwarf males which settle on the mantle margins of hermaphrodites. The eggs are normally retained in the mantle cavity until they hatch as nauplii. In the normal course of postembryonic development the naupliar stages are followed by a cypris stage which has a large bivalved carapace enclosing the limbs. Some of the parasitic and Arctic forms develop to the cypris stage before they leave the parental mantle cavity.

Subclass Malacostraca

Division Phyllocarida. *Nebalia bipes* carries its eggs between the thoracic limbs. After leaving the vitelline membrane the embryo moults three times before it is liberated. The newly released young resemble the adults, except that the last three pairs of limbs are rudimentary (Manton 1934).

Division Hoplocarida. The eggs are carried by the females, using their chelate legs, or, in other species, the eggs are deposited in the burrow inhabited by the female. The stage of development on the assumption of a free swimming mode of life is variable. The larvae of Coronida hatch with a large carapace, head appendages and five pairs of thoracic limbs, but the abdomen is unsegmented and lacks appendages. The larva of Squilla leaves the egg with a carapace, head appendages, only two thoracic appendages, but with four or five pairs of appendages on a segmented abdomen.

Division Syncarida. Order Anaspidacea. The eggs of Anaspides are not carried by the females, but are laid in a sheltered position in water. Development is direct, but the appendages of the newly emerged young are not fully developed (Hickman 1937).

Order Bathynellacea. Eggs are laid separately and are not carried by the females. The newly emerged larva of *Bathynella natans* bears antennules, antennae and mouthparts and the first thoracic limbs. Progressive additions are made to the limbs at successive moults until the adult condition is reached (Jacobi 1954).

Order Stygocaridacea. The members of this order produce relatively large eggs which are not carried by the females. Development appears to be direct, and the newly liberated young have a well-developed series of appendages (Noodt 1964).

Division Peracarida. The eggs are carried by the females, most usually in a ventral pouch formed by oostegites, which are flat plates arising at the bases of the thoracic limbs. Some isopods, such as Sphaeroma, have internal pouches projecting from the ventral surface into the body cavity. The Thermosbaenacea carry their eggs in a dorsal brood pouch between the trunk and the carapace. The young are released at an advanced stage of development, and resemble closely the adult. The work of Saudray (1954) indicates that the embryos of the isopod *Ligia oceanica* receive nutriment from the mother during embryonic development.

Division Eucarida. Order Euphausiacea. The eggs are usually carried between the thoracic legs. The larva emerging from the egg is a nauplius with an oval, unsegmented body and the usual three appendages.

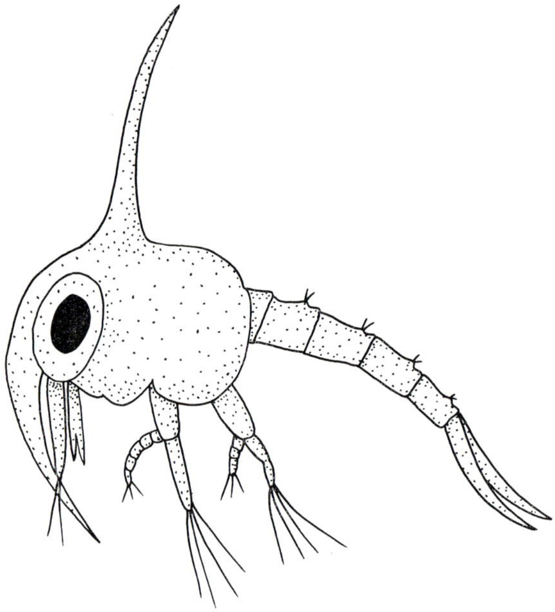

Fig. 2. Zoea larva of Carcinus (Decapoda: Brachyura). Note the presence of thoracic limbs and large lateral eye.

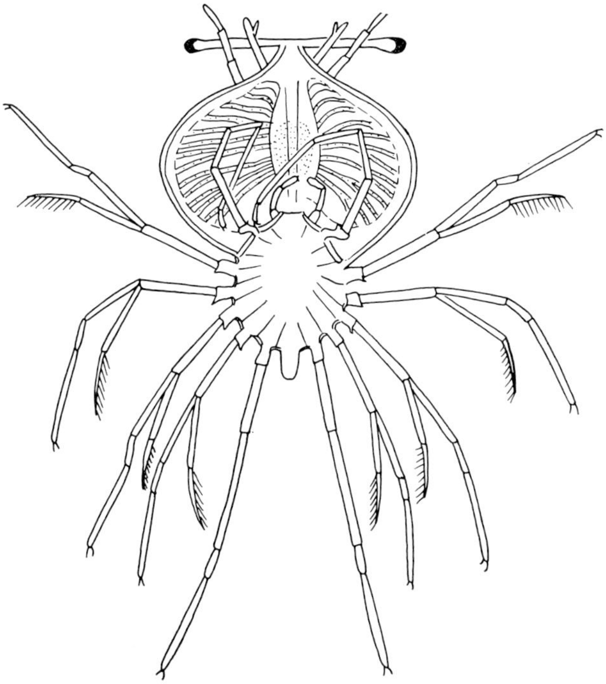

Fig. 3. Phyllosoma larva of Jassus (Decapoda: Palinura).

Order Decapoda. Most decapods carry their eggs by cementing them to the abdominal appendages, or pleopods. Recent work by Cheung (1966) indicates that the cement attaching the eggs to the pleopods is derived from the eggs themselves and is not secreted by the female. The members of the family Penaeidae do not carry their eggs but shed them freely into the water. The newly hatched penaeid is a nauplius, but most other decapods hatch at a later stage. The zoea is the most characteristic decapod larva, but specialised forms, such as the phyllosoma (fig. 3) are found in the tribe Palinura.

Many of the freshwater decapods lay large eggs, and development is more or less direct, so that there is no free living zoea. The young are then liberated in a form resembling a miniature adult, but differing in details of the appendages.

11.2. Oogenesis

In the developing oocytes of *Artemia salina* three phases were described by Fautrez-Firlefyn (1957).

(1) Young oocytes with a nucleus containing strongly Feulgen positive chromatic granules, and one or more strongly basophilic nucleoli. The cytoplasm takes up pyronine very readily, and the intensity of colouration is uniform throughout the cell.

(2) Previtellogenic oocytes. The chromatic granules and nucleoli do not stain as readily as in the previous stage. A region of perinuclear cytoplasm with strong basophilic characteristics is developed.

(3) Vitellogenic oocytes. The chromatin becomes strongly Feulgen positive again, but the nucleoli are not visible. Yolk granules appear in the cytoplasm.

Similar stages can be recognised in other Crustacea, with small variations in the details.

In recent years much interest has centered on the activities of the nucleus during the change from the previtellogenic to the vitellogenic stage. In particular there are many indications of the importance of the nucleoli. The number and form of the nucleoli show some variation, and they show differences in their relations with other parts of the nucleus. In Daphnia and Cyclops the nucleolus appears to be in contact with a chromatic granule after the first young stages (Fautrez and Fautrez-Firlefyn 1951), but in Artemia the nucleoli are surrounded by a ring of chromatic granules (Fautrez-Firlefyn 1951).

Nucleoli have been seen to extrude part of their contents into the nucleoplasm, and into the cytoplasm outside the nuclear membrane. In Cyclops there appears to be a definite connection between the nucleolus and the nuclear membrane; this feature is probably connected with the transfer of nucleolar material to the cytoplasm (Fautrez 1958, 1959).

In Calanus the extrusions from the nucleoli of early oocytes have been observed on the outside of the nuclear membrane. Fragmentation and dispersal of this material was found to coincide with the onset of vitellogenesis. It thus seems that liberation of material from the nucleolus is a primer for the formation of yolk.

The materials transferred from the nucleolus to the cytoplasm are a complex mixture including phospholipids and RNA. It has been observed in Artemia that the nucleoli are rich in these substances at the beginning of the previtellogenic phase, but lose them later, before the beginning of vitellogenesis (Fautrez and Fautrez-Firlefyn 1955). A similar decrease in the RNA content of the nucleolus is found in Daphnia and Cyclops, but in these forms the decrease is later in oogenesis, and occurs as vitellogenesis is taking place.

It is not certain whether all the material transferred from the nucleus to the cytoplasm comes from the nucleolus; other parts of the nucleus may also contribute. In Asellus an accumulation of protein and DNA forms on one side of the nucleus, and is discharged through the nuclear membrane at the onset of vitellogenesis (Montefoschi and Magaldi 1953). Part of this material is assumed to come from nucleolar extrusions, but this is not certain.

There is also evidence that mitochondria are involved in yolk formation. In the early oocytes of Artemia there is a perinuclear concentration of mitochondria, but in the later oocytes the main concentration of mitochondria is located peripherally just beneath the cortex. There are in the literature some descriptions of the direct transformation of mitochondria into yolk platelets, for instance in the prawn Palaemon (Bhatia and Nath 1931). Recent studies on the yolk nucleus of Artemia indicate that the process is likely to be complex. The yolk nucleus is an organelle with intense acid phosphatase activity which appears in previtellogenic oocytes. The electron microscope studies of Anteunis et al. (1964) have shown that this structure is a complex of several different types of micro-organelles. These include dense bodies about 500 mμ in diameter, multivesicular bodies about 300 mμ in diameter, microvesicles about 30 mμ in diameter, and osmiophilic granules or ribosomes. In the immediate vicinity of the yolk nucleus the cytoplasm includes mitochondria and Golgi elements. As vitellogenesis proceeds distinct yolk platelets and fat globules are formed. These reach a much greater size than the micro-organelles forming the yolk nucleus: the platelets reach 3 μ in diameter and the fat globules about 0.5 μ.

The formation of glycogen in oocytes seems to be independent of yolk formation. In Chirocephalopsis glycogen appears before the yolk platelets, and is thought to be synthesised endogenously from glucose passing into the oocyte from the haemolymph (Linder 1959).

The process of vitellogenesis is not purely internal to the oocyte. In most Crustacea one or more nurse cells become associated with the oocyte and transfer nutrient material to it. In Artemia the nurse cells become polyploid by endomitosis, and then extrude DNA from the nucleus into the cytoplasm before being phagocytised by the developing oocyte (Fautrez-Firlefyn 1950, 1951). In Daphnia groups of four cells are formed in the ovary. Three of these cells break down and are ingested by the fourth, which becomes the definitive oocyte. Sometimes all four cells break down and are ingested by a neighbouring oocyte. This process enables the rapid build-up of relatively large cells in a short time. The cycle of oogenesis in Daphnia is closely related to the moulting cycle. In good conditions an adult female Daphnia moults once every two or three days. After each moult a batch of eggs is laid into the brood pouch, and the ovary then has to regenerate rapidly ready for the next moult. The next batch of oocytes enters vitellogenesis about a day after the last batch of eggs has been laid.

A complication in Daphnia is that this cladoceran is capable of laying two different sorts of eggs. Normally the eggs are parthenogenetic; they form a single polar body and do not require fertilisation. In poor environmental conditions a second type of egg is produced. This type produces two polar bodies and requires fertilisation before it will develop. At each moult only one type of egg is laid, but during the intermoult period both types may develop in the ovary. This process has been beautifully illustrated in colour by Von Scharfenberg (1910). A struggle develops between the two types of egg, and according to the effect of the environment on the metabolism of the female one or the other prevails. If conditions are good the parthenogenetic eggs develop their characteristic colouration and large fat droplets, and several will be produced by each ovary. When this happens the fertilisable eggs degenerate within the ovary. If conditions are poor the parthenogenetic eggs do not develop, and each ovary produces a single fertilisable egg. When a fertilisable egg is produced in the ovary of Daphnia several four-cell groups may be absorbed by the oocyte.

11.3. The egg and its segmentation

Crustacean eggs show great variation in colour, size and shape. In general the shape is sphaeroidal, but occasionally there are bizarre departures from this form. This is particularly so with the resting eggs of the Conchostraca (fig. 4).

Some crustacean eggs are colourless, like those of *Leptodora kindti*, but most are pigmented. The most frequent pigments are carotenoid proteins

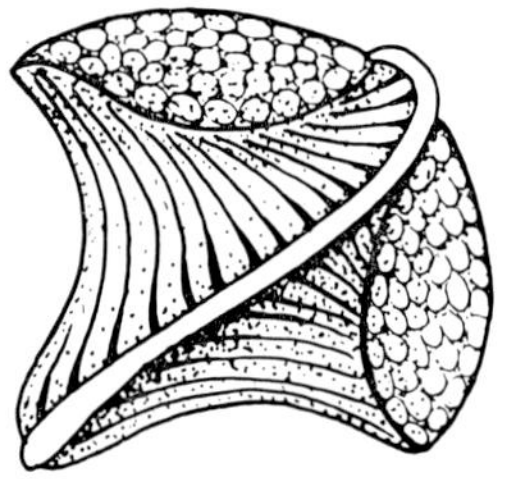

Fig. 4. Egg of *Limnadia lenticularis* (conchostraca).

(see p. 351), but haemoglobin is also found in the eggs of some Cladocera. Sometimes the pigment is located in the outer membranes. This is so in Artemia, where the outer case of the resting egg is often heavily impregnated with haematin. The amount of haematin varies with the salinity of the water in which the eggs are laid (Gilchrist and Green 1960; see p. 350).

The relative size of crustacean eggs shows ecological variation. In general the eggs of terrestrial Crustacea are larger than those of allied forms in fresh water, which in turn are larger than those of allied species in the sea. These differences can be related to the provision of nutrients and water for the developing embryo.

The amount of yolk in the egg influences the segmentation and further development of the embryo. Total cleavage may be found in crustacean eggs with little yolk. Some of the primitive decapods, such as the penaeid Lucifer, exhibit total cleavage (Brooks 1882), as well as some Euphasiacea (Taube 1909) and some Amphipoda (Heidecke 1904). The members of the parasitic isopod family Bopyridae produce relatively small eggs which exhibit unequal holoblastic segmentation (Bonnier 1900), while *Hemioniscus balani*, which belongs to a different family of parasitic isopods, shows equal holoblastic segmentation.

In Hemimysis the eggs are relatively large and yolky, and the central nucleus divides to produce several nuclei which lie in stellate masses of protoplasm deep in the yolk. Fine strands of protoplasm pass through the yolk, and some of these connect the stellate masses with a thin layer lying at the egg surface. By the time that 16 nuclei have been produced the stellate masses have moved nearer to the surface of the yolk. When 32 nuclei are present they begin to appear at the surface. This occurs first at the vegetal pole, and somewhat later at the animal pole. At the 128-cell stage the yolk is completely enclosed by blastomeres which are thick over the animal pole and very thin over the vegetal pole.

The early stages of *Nebalia bipes* resemble those of Hemimysis in that there is no external sign of cleavage furrows (Manton 1934), but the behaviour of the central stellate mass of protoplasm shows some differences. In Nebalia this mass divides and rises to the surface of the yolk as two or three patches or pseudoblastomeres, but only one of these contains a nucleus. This nucleus divides, and the products of subsequent divisions eventually form a layer over the whole surface of the yolk.

In most decapodas the central stellate mass of protoplasm divides as in Hemimysis, and the nucleated masses move towards the periphery, but radial cleavage furrows appear before the nuclei have reached the surface of the yolk (Weygoldt 1961; Sollaud 1923; Herrick 1894; Weldon 1892). The result of this process is a series of more or less equal primary yolk pyramids, or blastomeres, each containing yolk and a central mass of protoplasm with a nucleus.

11.4. Formation of the germ layers

One of the most thorough descriptions of gastrulation in a crustacean is that given by Manton (1928) for *Hemimysis lamornae*. This has been selected as an example because it is not a particularly specialised form, and passes through many of the stages exhibited by other Crustacea. *Hemimysis* develops to an advanced stage before emerging, and thus allows considera-

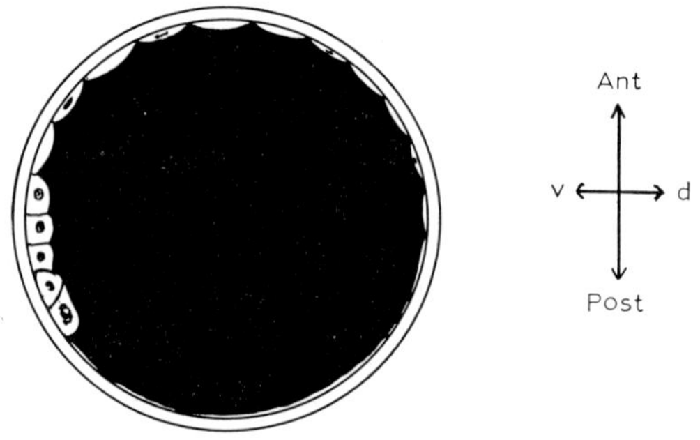

Fig. 5. Sagittal section showing a late segmentation stage of *Hemimysis lamornae*, with the ventral thickening forming the germinal disk. (After Manton 1928.)

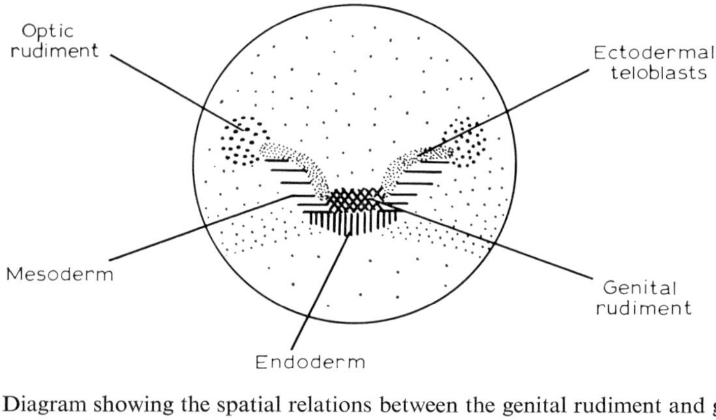

Fig. 6. Diagram showing the spatial relations between the genital rudiment and germinal layers of *Hemimysis lamornae* before gastrulation. (After Manton 1928.)

tion of some aspects of organogenesis which would not be possible if a form emerging as a nauplius larva was selected.

After the yolk has been enclosed by blastomeres the blastodermic disk shows an increase in the thickness of the cells at the posteroventral edge of the animal pole. From this region cells multiply and migrate over the yolk to form a transverse band. Within this band there is a central group of about twelve cells with small granular nuclei: these form the genital rudiment. On either side of the genital rudiment a curved band of cells extends laterally and forwards. Within each of these bands the most anterior row of cells is destined to form the ectodermal teloblasts, and the next row of cells migrates inwards to form the mesoderm. At the posterior edge of the blastodisk there are some cells which take yolk into vacuoles. These cells are the vitellophages, or yolk cells, which later form the midgut epithelium.

Gastrulation in Hemimysis is epibolic; only a slight blastoporal depression is formed (fig. 7). The cells migrate separately under the ectoderm, and there is never any formation of germ layers by delamination.

There are normally fifteen ectodermal teloblasts in a row just in front of the blastoporal depression, and their number remains constant while they are producing ectoderm. The whole row of cells undergoes division at approximately the same time, cutting off a row of smaller cells anteriorly. This is repeated, producing a regular series of rows of ectodermal cells. At the anterior end of these regular rows there is an irregular layer of ectodermal cells formed from the anterior part of the germinal disk.

While the ectodermal teloblasts are beginning these activities the genital rudiment sinks inwards and moves forwards under the ectoderm. Mesodermal cells also migrate inwards. A row of eight relatively large mesodermal cells comes to lie under the ectodermal teloblasts and forms the mesodermal teloblasts. After the mesodermal teloblasts have formed there is still some inward migration of mesoderm cells, which lie behind the teloblasts. Some of these cells migrate forwards under the teloblasts and come to lie on either side of the genital rudiment. These cells will eventually form the head mesoderm bands.

The eight mesodermal teloblasts divide in a manner similar to the ectodermal teloblasts, so that rows of mesodermal cells are formed beneath the rows of ectodermal cells. The ectodermal teloblasts start their activities earlier so that there is normally one more row of ectoderm than mesoderm.

While these changes are taking place in the region of the blastoporal depression other changes are taking place elsewhere in the blastodisk. The optic rudiments begin to form as lateral thickenings at the ends of the rows

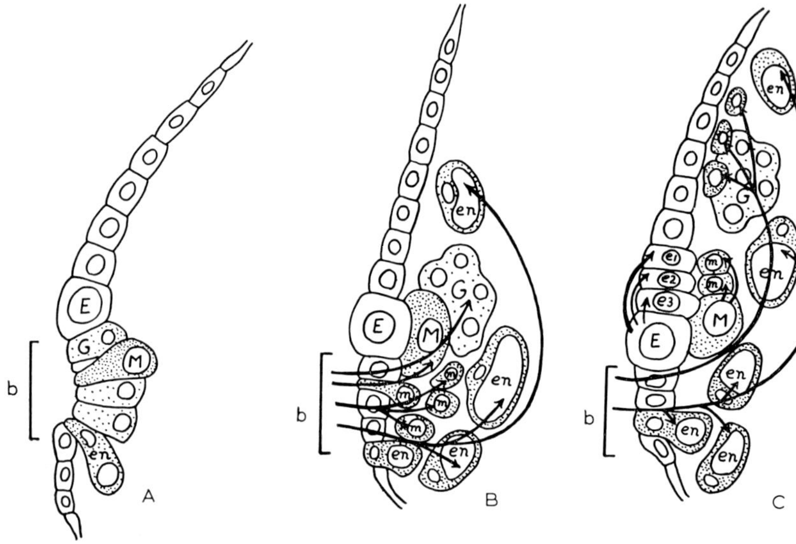

Fig. 7. Gastrulation in *Hemimysis lamornae:* diagrammatic sagittal views showing the process of gastrulation by immigration from a blastoporal area. A: germ layers and blastoporal area just defined. B: gastrulation in progress. Ectodermal teloblasts differentiated in the anterior lip of the blastopore; the genital rudiment has migrated forwards and the mesodermal teloblasts, head mesoderm and endoderm cells are passing inwards. C: Gastrulation is complete except for the continued formation of endoderm. Ectodermal and mesodermal teloblasts are now active and the head mesoderm bands are being formed. b: blastoporal area; E: ectodermal teloblast; en: endoderm cell (yolk cell or vitellophage); e1, e2, e3: descendants from ectodermal teloblasts; G: genital rudiment; M: mesodermal teloblast; m, head mesoderm band cell; m1, m2: descendants from mesodermal teloblasts. (After Manton 1928.)

of teloblasts. The teloblasts move backwards as they produce rows of cells anteriorly, but the optic rudiments move forwards, becoming less lateral in position. Small groups of cells migrate inwards at the posterior edges of the optic rudiments and form the preantennulary mesoderm.

The result of these movements is a Y-shaped region of embryonic material. The two arms of the Y are formed by the optic rudiments, and the stem mainly by the descendants of the teloblasts.

From the arms of the Y between the optic rudiments the anterior appendages begin to appear as projections from the surface. At first these appendages are simply hollow projections of ectoderm. Later the groups of mesodermal cells at the sides of the genital rudiment begin to elongate and form bands lying medial to the bases of the appendages. Some of these eventually gives rise to the dorsal longitudinal muscle, heart, pericardial floor and genital ducts (figs. 9, 10).

The gonads are formed from the genital rudiment which migrates into the median line of the first thoracic segment. This rudiment divides into two which lie in a ventrolateral position. Each of these two rudiments becomes covered with a thin sheet of cells developed from the inner side of the dorsal mesoderm. As the mesoderm travels upwards the genital rudiments are carried with it until they meet again below the pericardial floor. During this process the germ cells increase in number, and once they come to lie beneath the pericardial floor the gonads elongate to occupy a considerable length of the thorax. The gonoducts form from the floor of the pericardium. A pair of longitudinal ridges forms near the midline. These ridges become hollow and form ducts with walls that are continuous with the sheets of mesoderm covering the gonads (fig. 10).

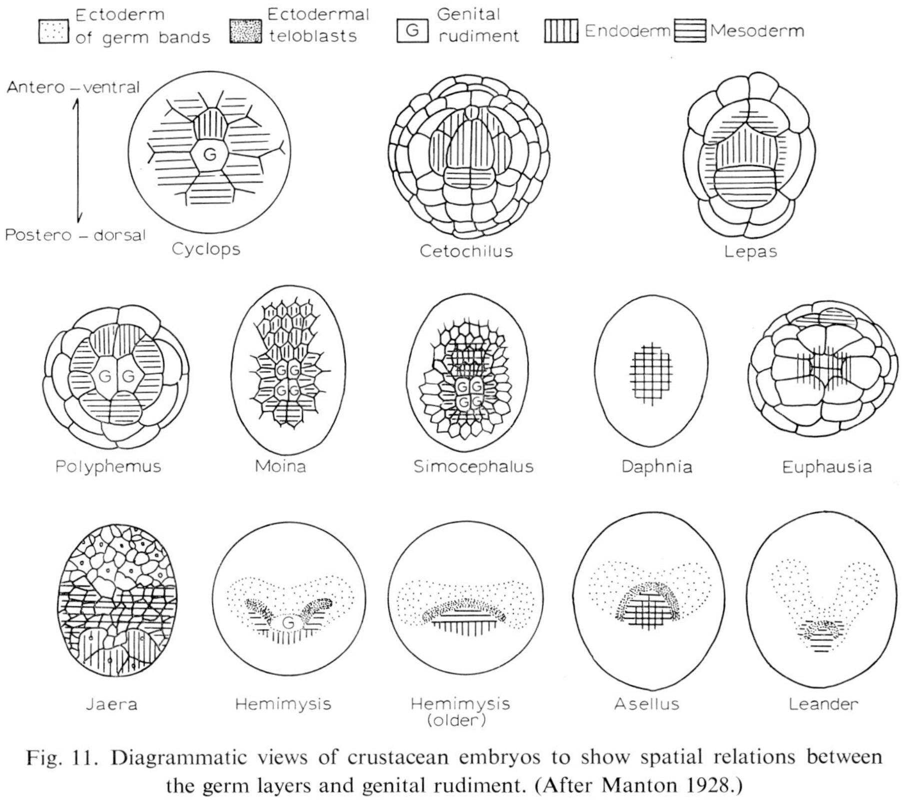

Fig. 11. Diagrammatic views of crustacean embryos to show spatial relations between the germ layers and genital rudiment. (After Manton 1928.)

The development of the excretory organs shows considerable variation. In some groups, such as the Amphipoda, the functional gland of the adult is in the antennal segment. But in the Isopoda and Tanaidacea the functional excretory organ of the adult lies in the maxillary segment. The primitive mysidaceans Eucopia and Lophogaster have both organs, although the antennal gland is better developed. In the forms which have only one pair of excretory organs the other pair may show a transient appearance during embryonic development, and may even persist as a reduced structure in the juveniles and adults.

TABLE 1

Systematic list of selected descriptions of the embryonic development of Crustacea. (Most of the references selected are later than 1880, many earlier papers are referred to by Balfour (1880).)

Subclass Branchiopoda
- Order Notostraca: Claus (1886)
- Order Anostraca: Brauer (1892), Claus (1886), Reynier (1959)
- Order Conchostraca: Cannon (1924)
- Order Cladocera: Agar (1908), Baldass (1937, 1942), Cannon (1921), Dohrn (1870b), Gerschler (1911), Grobben (1879, 1893), Kühn (1911, 1912), Lebedinsky (1891), Samter (1900), Sudler (1899), Vollmer (1912), Wotzel (1937), Warren (1901)

Subclass Ostracoda: Müller-Cale (1913), Weygoldt (1960a)

Subclass Cirripedia: Groom (1894), Delsman (1917)

Subclass Copepoda: Amma (1911), Fuchs (1914), Grobben (1881), Jacobs (1925), Stich (1950), Witschi (1934)

Subclass Malacostraca
- Division Phyllocarida: Manton (1934)
- Division Hoplocarida: Nair (1941), Shiino (1942)
- Division Syncarida: Hickman (1937)
- Division Peracarida
  - Order Mysidacea: Manton (1928), Nair (1939), Needham (1937)
  - Order Cumacea: Butschinsky (1893), Dohrn (1870a), Grschebin (1910)
  - Order Amphipoda: Weygoldt (1958)
  - Order Isopoda: Goodrich (1939), McMurrich (1895), Nair (1956), Needham (1941–2), Weygoldt (1960b), Strömberg (1965, 1967)
  - Order Tanaidacea: Dohrn (1870c), Scholl (1963)
  - Order Thermosbaenacea: Barker (1962)
- Division Eucarida
  - Order Euphausiacea: Taube (1909)
  - Order Decapoda: Aiyer (1949), Nair (1949), Reichenbach (1886), Shiino (1950), Sollaud (1923), Weygoldt (1961)

For the student wishing to study the range of variation in development of various Crustacea a selected list of references is given in table 1.

11.6. Experimental analysis of development

11.6.1. Centrifuge studies

The centrifuge has been used to elucidate several different aspects of crustacean development. This first type of information made available by this means is related to the structure and composition of the egg. Kaudewitz (1950) was able to demonstrate differences in the yolk content of Daphnia eggs from different cultures by centrifuging both for the same length of time and then noting the amount of yolk which separated out.

Centrifugation at different stages of development also yields information about changes in composition during development. Stich (1950a) found that Cyclops oocytes could recover from mild centrifugation and develop normally, but if early cleavage stages were centrifuged for the same duration and strength they were damaged to such an extent that development was abnormal. This is taken to indicate that in the oocyte there has been no determination of the parts of the future embryo, but in the early cleavage stages such determination has taken place, and is distorted or destroyed in the centrifuge.

The nucleus of the egg of Cyclops can be split by centrifugation into several karyomeres. These artificial subunits showed the remarkable capacity to reunite after centrifugation. If 12, 24 or 48 karyomeres were produced they would reunite into 1, 2 or 4 nuclei (Stich 1950b). It was also possible by means of centrifugation to dissociate nuclear division from cell wall formation. By this means an embryo could be produced with numerous nuclei, but without any complete cell walls, when a normal embryo of the same age would have each nucleus separated from the others by cell walls.

The centrifuge can also indicate the degree of structural organisation in the early cleavage stages. The studies of Kaudewitz (1950) on the eggs of Daphnia indicate that in the early cleavage stages the polarity and bilaterality of the embryo are already established. If an egg of Daphnia is centrifuged, even before cleavage has begun, and the direction of centrifugation is at an angle to the future axis of the embryo, then the embryo develops with a distortion at an angle between the normal axis and the centrifugal direction. The nature and extent of the abnormality naturally varies with the strength and duration of centrifugation.

Apart from redistributing the centre of organisation so that the embryo is distorted, the disorganisation caused by centrifuging also results in a retardation of development. An uncleaved centrifuged egg of *Daphnia pulex* takes 45 hr to reach the same stage of development as a normal embryo aged 36 hr.

11.6.2. Ablation studies

Another approach to analysis of development is the removal or inactivation of part of the embryo. Various techniques have been applied to crustacean embryos. The early cleavage stages may be separated mechanically, or blastomeres can be destroyed by ultra-violet radiation, or by electrical heating of a platinum needle.

A study of this type has been made by Kajishima (1952) on the embryos of the isopod *Megaligia exotica*. When the embryo was damaged with a heated platinum needle the effect varied with age of the embryo and the location of the damage. If more than one-tenth of the embryo was damaged in the very early stages, such as uncleaved eggs or up to 32 nuclei, then no embryos developed normally. If similar damage was inflicted at later stages then some perfect embryos did develop, but if they were damaged ventrally the injury was more likely to be fatal. This ventral location of the most vulnerable site indicates that the main determining or organising centre lies in the middle of the germinal disk. In general it was found that if the sphere of injury was kept constant, the influence of the operation was stronger in the early stages and the effect decreased when later stages were operated on.

The experiments with a heated needle were followed by experiments in which part of the embryo was pricked with a fine glass tube and some of the

TABLE 2

Effects of median constriction of embryos of *Megaligia exotica* with silk loops. (After Kajishima 1952.)

Operated stage*	Total no.	Successful no.	No. failing to differentiate	Perfect embryos from one half	Partial embryos from one half
2	18	9	5	4	0
3	25	15	7	6	2
4	19	14	7	2	5

* Stage 2, 2 to 32 nuclei; stage 3, blastula; stage 4, migration, 250 nuclei, denser in ventral region; endomesoderm detectable.

material removed. When the germinal disk was formed it was found that removal of the central region resulted in the death and disintegration of all the embryos treated in this way. If a similar injury was made to an extra-germinal region about 50% of the embryos developed normally. The important area removed from the centre of the germinal disk was that concerned in the formation of the endomesoderm. This region thus seems to be involved in the determination of the development of the rest of the embryo.

A further experiment performed by Kajishima (1952) involved the constriction of embryos with silk loops. The loops were arranged so that they cut through the median plane of the embryo. Table 2 shows the results obtained with early embryos. An important point to note is that the embryos developing from one half were perfect, but half the normal size. In stage 4 all the germinal cells are gathered in the ventral region of the embryo, and it seems that as long as a substantial part of the endomesoderm is included in a constricted half there is a good chance of the embryo developing normally.

There is some variation in the Crustacea in the degree to which determination of the future embryo has been developed in the early stages. Some, such as some parasitic copepods and the fertilised eggs of Cladocera, are not very strongly determined, but others, such as the parthenogenetic eggs of Cladocera, and the fertilised eggs of Cyclops, are strongly determined (Jacobs 1925).

11.6.3. Chemical inhibitor studies

A third technique developed to study the mechanics of development involves the use of chemical substances which interfere with nuclear division and with cell wall formation. There have been comparatively few studies of this type on crustacean embryos. The most significant are those made on the early stages of Artemia.

During the early development of Artemia the division between the blastomeres is made by means of a superficial furrow and a deep partition. The superficial furrow is preceded by a thickening of the cortical cytoplasm, which then disappears into the perivitelline space after the formation of a new membrane (Fautrez-Firlefyn and Fautrez 1962). The deepening of the superficial furrow seems to be produced largely by autonomic cortical activity. The formation of the deep partition is dependent on the separation of the daughter nuclei. The difference between the furrow and partition is shown in eggs when the daughter nuclei fail to separate; the superficial furrow will still appear but the partition does not (Fautrez 1963).

Beta-mercaptoethanol is a strong reducing agent which is capable of breaking the disulphide linkages of proteins. This substance produces blockage of the early cleavage stages of Artemia, and causes an increase in the perinuclear cytoplasm (Fautrez and Fautrez-Firlefyn 1963). A similar effect is produced by dithiodiglycol, which is an oxidation product of β-mercaptoethanol (Fautrez-Firlefyn and Fautrez 1963). If these substances are applied in the later stages of a division that division is not blocked, but the next one is. There are some differences in the behaviour of daughter nuclei in the presence of these two substances. With β-mercaptoethanol the nuclei move apart, and there is a lowering of the interblastomere adhesion, but with dithiodiglycol the nuclei lie closer together and the blastomeres are closely pressed together. It is thought that these different effects are results of the action of the two substances on the spindle. Beta-mercaptoethanol disrupts the -SS- bonds and causes the spindle to become more fluid, while dithiodiglycol stabilises the bonds and prevents the spindle from elongating. Now, since both substances produce an increase in the perinuclear cytoplasm this particular phenomenon cannot be caused by the breaking of -SS- bonds, but must be due to some other cause. This perinuclear cytoplasm is rich in ribonucleic acid, but the details of the mechanism causing the increase are unknown (Roels et al. 1964).

11.7. Temperature and rate of development

The length of time between oviposition and the emergence of young shows great variation between different Crustacea. Two extreme examples will be sufficient to illustrate this. The development of *Homarus americanus* from laying to hatching takes 11 to 12 mth in Canadian waters, while the egg of a small cladoceran in a small desert pool may hatch in less than two days. Within one species the duration of embryonic development varies inversely with temperature over the normal viable range. The early stages of the Canadian Homarus mentioned above were studied by Templeman (1940), who found that the time taken to reach the 16-cell stage varied from 25 days at 4.7 °C to 2 days at 18.5 °C.

A detailed comparison of the rates of development of the embryos of several species of barnacles has been made by Patel and Crisp (1960). Embryos of species from northern waters, such as *Balanus balanoides* and *B. balanus*, that breed in winter in temperate regions, could not withstand temperatures higher than 15 °C, and failed to develop at temperatures above this level. Intermediate forms such as *Verruca stroemia* and *B. crenatus*, could develop normally at temperatures up to 23 °C, and southern forms such as *B. perforatus*, *B. amphitrite*, and *Chthamalus stellatus* developed normally up to 30 °C. The southern species failed to develop at low temperatures. For example, it seemed that temperatures above 14 °C were necessary for the development of *B. amphitrite*. One species, *Elminius modestus*, which breeds throughout the year in temperate regions, was found to be capable of developing normally at all temperatures between 3 and 30 °C.

The actual rate of embryonic development of all the barnacle species increased 3 or 4 times for a 10 °C rise in temperature over the lower part of the viable temperature range. For example, the embryos of *Elminius modestus* took 40 days for complete development at 6 °C, but only 10 days at 15 °C. In the upper part of the viable temperature range the rate of development did not increase so quickly, so that there was a levelling off, and above the upper limit the embryos become abnormal and eventually disintegrated.

11.8. Changes in dry weight during embryonic development

Measurements of dry weight provide a simple means of studying changes in constitution during development. This can give, under certain conditions, an indication of the amount of material used by the developing embryo. The method is most likely to be successful with freshwater and terrestrial Crustacea, which have limited opportunities for uptake of salts. The dry weight of the embryos of such Crustacea should diminish as respiratory substrates are oxidised. The embryos of Crustacea in saline habitats may be able to maintain their dry weights by taking up salts. Dutrieu (1960) found that the dry weight of Artemia embryos did not change during the course of development, and attributed this to uptake of salts. In contrast the dry weight of the parthenogenetic egg of the freshwater cladoceran *Daphnia magna* decreases by 16–25 % during embryonic development, and a similar decrease is found in *D. curvirostris* (Green 1956a).

The eggs of the lobster *Homarus vulgaris* lose 7 % of their dry weight by the time the larvae are ready to hatch (Saudray 1954). This lower percentage loss compared to freshwater Cladocera may be due to salt uptake.

A special feature is encountered when eggs are enclosed in a brood pouch and receive nutrients secreted by the mother. The dry weight can then give an estimate of the material secreted by the mother, although in marine Crustacea there is always the complicating factor of salt uptake. Newly hatched young of *Ligia oceanica* have a dry weight 36 % in excess of the initial dry weight of the egg (Saudray 1954). If one accepts that salt uptake may compensate for the loss of dry weight caused by utilisation of respiratory substrates, as in Artemia, then the extra 36% dry weight of the neonate Ligia represents material supplied by maternal secretions.

11.9. Osmotic relations during embryonic development

The eggs of Daphnia, Simocephalus, Scapholeberis and Eurycercus can be dissected from the maternal brood pouch and will develop in freshwater (Rammner 1933). If the eggs of *Daphnia pulex* are removed from the brood pouch several hours after being laid, and are then placed in distilled water, they will develop normally (Ramult 1914). But if the eggs are placed in distilled water during the first hour after laying they fail to develop and often burst.

When first laid into the brood pouch the eggs of Daphnia swell rapidly and then develop a tough membrane that prevents any further swelling for a period which varies according to temperature, but may be as long as 30 hr at 10 °C. After this period the membrane normally bursts and then the embryo increases in size by uptake of water. Solutions with concentrations over 100 mM prevent the egg membrane from bursting, so that instead of expanding in the usual manner the embryo continues its development inside the membrane. All the limbs develop, but their shapes are distorted by the pressure of the egg membrane. Such enclosed embryos eventually die after failing to burst out of the egg membrane. The power to resist high external osmotic concentrations increases with the age of the embryo, and is greatest after the egg membrane has been shed. When the eggs of *D. pulex* are newly laid they will not develop in solutions with concentrations above $\frac{1}{12}$ N NaCl, but after the egg membrane has been shed they can develop in concentrations up to $\frac{1}{6}$ N NaCl (Ramult 1925).

The osmotic pressures of the egg and embryo show an increase as development proceeds (Przylecki 1921). Parthenogenetic embryos of *Simocephalus vetulus* at an age of 6 hr show a freezing point depression ($\triangle$) of -0.245 °C, while embryos that have reached the age of 54 hr give a value of -0.752 °C. These data were obtained by determining the strength of a glucose solution which did not cause any shrinkage or swelling of the egg or embryo.

The fertilised eggs of *D. pulex* normally remain dormant through the winter. The osmotic pressure in these eggs rises rapidly to give a freezing point depression of -0.74 °C (Przylecki 1921). This value remains constant while the egg is dormant for about 4 mth. The osmotic pressure then falls by about a third and then rises again just before the neonate emerges.

The hatchability of the parthenogenetic eggs of *S. vetulus* in artificial media has been studied by Hoshi (1951). In pure solutions of sodium chloride the highest percentage hatch was found at a concentration of 0.01 M, with a progressive decrease in the hatch at higher concentrations. About 60% of the embryos hatch in distilled water, but in natural pond water a 100% hatch was recorded. Using 10 ml of a 0.01 M solution of NaCl as a base varying volumes of an isotonic solution of calcium chloride were added to give a series of solutions differing in calcium content. The greatest hatch of eggs was found when 0.2 ml of 0.007 M $CaCl_2$ was added. This gave a calcium-sodium ratio similar to that of sea water, but at much greater dilution. Lower and higher calcium-sodium ratios decreased the percentage of embryos that hatched. Hoshi also confirmed that the embryos show an increase in resistance to high external osmotic pressures as they get older.

Absorption of water during the course of embryonic development appears to be a general phenomenon among Crustacea. This has been demonstrated in a wide range of crustaceans from Cladocera (Ramult 1925) to sand-crabs (Needham and Needham 1930). A good example is given by Barnes (1965) who estimated the water content of two species of Balanus. A million newly deposited eggs of *B. balanoides* contain just under 2 g of water, but by the time the nauplii are ready to become free swimming this quantity has increased to about 3.5 g. Similarly with *B. balanus* a million eggs contain just under 3 g of water, and a million nauplii contain about 4.7 g. The increase in water content is most rapid during the middle stages of development when there is marked cellular differentiation and metabolic activity.

Osmotic phenomena are also of importance in the physical process of hatching from the crustacean egg. This aspect is particularly important in forms dwelling in saline waters. The brine shrimp, *Artemia salina*, has a modified carbohydrate metabolism which leads to the production of glycerol and increases the internal osmotic pressure. This phenomenon is discussed more fully in section 11.11.

The hatching process in the decapod Palaeomonetes involves bursting the outer membrane by osmotic swelling of an inner membrane (Davis 1965). The outer membrane slowly slips off the inner membrane, but the two remain connected at the point where the egg-stalk is attached. A space develops between the inner membrane and the larva, but in the later stages of hatching the fluid in this space is removed by rectal swallowing. This process involves antiperistaltic movements of the rectum, resulting in the intake of fluid to the gut. This uptake of water increases the size of the larva so that it comes to be pressed tightly against the inner membrane. The first abdominal segment of the larva bears a sharp border dorsally, and it is in this region that the inner membrane first splits. The final escape of the larva is aided by water currents produced by movements of the pleopods of the female to which the eggs are attached.

Hatching of the American lobster, *Homarus americanus*, shows some similarities to the process described above, but the initial swelling that causes rupture of the outer membrane is not caused by osmotic swelling of an inner membrane, but by osmotic swelling of the larva itself (Davis 1964). There does not appear to be any part of the larval body modified for bursting the inner membrane, and the young larva is dependent on vigorous movements of the pleopods made by the female. When making these movements the female extends the abdomen and legs so that the larvae are swept posteriorly. The larvae, or prelarvae, are at first immotile, but after a variable period of time the larval cuticle splits over the dorsal region of the carapace, and the larva escapes by means of leg movements and stretching movements of the abdomen. When the cuticle has been freed as far as the telson the abdomen is flicked vigorously to finally cast the cuticle away.

11.10. Oxygen consumption during development

The amount of oxygen consumed by an embryo provides a measure of its metabolic rate, and can be used to study the effects of environmental conditions on embryonic metabolism.

The effects of temperature on the oxygen uptake by embryos of two species of cirripedes have been studied by Barnes and Barnes (1959). They found that all stages increased their oxygen consumption with increasing temperature, but the rate of increase shown by the young stages of *Balanus balanoides* was lower than that shown by the later stages. At any one temperature the oxygen uptake of young embryos was lower than that of older embryos. This was attributed to the larger amount of active protoplasm present in the older embryos when compared with the younger stages, which were richer in yolk.

The effect of size on oxygen consumption was shown well by a comparison of the eggs of *B. balanoides* and *Pollicipes polymerus* (Barnes and Barnes 1959). The volume of one egg of *P. polymerus* was found to be one twelfth that of an egg of *B. balanoides*, but the rate of oxygen consumption lay between one third and one sixth of the oxygen consumed by an egg of *B. balanoides*. In terms of oxygen consumed per unit wet weight the amount used by eggs of *P. polymerus* lay between two and four times that consumed by eggs of *B. balanoides*.

The eggs of *B. balanus* are slightly larger than those of *B. balanoides*, but their consumption of oxygen is disproportionately greater. This difference appears to be correlated with the rates of development of these two species. The eggs of *B. balanoides* are laid in the autumn and the nauplii are liberated in the spring, while the eggs of *B. balanus* are not laid until February, but the nauplii are also liberated in time to catch the spring outburst of phytoplankton. The difference in rate of development is particularly marked in the later stages. Some stages of *B. balanus* are passed through in a fifth or a ninth of the time taken by *B. balanoides*, but the consumption of oxygen on a per egg basis does not rise much above twice that of *B. balanoides* (Barnes 1965). This means that the correlation between oxygen consumption and rate of development, at least as judged by visible stages of development, is not particularly close.

The relationship between the osmotic pressure of the external medium and the consumption of oxygen by embryos of *Artemia salina* has been studied by Clegg (1964). When incubated in 0.25 M sodium chloride solution at 30 °C the embryos consumed oxygen at a rate of 60 μl/30 mg of cysts between the third and fourth hours after being wetted. If the concentration of sodium chloride was increased, the consumption of oxygen fell. A similar decrease of oxygen consumption was found when other substances such as glucose, sucrose and mannitol were used to increase the external osmotic pressure. As the external osmotic pressure increased up to 30 atm the decrease in oxygen consumption was not caused by any reduction in the number of cysts respiring or by any decrease in the final percentage of emergence, but by a reduction in the rate of embryonic development. At external osmotic pressures above 30 atm the percentage emergence began to fall, and at 65 atm no emergence took place.

Oxygen consumption by embryos of *Simocephalus vetulus* has been measured by Hoshi (1950a). In the early stages, when the embryos were 10 or 12 hr old, the amount of oxygen consumed at 27 °C was 0.019 μl per embryo per hour. In the later stages the figures increased to 0.056 μl per embryo per hour. The latter figure was doubled when the young were released from the brood pouch and became free swimming, so that the overall increase in oxygen consumption was approximately sixfold. In a similar study of Artemia it was found that the encysted embryos of *Artemia salina* consumed oxygen at a rate of 200 μl/hr/100 mg dry weight at 30 °C 4 hr after being wetted. The free swimming nauplii consumed oxygen at a rate of 1740 μl/hr/100 mg dry weight (Dutrieu 1960). This increase of over eightfold is somewhat greater than that found by Hoshi (1950a) for Simocephalus. It is of interest in relation to these results that after surveying the literature Hemmingsen (1960) reached the conclusion that the metabolism of unicellular organisms, including the eggs of marine invertebrates, is on the average about one-eighth of that of equal sized metazoans. This finding is in remarkable agreement with the increase in oxygen consumption found during the development of Artemia and Simocephalus. As the embryo changes from a basically unicellular state to the multicellular or metazoan condition its metabolic rate changes from that characteristic of a protozoan to the higher level found in the metazoa.

If the measurement of oxygen consumption is combined with an estimate of carbon dioxide output a respiratory quotient (RQ) can be found which may indicate the substrate used to supply energy to the developing embryo. The RQ can be modified by factors which influence the release of carbon dioxide, so that caution is always necessary in drawing conclusions from respiratory studies. Nevertheless, the RQ can provide information which may help to confirm conclusions drawn from other types of data.

The RQ of embryos of Simocephalus has been measured by Hoshi (1950a). The RQ rose from 0.74 when the embryos were 10 hr old to 0.99 when the young were liberated from their mothers' brood pouch. These results suggest that fat is utilised to supply energy in the early stages of development and carbohydrates are used in the later stages. An important point to note when comparing Hoshi's results with those of other workers is that the earliest stages were 10 hr old. This is about 20% of the duration of development of Simocephalus at the temperature used by Hoshi, so that the earliest stages were not studied. Needham (1933) studied embryos of the shore crab, *Carcinus maenas*. His earliest stages were in the process of cleaving, and had an RQ close to unity. By the time stage 2 was reached the RQ had fallen to 0.72. The later stages showed a gradual increase in RQ to 0.83 at the time of hatching. The durations of the stages used by Needham are not known, but it seems likely that the low RQ found by him in stage 2 corresponds to the similar figure found by Hoshi in Simocephalus when one-fifth of the duration of development had passed.

11.11. Carbohydrates

During the development of the parthenogenetic egg of *Simocephalus vetulus* glycogen is confined to the cytoplasmic parts of the developing embryo and is not found in the yolk (Hoshi 1951, 1953, 1954). As development proceeds glycogen appears in the muscles, the wall of the alimentary canal and in the wall of the maxillary gland. Glycogen is formed rather than utilised in the early stages of development. The normal glycogen content of the gastrula is 0.7% of the wet weight. Later the content rises to 0.91% of the wet weight, and when the neonatae emerge from the maternal brood pouch the glycogen content falls to 0.71% of the wet weight.

In anaerobic conditions at 20°C the embryos of *S. vetulus* die after 11 hr, when about 28% of the total glycogen has been utilised. When a small amount of oxygen (0.02 ml/l) is present in the medium the amount of glycogen used in 11 hr is increased to 47% of the total available. At higher concentrations of oxygen less glycogen is used, and the metabolism of the embryo is more like that of embryos in well aerated water. Hoshi (1951, 1953, 1954) makes it clear that in well aerated conditions the embryos of *S. vetulus* make glycogen in the early stages and utilise a small amount in the later stages. This provides a contrast with the barnacles *Balanus balanoides* and *B. balanus*. In embryos of both these species the total glycogen decreases steadily during development. Glucose in these embryos increases during the first half of development, then decreases again. The decrease in glycogen during the early stages is greater than the increase in glucose, so that the overall change in carbohydrates is a steady decrease (Barnes 1965).

The formation of glycogen in Simocephalus has a parallel in the encysted embryos of Artemia, where the main carbohydrate reserve is trehalose (Dutrieu 1960; Clegg 1964). During the course of development the trehalose content falls from 15.3% of the dry weight of the egg to 0.6% of the dry weight of the newly hatched nauplius. At the same time the glycogen content shows an increase from 1.5 to 5.0% of the dry weight.

The osmotic pressure of the external medium has a strong influence on the metabolism of carbohydrates by the developing embryo of Artemia. The normal desiccated cysts contain 34 μg glycerol per mg dry weight. When incubated in 0.25 M sodium chloride the glycerol content increases to 56 μg per mg dry weight just before emergence. If the external medium during incubation is 0.75 M sodium chloride the glycerol content increases to 80 μg per mg dry weight (Clegg 1964). In contrast less glycogen is formed at high than at low external osmotic pressures.

High external osmotic pressures thus appear to stimulate the formation of glycerol at the expense of glycogen. The source of both these substances is trehalose, which forms the main carbohydrate reserve of Artemia embryos. The formation of glycerol is an adaptation to increase the internal osmotic pressure above the external osmotic pressure so that osmotic rupture of the hard outer shell may be facilitated. The presence of glycerol in the desiccated cysts may also be an adaptation enabling the embryos to resist freezing and drying. The hygroscopic properties of glycerol may aid in the retention of water when the cysts are dried. The role of glycerol in the cold hardiness of insects now seems well established (Salt 1961), but Artemia may also gain advantage by utilising trehalose as the main carbohydrate reserve. Asahina and Tanno (1964) have shown that the frost-resistant prepupal larva of the sawfly *Trichiocampus populi* contains a high concentration of trehalose, but no glycerol. It is probable that the combined high concentrations of trehalose and glycerol are involved in the remarkable resistance to cold and heat shown by the encysted embryos of Artemia.

Trehalose in the encysted egg of Artemia is formed within the egg and not supplied as trehalose by the mother. Glucose is the only sugar in the blood of female Artemia (Clegg 1965). The second type of egg, which is retained in the brood pouch of the female Artemia until a nauplius larva emerges, shows a difference in carbohydrate metabolism from the egg which becomes encysted and dormant. The glycogen contents of both types are similar after 10 hr of development, but the egg destined to become dormant then begins to show a decrease in glycogen and an increase in trehalose. The nondormant embryos develop a higher glycogen content, but their trehalose content is hardly detectable. This difference between the dormant and non-dormant embryos of Artemia is another indication that trehalose plays a significant part in the dormant state (Clegg 1965).

11.12. Lipids

Direct observation of the centrifuged egg of *Simocephalus vetulus* reveals that the oil droplet which is so conspicuous near the beginning of development is broken up and decreases in volume as development proceeds. Utilisation of lipids seems to be greatest early in development and this is supported by the respiratory quotient (Hoshi 1950a, b). The fat droplets in the embryos of Simocephalus and many other Cladocera become restricted to the cytoplasm of the fat body as soon as the large cells which constitute this body are formed. The droplets are often coloured by carotenoid pigments and form a strong contrast with the cytoplasm.

There is some disagreement concerning the changes in the lipid content of developing Artemia embryos. Dutrieu (1960) states that the total lipids increase slightly during the transition from egg to nauplius, from 20 to 24% of the dry weight. Urbani (1962) shows the total lipid content decreasing during the course of embryonic development, and from the slope of the line on the graph it would appear that the rate of decrease is more rapid towards the end of development. Bellini and Lavizzari (1958) measured lipase activity in Artemia embryos, and found it to be greatest just before the nauplius became free swimming; the increase was threefold or fourfold between immersion and emergence. This may not necessarily be connected with the utilisation of lipids, but could indicate a reorganisation of lipids. Dutrieu (1960) records a change in the composition of lipids during the development of Artemia; the unsaponifiable fraction increases from 4.8 to 12% of the total lipids.

A detailed study of the lipids in embryos of *Balanus balanoides* and *B. balanus* has been made by Dawson and Barnes (1966). The main lipid components of the eggs of both species are triglycerides and phospholipids; small amounts of free fatty acids, sterols and a nonpolar material (possibly hydrocarbon) were also present. The range of fatty acids found in the neutral lipid fraction (mainly triglyceride) was wide, extending from C13 to C22, and was dominated by highly unsaturated fatty acids, of which eicosapentaenoic and docosahexanoic acids were the most important. The only saturated fatty acid of any importance was palmitic acid. The high degree of polyunsaturation in the lipids of these barnacle eggs may be related to the fact that in their natural environment they are exposed to low temperatures. A tendency to increase the degree of unsaturation of lipids in cold conditions has been noted in a wide range of animals. It is noteworthy that the degree of unsaturation is higher in *B. balanoides* which is exposed on the shore, than in *B. balanus* which remains permanently submerged and so does not suffer the same exposure to low winter temperatures as its congener.

During the course of development the triglyceride content of the Balanus embryos decreased markedly, so that the final figure was less than a third of the original content. The phospholipid, lecithin, also showed a decrease in the middle stages, but showed a slight increase in the late stages of development. Another phospholipid, cardiolipin, showed a consistent increase of about seven- or eightfold during development. Cardiolipin is a mitochondrial lipid which appears to be associated with the cytochrome oxidase complex. It might thus be expected to show some relationship with respiratory processes. There is good agreement between the seven- or eightfold increase in cardiolipin and the six- or eightfold increase in oxygen consumption that has been found in crustacean embryos during development (see p. 339). Unfortunately the data available at present can only indicate the possibility of such a relationship. It would be most interesting to have information concerning cardiolipin content and oxygen consumption of a single species from egg to newly hatched young.

The embryos of higher Crustacea such as *Ligia oceanica* and *Homarus vulgaris* also utilise considerable amounts of lipid during development. In Ligia 32% of the initial lipid content of the egg disappears by the time the neonate emerges, and in Homarus the corresponding figure is 60% (Saudray 1954). The lower figure for Ligia may be caused by the embryos receiving nutrients while in the maternal brood pouch.

11.13. Amino acids and proteins

The amino acid composition of the proteins in ovarian eggs of the Japanese Spiny lobster, *Panulirus japonicus*, has been analysed by Suyama (1959). Glutamic acid and aspartic acid each form about 12% of the total. Arginine, alanine, glycine, histidine, isoleucine, leucine, lysine, phenylalanine, proline, serine, threonine, tyrosine and valine are all present in percentages ranging from approximately 4 to 7. Much smaller amounts of cystine, hydroxylysine, hydroxyproline, methionine and tryptophan are also present. The possibility of changes in these percentages during embryonic development has not yet been studied in Panulirus, but in Artemia some detailed information is available concerning the metabolism of amino acids (Dutrieu 1960; Emerson 1967a).

There are very small changes in the total nitrogen and protein nitrogen during the course of embryonic development in Artemia, so that protein is not apparently used to any significant extent as a respiratory substrate. There is however considerable evidence for the reorganisation of yolk protein into new protein components of embryonic tissues. Such a reorganisation probably involves the amino acid pool as an intermediate.

The evidence for the reorganisation of proteins takes several forms. There are marked changes in the concentrations of some of the amino acids. Dutrieu (1960) found that the six most abundant free amino acids in Artemia eggs were, in order of abundance, aspartic acid, cysteine, serine, glycine, alanine and glutamic acid. In the hatched nauplius the order became: alanine, serine, glycine, proline, tyrosine and glutamic acid. There was an increase in the proportion of proline and a marked decrease in aspartic acid. Of the less abundant amino acids, arginine and threonine were scarcely detectable in eggs, but were much more in evidence in nauplii.

More detailed quantitative studies have been made by Emerson (1967) who found that alanine, glycine, threonine, histidine and lysine increased in quantity during development. Alanine showed a rapid rise in concentration, then a fall to a lower level. This resembled the pattern of activity of the enzyme glutamic-pyruvic transaminase. It was suggested that the following transamination was taking place: glutamate + pyruvate = α ketoglutarate + alanine.

The pyruvate for this reaction could be derived from the metabolism of trehalose, which forms the main respiratory substrate at the time when alanine is formed rapidly. There is thus a linkage between the metabolism of carbohydrates and the metabolism of amino acids in the embryo of Artemia.

When the concentration of alanine begins to fall there is a marked increase in oxygen consumption (Emerson 1967a). This increase may reduce the availability of pyruvic acid, and further there is a change from the utilisation of trehalose as the main respiratory substrate to the utilisation of lipids. The evidence for this change is based on a lowering of the respiratory

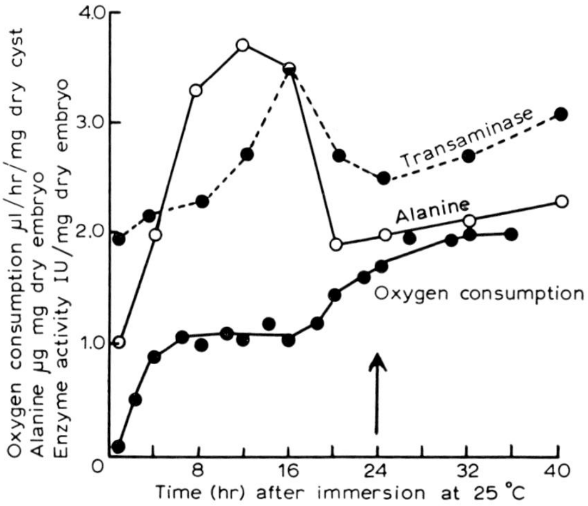

Fig. 12. Relationship between glutamic-pyruvic transaminase activity, alanine levels and oxygen consumption in Artemia embryos incubated in 0.5 M NaCl. The time for 50% emergence is indicated by an arrow. (After Emerson 1967a.)

quotient during the later stages of development (Emerson 1963) and the work of Urbani (1962) indicates that lipids are utilised more rapidly towards the end of embryonic development.

A rapid fall in the concentration of aspartic acid during the first 8 hr after hydration of Artemia cysts may have two causes. It is possible that aspartic acid could be used in the synthesis of pyrimidine, or it may have its amino group transferred to other amino acids, leaving the α-keto acid skeleton to be metabolised, or used in the synthesis of other amino acids.

Glutamic acid appears to be important in the exchange of amino groups between amino acids. Emerson (1967a) found two transaminases and a dehydrogenase in the encysted embryos of Artemia. All three of these enzymes catalysed reactions involving glutamic acid. Rapid turnovers of glutamic and aspartic acids are also indicated by the fact that these substances become highly labelled when Artemia cysts are incubated in an atmosphere containing $^{14}CO_2$.

11.14. Enzyme activity during embryonic development

Most of the work on enzymes in crustacean embryos has been concerned with the encysted embryos of Artemia. A list of most of the enzymes studied so far is given in table 3.

TABLE 3

Enzymes in encysted embryos of *Artemia salina*.

Enzyme	Authorities
Proteinase	Urbani and De Cesaris-Coromaldi (1953)
Dipeptidase (alanine-glycine)	Urbani et al. (1952), Bellini (1957)
Glutamic-oxalacetic transaminase	Emerson (1967a)
Glutamic-pyruvic transaminase	Emerson (1967a)
Glutamic dehydrogenase	Emerson (1967a)
Lipase	Bellini and Lavizzari (1958)
Amylase	Urbani et al. (1953), Bellini (1958)
Acid ribonuclease	Urbani and Bellini (1958)
Alkaline phosphatase	Urbani and Urbani Mistruzzi (1953)

Studies on enzymes concerned with amino acids and proteins indicate that there is a peak of activity at the time when the embryo is emerging from the outer shell. Dipeptidase activity increases fourfold in the first hour after the egg has been immersed in water, then rises by half as much again at the time of emergence. Proteinase does not increase in the first hour after immersion, but later rises to a peak at the same time as dipeptidase activity (Bellini 1957). The latter activity is greater in tetraploid than in diploid nauplii (De Cesaris-Coromaldi and Urbani 1959). It is possible that the greater dipeptidase activity in the tetraploids may be related to an overall increase in nitrogen metabolism. The end product of nitrogen metabolism in both diploids and tetraploids is ammonia, and the tetraploids have the higher rate of ammonia elimination (Bellini and De Vincentiis 1960).

Variations in the activities of enzymes during the development of Artemia are shown in fig. 13. It is clear that the enzymes concerned with proteins and

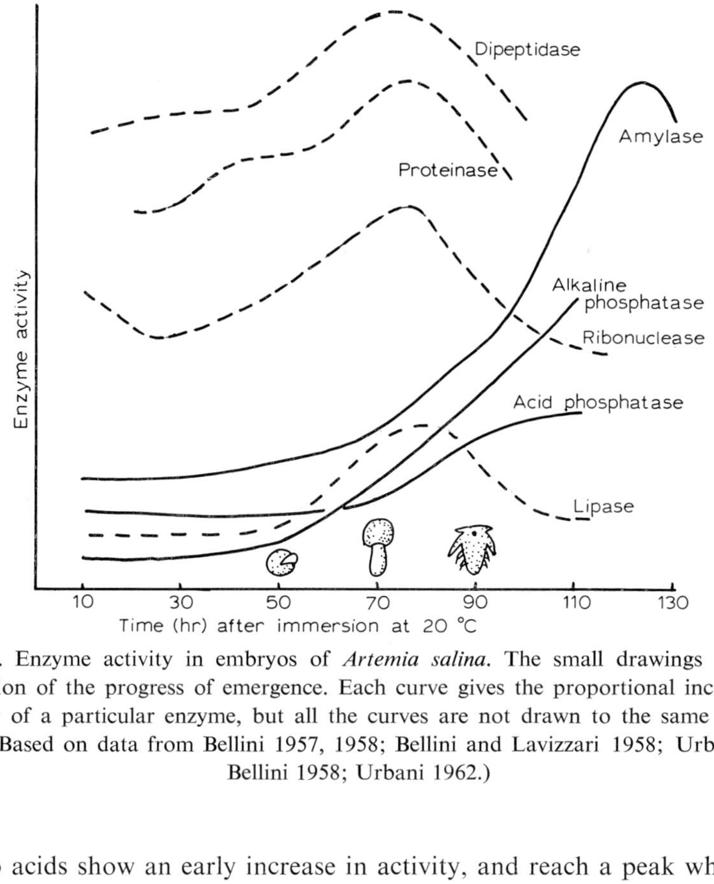

Fig. 13. Enzyme activity in embryos of *Artemia salina*. The small drawings give an indication of the progress of emergence. Each curve gives the proportional increase in activity of a particular enzyme, but all the curves are not drawn to the same vertical scale. (Based on data from Bellini 1957, 1958; Bellini and Lavizzari 1958; Urbani and Bellini 1958; Urbani 1962.)

amino acids show an early increase in activity, and reach a peak when the embryo is emerging from the shell and is still enclosed in a thin membrane. Acid ribonuclease shows a peak at the same time as the proteinase, but in the early stages it shows a slight decrease. The concentration of ribonucleic acid in the embryo shows a peak which coincides with the beginning of the free swimming nauplius stage, and a peak of DNA is found about 10 hr later (Bellini 1960).

Lipase activity in the embryo of Artemia is greatest at a somewhat later stage than the peak proteinase activity. The metabolism of lipids shows a considerable increase between emergence into the thin membrane and the free swimming nauplius. There is an even later peak in amylase activity, which becomes greatest about 10 hr after the nauplius has become free swimming. In contrast to these enzymes which show peaks of activity the two phosphatases show a steady increase, which begins as the embryos emerge through the cyst wall.

There are no comparable data on the enzymes in other crustacean embryos. One might expect considerable variation between embryos of different Crustacea. Within the genus Marinogammarus it has been shown that *Marinogammarus obtusatus* maintains a constant high level of esterase activity throughout embryonic development, while *M. finmarchicus* starts with a low esterase activity which suddenly increases tenfold when the embryo is 8 days old (at 15 °C, the total duration of development being 19 days). This sudden increase is coincident with an acceleration of organogenesis (Doyle et al. 1959).

11.15. Haem pigments

Haem pigments are present in most animal tissues in the form of cytochromes, but the concentration is normally not sufficient to cause any colouration. The occurrence and function of cytochromes in crustacean embryos does not appear to have been studied. In contrast the red respiratory pigment haemoglobin often occurs in concentrations high enough to cause a strong colouration and has been studied in some detail. Among the Crustacea this pigment occurs only in some of the nonmalacostracan groups, and among these is most conspicuous in the Branchiopoda. The presence of haemoglobin in the parthenogenetic eggs of Daphnia was recorded by Teissier (1932) and confirmed by Fox (1948) who found that the amount diminishes as development proceeds. Haemoglobin passes from the maternal blood into the eggs while still in the ovary during a few hours before the eggs are laid (Dresel 1948). A female may pass one-third of her haemoglobin from the blood to the ovaries. The amount of haemoglobin passed into each egg depends on the state of nutrition of the mother and the oxygen content of the water in which she lives (Green 1956b). High concentrations of haemoglobin are found in eggs laid by females of *Daphnia magna* producing few eggs in poorly aerated water when food is abundant. Low concentrations are found in eggs laid by females in well aerated water. When the concentration of haemoglobin in the eggs is high most of the pigment is taken into the fat cells of the embryo as soon as these are formed. The haemoglobin in these cells is then broken down during the course of embryonic development (Green 1955). When the concentration of haemoglobin in the egg is not so high the fat cells are not coloured by haemoglobin, but the concentration still diminishes during the course of embryonic development (Fox 1948; Phear 1955), so that the neonatae are not visibly coloured by the pigment.

When embryos of *Simocephalus vetulus* are treated with carbon monoxide to inactivate haemoglobin their rate of oxygen consumption does not differ significantly from that of normal embryos (Hoshi 1957). This respiratory behaviour of the embryos contrasts with that of the adults, which consume less oxygen when their haemoglobin is rendered non-functional, especially when the oxygen content of their medium is reduced. The lack of difference in oxygen uptake between embryos treated with carbon monoxide and normal embryos suggests that the haemoglobin does not function in embryos as a respiratory pigment, but plays some other role. Embryos treated with carbon monoxide take a longer time to complete their development, so that haemoglobin serves a function in accelerating development, particularly in poorly aerated water (table 4). A similar effect has been found in Daphnia (Fox 1948).

It may be that haemoglobin is passed into the eggs of Cladocera merely as a supply of protein, and carbon monoxide may render it unsuitable for

TABLE 4

Effect of carbon monoxide on the development of embryos of *Simocephalus vetulus*. (After Hoshi 1957.)

Temperature °C	Oxygen concentration cc/l	Medium	Time in hr for 50% hatch
25	5.4	without CO	18.5
25	5.4	with CO	22.1
25	1.5	without CO	20.6
25	1.5	with CO	25.0
28	1.9	without CO	19.2
28	2.0	with CO	23.5

use by the embryos. The greater difference in rate of development between treated and untreated embryos in poorly aerated water may be due to the slower rate of decomposition of carboxyhaemoglobin in these conditions. The decrease in haemoglobin during the course of development supports the view that its function is to supply proteins for embryogenesis. Haemoglobin may have an advantage in this respect over proteins not attached to a prosthetic group in that it may be stabilised against the action of the less specific proteases and so have the possibility of being retained until a specific stage of development is reached. This suggestion is the same as that made in relation to carotenoproteins in crustacean embryos (p. 353).

The gut of late stage embryos of Daphnia contains a haemochromogen, named daphniarubin by Fox (1948). In freshly laid eggs and early embryos no haemochromogen can be detected, so that the pigment found in the later stages must be formed during the course of embryonic development. The total haem contents of early and late stages are similar, suggesting that the haemochromogen is formed from some other haem pigment (Phear 1955). There is a decrease in the haemoglobin content of the embryo during development (Fox 1948; Phear 1955) and this suggests that part at least of the haemochromogen formed in late embryos is derived from haemoglobin present in the egg.

The embryos of *Artemia salina* do not contain haemoglobin, but the tough shell which coats the resting egg contains a haem pigment in the form of haematin (Needham and Needham 1930). The colour of the shell of the resting eggs of Artemia can vary from pale cream to a very dark brown. This variation in colour has been shown to be caused by variation in the haematin content (Gilchrist and Green 1960). More haematin is deposited in the shells of eggs laid in concentrated brines with a low oxygen content than in more dilute media with higher oxygen contents.

11.16. Bile pigments

Although the developing embryos of Daphnia can destroy haemoglobin during the course of development, no bile pigments have been detected in embryos or adults (Fox 1955; Smaridge 1956). This lack of bile pigments indicates that the metabolic pathway of haemoglobin breakdown in Daphnia is different from that found in the vertebrates where bile pigments are the normal end product. It has been suggested that the breakdown of haemoglobin in Daphnia involves a coupled oxidation with unsaturated fatty acids such as linoleic acid (Green 1957a). The fat cells, which are the main site of haemoglobin breakdown, usually contain droplets with a complex mixture of lipids (Jager 1935), so that the materials for such an oxidation are available in both embryos and adults.

The normal parthenogenetic eggs of *Polyphemus pediculus* are colourless, but as the embryo develops the eye rudiments are seen to be green in colour. This colour is due to biliverdin (Green 1961). The pigment seems to be synthesised in the developing eyes and is not found elsewhere in the body. Later in development the green pigment is obscured by a high concentration of an ommochrome pigment which progressively darkens the compound eye until it appears black.

Bile pigments are known to be present in the eggs and embryos of some parasitic cirripedes. Bloch-Raphael (1948) found such pigments in *Septosaccus cuenoti*, and took the view that these pigments were related in some way to the metabolism of haem compounds in the embryos.

11.17. Carotenoid pigments

The eggs of many Crustacea are pigmented by carotenoids. These pigments may be free, or linked to a protein. When free the colours range from yellow to red, but when linked to a protein the colour range is greatly increased. Green, blue, brown and purple colours are produced by linkage of carotenoids to proteins. The colour of the carotenoprotein in the eggs of *Cyclops vernalis* varies according to the food eaten by the copepod. This variation appears to be due to differences in proportions of carotenoprotein linkages giving purple and red colours. The carotenoid does not vary, but is linked in at least two ways so that a range of colours can be produced by varying the relative proportions of each linkage (Dupraw 1958).

The eggs of lobsters (Homarus spp.) contain a green carotenoprotein named ovoverdin by Stern and Salomon (1937). When this pigment is treated with agents to denature protein the colour changes to red, owing to liberation of astaxanthin (Kuhn and Sorensen 1938a, b). This change in colour also occurs naturally towards the end of embryonic development. The total carotenoid content of the embryo does not change, but the link between carotenoid and protein is broken (Goodwin 1951).

An intense purple carotenoprotein is found in the eggs of the hermit crab, *Eupagurus bernhardus*. This pigment contains an astaxanthin ester (Cheesman and Prebble 1966).

Carotenoproteins differing in colour are also found in the eggs of the cirripede Lepas (Ball 1944), the copepods *Idya furcata* and Cyclops (Lwoff

1925, 1927; Dupraw 1958) and in a wide range of Cladocera (Green 1957b). In the Cladocera the carotenoprotein and the free carotenoid in fat globules become restricted to the embryonic fat cells as soon as these are formed. At the end of embryonic development the link with the protein is broken. The cytoplasm of the fat cells becomes paler and the freed carotenoid passes into the globules so that their colour is intensified. In some embryos of *Simocephalus vetulus* it has been observed that when the concentration of carotenoprotein is high carotenoids are excreted through the gut. The pigment disappears from the fat cells and appears in the gut lumen (Green 1966). Sometimes the concentration in the gut of late embryos is so high that distinct crystals are formed, and their progress down the gut can be followed.

The eggs of *Artemia salina* contain two carotenoids. Canthaxanthin is the major pigment, forming about 95% of the total, and echinenone forms the remainder (Krinsky 1965). Earlier work had indicated that there may be a net synthesis of carotenoid during the development of Artemia (Needham and Needham 1930). Reinvestigation revealed that the difference between quantitative extracts from eggs and nauplii was approximately 8% and it was considered that this could be caused by differing degrees of adsorbtion on to embryonic tissues and by slightly different extinction coefficients of the pigments from eggs and nauplii (Gilchrist and Green 1960), so that the synthesis of carotenoid during embryonic development of Artemia must be considered unproven.

Eggs of the crab *Carcinus maenas* contain a much wider range of carotenoids than the eggs of Artemia. Lenel (1961) found β-carotene, cryptoxanthin, a free xanthophyll, hypophasic astaxanthin and traces of a monohydroxy ketocarotenoid. A recent re-study of the carotenoids of Carcinus indicates that isocryptoxanthin (4-hydroxy β-carotene) is present rather than cryptoxanthin (3-hydroxy β-carotene). Gilchrist and Lee (1967) found that the ovaries of Carcinus contained β-carotene, δ-carotene, echinenone, isocryptoxanthin, lutein, zeaxanthin and astaxanthin.

The rhizocephalan parasite, *Sacculina carcini*, which occurs commonly on Carcinus, and has rootlike processes which extend throughout the body of its host, accumulates only β-carotene, and this is the only carotenoid in its embryos.

It is now generally accepted that crustaceans do not synthesise carotenoids from noncarotenoid precursors, but need a source of carotenoid in their food. The carotenoids taken in with the food can be modified and changed into other carotenoids. There is some evidence that β-carotene is the main precursor of other carotenoids in the Crustacea. For instance, Teissier (1932) fed Daphnia with a variety of carotenoids, but only carotene – presumably mostly β-carotene – was effective in producing green carotenoprotein pigmentation of the eggs. Recent investigations indicate that the transition from β-carotene to astaxanthin follows a route of the following nature (Lee 1966a, b; Gilchrist and Lee 1967):

β-carotene
↓
monohydroxy β-carotene
(isocryptoxanthin)
↓
monoketo β-carotene
(echinenone)
↓
monohydroxy-monoketo β-carotene
↓
diketo β-carotene
(canthaxanthin)
↓
monohydroxy-diketo β-carotene
↓
dihydroxy-diketo β-carotene
(astaxanthin).

There are still many details of this process to be settled, but it now seems evident that the first substitution into the β-carotene molecule results in the formation of isocryptoxanthin, and in some crustaceans, such as Idotea, the process stops at canthaxanthin.

The function of carotenoids in crustacean embryos is not known. Carotenoproteins may function by absorbing light and so protect embryos from injury by solar radiation. This function is suggested by the fact that there is an increased deposition of carotenoprotein in the eggs laid by females of *Daphnia magna* exposed to light when compared to females kept in the dark (Green 1957b). The breakdown of carotenoproteins towards the end of embryonic development suggests that if there is a function it is more important in the early stages than later. There is the possibility that linking a protein with a carotenoid may remove the protein from the possibility of attack by certain enzymes. Stabilisation of a protein by linkage to a carotenoid is found in an extreme form in the eggs of the tropical snail *Pomacea canaliculata*. The carotenoprotein, ovorubin, is not readily coagulated by heat and resists attack by trypsin, but when separated from the carotenoid the apo-protein is readily coagulated by temperatures above 70 °C and can be digested by trypsin (Cheesman 1958). In a less extreme form such a linkage with a carotenoid might enable a protein to be held in reserve until a specific stage of development.

Further support for the view that carotenoid-protein complexes form food reserves in crustacean eggs comes from the work of Zagalsky et al. (1967). The ovaries of the crab *Cancer pagurus* and the eggs of the prawn *Plesionika edwardsi* contain high concentrations of glycolipoproteins of large molecular size, rich in phospholipids and cholesterol, and containing mannose and glucosamine as the carbohydrate component. The overall composition of these complexes is roughly equivalent to a combination of the yolk and white proteins of the hen's egg into a single molecular species. In Plesionika the only carotenoid associated with the lipoprotein was astaxanthin, but in Cancer the protein complex was orange red and yielded a mixture of carotenoids.

References

AGAR W. E., 1908. Note on the early development of a cladoceran *(Holopedium gibberum)*. Zool. Anz. *33*, 420–7.

AIYER P., 1949. On the embryology of *Palaemon idae* (Heller). Proc. Zool. Soc. Bengal *2*, 101–47.

AMMA K., 1911. Ueber die Differenzierung der Keimbahnzellen bei den Copepoden. Arch. Zellforsch. *6*, 479–576.

ANTEUNIS A., N. FAUTREZ-FIRLEFYN, J. FAUTREZ and A. LAGASSE, 1964. L'ultra-structure du noyau vitellin de l'œuf d'*Artemia salina*. Exptl. Cell Res. *35*, 239–47.

ASAHINA E. and K. TANNO, 1964. A large amount of trehalose in a frost resistant insect. Nature *204*, 1222.

BALFOUR F. M., 1880. A treatise on comparative embryology. MacMillan & Co., London.

BALL E. G., 1944. A blue chromoprotein found in the eggs of the goose-barnacle. J. Biol. Chem. *152*, 627–34.

BARKER D., 1962. A study of *Thermosbaena mirabilis* Monod (Malacostraca, Peracarida) and its reproduction. Quart. J. Microscop. Sci. *103*, 261–86.

BARNES H., 1965. Studies in the biochemistry of cirripede eggs. 1. Changes in the general biochemical composition during development of *Balanus balanoides* and *B. balanus*. J. Marine Biol. Assoc. U.K. *45*, 321–39.

BARNES H. and M. BARNES, 1959. The effect of temperature on the oxygen uptake and rate of development of the egg masses of two common cirripedes, *Balanus balanoides* (L.) and *Pollicipes polymerus* J. B. Sowerby. Kieler Meeresforsch. *15*, 242–51.

BELLINI L., 1957. Studio delle dipeptidasi e proteinasi nello sviluppo di *Artemia salina* Leach. Atti Accad. Naz. Lincei Rc. (8) *22*, 340–6.

BELLINI L., 1958. Studio delle amilasi nello sviluppo di *Artemia salina* Leach. Atti Accad. Naz. Lincei Rc. (8) *23*, 303–7.

BELLINI L., 1960. Osservazione sugli acidi nucleici nello sviluppo di *Artemia salina* Leach. Ric. Sci. *30*, 816–22.

BELLINI L. and D. M. DE VINCENTIIS, 1960. Observations on the end products of protein metabolism in diploid and tetraploid *Artemia salina* (Leach). Exptl. Cell Res. *21*, 239–41.

BELLINI L. and G. S. LAVIZZARI, 1958. Studio delle lipasi nello sviluppo di *Artemia salina* Leach. Atti Accad. Naz. Lincei Rc. (8) *24*, 92–5.

BHATIA D. R. and V. NATH, 1931. Studies in the origin of yolk. VI. The crustacean oogenesis. Quart. J. Microscop. Sci. *74*, 669–99.

BIGELOW M. A., 1902. The early development of *Lepas*. Bull. Mus. Comp. Zool. Harvard Coll. *40*, 1–144.

BLOCH-RAPHAEL C., 1948. Évolution de l'hémoglobine et de ses dérivés au cours du développement de *Septosaccus cuenoti* (Duboscq). Compt. Rend. Soc. Biol. *142*, 67–8.

BONNIER J., 1900. Contribution à l'étude des Épicarides: Les Bopyrides. Trav. Stat. Zool. Wimereux *8*, 1–475.

BRAUER A., 1892. Ueber das Ei von *Branchipus grubei* von der Bildung bis zur Ablage. Abhand. Preuss. Akad. Wiss. *1892*, 1–66.

BROOKS W. K., 1882. Leucifer, a study in morphology. Phil. Trans. Roy. Soc. B *173*, 57–137.

BUTSCHINSKY P., 1893. Zur Embryologie der Cumaceen. Zool. Anz. *16*, 386–7.

CANNON H. G., 1921. The early development of the summer egg of a cladoceran *(Simocephalus vetulus)*. Quart. J. Microscop. Sci. *65*, 627–42.

CANNON H. G., 1924. On the development of an estherid crustacean. Phil. Trans. Roy. Soc. B *212*, 395–430.

CHEESMAN D. F., 1958. Ovorubin, a chromoprotein from the eggs of the gastropod mollusc *Pomacea canaliculata*. Proc. Roy. Soc. B *149*, 571–87.

CHEESMAN D. F. and J. PREBBLE, 1966. Astaxanthin ester as a prosthetic group: a carotenoprotein from the hermit crab. Comp. Biochem. Physiol. *17*, 929–36.

CHEUNG T. S., 1966. The development of egg-membranes and egg attachment in the shore crab, *Carcinus maenas*, and some related decapods. J. Marine Biol. Assoc. U.K. *46*, 373–400.

CLAUS C., 1886. Untersuchungen über die Organisation und Entwicklung von *Branchipus* und *Artemia*. Arb. Zool. Inst. Univ. Wien *6*, 267–370.

CLEGG J. S., 1964. The control of emergence and metabolism by external osmotic pressure and the role of free glycerol in developing cysts of *Artemia salina*. J. Exptl. Biol. *41*, 879–92.

CLEGG J. S., 1965. The origin of trehalose and its significance during the formation of encysted dormant embryos of *Artemia salina*. Comp. Biochem. Physiol. *14*, 135–43.

DAVIS C. C., 1964. A study of the hatching process in aquatic invertebrates. XIII. Events of eclosion in the American lobster, *Homarus americanus* Milne-Edwards (Astacura, Homaridae). Am. Midland Naturalist *72*, 203–10.

DAVIS C. C., 1965. A study of the hatching process in aquatic invertebrates. XIV. An examination of hatching in *Palaemonetes vulgaris* (Say). Crustaceana *8*, 233–8.

DAWSON R. M. C. and H. BARNES, 1966. Studies in the biochemistry of cirripede eggs. II.

Changes in lipid composition during development of *Balanus balanoides* and *B. balanus*. J. Marine Biol. Ass. *46*, 249–61.

DE CESARIS-COROMALDI O. and E. URBANI, 1959. Dipeptidasi e lipasi in *Artemia salina* (Leach) diploide e tetraploide. Atti Accad. Naz. Lincei Rc. (8) *26*, 801–6.

DELAMARE-DEBOUTTEVILLE C., 1954. Le développement postembryonaire des Mystacocarides. Arch. Zool. Exptl. Gén. *91*, 25–34.

DELSMAN H. C., 1917. Die Embryonalentwicklung von *Balanus balanoides*. Tijdschr. Nederl. Dierk. Ver. (2) *15*, 419–520.

DOHRN A., 1870a. Ueber den Bau und die Entwicklung der Cumaceen. Jena Z. Med. Naturwiss. *5*, 54–80.

DOHRN A., 1870b. Untersuchungen über Bau und Entwicklung der Arthropoden. 3. Die Schalendrüse und die embryonale Entwicklung der Daphniden. Jena Z. Med. Naturwiss. *5*, 277–92.

DOHRN A., 1870c. Zur Kenntnis vom Bau und der Entwicklung von Tanais. Jena Z. Med. Naturwiss. *5*, 293–306.

DOYLE W. L., R. RAPPAPORT and M. E. DOYLE, 1959. Formation and distribution of esterase in gammarids. Physiol. Zool. *32*, 246–55.

DRESEL E. I. B., 1948. Passage of haemoglobin from blood into eggs of Daphnia. Nature *162*, 736.

DUPRAW E. J., 1958. Analysis of egg colour variation in *Cyclops vernalis*. J. Morphol. *103*, 31–63.

DUTRIEU J., 1960. Observations biochimiques et physiologiques sur le développement d'*Artemia salina* Leach. Arch. Zool. Exptl. Gén. *99*, 1–134.

EMERSON D. N., 1963. The metabolism of hatching embryos of the brine shrimp *Artemia salina*. Proc. S. Dakota Acad. Sci. *42*, 131–5.

EMERSON D. N., 1967a. Some aspects of free amino acid metabolism in developing encysted embryos of *Artemia salina*, the brine shrimp. Comp. Biochem. Physiol. *20*, 245–61.

EMERSON D. N., 1967b. Surface area respiration during the hatching of encysted embryos of the brine shrimp, *Artemia salina*. Biol. Bull. *132*.

FAUTREZ J., 1959. Le nucléole et l'anabolisme protéique. Biol. Jaarb. *27*, 17–20.

FAUTREZ J., 1963. Dynamisme de l'ana-télophase et cytodiérèse. Symp. Int. Soc. Cell Biol. *2*, 199–213.

FAUTREZ J. and N. FAUTREZ-FIRLEFYN, 1951. À propos de la chromatine et des nucléoles dans la vésicule germinative de l'oocyte de quelques Crustacés. Biol. Jaarb. *18*, 27–40.

FAUTREZ J. and N. FAUTREZ-FIRLEFYN, 1955. Sur la présence de phospholipines dans les nucléoles de l'oocyte d'*Artemia salina*. Compt. Rend. Ass. Anat. *42*, 506–9.

FAUTREZ J. and N. FAUTREZ-FIRLEFYN, 1961. La métachromasie au bleu de toluidine dans l'œuf d'*Artemia salina*. J. Embryol. Exptl. Morphol. *9*, 60–7.

FAUTREZ J. and N. FAUTREZ-FIRLEFYN, 1963. À propos de la localisation du noyau dans les blastomères de l'œuf d'*Artemia salina*. L'influence du β-mercapto-ethanol et du dithiodiglycol sur cette localisation. Develop. Biol. *6*, 250–61.

FAUTREZ-FIRLEFYN N., 1950. Expulsion d'acide thymonucléique hors de noyeau de certaines cellules de l'ovaire d'*Artemia salina*. Compt. Rend. Soc. Biol. *144*, 1127–8.

FAUTREZ-FIRLEFYN N., 1951. Étude cytochimique des acides nucléiques au cours de la gametogénèse et des premiers stades du développement embryonnaire chez *Artemia salina* L. Arch. Biol. *62*, 391–438.

FAUTREZ-FIRLEFYN N., 1957. Protéines, lipides et glucides dans l'œuf d'*Artemia salina*. Arch. Biol. *68*, 249–96.

FAUTREZ-FIRLEFYN N. and J. FAUTREZ, 1962. La cytodiérèse au cours des divisions de segmentation de l'œuf d'*Artemia salina*. Biol. Jaarb. *30*, 81–5.

FAUTREZ-FIRLEFYN N. and J. FAUTREZ, 1963. Développement anormal du plasme périnucléaire dans l'œuf en segmentation d'*Artemia salina* sous l'effet du β-mercapto-ethanol et du dithiodiglycol. Exptl. Cell Res. *31*, 1–7.

FOX H. M., 1948. The haemoglobin of Daphnia. Proc. roy. Soc. B *135*, 195–212.

FOX H. M., 1955. L'hémoglobine de la Daphnie et les problèmes qu'elle soulève. Bull. Soc. Zool. Fr. *80*, 288–98.

FUCHS K., 1914. Die Kiemblätterentwicklung von *Cyclops viridis* Jurine. Zool. Jahrb., Abt. Anat. *38*, 103–56.

GERSCHLER M. W., 1911. Monographie der *Leptodora kindtii* (Focke). Arch. Hydrobiol. *6*, 415–66, *7*, 63–118.

GILCHRIST B. M. and J. GREEN, 1960. The pigments of Artemia. Proc. Roy. Soc. (London), Ser. B *152*, 118–36.

GILCHRIST B. M. and W. L. LEE, 1967. Carotenoids and carotenoid metabolism in *Carcinus maenas* (Crustacea: Decapoda). J. Zool. Lond. *151*, 171–80.

GOODRICH A. L., 1939. The origin and fate of the endoderm elements in the embryology of *Porcellio laevis* Latr. and *Armadillidium nasutum* B.L. (Isopoda). J. Morphol. *64*, 401–26.

GOODWIN T. W., 1951. Carotenoid metabolism during development of lobster eggs. Nature *167*, 559.

GRAVIER C., 1931. La ponte et l'incubation chez les Crustacés. Ann. Sci. Nat. Zool. *14*, 303–419.

GREEN J., 1955. Haemoglobin in the fat cells of Daphnia. Quart. J. Micr. Sci. *96*, 173–6.

GREEN J., 1956a. Growth, size and reproduction in Daphnia (Crustacea: Cladocera). Proc. Zool. Soc. Lond. *126*, 173–204.

GREEN J., 1956b. Variation in the haemoglobin content of Daphnia. Proc. Roy. Soc. (London), Ser. B *145*, 214–32.

GREEN J., 1957a. *Daphnia*, the water flea. New Biol. *23*, 48–64.

GREEN J., 1957b. Carotenoids in *Daphnia*. Proc. Roy. Soc. (London), Ser. B *147*, 392–401.

GREEN J., 1961. Biliverdin in the eyes of *Polyphemus pediculus* (L.) (Crustacea: Cladocera). Nature *189*, 227–8.

GREEN J., 1966. Variation in carotenoid pigmentation of *Simocephalus vetulus* (O.F. Muller) (Crustacea: Cladocera). J. Zool. Lond. *149*, 174–87.

GROBBEN C., 1879. Die Entwicklungsgeschichte der *Moina rectirostris*. Arb. Zool. Inst. Univ. Wien *2*, 1–66.

GROBBEN C., 1881. Die Entwicklungsgeschichte von *Cetochilus septentrionalis* Goodsir. Arb. Zool. Inst. Univ. Wien *3*, 1–40.

GROOM T. T., 1894. On the early development of Cirripedia. Phil. Trans. Roy. Soc. London, Ser. B *185*, 119–232.

GRSCHEBIN S., 1910. Zur Embryologie von *Pseudocuma pectinata* Sowinsky. Zool. Anz. *35*, 808–13.

HANSEN H. J., 1897. The Choniostomatidae. Høst and Son, Copenhagen.

HEIDECKE P., 1904. Untersuchungen über die ersten Embryonalstadien von *Gammarus locusta*. Jena Z. Med. Naturwiss. *38*, 505–52.

HEMMINGSEN A. M., 1960. Energy metabolism as related to body size and respiratory surfaces, and its evolution. Rep. Steno Mem. Hosp. Nord. Insulin Lab. *9*, 7–110.

HERRICK F. H., 1894. The reproduction of the lobster. Zool. Anz. *17*, 289–92.

HICKMAN V. V., 1937. The embryology of the syncarid crustacean *Anaspides tasmaniae*. Pap. Proc. Roy. Soc. Tasmania *1936*, 1–36.

HINTON H. E., 1954. Resistance of the dry eggs of *Artemia salina* L. to high temperatures. Ann. Mag. Nat. Hist. (12) *7*, 158–60.

HOSHI T., 1950a. Studies on physiology and ecology of plankton. III. Changes in respiratory quotient during embryonic development of a daphnid, *Simocephalus vetulus* (O.F. Muller). Sci. Rep. Tohoku Univ., (Fourth Ser.) *18*, 316–23.

HOSHI T., 1950b. Studies on physiology and ecology of plankton. V. Fatty substance in development of *Simocephalus vetulus*, with reference to behaviour of yolk granule. Sci. Rep. Tohoku Univ., (Fourth Ser.) *18*, 464–6.

HOSHI T., 1951. Studies on physiology and ecology of plankton. VI. Glycogen in embryonic life of *Simocephalus vetulus*, with some notes on the energy source of development. Sci. Rep. Tohoku Univ., (Fourth Ser.) *19*, 123–32.

HOSHI T., 1953. Studies on physiology and ecology of plankton. IX. Changes in the glycogen content during development of the daphnid, *Simocephalus vetulus* under aerobic and anaerobic conditions. Sci. Rep. Tohoku Univ., (Fourth Ser.) *20*, 6–10.

HOSHI T., 1954. Studies on physiology and ecology of plankton. X. Relation between oxygen content in medium, incubation period of animal and glycogen content of the daphnid, *Simocephalus vetulus*, kept under anaerobic conditions. Sci. Rep. Tohoku Univ., (Fourth Ser.) *20*, 260–4.

HOSHI T., 1957. Studies on physiology and ecology of plankton. XIII. Haemoglobin and its role in the respiration of the daphnid, *Simocephalus vetulus*. Sci. Rep. Tohoku Univ., (Fourth Ser.) *23*, 35–58.

JACOBI H., 1954. Biologie, Entwicklungsgeschichte und Systematik von *Bathynella natans* Vejd. Zool. Jahrb. (Syst.) *83*, 1–63.

JACOBS M., 1925. Entwicklungsphysiologische Untersuchungen am Copepodenei (*Cyclops viridis* Jurine). Z. Wiss. Zool. *124*, 487–541.

JÄGER G., 1935. Ueber den Fettkörper von *Daphnia magna*. Z. Zellforsch. *22*, 89–131.

KAJISHIMA T., 1952. Experimental studies on the embryonic development of the isopod crustacean *Megaligia exotica*. Annotationes Zool. Japon. *25*, 172–181.

KAUDEWITZ F., 1950. Zur Entwicklungsphysiologie von *Daphnia pulex*. Arch. Entwicklungsmech. Organ. *144*, 410–47.

KRINSKY N. I., 1965. The carotenoids of the brine shrimp, *Artemia salina*. Comp. Biochem. Physiol. *16*, 181–7.

KÜHN A., 1911. Über determinierte Entwicklung bei Cladoceren. Zool. Anz. *38*, 345–57.

KÜHN A., 1912. Die Sonderung der Keimbezirke in der Entwicklung der Sommereier von *Polyphemus pediculus* De Geer. Zool. Jahrb., Abt. Anat. *35*, 243–340.

KUHN R. and N. A. SORENSEN, 1938a. Über die Farbstoffe des Hummers (*Astacus gammarus* L.). Angew. Chem. *51*, 465–8.

KUHN R. and N. A. SORENSEN, 1938b. Über Astaxanthin und Ovoverdin. Chem. Ber. td. *71*, 1879–88.

LEBEDINSKY J., 1891. Die Entwicklung von Daphnia aus dem Sommerei. Zool. Anz. *14*, 149–52.

LEE W. L., 1966a. Pigmentation of the marine isopod *Idothea montereyensis*. Comp. Biochem. Physiol. *18*, 17–36.

LEE W. L., 1966b. Pigmentation of the marine isopod *Idothea granulosa* (Rathke). Comp. Biochem. Physiol. *19*, 13–27.

LENEL R., 1961. Sur le métabolisme des pigments caroténoïdes de *Carcinus maenas* Linné. Thesis, University of Nancy.

LINDER H. J., 1959. Studies on the fresh water fairy shrimp *Chirocephalopsis bundyi*. 1. Structure and histochemistry of the ovary and accessory reproductive tissues. J. Morphol. *104*, 1–60.

LWOFF A., 1925. Un carotinoïde, pigment oculaire de Copépodes, son origine et son évolution pendent l'ontogénèse. Compt. Rend. Soc. Biol. *93*, 1602–4.

LWOFF A., 1927. Le cycle du pigment carotinoïde chez *Idya furcata* (Baird) (Copépode harpacticide). Nature, origine, évolution du pigment et des réserves ovulaires au cours de la segmentation. Structure de l'œil chez les Copépodes. Bull. Biol. *61*, 193–204.

MANTON S. M., 1928. On the embryology of a mysid crustacean *Hemimysis lamornae*. Phil. Trans. Roy. Soc. London, Ser. B *216*, 363–463.

MANTON S. M., 1934. On the embryology of the crustacean *Nebalia bipes*. Phil. Trans. Roy. Soc. London, Ser. B *223*, 163–238.

MATHIAS P., 1934. Résistance au froid et à la chaleur de l'œuf d'*Artemia salina* L. (Crustace phyllopode). Compt. Rend. Congr. Soc. Savantes Paris Dept., Sect. Sci. 157–61.

MCMURRICH J. P., 1895. Embryology of the isopod Crustacea. J. Morphol. *11*, 63–154.

MONTEFOSCHI S. and A. MAGALDI, 1953. Comportamento degli acidi nucleinici nella ovogenesi di *Asellus aquaticus*. Pubbl. Staz. Zool. Napoli *24*, 167–87.

MÜLLER-CALÉ C., 1913. Ueber die Entwicklung von *Cypris incongruens*. Zool. Jahrb., Abt. Anat. *36*, 113–70.

NAIR K. B., 1939. The reproduction, oogenesis and development of *Mesopodopsis orientalis* Tattersall. Proc. Indian Acad. Sci., Sect. B *9*, 175–223.

NAIR K. B., 1941. On the embryology of Squilla. Proc. Indian Acad. Sci., Sect. B *14*, 543–76.

NAIR K. B., 1949. The embryology of *Caridina laevis* Heller. Proc. Indian Acad. Sci., Sect. B *29*, 211–88.

NAIR S. G., 1956. On the embryology of the isopod Irona. J. Embryol. Exptl. Morphol. *4*, 1–33.

NAKANISHI Y. H., T. IWASAKI, T. OKIGAKI and H. KATA, 1962. Cytological studies of *Artemia salina*. 1. Embryonic development without cell multiplication after the blastula stage in encysted dry eggs. Annotationes Zool. Japon. *35*, 223–8.

NEEDHAM A. E., 1937. Some points on the development of *Neomysis vulgaris*. Quart. J. Microscop. Sci. *79*, 559–89.

NEEDHAM A. E., 1941–2. The structure and development of the segmental excretory organs of *Asellus aquaticus* (Linné). Quart. J. Microscop. Sci. *83*, 205–43.

NEEDHAM J., 1933. The energy sources in ontogenesis. VIII. The respiratory quotient of developing crustacean embryos. J. Exptl. Biol. *10*, 79–87.

NEEDHAM J. and D. M. NEEDHAM, 1930. On phosphorus metabolism in embryonic life. 1. Invertebrate eggs. J. Exptl. Biol. *7*, 317–48.

NOODT W., 1964. Natürliches System und Biogeographie der Syncarida (Crustacea Malacostraca). Gewäss. Abwäss. 37–8, 77–186.

PATEL B. and D. J. CRISP, 1960. Rates of development of the embryos of several species of barnacles. Physiol. Zool. *33*, 104–19.

PHEAR E. A., 1955. Gut haems in the invertebrates. Proc. Zool. Soc. London *125*, 383–406.

PRYZLECKI S., 1921. Récherches sur la pression osmotique chez les embryons de Cladocères. Trav. Inst. Nencki 1.

RAMMNER W., 1933. Wird der Cladoceren-Embryo vom Muttertier ernährt? Arch. Hydrobiol. *25*, 692–8.

RAMULT M., 1914. Untersuchungen über die Entwicklungsbedingungen der Sommereier von *Daphnia pulex* und anderen Cladoceren. Bull. Int. Acad. Cracovie *1914*, 481–514.

RAMULT M., 1925. Development and resisting power of Cladocera embryos in the solutions of certain inorganic salts. Bull. Int. Acad. Cracovie *1925*, 135–94.

REICHENBACH H., 1886. Studien zur Entwicklungsgeschichte des Flusskrebses. Abhandl. Senckenberg. Naturforsch. Ges. *14*, 1–137.

REYNIER M., 1959. Recherches sur le développement et la reproduction d'*Artemia salina*. Bull. Soc. Sci. Nancy *18*, 155–75.

ROELS M. P., N. FAUTREZ-FIRLEFYN and J. FAUTREZ, 1964. Étude biochimique de l'effet du β-mercapto-ethanol sur la teneur en acide ribonucléique de l'œuf d'*Artemia salina*. Exptl. Cell Res. *35*, 248–54.

SALT R. W., 1961. Principles of insect cold hardiness. Ann. Rev. Entomol. *6*, 55–75.

SAMTER M., 1900. Studien zur Entwicklungsgeschichte der *Leptodora hyalina*. Z. Wiss. Zool. *68*, 169–260.

SANDERS H. L., 1963. The Cephalocarida. Functional morphology, larval development, comparative external anatomy. Mem. Conn. Acad. Arts Sci. *15*, 1–80.

SARS G. O., 1887. On *Cyclestheria hislopi* (Baird) a new generic type of bivalve Phyllopoda: raised from dried Australian mud. Christ. Vidensk-Selsk. Forh. *1887*, 1–65.

SAUDRAY Y., 1954. Utilisation des réserves lipidiques au cours de la ponte et du développement embryonnaire chez deux Crustacés: *Ligia oceanica* Fab. et *Homarus vulgaris* Edw. Compt. Rend. Soc. Biol. *148*, 814–6.

SCHOLL G., 1963. Embryologische Untersuchungen an Tanaidaceen (*Heterotanais oerstedi* Kroyer). Zool. Jahrb., Abt. Anat. *80*, 500–54.

SCHRADER F., 1926. The cytology of pseudo-sexual eggs in a species of Daphnia. Z. Induktive Abstammungs-Vererbungslehre *40*, 1–27.

SHIINO S. M., 1942. Studies on the embryology of *Squilla oratoria*. Mem. Coll. Sci. Kyoto *17B*, 77–174.

SHIINO S. M., 1950. Studies on the embryonic development of *Panulirus japonicus* (Von Siebold). Contr. Fac. Fish. Univ. Mie. *1*, 1–168.

SMARIDGE M. W., 1956. Distribution of iron in Daphnia in relation to haemoglobin synthesis and breakdown. Quart. J. Microscop. Sci. *97*, 205–14.

SOLLAUD S., 1923. Recherches sur l'embryogénie des Crustacés décapodes de la sous-famille des Palaemoninae. Bull. Biol. Suppl. *5*, 1–234.

STERN K. G. and K. SALOMON, 1937. Ovoverdin, a pigment chemically related to visual purple. Science *86*, 310–11.

STICH H., 1950a. Histochemische Untersuchungen frühembryonaler Sonderungsprozesse normaler und zentrifugierter Eier von *Cyclops viridis* J. Arch. Entwicklungsmech. Organ. *144*, 364–80.

STICH H., 1950b. Karyologische Untersuchungen an zentrifugierten Keimen von *Cyclops viridis*. J. Biol. Zentralbl. *69*, 197–209.

STRÖMBERG J., 1965. On the embryology of the isopod *Idotea*. Arkiv. Zool. (2) *17*, 421–73.

STRÖMBERG J., 1967. Segmentation and organogenesis in *Limnoria lignorum* (Rathke) (Isopoda). Arkiv. Zool. (2) *20*, 91–139.

SUDLER M. T., 1899. The development of *Penilia schmackeri* Richard. Proc. Boston Soc. Nat. Hist. *29*, 109–29.

SUYAMA M., 1959. Biochemical studies on the eggs of aquatic animals. Bull. Jap. Soc. Sci. Fish. *25*, 48–51.

TAUBE E., 1909. Beitrage zur Entwicklungsgeschichte der Euphausiden. 1. Von der Furchung des Eies bis zur Gastrulation. Z. Wiss. Zool. *92*, 427–64.

TEISSIER G., 1932. Origine et nature du pigment carotinoïde des œufs de *Daphnia pulex* (de Geer). Compt. Rend. Soc. Biol. *109*, 813–5.

TEMPLEMAN W., 1940. Embryonic developmental rates and egg laying of Canadian lobsters. J. Fish. Res. Bd. Canada *5*, 71–83.

URBANI E., 1962. Comparative biochemical studies on amphibian and invertebrate development. Advan. Morphogenesis *2*, 61–108.

URBANI E. and L. BELLINI, 1958. Studio della ribonucleasi acida nello sviluppo di *Artemia salina* Leach. Atti Accad. Naz. Lincei Rc. (8) *25*, 198–201.

URBANI E. and L. DE CESARIS-COROMALDI, 1953. Osservazioni sulla vita latente. II. Studio della proteinasi di *Artemia salina* L. Atti Acad. Naz. Lincei Rc. (8), *14*, 144–9.

URBANI E., L. ROGNONE and S. RUSSO, 1952. Osservazioni sulla vita latente. I. Studio della dipeptidasi di *Artemia salina* L. Atti Accad. Naz. Lincei Rc. (8), *13*, 300–8.

URBANI E., S. RUSSO and L. ROGNONE, 1953. Osservazioni sulla vita latente. III. Studio della amilasi di *Artemia salina* L. Atti Accad. Naz. Lincei Rc. (8), *14*, 697–701.

URBANI E. and L. URBANI-MISTRUZZI, 1953. Osservazioni sulla vita latente. IV. Studio della fosfatasi alcalina di *Artemia salina* L. Atti Accad. Naz. Lincei Rc. (8), *15*, 126–31.

VOLLMER C., 1912. Zur Entwicklung der Cladoceren aus dem Dauerei. Z. Wiss. Zool. *102*, 646–700.

VON BALDASS F., 1937. Entwicklung von *Holopedium gibberum*. Zool. Jahrb., Abt. Anat. *63*, 399–454.

VON BALDASS F., 1942. Entwicklung von *Daphnia pulex*. Zool. Jahrb., Abt. *67*, 1–60.

VON SCHARFENBERG U., 1910. Studien und Experimente über die Eibildung und den Generationszyklus von *Daphnia magna*. Int. Rev. Hydrob. Hydrog., Biol. Suppl. *2*, 1–42.

WARREN E., 1901. A preliminary account of the development of the free swimming nauplius of *Leptodora hyalina* Lillj. Proc. Roy. Soc. (London) Ser. B *68*, 210–8.

WELDON W. F. R., 1892. Formation of the germ layers in Crangon. Quart. J. Microscop. Sci. *33*, 343–63.

WEYGOLDT P., 1958. Die Embryonalentwicklung des Amphipoden *Gammarus pulex pulex* (L.). Zool. Jahrb., Abt. Anat. *77*, 51–110.

WEYGOLDT P., 1960a. Embryologische Untersuchungen an Ostrakoden: die Entwicklung von *Cyprideis litoralis* (G. S. Brady) (Ostracoda, Podocopa, Cytheridae). Zool. Jahrb., Abt. Anat. *78*, 369–426.

WEYGOLDT P., 1960b. Beitrag zur Kenntnis der Malakostrakenentwicklung: die Keimblätterbildung bei *Asellus aquaticus* (L.). Z. Wiss. Zool. *163*, 340–54.

WEYGOLDT P., 1961. Beitrag zur Kenntnis der Ontogenie der Dekapoden: embryologische Untersuchungen an *Palaemonetes varians* (Leach). Zool. Jahrb., Abt. Anat. *79*, 223–70.

WHITAKER D. M., 1940. The tolerance of *Artemia* cysts for cold and high vacuum. J. Exptl. Zool. *83*, 391–9.

WITSCHI E., 1934. On determinative cleavage and yolk formation in the harpacticoid copepod *Tisbe furcata* (Baird). Biol. Bull. *67*, 335–40.

WOTZEL F., 1937. Zur Entwicklung des Sommereies von *Daphnia pulex*. Zool. Jahrb., Abt. Anat. *63*, 455–70.

ZAGALSKY P. F., D. F. CHEESMAN and H. J. CECCALDI, 1967. Studies on carotenoid containing lipoproteins isolated from the eggs and ovaries of certain marine invertebrates. Comp. Biochem. Physiol. *22*, 851–71.

G. Reverberi (ed.), Experimental embryology of marine and fresh-water invertebrates

CHAPTER 12

Echinoids

G. CZIHAK

Max-Planck-Institut für Zellbiologie, Tübingen

12.1. Introduction

Echinids have been famous objects for embryological and experimental biological research for about a hundred years. Much celebrated work of Driesch, Boveri, Morgan, Runnström, Hörstadius, Brachet and many others has been performed on sea urchin eggs. This is due to the fact that a ripe female often contains several million easily accessible eggs which are highly transparent because of the paucity of yolk, and which develop rather quickly and synchronously after fertilization. Furthermore, the maturity periods of several species are long, as for instance in *Paracentrotus lividus* (October to May), and sometimes overlap with those of other species, so that one can work with this material throughout the year.

Several methods of obtaining gametes are used. The old, cruel one is to cut the sea urchins open after rinsing them with tap water. The oral side with the lantern of Aristotle is removed, the coelomic fluid washed off with filtered sea water and the gonads removed with forceps or a spoon. They are placed in a beaker and the eggs which gush out are washed several times by pouring filtered sea water over them. Then they are allowed to settle by gravity or centrifugation. The so-called 'dry sperm' is obtained by puncturing the male gonads. A small quantity of dry sperm is transferred by pipette into a beaker containing filtered sea water and then mixed. Insemination is brought about by adding some drops of sperm suspension of slightly milky appearance to the well-stirred egg suspension. Further drops of sperm suspension may be added one or two minutes later. In this way only eggs that are not yet fertilized are inseminated and thus the highest possible fertilization rate is achieved.

Gametes can also be obtained by injecting isotonic (0.6 mol) KCl solution into the body cavity after pricking the soft body wall of the mouth field with the needle of a syringe. The animals will shed their gametes after a few minutes or even seconds. If not injected too much, the animals survive and can be used again. The eggs must be washed as above. This method is to be preferred unless a maximum quantity of eggs is needed. In this case electric stimulation can be used as well. Application of 8–12 volt alternating or direct current with electrodes at two opposite points of the shell is generally successful.

The fertilization membrane, which becomes elevated about 60 sec after addition of sperm, can be removed by sucking the egg suspension vigorously into a narrow pipette or pouring it through a nylon net or silk with apertures slightly smaller then the egg diameter. In both cases the eggs become somewhat compressed so that the rising fertilization membrane bursts.

12.2. Normal development

From their maturation in the ovary the eggs of sea urchins have a polar differentiation, which in some species, however, is not morphologically evident. No polarity can be seen in the oocytes of Paracentrotus, which are characterized by a smaller size, an ovoid form, and a large nucleus with a large nucleolus (fig. 1), whereas in the ripe eggs of *Paracentrotus lividus* polar-

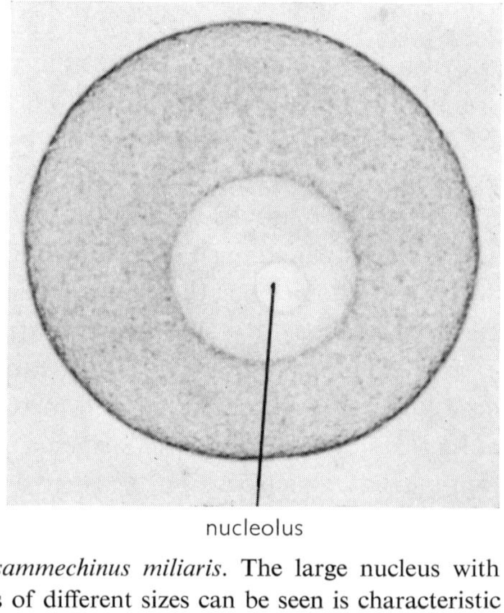

Fig. 1. Oocyte of *Psammechinus miliaris*. The large nucleus with a large nucleolus in which vacuoles of different sizes can be seen is characteristic for the oocyte.

ity becomes visible in that the previously uniformly scattered pigment granules appear concentrated in a belt-like zone near the vegetal pole after the second maturation division (Selenka 1883; Boveri 1901). The Paracentrotus eggs of Villefrance (France) nearly always have a well-developed pigment ring, but those of Naples are poor in this respect. In other species such as *Echinocyamus pusillus* the appearance of the animal cytoplasm is different from that of the vegetal pole. In dark-field examination of fertilized Paracentrotus eggs, Runnström (1928a) found a clear, orange, ring-shaped zone between the equator and the pigment ring which is lacking in the unfertilized egg: this is further evidence of the polarity of the egg cortex and its changes after fertilization.

The eggs are normally shed after the second maturation division. After artificial spawning, however, the egg batches may contain a considerable number of all the stages between the oocyte and the mature egg. The latter are surrounded by a jelly-coat of about 20 μm. It can be removed by brief treatment of the eggs in acidified sea water of pH 5.0. The jelly-coat is per-

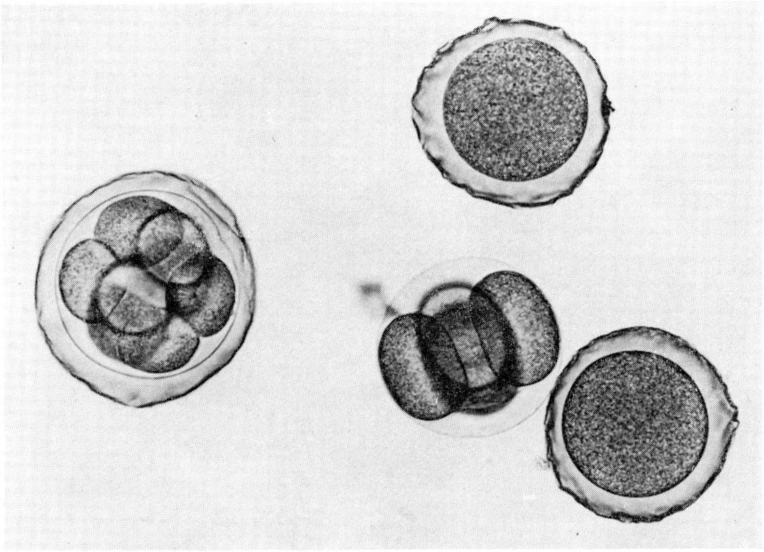

Fig. 2. Group of eggs of Paracentrotus after vital staining with Janus green B diluted 1 : 100,000. By this treatment the jelly-coat of the egg, present in nearly all the eggs shown, becomes visible. At the right, two unfertilized eggs; near the centre, a four-cell stage surrounded by the tiny fertilization membrane; at the left an eight-cell stage; the four-cell stage has lost the jelly-coat as a consequence of the mechanical treatment in making the preparation.

forated by a narrow channel situated above the animal pole opposite the previous attachment (vegetal) pole at the ovary wall. Sometimes one of the polar bodies can be found in the channel after natural spawning. The jelly-coat is so transparent that in spite of its thickness it can only be seen by the fact that particles in the sea water are kept away from the egg surface. But it can be stained, e.g. with Janus green (fig. 2). The pole opposite the animal pole is called vegetal because the blastomeres arising from the vegetal ooplasm differentiate into 'vegetal' organs such as gut (intestine), coelomic cavities, sex organs and so on.

If the concentration of the sperm suspension is not too high, only one sperm succeeds in penetrating the jelly-coat and egg surface, after which the outer membrane of the double plasmalemma becomes elevated to form a fertilization membrane some microns away from the egg membrane. The lifting begins at the penetration point and in *Psammechinus miliaris* and *Paracentrotus lividus* spreads over the egg surface within about 90 sec. The fertilization membrane is smooth (in other echinoderms sometimes spiny), transparent, and about 0.1 μm thick. With the elevation of this membrane the cortical bodies dissolve. These bodies are species-specific ovoid particles, 0.4–0.8 μm in diameter, with a granular or lamellar inner structure. They have been studied in different species by Afzelius (1956) and

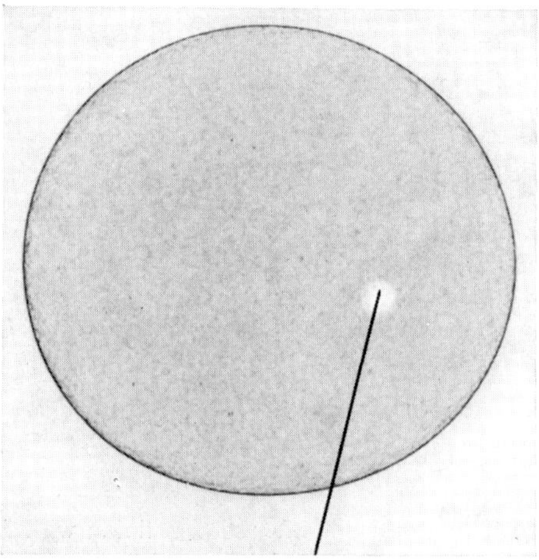

Fig. 3a. Ripe, unfertilized living egg of *Psammechinus miliaris*. The pronucleus (haploid nucleus of the egg) is much smaller than the oocyte nucleus and has completed the second maturation division.

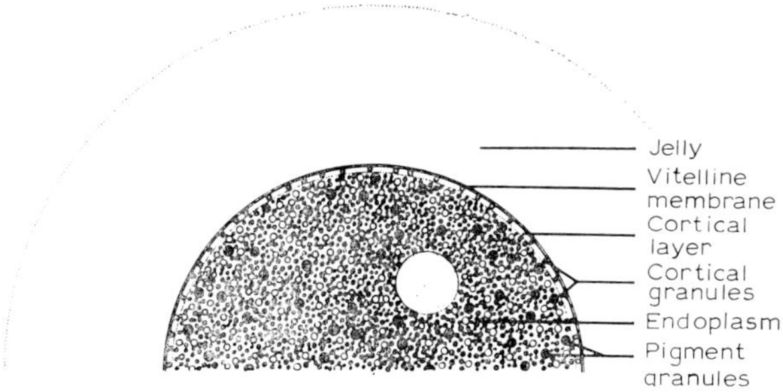

Fig. 3b. Schematic drawing of an unfertilized egg of Arbacia. (After Harvey, 1956.)

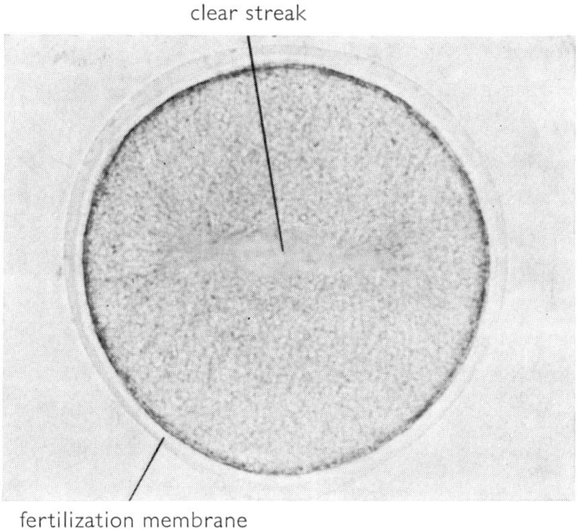

Fig. 4a. Fertilized, living egg of *Paracentrotus lividus*, 40 min after insemination. The fertilization membrane rose about 90 sec after the addition of sperm. It will remain until hatching. In the centre of the egg a 'clear streak' of cytoplasm is visible, a protoplasmatic figure indicating the centriole movement to opposite poles of the fused male and female pronucleus, the membrane of which has already dissolved.

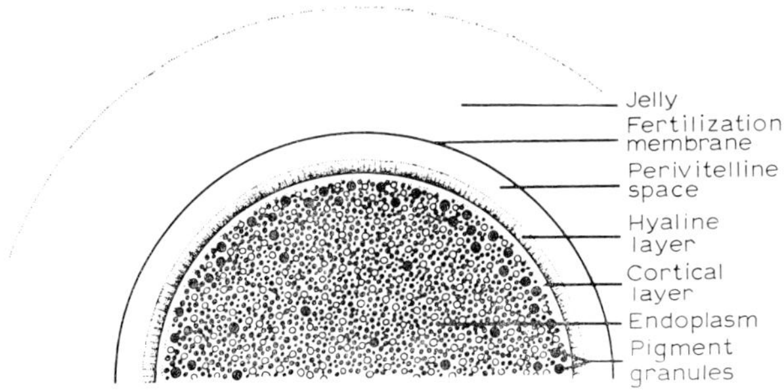

Fig. 4b. Schematic drawing of the Arbacia egg after fertilization. The female nucleus has disappeared.

by electron microscope by Runnström. Their breaking up probably contributes to the formation of the hyaline layer, the outmost layer of the egg-plasm, and to the filling of the perivitelline space between the fertilization and the egg-membrane. Fertilized eggs can easily be distinguished from unfertilized ones by the elevated fertilization membrane (figs. 3 and 4). Under unfavourable circumstances, however, for instance in polluted sea water, the fertilization membrane lifts only a few μm or less.

Normally the entire sperm, with the tail, enters the cytoplasm of the egg; sometimes, however, the tail is found in the perivitelline space. The sperm then migrates towards the egg nucleus or is transported there by cytoplasmic streamings and the sperm nucleus swells by a loosening of the densely packed chromatin. But the egg nucleus also moves towards the sperm nucleus and according to W. and G. Kuhl (1950) the male and female pronuclei begin to fuse in *Psammechinus miliaris* as little as 3 min after fertilization. The resulting zygote (fertilized egg) nucleus then moves back to the egg centre. Around the centriole of the sperm, formerly enclosed in the middle-piece, the astrosphere develops, the rays of which can be recognized as they force cytoplasmic granules to line up in straight rows (cf. fig. 5). At the beginning of the prophase of the first cleavage the centriole has divided into two parts which move to opposite points of the zygote nucleus in the equator of the egg. The astrosphere becomes stretched by the

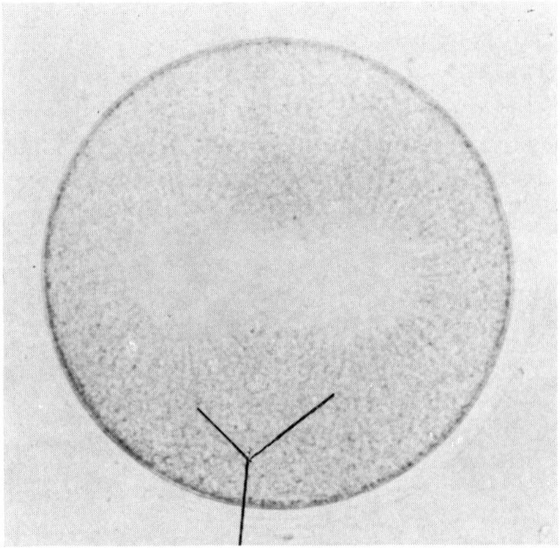

Fig. 5. Egg of Paracentrotus in the metaphase of the first division characterized by maximal development of astral rays, 60 min after fertilization.

centriole movement forming the 'clear streak' (fig. 4), conspicuous even with low magnification as a stripe of homogeneous granula-free cytoplasm almost in the centre of the egg. This is furthermore a sign of the coming metaphase and is often used as a mark for successful fertilization if the

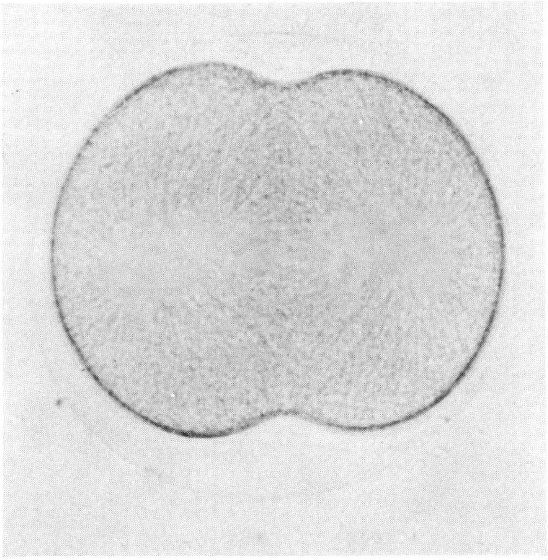

Fig. 6. Anaphase of the first division, 65 min after fertilization. The cleaving furrow cuts through within a few minutes.

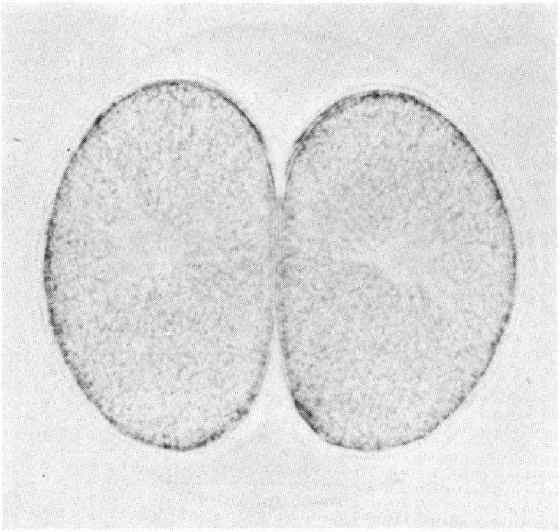

Fig. 7. Completed first cleavage of Paracentrotus. The nuclei of the two-cell stage became visible after the fusion of several small 'Kernbläschen'.

fertilization membrane is not well elevated. After condensation of the chromosomes and disappearance of the nuclear envelope, spindle fibres – attached to the chromosomes – become visible. They are oriented towards the centrioles, thus lying parallel to the equator. Hence the subsequent division plane cuts between the animal and vegetal poles (figs. 6, 7), and the two blastomeres of the two-cell stage are equivalent: each of them contains the same amount of animal and vegetal cytoplasm.

When the chromosomes begin to decondense after anaphase, temporary nuclear membranes develop around single chromosomes or groups of chromosomes (fig. 16). Later in telophase these 'Kernbläschen' unite to a single nucleus (figs. 7, 16).

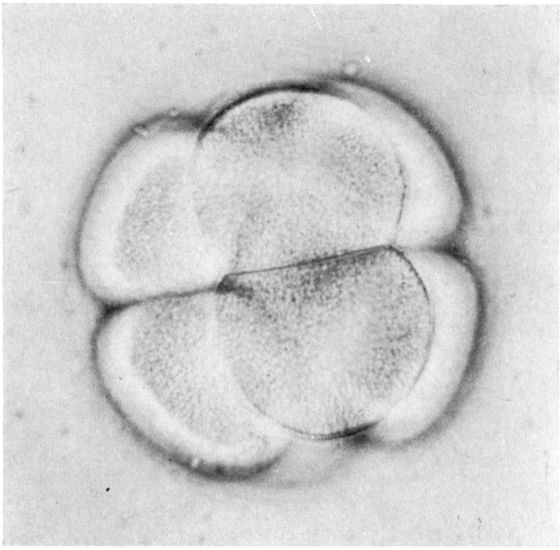

Fig. 8. Eight-cell stage of Paracentrotus. If the pigment ring is not well developed, animal and vegetal cells cannot be discerned. Like all the previous photographs a microphotograph of a living egg.

During the second cleavage the spindle is once again oriented in the equatorial plane, but perpendicular to the first spindle axis. Thus the cleavage plane is of course at right angles to the previous one. The resulting 4 blastomeres are still equivalent (fig. 2). In the third cleavage however the spindle axes are perpendicular to the equatorial plane, and the 8 blastomeres (fig. 8) formed have the same size but are different: 4 are animal blastomeres containing mostly animal cytoplasm and 4 are vegetal ones

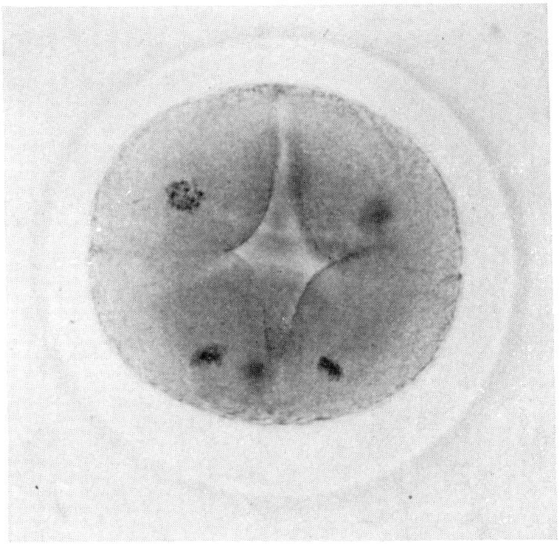

Fig. 9a. Egg of Paracentrotus in the metaphase of the fourth cleavage fixed in Carnoy and stained with acetocarmine. The metaphase plate of the animal cells is seen in part from above (blastomere at upper left) so that the chromosomes can be distinguished at certain points. The metaphase plate of the vegetal blastomeres does not lie in the centre but nearer to the vegetal pole and is seen in side-view here. Note the contraction of the fertilization membrane by fixation.

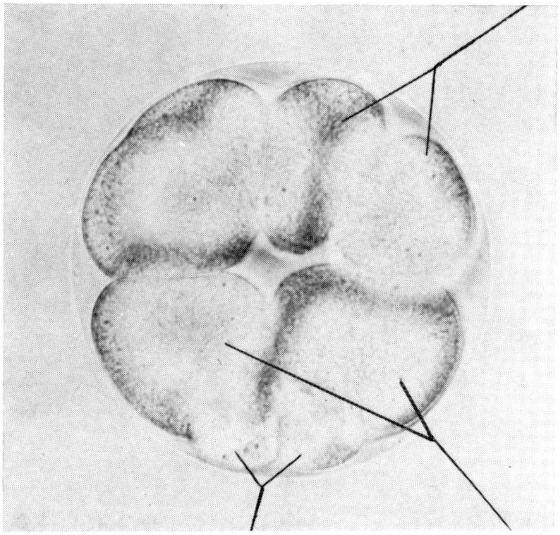

Fig. 9b. Corresponding living egg: two of the visible animal cells have already divided into four mesomeres and the vegetal cells into two macromeres and two micromeres.

with vegetal plasm (cf. p. 426). They can only be discerned if the pigment-ring is well developed or if there is a difference in the structure of the cytoplasm from the beginning. From this 8-cell stage the 16-cell stage results by

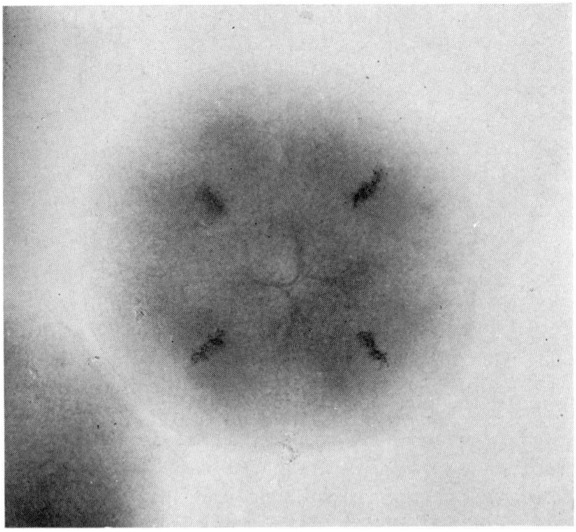

Fig. 10a. The four animal cells of the same stage as in fig. 9a seen from the animal pole, showing the tangential spindle and radial metaphase plate orientation.

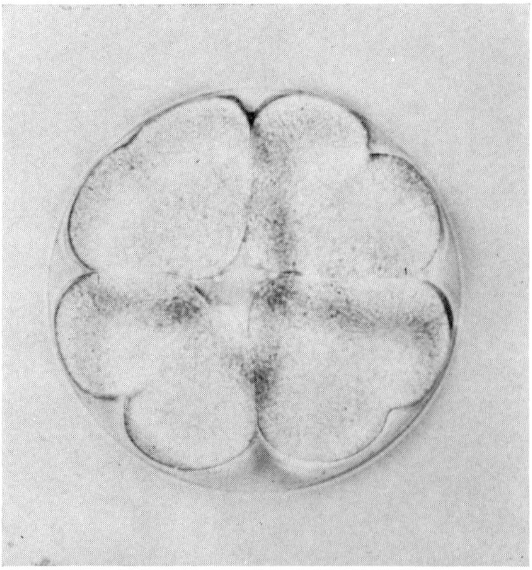

Fig. 10b. The same top-view of a living egg about to complete the fourth division.

cleavage of the animal blastomeres with spindle axis parallel to the equator into 8 mesomeres (figs. 9b, 10a, 10b) and simultaneous cleavage of the vegetal blastomeres by eccentric and oblique spindles near the vegetal pole (figs. 9a-b, 11) into 4 macro- and 4 micromeres. Further cleavage is often

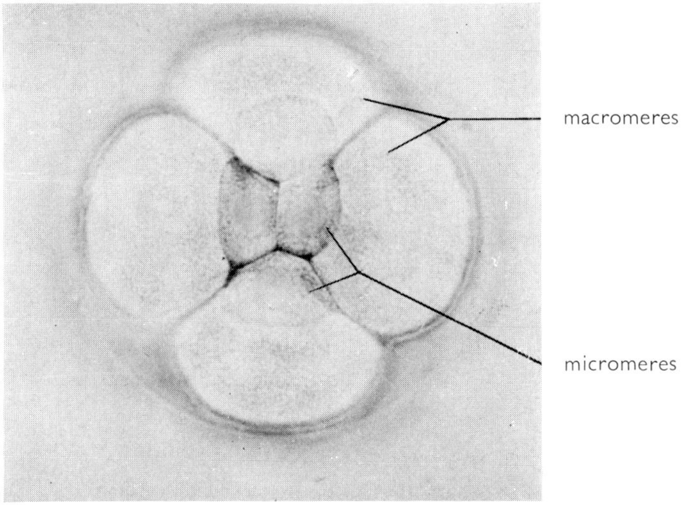

Fig. 11. Sixteen-cell stage of Paracentrotus seen from the vegetal pole, i.e. the opposite view to that shown in fig. 10. The four macromeres and micromeres can be seen. The position of the micromeres is usually more regular: compare fig. 12.

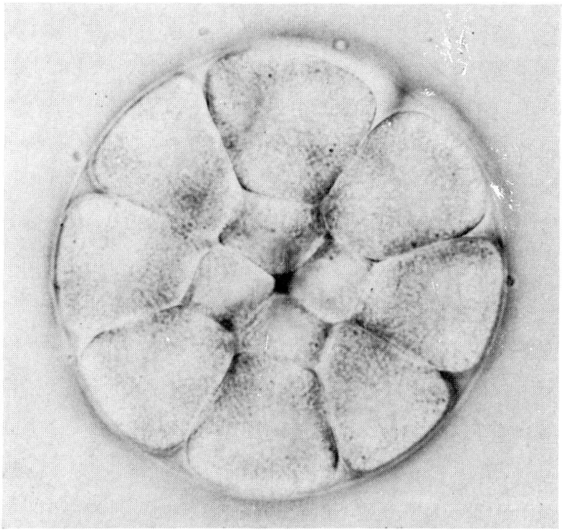

Fig. 12. 28-cell stage of Paracentrotus seen from the vegetal pole: the four macromeres have divided into eight, the four micromeres remain uncleaved for a short time.

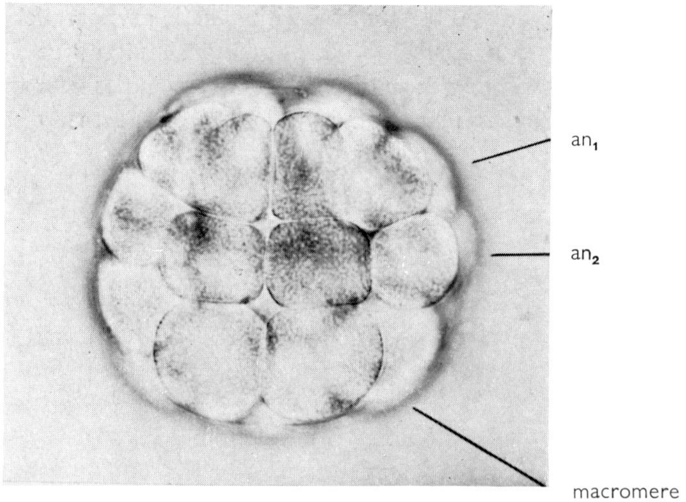

Fig. 13. Side-view of the same stage showing the two circles of mesomeres (an$_1$ and an$_2$)

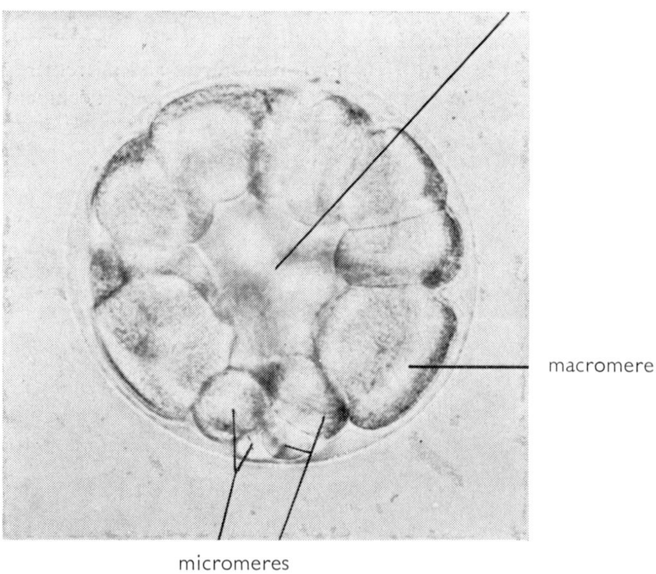

Fig. 14. Optical section through a 32-cell stage of the Paracentrotus egg viewed from the side. The micromeres have now completed their first, unequal division, resulting in four larger and four smaller micromeres.

no longer simultaneous, as the micromeres have a prolonged interphase (Hagström and Lönning 1965). Thus in the 5th cleavage a 28-cell stage (figs. 12, 13) is passed through before the 32-cell stage (fig. 14) is reached when the micromeres have completed their retarded cleavage. The next stages are a 56- and a 64-cell stage.

As cleavage progresses, the central cavity between the blastomeres, perceptible for the first time in the 8-cell stage, becomes larger, and is then called the blastocoel (fig. 15) of the blastula stage. According to Agrell and

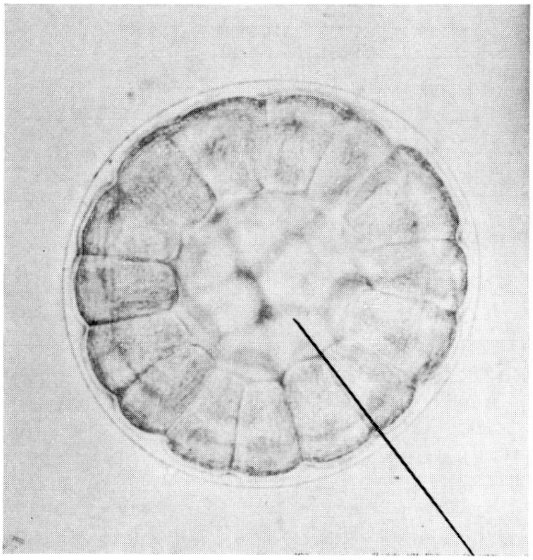

Fig. 15. Optical section through a 5-hr blastula of Paracentrotus: it is very difficult here to discern descendents of meso-, macro- and micromeres.

Persson (1956) the synchrony of cleavage lasts for different periods of time; Agrell (1956) observed waves of mitosis in the later cleavage of *Paracentrotus lividus*, starting from the macromeres and finally reaching the blastomeres of the animal pole.

The most accurate time-table of the mitotic events has been published by Fry (1936) for *Arbacia punctulata*. His results are presented in fig. 16.

The duration of the different mitotic phases is very similar in *Psammechinus miliaris*, *Paracentrotus lividus* and *Arbacia punctulata*. Great differences have been found between other species. According to Callan (1949) *Psammechinus microtuberculatus* cleaves more rapidly than three other Mediter-

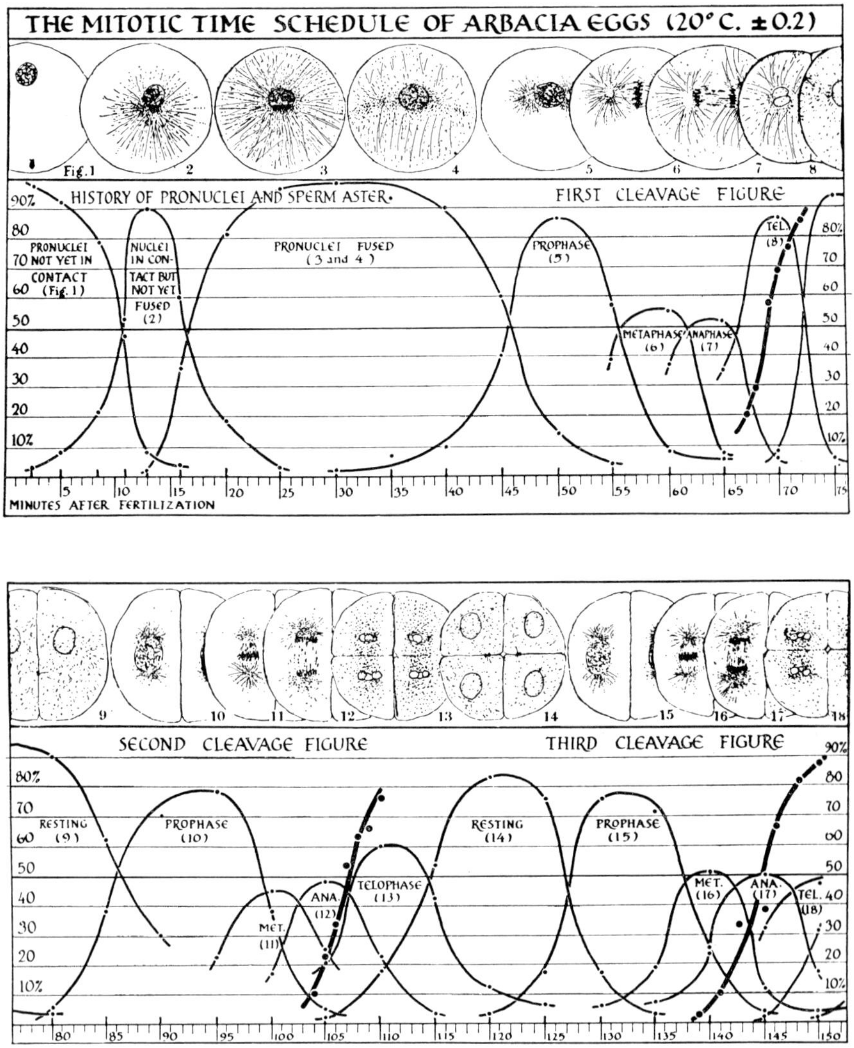

Fig. 16. Diagram of the first division cycles of Arbacia eggs. Abscissa: time after fertilization in minutes. Ordinates: percentage of eggs found at a given time in the indicated stage. For instance 40 min after fertilization the pronuclei have fused in 90% of the eggs and the prophase has already begun in 10%; 65 min after fertilization nearly all the eggs have completed prophase and metaphase, more than 50% are found in anaphase and as many as 35% in telophase of the first cleavage. The steps of the cleavage cycle are numbered and drawn over the graphs schematically. Three thicker lines indicate the transition from the one-cell stage to the two-cell stage, the two- to the four- and the four- to the eight-cell stage respectively. Note the 'Kernbläschen' in steps 8, 13 and 18. (After Fry 1936.)

ranean species, of which Sphaerechinus is the slowest. This is shown in the following table.

	1st mit.ph. (min)	2nd mit.ph. (min)	3rd mit.ph. (min)
Psammechinus microtuberculatus	61	37.5	34.8
Paracentrotus lividus	76	42.7	44.3
Arbacia lixula	99.4	54.1	55.1
Sphaerechinus granularis	100.5	57.7	55.3

The temperature was 18 °C. The appearance of the furrow was chosen as limit between two stages.

The velocity of cleavage is determined by factors of the cytoplasm, because in hybrids it corresponds to the timetable of the female (Moore 1933). Parts of the egg and fragments harvested after centrifugation have different cleavage intervals after fertilization. So it seems that the responsible cytoplasmic factors can be dislocated by centrifugation.

Hagström and Lönning (1965) analyzed the cleavage sequence of Echinocyamus after the 16-cell stage and presented the following scheme (fig. 17).

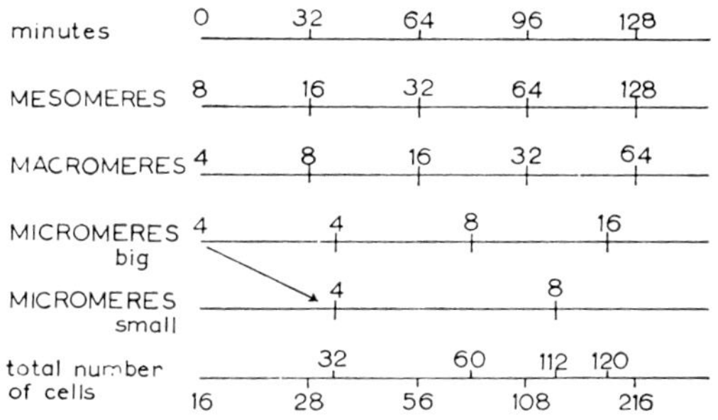

Fig. 17. Timetable of cleavage of the blastomeres of *Echinocyamus pussillus* beginning with the 16-cell stage. The mean time of cleavage of the meso- and macromeres is 32 min, of the large micromeres 40 and of the smaller ones 65 min.

The mesomeres have the shortest interphase, that of the macromeres is scarcely longer, and the micromeres exhibit a long delay when cleaving into large and small micromeres. The latter are still slower, as they remain uncleaved for 65–90 minutes. These different cleavage rates are plausible if it is assumed that a mitotis-triggering substance is produced in the cytoplasm, a process which may last longer in cells poor in cytoplasm. According to Agrell (1956) chromosome-elimination occurs in the micromeres, but the primary mesenchyme cells, differentiating from micromeres, have the normal chromosome number once again.

The shape of the blastula with one layer of cells is the consequence of the tangentially oriented spindle axis in blastomere cleavage. The spindle axis of each cleavage is always perpendicular to the previous one (Wheeler 1963), and thus follows the scheme of Costello (cf. p. 413).

Cilia develop on the surface of the blastomeres even before hatching. Late blastula-stages rotate within the fertilization membrane. At the thicker blastoderm-epithelium of the animal pole the cilia are longer, denser and stiff; this is known as the ciliary tuft (fig. 49H).

As early as in the blastula, size differences of cells in different regions are recognizable: the blastomeres of the vegetal region are higher than the animal ones (fig. 18), with the exception of the animal pole where higher cells bear the ciliary tuft (figs. 19, 20).

The blastula produces a hatching enzyme capable of dissolving the fertilization membrane. After hatching the blastulae swim in spiral courses. The round or ovoid form of the blastula is then changed by a flattening of

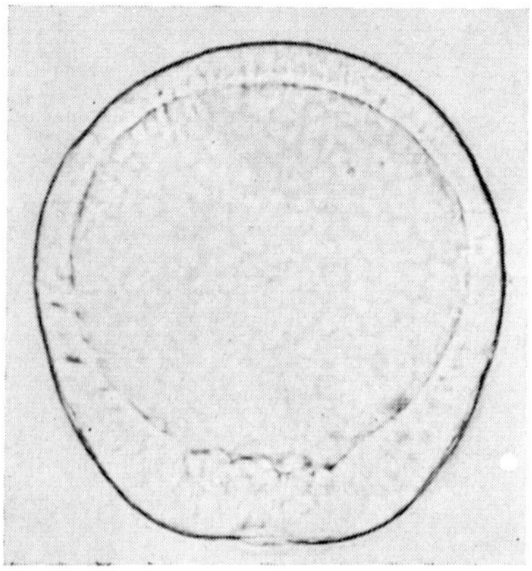

Fig. 18. Nine-hour blastula of Psammechinus after hatching. The size of the blastomeres can be appreciated in this microphotograph, especially in the animal region. The cells of the vegetal pole are higher, the vegetal pole itself is slightly flattened. The micromeres have developed to primary mesenchyme cells and are beginning to separate in the centre of the vegetal pole.

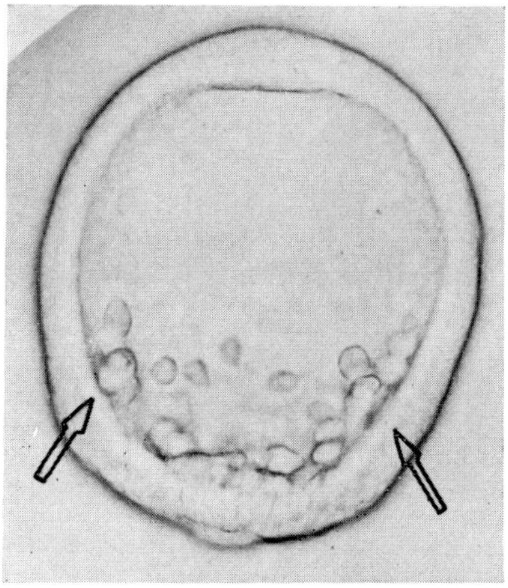

Fig. 19. Twelve-hour blastula of Psammechinus. After the cells of the primary mesenchyme have spread in the blastocoel they begin to gather in two clusters (arrows). Note the high cells at the apical pole. The ciliary tuft is hardly visible in living embryos without phase contrast.

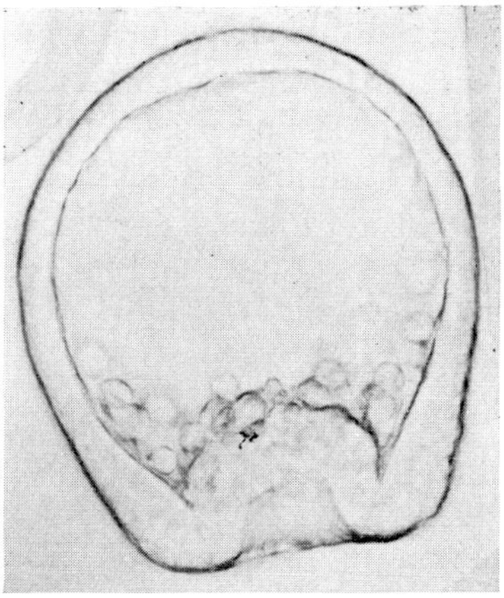

Fig. 20. Beginning of gastrulation in a 14-hour embryo of Psammechinus.

the vegetal region (fig. 18). The cells in the centre of the vegetal pole, originating from the micromeres, begin to move towards the blastocoel with pulsations of the centripetal pole (fig. 22). The immigrating cells differentiate into the primary mesenchyme cells, characterized by the production of long protoplasmic filaments (Okazaki 1960). They first form a little cluster and can

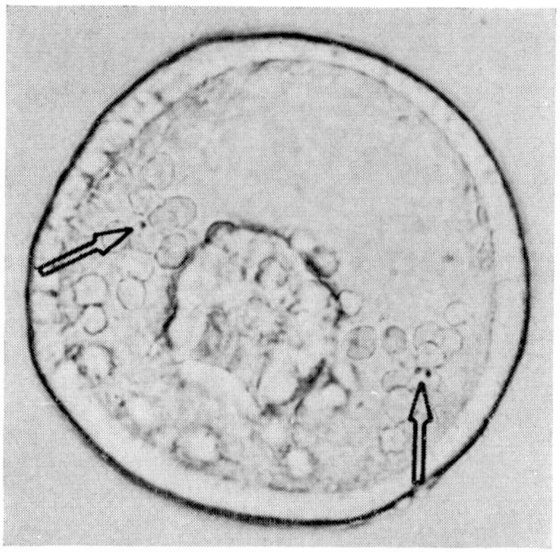

Fig. 21. A further step in the process of archenteron invagination. Mid-gastrula stage of Psammechinus, 17 hr after fertilization. Within the two heaps of primary mesenchyme cells the first granules of the skeleton had been secreted: they appear here as dark grains (arrows).

later be found spreading to the border of the flattened vegetal region. Finally they are arranged in a circle, connected to the attachment zone of the ectoderm and connected among themselves by their long filaments. The blastula also then flattens in the region of the future oral field. At the same time the central part of the vegetal pole region begins to form the invagination of the archenteron (fig. 20). Gustafson and several coworkers have studied the cell movements very carefully by time-lapse cinematography (fig. 22). It was found that the cells of the future archenteron also make pulsatory movements towards the blastocoel, bringing about the invagination of this cell group (Kinnander and Gustafson 1960). Gustafson and Wolpert summarized these results in 1967.

By the flattening of the oral region (fig. 20), the radially symmetric blastula is transformed into a bilaterally symmetric larva. A dorsal and ventral part of the ectoderm can now be distinguished. Along the borderline of the oral field (fig. 68) the epithelium becomes thicker, and thinner dorsally (fig. 23). In the ventral region the rim of the oral field is thickest at two points (fig. 27 F_2, H_2), the so-called attachment zones (Runnström) or lateral ectoderm thickenings. These parts of the ectoderm obviously exert an attraction on the primary mesenchyme cells, which in part crowd beyond them (fig. 19, 21, 26). Some observations support the assumption that differentiation of the oral field (fig. 68) starts in its centre, the future mouth, and that the points attractive to the cells of the primary mesenchyme differentiate where

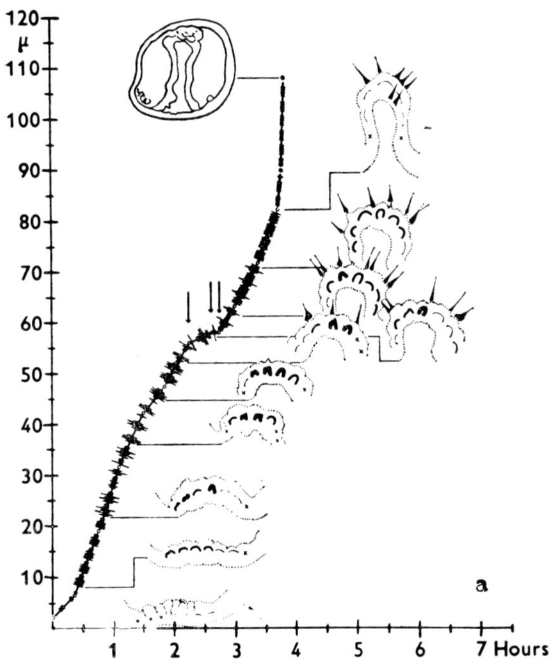

Fig. 22. Diagram showing the time dependence of gastrulation. From the beginning of gastrulation until 2 hr later, invagination of the archenteron is brought about by pulsatory (centripetal) movements of the most vegetal cells towards the blastocoel. Invagination then stops for about 40 min. After the protrusion of long filaments from the tip of the archenteron across the blastocoel and establishment of contact with the blastocoel wall, invagination continues until full length. Abscissa: time after beginning of gastrulation in hours; ordinate: length of the archenteron in microns. (After Kinnander and Gustafson 1960.)

the border of the oral field comes into contact with a certain level of the animal and vegetal gradients (Czihak 1961, 1962b; see p. 434).

The autonomous invagination of the archenteron stops at about $\frac{1}{3}$ of the maximum length of the archenteron (cf. fig. 22). Cells in the tip of the archenteron then protrude long filaments which finally adhere to the inner side of the apical pole and the dorsal side. After a short standstill, gastrulation (= invagination of the archenteron) proceeds further by contraction of these filaments. When the archenteron has reached its maximum length, its tip seems to dissolve (fig. 24) during the emigration of the cells of the secondary mesenchyme. Later the archenteron tip once again has a smooth contour (fig. 29).

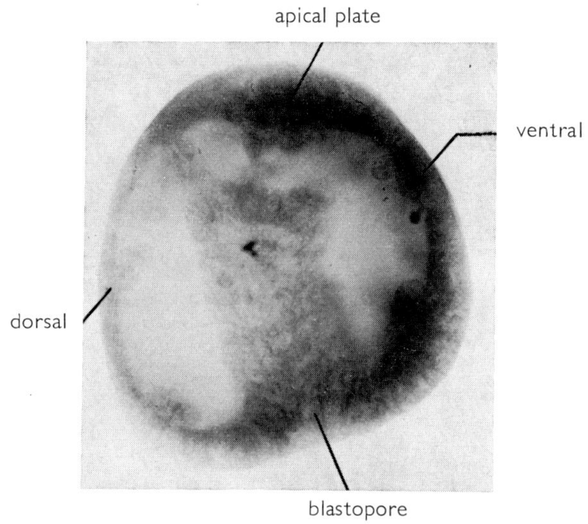

Fig. 23. Gastrula of Psammechinus, 23 hr after fertilization, side-view. Fixed in Bouin-Allen, stained with azur B. The future oral field on the ventral side is more strongly stained and the cells are higher than those of the dorsal side.

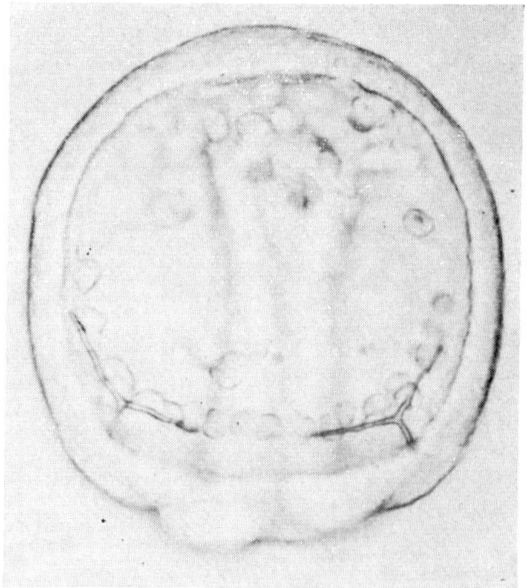

Fig. 24. A 16-hr gastrula of Paracentrotus at the end of primary mesenchyme formation, seen from ventral side. The tip of the archenteron seems to have dissolved into single cells. Note the assemblage of primary mesenchyme cells following the ventral border line of the future oral field and the triradiate spicules.

During gastrulation the anlagen of the skeleton are formed within the two clusters of primary mesenchyme. The development starts with very small grains of inorganic material (figs. 21, 27G, H) produced by the cells, subsequently 'growing out' to triradiate spicules (fig. 24, 27I) by further deposition of mineral substances in three corners (Woodland 1906) from the surface of the cells. The cells of the primary mesenchyme must be arranged in a certain pattern. It was already mentioned that they are connected by protoplasmic threads. If this connection is partially interrupted, supernumerary spicules are formed, which later disappear as soon as the connection is reestablished (Okazaki 1960). Thus information concerning skeleton formation must be transported along the protoplasmic filaments.

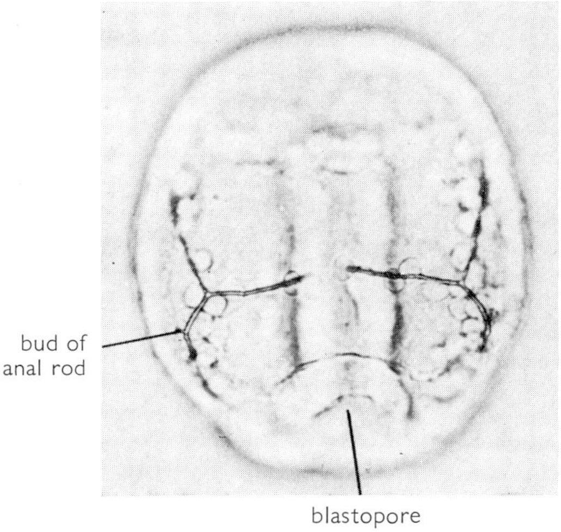

Fig. 25. Oblique view of an 18-hr gastrula of Paracentrotus showing the arrangement of primary mesenchyme cells during the formation of the skeleton. Note the obtuse angle in the vertex rod at which the anal rod buds.

With the complete flattening of the oral and anal field the gastrula has reached the tetrahedron or prism-stage (figs. 25–32). The tip of the archenteron becomes asymmetric: at the right edge the archenteron is longer and appears obliquely cut, subterminally two sacs grow out which are finally separated from the archenteron (figs. 29, 30, 32). Normally the left coelomic sac is larger than the right (fig. 32). The cell number ratio of the right to the left coelom is often 3:2 and there is some evidence that an anlage of five parts is divided unequally during the formation of the coelomic sacs (Czihak

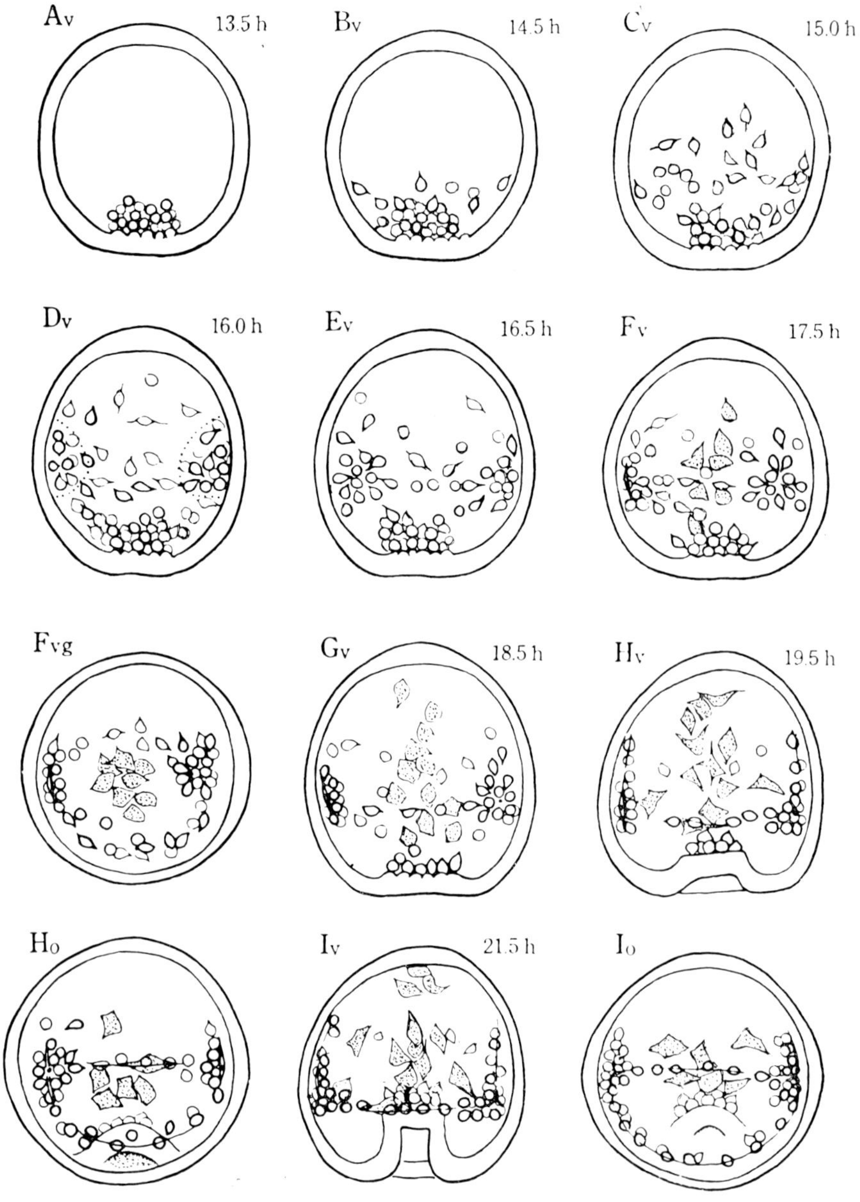

Fig. 26. Immigration and movements of primary and secondary mesenchyme cells of *Clypeaster japonicus*. The time after fertilization is indicated at the top right. In F the first secondary mesenchyme cells, much larger than the primary ones. v = ventral view, o = oblique view. (After Okazaki 1960.)

1962b). During and after this process the cells of the secondary mesenchyme spread in the blastocoel and the tip of the archenteron bends towards the oral field to come into contact with the mouth groove (fig. 31) with the aid of plasmatic protrusions. The mouth opening (deuterostomium) breaks through after the contact between entoderm and ectoderm has been established. The opening of the archenteron invagination (blastoporus) differentiates into the anus. The mouth and a small part of the oesophagus are built up out of ectodermal material. The archenteron differentiates into the intestine, first by two constrictions separating the oesophagus from the stomach and the stomach from the rectum (fig. 31).

The rods of the triradiate spicules grow further by deposition of mineral substances within the syncytium of primary mesenchyme cells. No cell boundaries were found electronmicroscopically (Hagström und Lönning 1968) in the syncytium of primary mesenchyme cells.

An oral rod grows out towards the apical pole (fig. 33) and later into the oral arms (fig. 34), and the vertex-rod grows towards the dorsal side at right angles to both the transverse rod and the anal rod growing out into the anal arms (figs. 33, 34; Von Ubisch 1937).

Some species, for instance Sphaerechinus and Echinarachnius, have fenestrated anal rods. Their development is shown in figs. 35–37.

A ciliary band develops at the border line of the oral field and at 4 points we find buds representing the anlagen of the arms. They at first develop autonomously, even when the skeleton is absent. But subsequent growth is dependent on the skeleton: the arms become longer only if the rods grow into them (Von Ubisch 1933). The larva is called the pluteus as soon as the arms have begun to develop. The apical ciliary tuft, still present in late prisms (fig. 31), disappears with the outgrowth of the oral arms.

After their separation from the archenteron, the coelomic sacs remain for a short time unchanged at the sides of the differentiating oesophagus. Afterwards they stretch, their cavity temporarily disappears, and they begin to migrate along the oesophagus towards the stomach, attracted by unknown forces. From the larger coelomic sac, normally the left one, a process (the hydroporus channel) is developed (fig. 38), growing out towards the hydroporus, the latter primarily being a tiny invagination of the dorsal integument. After contact has been established, there is an open connection between the coelomic cavity and the surrounding medium. When the coelom is lacking, the hydroporus-anlage is formed as well, showing that it is independent of coelom differentiation.

As soon as about the half of the coelomic cells have reached the stomach,

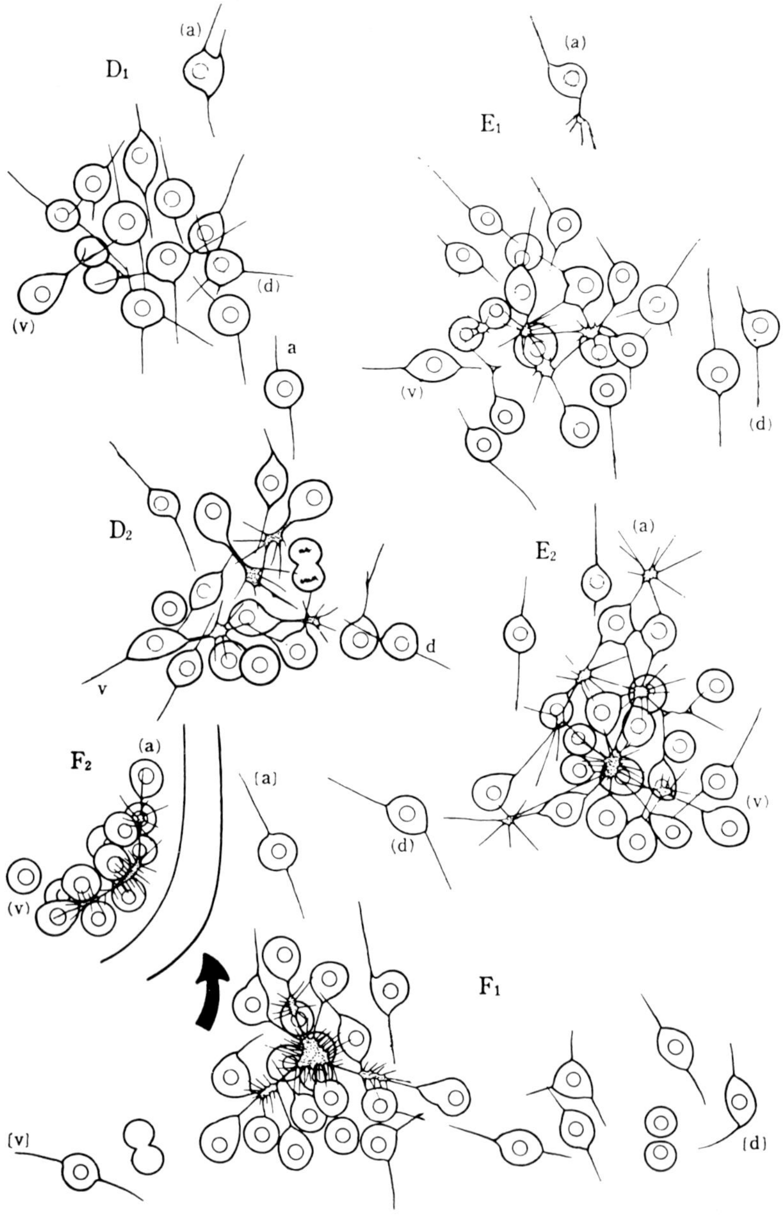

Fig. 27. Subsequent steps of cluster formation in the primary mesenchyme of Clypeaster. The lettering corresponds to that in fig. 26: G for instance is an 18.5 hr blastula. Different small syncytia (E, F) fuse and form a single triangular syncytium below each attachment

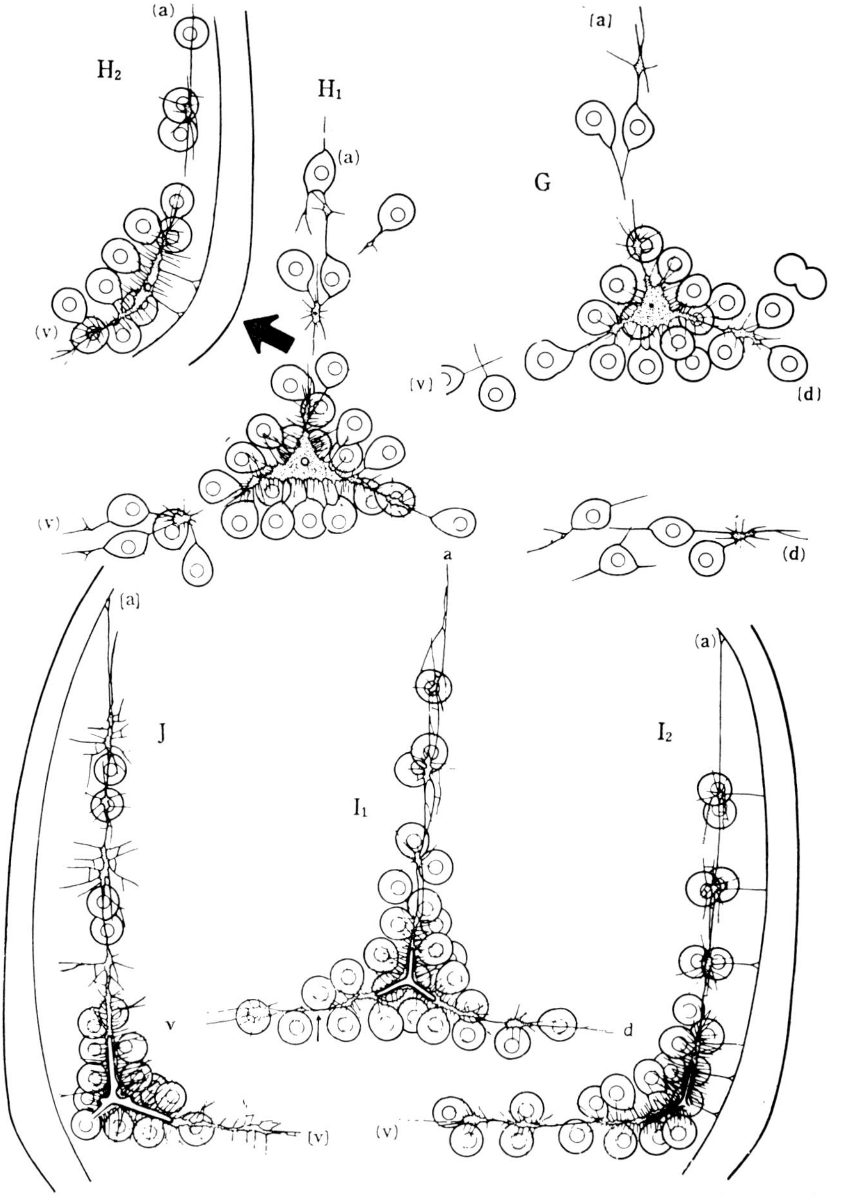

zone (lateral ectoderm thickening, indicated by arrow). In the centre of the syncytium first a grain of inorganic material is produced (G, H), growing to the triradiate form by further deposition of material (I). Note the strands by which the primary mesenchyme cells are connected and the thin threads which connect the strands with the blastula wall (I_2). (After Okazaki 1960.)

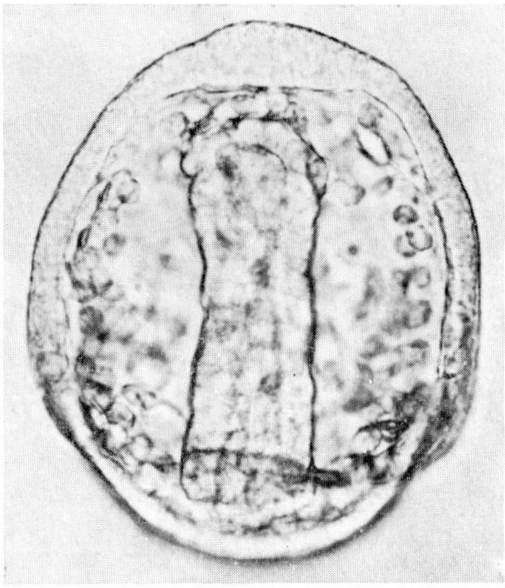

Fig. 28. Gastrula of Psammechinus, 23 hr after fertilization, seen from the dorsal side. Note the slightly asymmetric tip of the archenteron.

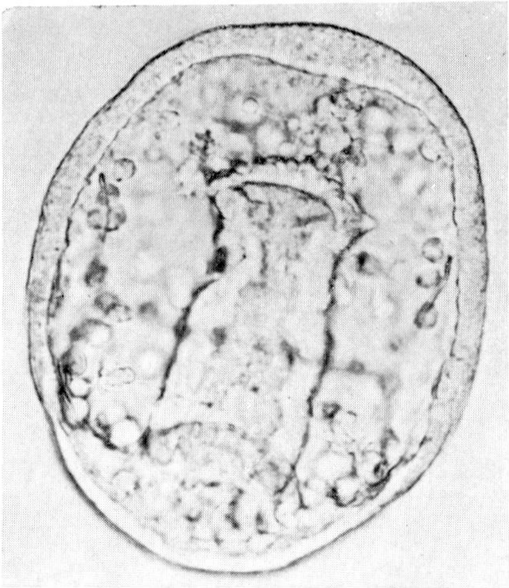

Fig. 29. Beginning formation of the coelomic sacs in a 25-hr-old gastrula of the same species.

the mass of coelomic cells beside the oesophagus becomes divided into two parts lying one after the other (figs. 38, 39). The parts near the mouth remain beside the oesophagus (fig. 39), but those on the stomach do not stop migrating and the cavities within them reappear. In the larger coelomic

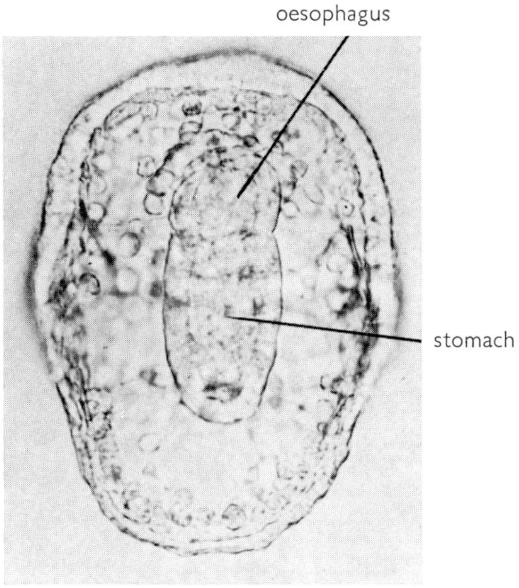

Fig. 30. Gastrula of 27 hr (prism-stage) after termination of evagination of the coelomic sacs, seen from the dorsal side. Slight indication of the future constriction between oesophagus and stomach.

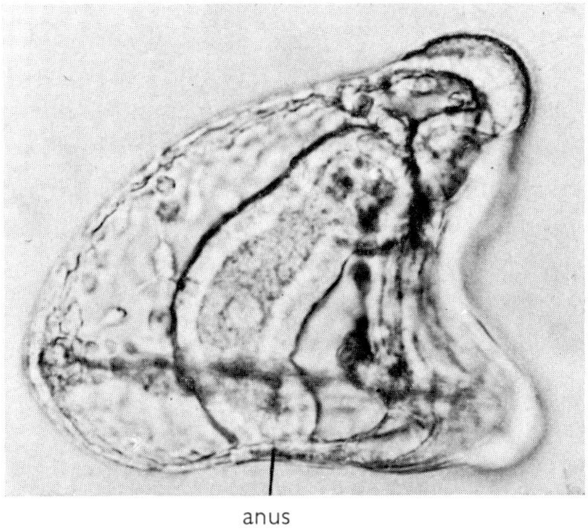

Fig. 31. Twenty-seven-hour-old prism of Psammechinus, side-view. The ciliary tuft is still present, the archenteron bent towards the oral field, where the mouth groove is sinking in and is already in contact with the tip of the archenteron. A slight constriction indicates the place where the oesophagus will become separated from the stomach.

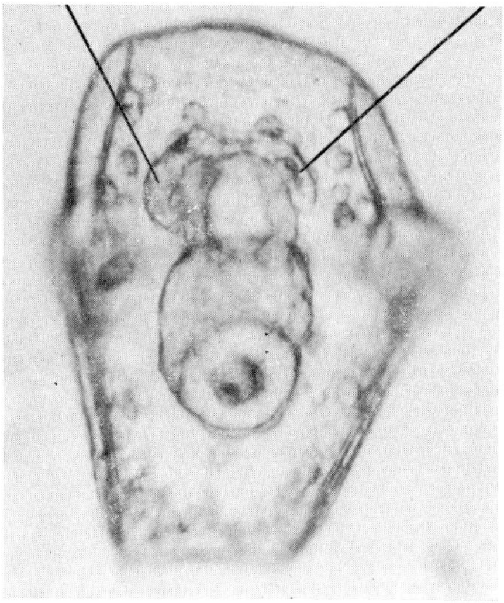

Fig. 32. Young pluteus of 36 hr, same species, seen from below. The oral rods immediately before outgrowth of the oral arms. The separation of oesophagus and stomach is more pronounced and the size difference of the coelomic sacs is striking. Here the right coelomic sac is the larger (situs inversus).

sac, normally the left one, another fission occurs: three parts can then be distinguished; one, nearest to the mouth, remains on the oesophagus, the second, the middle part, becomes the hydrocoel and the posterior one the somatocoel (fig. 40). The middle part becomes subdivided into hydrocoel-vesicle, stone channel, ampulle and hydroporus channel (fig. 41). The two parts of the right coelom remain without further division: an anterior part on the oesophagus and the posterior somatocoel. The somatocoels of both sides spread over (figs. 41, 42) the stomach by flattening and finally touch each other in the median plane (fig. 44), where the mesentery is formed by the two adjacent walls of the somatocoels. The hydrocoel-vesicle is now nearly spherical and moves above the left somatocoel (figs. 41, 43). Opposite the hydrocoel-vesicle the integument sinks in to form the vestibulum (fig. 41). It seems that at least in *Genocidaris maculata* the vestibulum develops independently of the hydrocoel (Von Ubisch 1959). When the hydrocoel-vesicle touches the vestibulum, both grow broader, creating a large contact zone (figs. 42, 43). From the hydrocoel disk three finger-like processes grow out into the cavity of the vestibulum, followed by two more (figs, 43a, 43b). These are the five primary tentacles, composed of one layer of coelom epithelium covered by ectoderm epithelium of the vestibulum. We then have a young imaginal disk, developing further to the ventral side of the sea urchin (figs. 44, 45). Opposite the imaginal disk we find the anlagen of some

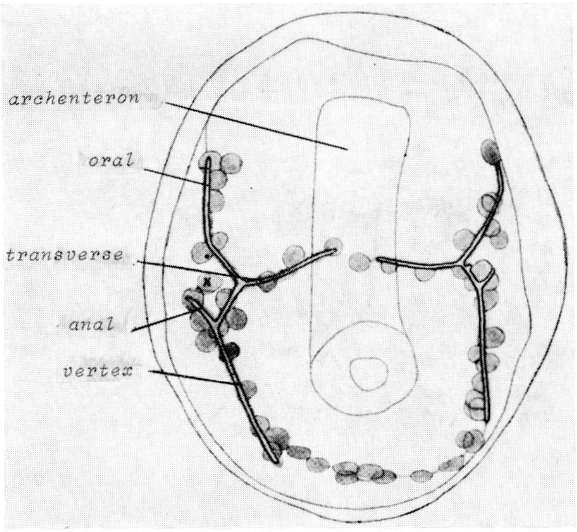

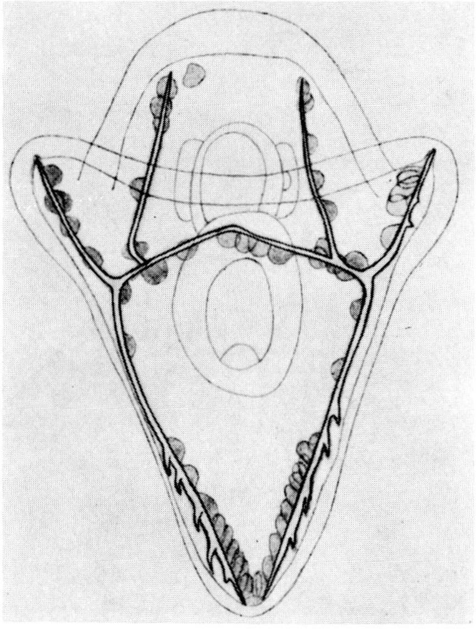

Figs. 33 and 34. Two steps of the development of the skeleton from the prism to pluteus of Psammechinus. (After Von Ubisch 1937.)

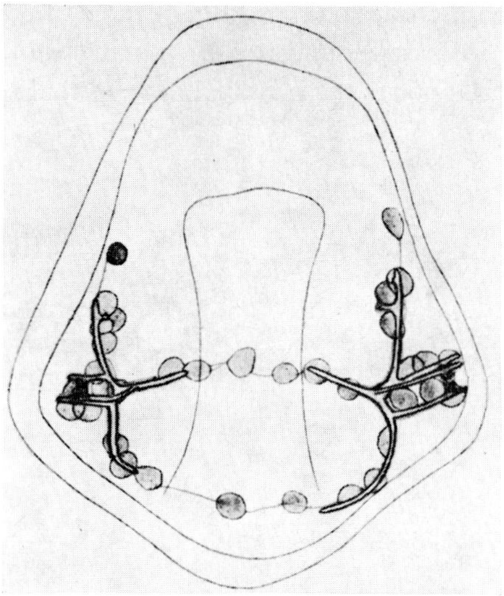

Fig. 35. (For the legend see fig. 37.)

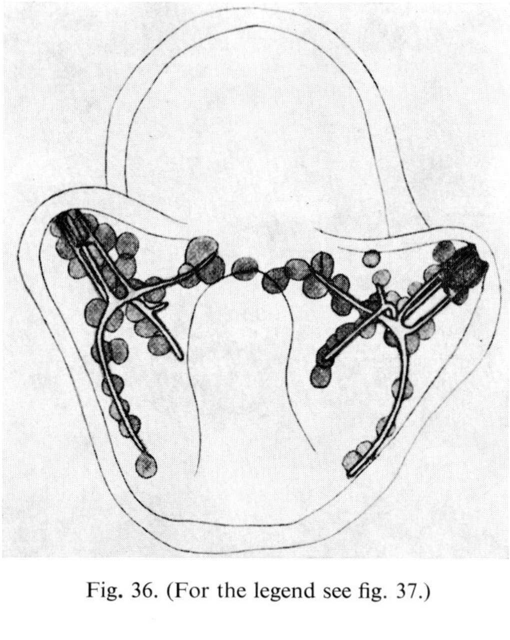

Fig. 36. (For the legend see fig. 37.)

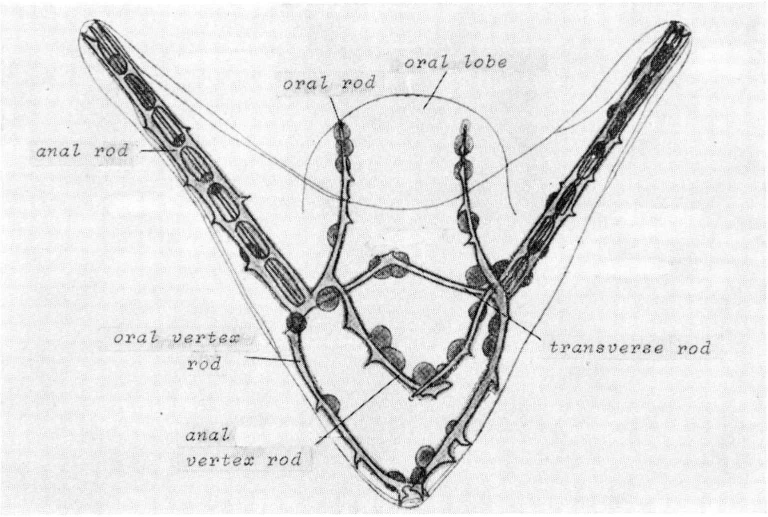

Fig. 37

Figs. 35, 36, 37. Three steps in the skeletal development of Echinocyamus, which is characterized by fenestrated anal rods. (After Von Ubisch 1937.)

spines and pedicellaria (fig. 44). The further development beyond metamorphosis is shown in figs. 44–47.

'Situs inversus', that is the occurrence of a larger coelomic sac at the right side and development of the hydrocoel at the right side, has frequently been observed. In these cases the vestibulum also forms at the right side and development therefore can proceed in a normal way. Pedicellaria are then formed first at the left side.

In some cases development of a hydrocoel was found on both sides. This can be interpreted as overdevelopment of the coelomic sacs on both sides or fission of a uniform anlage into two equal parts. The two hydrocoels develop normally into two imaginal disks and two ventral sides of sea urchins form Siamese twins after metamorphosis (fig. 48b). In this case of course no pedicellaria-anlagen can be found at the place of the imaginal disks. But two of them are formed when the imaginal disk is lacking on both sides (fig. 48c); this may be considered the consequence of the presence of two small coelomic sacs on both sides. According to the occurrence of the hydrocoel, + – (normal), – + (situs inversus), + + (twins) and – – types of plutei have been distinguished (Czihak 1960a).

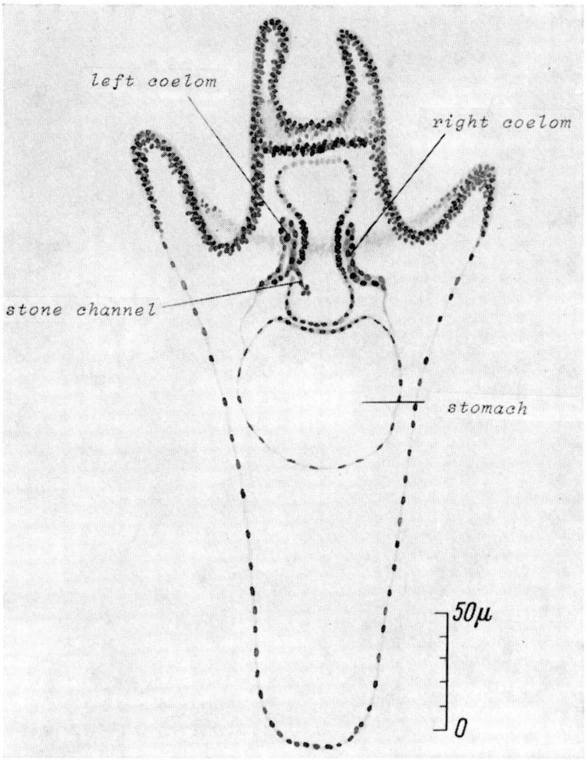

Fig. 38. Coelom development in a 4-day pluteus of Psammechinus. The skeleton has been dissolved by fixation; stained with borax-carmine. A cellular strand between the left coelom and the hydropore has developed; it will later be the stone channel.

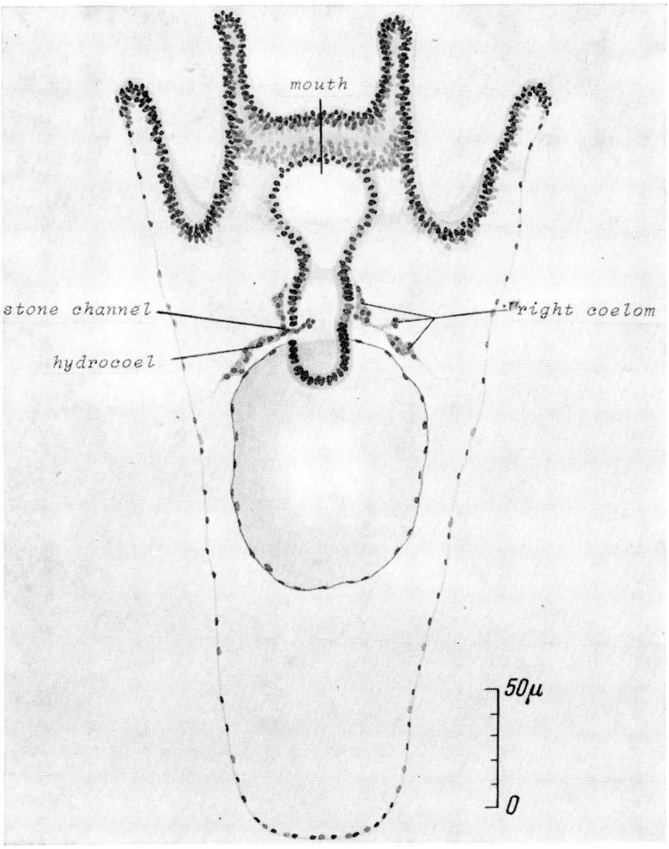

Fig. 39. Movement of the coelom-cells towards the stomach in a 7-day-old pluteus. Species and preparation as in fig. 38.

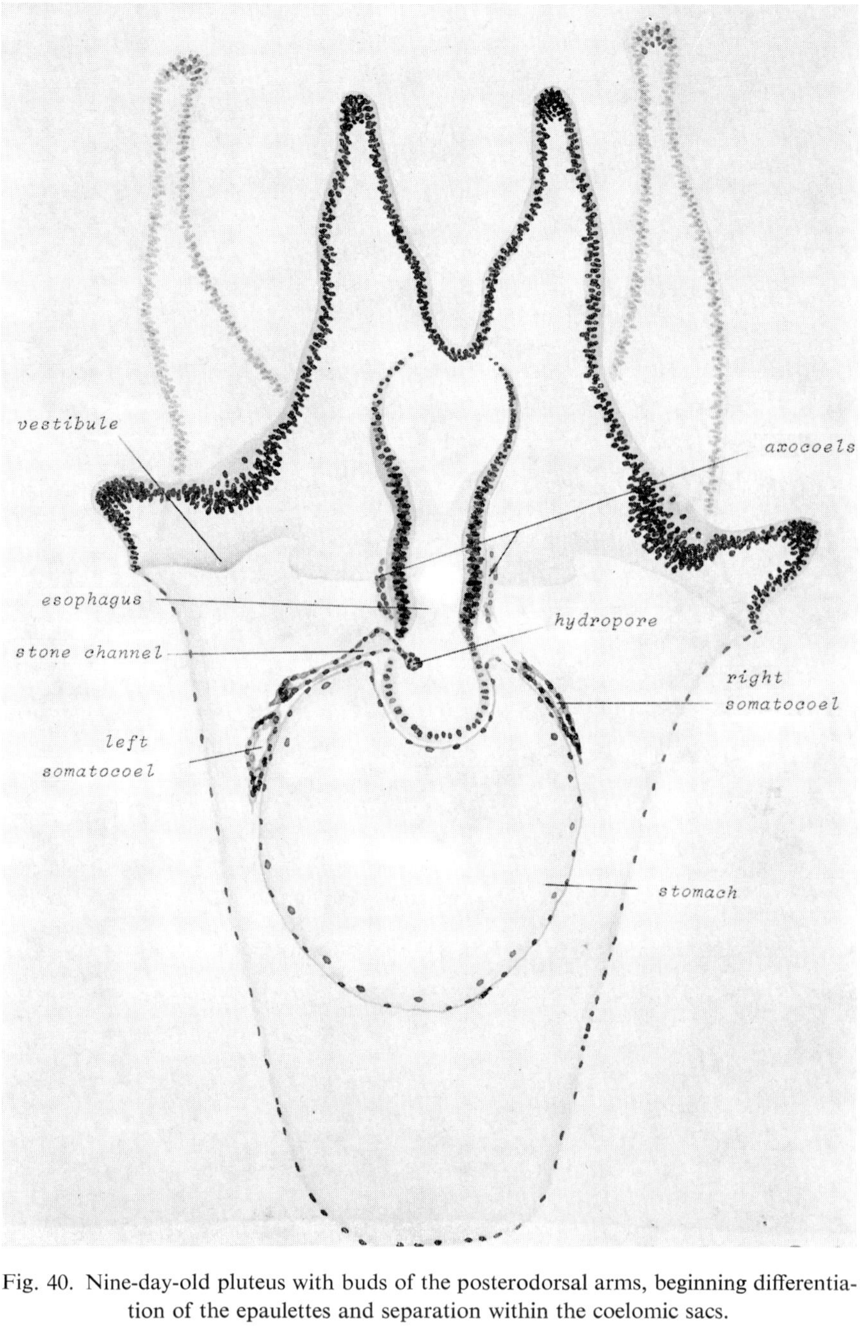

Fig. 40. Nine-day-old pluteus with buds of the posterodorsal arms, beginning differentiation of the epaulettes and separation within the coelomic sacs.

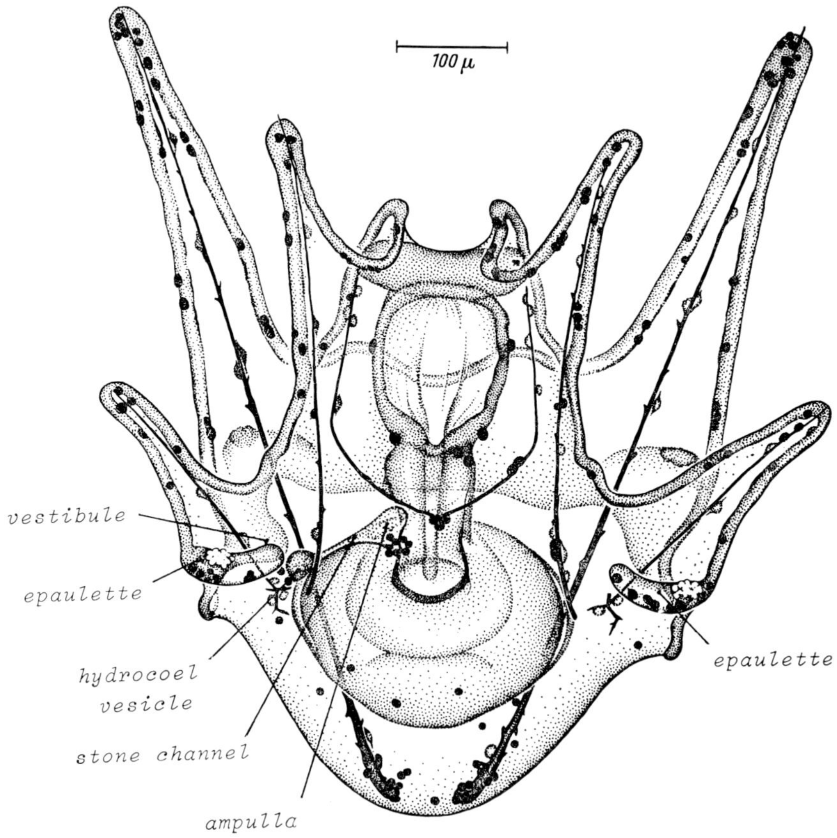

Fig. 41. Psammechinus-pluteus, 12 days old. The acute apex has been withdrawn, the apical rods are partially dissolved and near the mouth new arms with supporting skeletal rods (the preoral arms) have been formed. The epaulettes are now separated into band-like groups of ciliated cells. In the left coelom the hydrocoel-vesicle has been formed and opposite to it the ectodermal epithelium begins to sink in to form the vestibule. Beyond the hydropore the stone channel is dilated and called the ampulla.

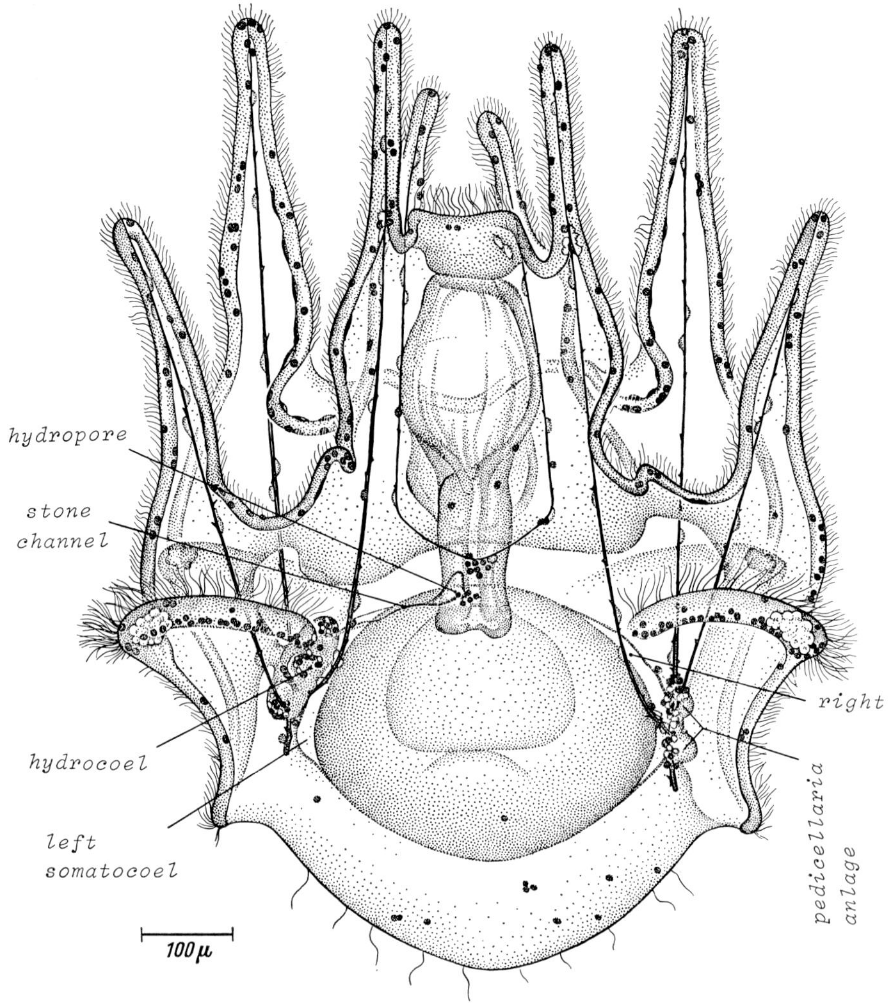

Fig. 42. A pluteus of Psammechinus, 14 days old, with 8 arms and apex rods completely withdrawn. The hydrocoel has developed 5 protrusions growing further to form the primary tentacles. Opposite the anlage of the imaginal disk (echinus rudiment) the anlagen of the pedicellaria have been formed.

(a)

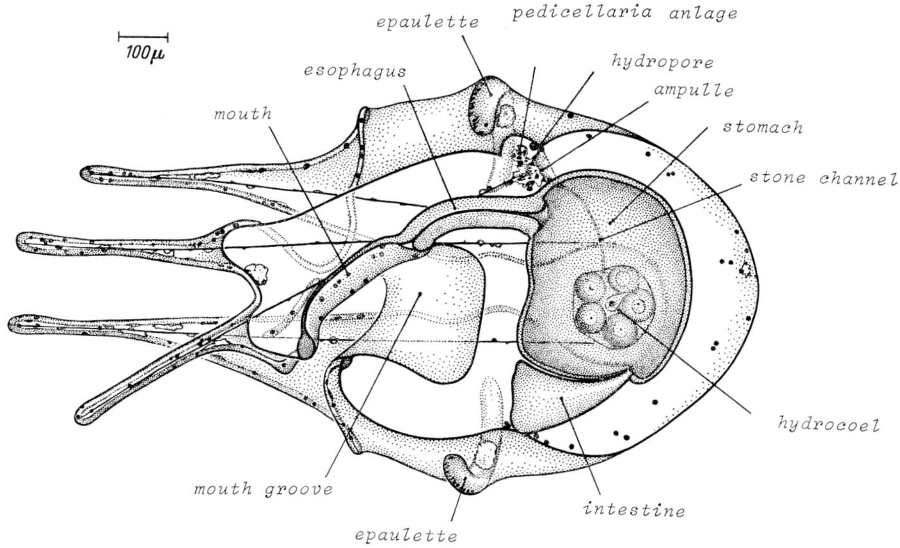

(b)

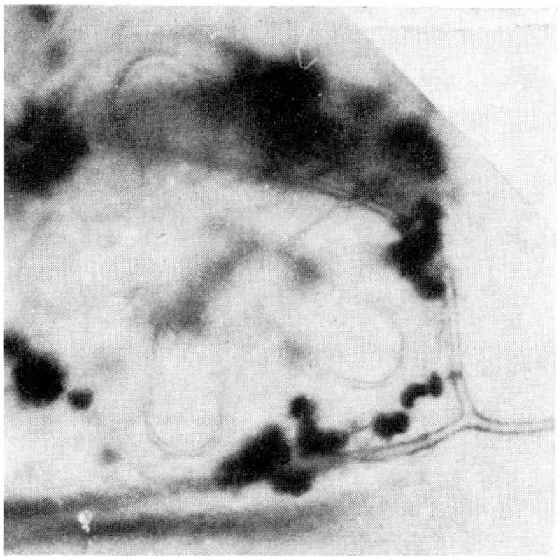

Fig. 43a, b. Psammechinus-pluteus of the same stage as the previous one in longitudinal section. The imaginal disk of this individual is located on the right side, and thus the pluteus is of the $-$ $+$ type (situs inversus, cf. fig. 32). Near the hydropore is an anlage of pedicellaria.

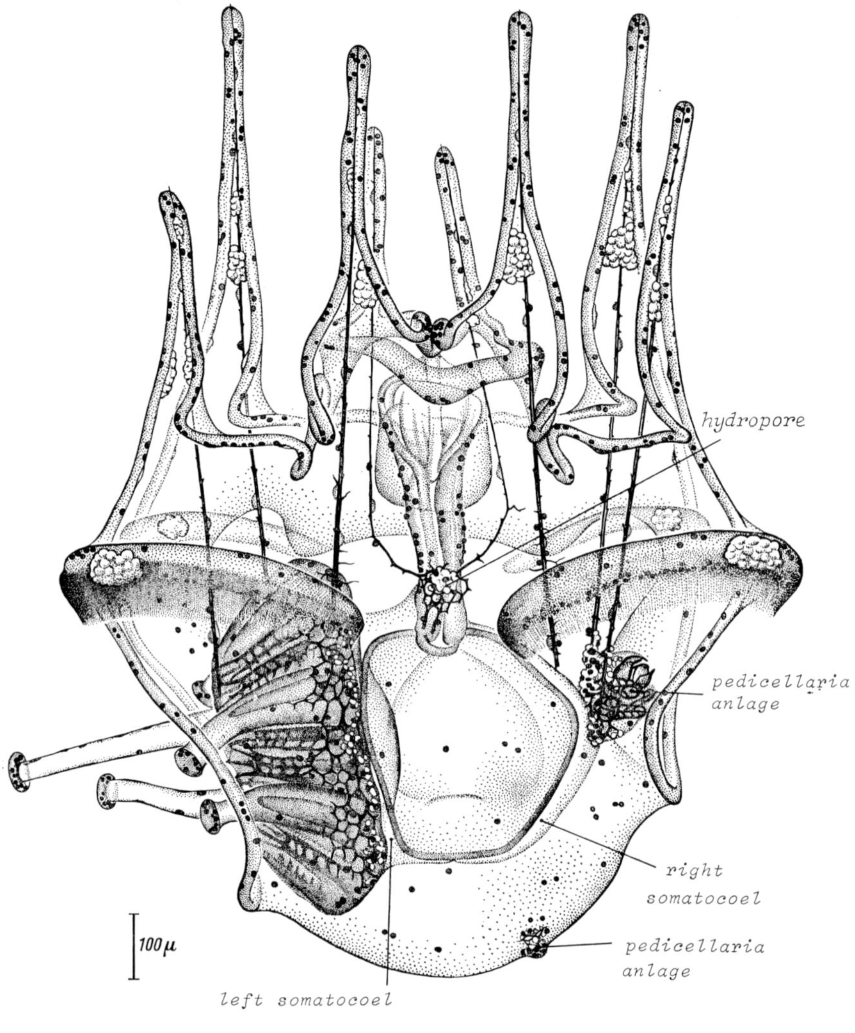

Fig. 44. Late pluteus stage of Psammechinus, 28 days old, with full developed imaginal disk. The long primary tentacles often come out from the vestibule in which the fenestrated larval spines with four tips are also visible. The first groups of pedicellaria together with larval spines opposite the imaginal disk. Another group of pedicellaria has developed near the apex. In the sagittal plane the two somatocoels are in contact, forming the mesentery.

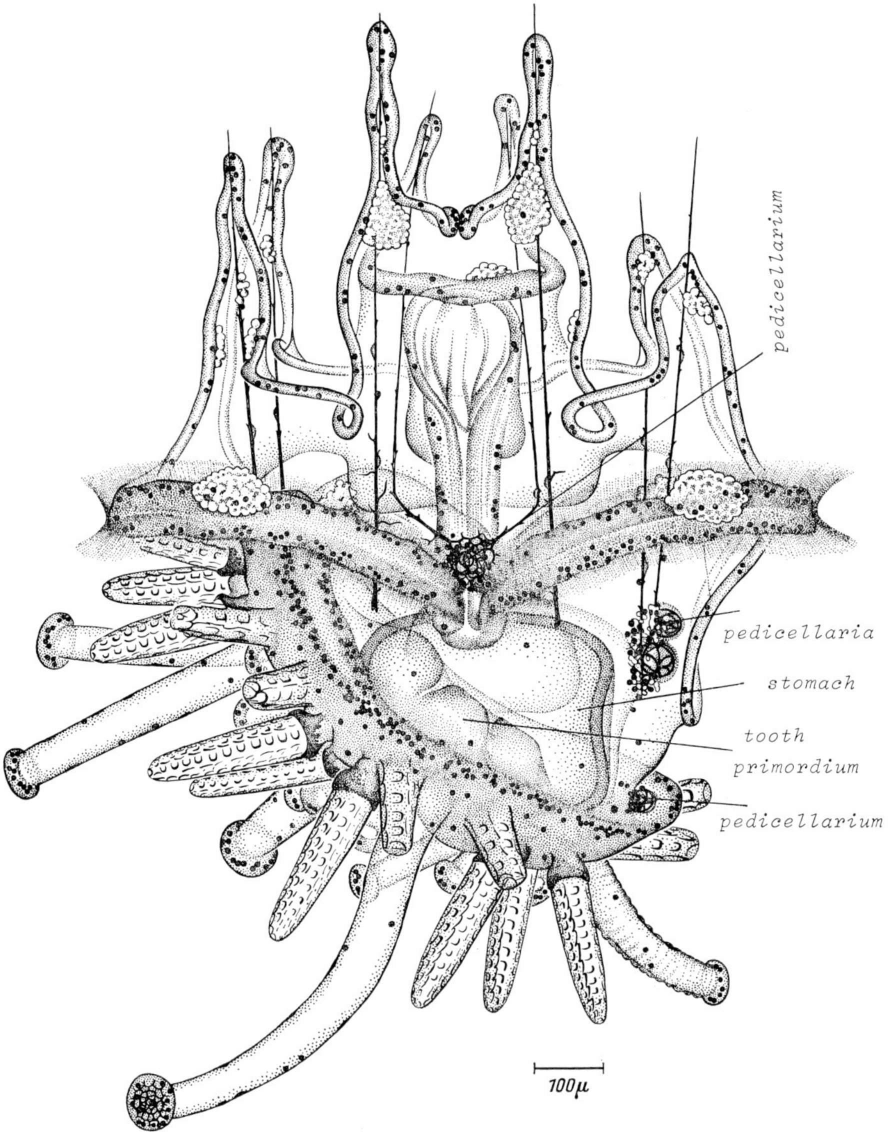

Fig. 45. First stage of metamorphosis in a 33-day-old pluteus of Psammechinus. The imaginal disk has begun to unroll and the ectodermal epithelium of the pluteus is beginning to shrink, first recognizable by its withdrawing from the arm rods. Besides the larval spines with four tips along the border line of the imaginal disk some of the definitive spines with acute tip have already been formed. The four tips of the larval spines are not yet developed to the extent of those in fig. 44: this is one of the frequent individual variations.
Note the teeth primordia visible by the impressions on the stomach.

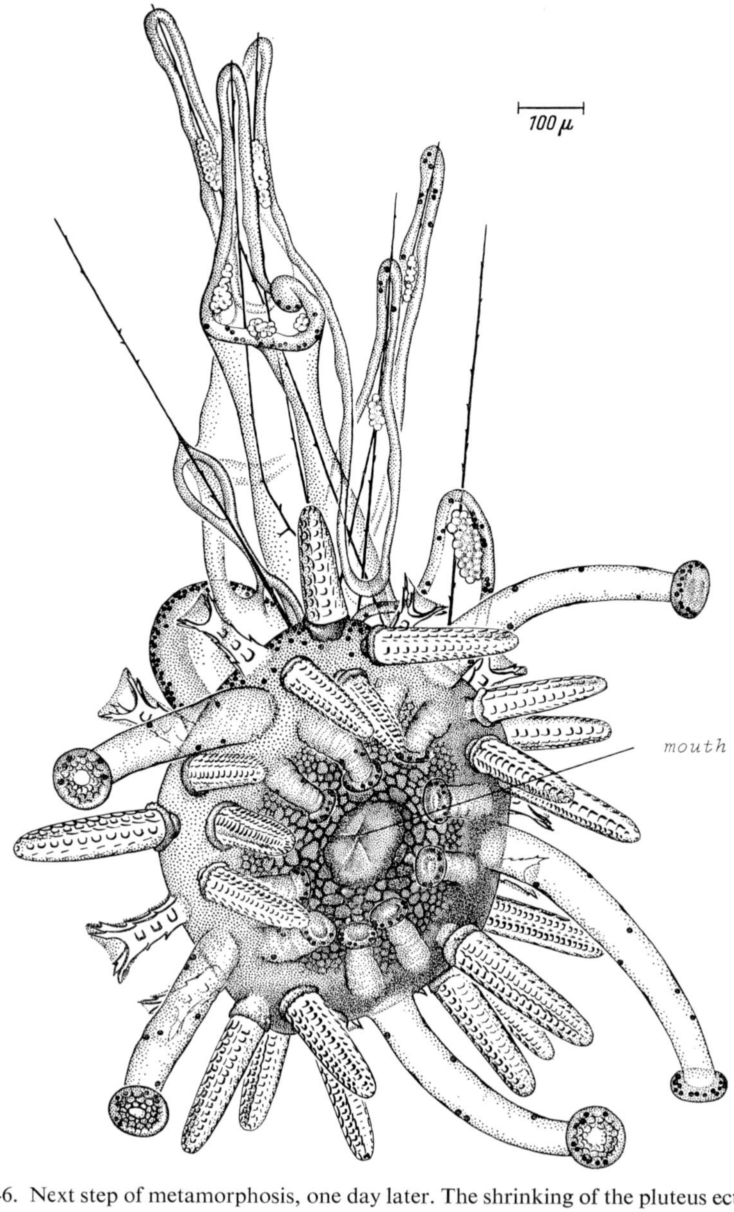

Fig. 46. Next step of metamorphosis, one day later. The shrinking of the pluteus ectoderm has proceeded. The former imaginal disk is seen from the new ventral side. Between each primary tentacle and the mouth two definitive podia develop and besides the two larval spines in each sector four definitive spines develop.

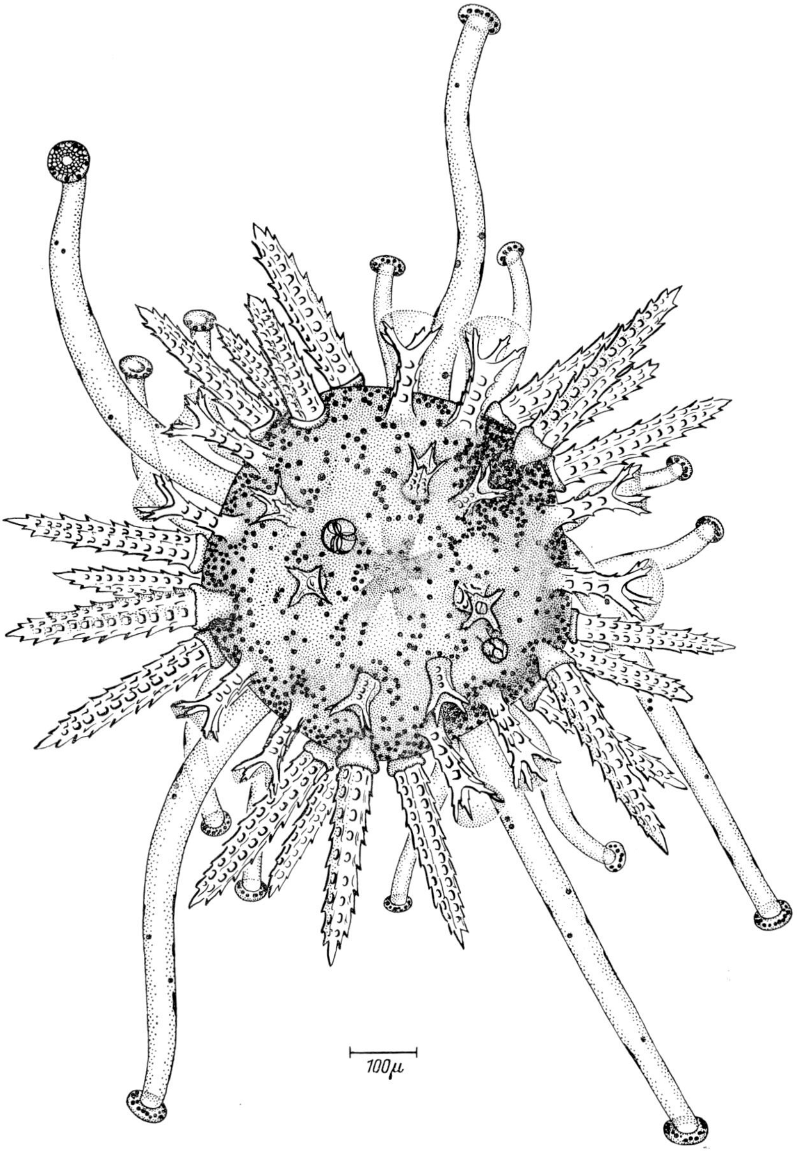

Fig. 47. Young sea urchin of the same species immediately after metamorphosis, dorsal view. By the shrinkage of the ectodermal epithelium of the pluteus 2 pedicellaria and 7 larval spines developed in the same anlagen have become situated on the dorsal half of the sea urchin. Figs. 41–47 were drawn by Dierl, Freiberg and Czihak from living specimens.

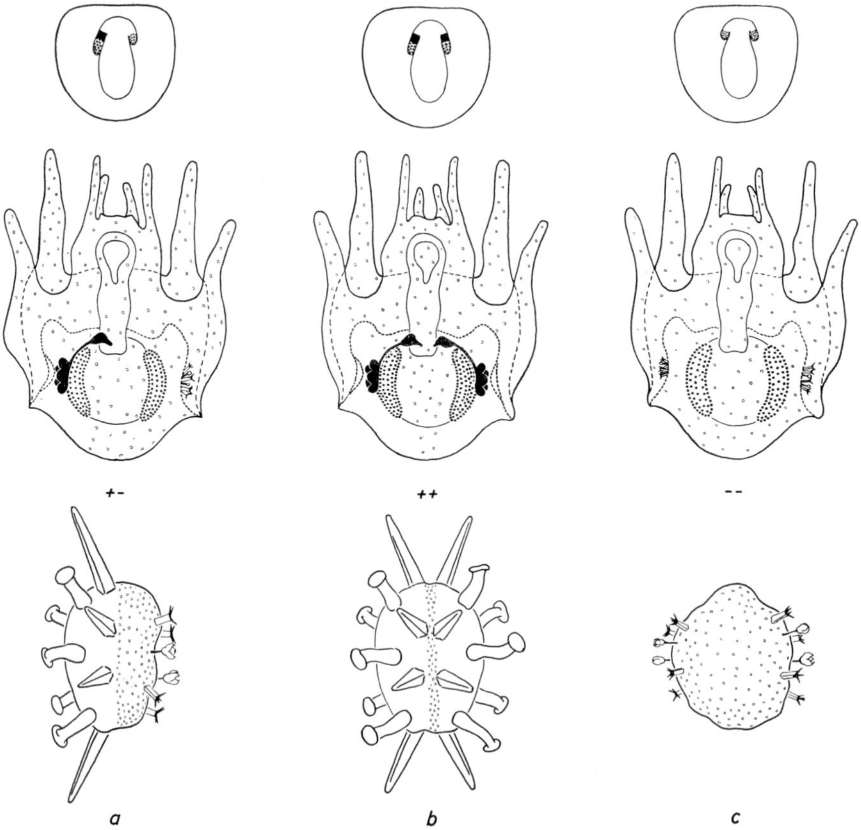

Fig. 48. Schematic representation of the development of the coelomic sacs and imaginal disks (echinus rudiment). (a) shows the most frequently observed case with the bigger coelomic sac in the left half, leading to development of the hydrocoel and imaginal disk on the left side and formation of pedicellaria and anlagen of larval spines on the opposite side. The imaginal disk forms the ventral, the shrinked pluteus ectoderm the dorsal side of the young sea urchin (below). This is the $+-$ type. There is also found a $-+$ type, the mirror-image of the previous one, with the imaginal disk on the right side. In (b) the $++$ type is shown, having two imaginal disks on either side, thought to be a consequence of two large coelomic sacs. After metamorphosis an individual is formed consisting of two ventral halves connected like Siamese twins. No pedicellaria anlagen are found. (c) represents the case of complete lack of hydrocoel and imaginal disk, probably due to a shortage of coelomic cells on either side. Thus on both sides anlagen of larval spines and pedicellaria only are formed. During metamorphosis the pluteus retracts to a sphere of two dorsal sea urchin halves bearing only pedicellaria and larval spines and not even having a mouth opening: this is called the $--$ type. The ectodermal epithelium of the pluteus, which after metamorphosis forms the dorsal skin of the sea urchin, is designated by small circles.

12.3. Cell lineage

Boveri (1901) already observed the development of blastomeres bearing pigment granules of the pigment ring. The pigment granules served as a natural marker and were later found in cells of the archenteron. Further work in this field was done by Von Ubisch using Vogt's method of vital staining of parts of embryos. Hörstadius succeeded in elucidating the prospective assignment of all the blastomere circles. As fig. 49 shows schematically, the mesomeres of the 16-cell stage divide into two circles of animal cells and reach the 32-cell stage (an_1 and an_2, figs. 13, 49E). The mac-

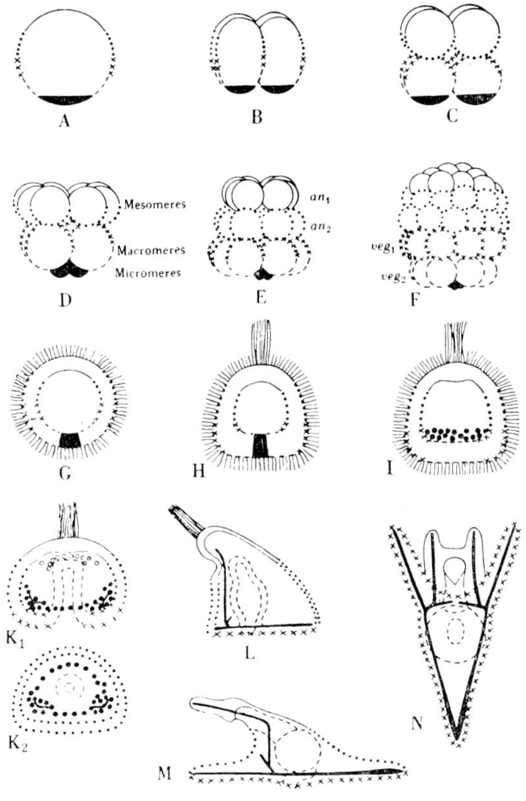

Fig. 49. Cell lineage of Paracentrotus demonstrating the future development of the egg parts. The blastomeres of veg_2 for instance, indicated by a dashed line, will form the archenteron and intestine, the black micromeres the primary mesenchyme and skeleton. (After Hörstadius 1939.)

romeres divide into two circles of vegetal cells in the 64-cell stage (veg_1 and veg_2, fig. 49F). In the blastula stage the offspring of the mesomeres form the animal half of the blastula and the cells developed from macro- and micromeres the vegetal half. The cells derived from the micromeres (black) immigrate into the blastocoel and differentiate into primary mesenchyme cells which later produce the skeleton. The descendants of the veg_2-circle blastomeres invaginate during gastrulation and form the intestine, secondary mesenchyme and coelomic sacs. From the blastomeres of the veg_1-circle the ventral or anal side of the pluteus is formed. The blastomeres of the an_2-circle differentiate into the oral and dorsal side near the ventral side and those of the an_1-circle into the other parts of the oral and dorsal side and the apical plate with the stiff cilia.

12.4. Centrifugation of eggs and change of cleavage pattern

We owe the most extensive research to E. B. Harvey. Her numerous original papers are summarized in two reviews (1949, 1956) which can be highly recommended for further reading.

12.4.1. Change of stratification of the egg content, change of egg shape and change of cleavage pattern

In Harvey's experiments the eggs of the American *Arbacia punctulata* were centrifuged in a density gradient, made up of 0.85 mol sucrose (iso-osmotic to the egg plasm) and sea water. The plasm of unfertilized eggs can be stratified at 3000 *g* in two minutes without changing the form. At 10,000 *g* they become stretched and after 4 min dumb-bell-shaped (fig. 50A) and torn into parts (figs. $50b_1$ and b_2). The egg nucleus is found in the lighter half. The following layers can be discerned after centrifugation of *Arbacia lixula* and *punctulata* eggs: 1st: oil droplet; 2nd: clear cytoplasm containing the egg nucleus near the oil droplet; 3rd: layer of mitochondria; 4th: yolk platelets; and 5th: pigment granules (fig. 50). The stratification is different in other species, as the yolk platelets and mitochondria often have other sizes (see table in Harvey 1956). The stratification of the egg content of some European sea urchin species after centrifugation was investigated by Callan (1949). In *Paracentrotus lividus* eggs (fig. 51) he found: 1st: oil droplet; 2nd: clear layer with nucleus; 3rd: yolk layer; 4th: clear cytoplasm; 5th: mitochondria (there is no mention of pigment), and in *Psammechinus miliaris*: 1st: oil; 2nd: yolk platelets with nucleus; 3rd: clear cytoplasm; 4th: mitochondria. In *Sphaerechinus granularis* the same stratification as in

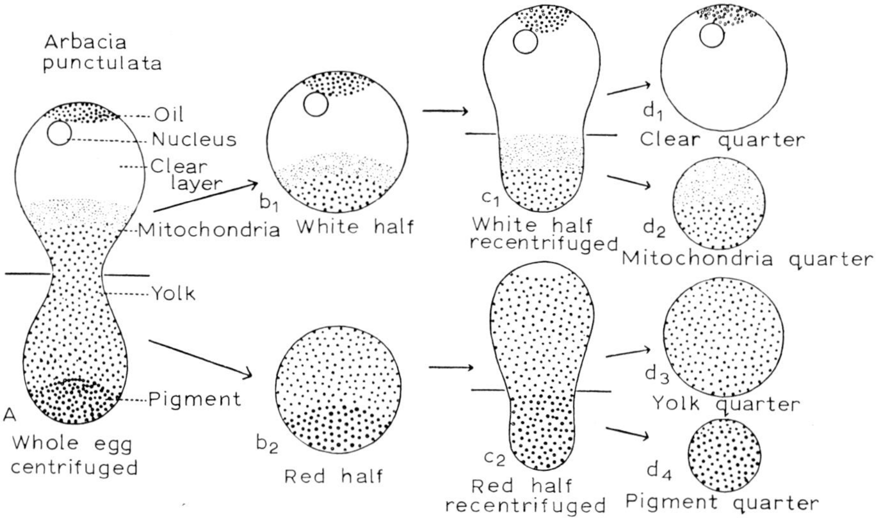

Fig. 50. Centrifugation of the Arbacia egg. In A the stratification of the content of the egg is shown, in b the separation of the egg halves by centrifugal forces: b_1 is the lighter egg half containing the nucleus, b_2 the heavier one without nucleus. If the halves of b are recentrifuged (c), they can be separated into the quarters shown in d. (After Harvey 1956.)

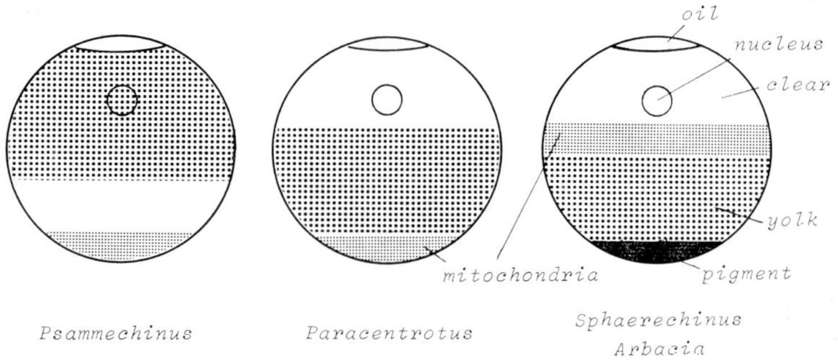

Fig. 51. Stratification of the egg content by centrifugation in three different species. (After Callan 1949.)

Arbacia was found with one exception: after the yolk platelets at the centrifugal pole there was clear cytoplasm instead of pigment (fig. 51).

Unfertilized eggs centrifuged at 3000 *g* gradually lose the stratification after centrifugation has been stopped. Some hours afterwards the layers are no longer separated and only the oil droplet is found up to 24 hr. Dumb-bell-shaped eggs can recover their rounded form some time after centrifugation has been stopped and the stratification also becomes lost after some hours. It may be assumed that there is a cortical pattern which remains unchanged and which can restore the normal stratification of the egg. When dumb-bell-shaped eggs are fertilized after centrifugation, however, they preserve this form up to the blastula stage.

The egg halves obtained by centrifugation (fig. 50b) are often of different sizes even within the same egg batch.

After prolonged centrifugation the eggs become divided into quarters as shown in fig. 50d. In this illustration it can be seen how the layers are shared among the different parts of the egg.

Freshly fertilized eggs can be divided into two parts if the fertilization membrane has been removed shortly after insemination (see p. 364). The stratification is then as in unfertilized eggs but the layers are less distinct. One rarely succeeds in dividing a fertilized egg within the fertilization membrane by centrifugal forces. The heavier egg half with the sperm nucleus develops more quickly than the centripetal egg nucleus.

Unfertilized eggs stratified by centrifugation develop normally after fertilization. According to Harvey the first cleavage plane is often but not always perpendicular to the stratification, the second, however, parallel and thus perpendicular to the previous one. The third cleavage plane is once again perpendicular to the stratification and to all previous ones. In comparison with the succession of normal cleavage the planes of the second and third cleavages are interchanged (fig. 53A). In normal development the blastomeres of the four-cell stage contain qualitatively the same parts of the egg plasm, whereas the blastomeres of centrifuged eggs are different in this respect after the second cleavage, as two contain plasm with no yolk but an oil droplet and the other two plasm rich in yolk. It is very interesting that the point at which the micromeres cleave off is unaffected by centrifugation: they always arise at the former vegetative pole (checked by the position of the polar bodies) or in its immediate neighbourhood (fig. 52) (Morgan and Spooner 1909). Probably the spindle orientation leading to micromere formation is directed by the egg cortex (see pp. 371, 412).

The cleavage pattern can also be changed by other techniques (see p. 411).

Eggs stretched by centrifugal forces and then fertilized before they can recover their spherical shape retain their dumb-bell form. The plasma obviously becomes stiff after fertilization. The spindle of the first cleavage lies in the long axis of stretched eggs. Thus two different blastomeres arise; one smaller with clearer plasma and a larger one with more mitochondria and yolk. In the second cleavage the small blastomere divides earlier. The cleavage plane is perpendicular to the first, i.e. parallel to the long axis. The other yolk-rich blastomere cleaves with the spindle axis parallel or perpendicular to the first axis. The further cleavages of the centrifugal and centripetal blastomeres become increasingly asynchronous, because the yolk-rich blastomeres continue to cleave more slowly. The micromeres can arise at different points of the stretched egg. There is no reason to deny that they arise at the previous vegetal pole, as Morgan and Spooner found in less centrifuged eggs. In spite of these profound changes the development proceeds normally!

12.4.2. Development of eggs divided into halves by centrifugation. Parthenogenetic development

Centripetal halves cleave with the plane in a different position at about the same velocity as normal uncentrifuged eggs. Micromeres do not arise in these halves and development generally stops at the blastula stage. These 'Dauerblastulae' (permanent blastulae) lack pigment and their skeleton is irregular. Development often stops even earlier.

They are rarely able to develop further to plutei which have regulated(?) the temporary loss of pigment.

There are several methods of initiating parthenogenetic development. Harvey (1956) states that hypertonic treatment is best for *Arbacia punctulata*, '... hypertonic sea water for about 20 min. The sea water is made hypertonic either by boiling to half its volume, or by adding 30 g NaCl per liter' Loeb published dozens of papers in this field. His 'double method' (1916) has often been used. Eggs are treated for 2–4 min with 50 ml sea water + 8 ml 0.1 n butyric acid, transferred to normal sea water, from which after 10 to 15 min they are put into hypertonic sea water containing 8 ml 2.5 mol NaCl per 50 ml. After 17.5–22.5 min they are transferred to normal sea water. Ishikawa replaced the butyric acid by thymol. Eggs are treated for about 30 sec in diluted thymol solution (2 parts of sea water saturated with thymol + 8 parts sea water) and then washed with normal sea water. After 20 min they are transferred for 20 min to hypertonic sea water (40 ml 2.5 n NaCl + 250 ml sea water) and afterwards again placed in normal sea water, where they remain for further development. Some biochemical processes connected with activation of eggs have been studied by Ishikawa (1962). Also D_2O was used for parthenogenetic activation of the egg (Gross et al. 1964).

After such treatments the eggs have a fertilization membrane, often start cleaving and develop to the blastula stage or in some cases even to the pluteus.

Centrifugal, yolk-rich halves contain no nucleus. After fertilization the sperm nucleus divides with small spindles but the formation of cell membranes is irregular so that some multinuclear blastomeres may result. Sometimes development gives rise to a small, heavily pigmented pluteus. Parthenogenetic development can also be induced in yolk-rich halves without a nucleus by hypertonic sea water. It is interesting to note that the cytoplasm behaves as if there were a nucleus: a monaster and an amphiaster are formed around a spot of clear plasm, and bisection of the plasm is certainly retarded though nearly normal. The subsequent transections of the plasm can give nucleus-less creatures with some hundred plasm compartments which may survive up to one month. Unfertilized eggs without parthenogenetic activation die after a few days, however.

As already mentioned, egg halves separated by centrifugation can be further divided by centrifugation at 10,000 *g* for 20 to 30 min; the resulting quarters in *Arbacia punctulata* are: a clear one with the nucleus, a mitochondria-rich, a yolk-rich and a pigmented quarter. Fertilized clear quarters divide slowly with the cleavage plane 'at random'; the second cleavage plane is always perpendicular to the first. Thus 4 equal-sized blastomeres arise. The hatching of the extremely transparent blastula is likewise retarded. Development of these quarters may continue and lead to small plutei which have an almost normal skeleton with very short arms. This is remarkable as there was no micromere formation. The quarters seem (!) to be free of mitochondria, but later mitochondria can be recognized in the archenteron wall; the lacking pigment is also restored after some days. In the quarters rich in mitochondria, which contain not only all of the egg's mitochondria but also some yolk platelets, the sperm nucleus divides somewhat more slowly than the nuclei of whole eggs. At the 16-cell stage the blastomeres are sometimes of equal size. Cleavage of the plasm is often retarded or lacking, however. Development never reaches gastrulation and there are no skeleton anlagen. The cleavage of fertilized yolk-rich quarters is similar to that of mitochondria-rich quarters. Such egg parts can, however, develop up to pluteus larvae with an irregular skeleton. The cleavage of fertilized pigment-rich quarters which contain some yolk, too, is also greatly retarded. These quarters are smaller than the others and cleavage seldom proceeds to the blastula stage.

Eggs centrifuged at 3000 *g* shortly after fertilization develop normally.

The spindle of the first cleavage often lies parallel to the stratification, but sometimes is found perpendicular to it. Halves of eggs divided after fertilization develop less well than fertilized ones which were divided before fertilization! The centripetal halves, which often contain the male and female pronuclei, give rise to dauerblastulae, rarely to poorly developed plutei. The centrifugal halves do not develop at all. This is in striking contrast to development of the same halves obtained before fertilization and activated to parthenogenetic development by hypertonic treatment.

It is striking that eggs centrifuged and bisected after fertilization should develop less well than those divided before fertilization and fertilized afterwards. Probably in the unfertilized egg development-controlling substances are distributed in an ordered manner but loosely, so that they may become more uniformly mixed, while heavier particles are translocated by centrifugation. Thus the halves may contain approximately the same substances and this could explain why development then is rather normal. After fertilization the ground plasm seems to become more solid than in centrifuged unfertilized eggs, as the retention of the stretched shape of centrifuged fertilized eggs indicates. A more solid plasm cannot of course be so easily mixed and thus egg parts obtained by centrifugation may contain different proportions of the above-mentioned substances. These results can be better understood and evaluated in the light of the double gradient hypothesis (cf. p. 425).

12.5. Alteration of cleavage patterns by other methods

As Boveri showed in 1895, the centrioles move to the future spindle poles in early prophase. The spindle fibres connected with the chromosomes become orientated towards the centrioles, which have clearly visible astral rays in sea urchin eggs. The orientation of the spindle seems to be predetermined by the situation of the centrioles. In normal development the centrioles lie within the equatorial plane during the first two cleavages. During the third cleavage the centrioles are found in the animal and vegetal plasms respectively. In the previous chapter it was pointed out that in centrifuged eggs the sequence of spindle orientations of the second and third cleavages can be reversed, probably because the altered stratification of the cytoplasm slowly returns to normal. Unaltered, even in centrifuged eggs (fig. 52), however, is the point at which micromere cleavage occurs. This obviously implies that one of the centrioles has a specific relation to the cortex of the vegetative pole.

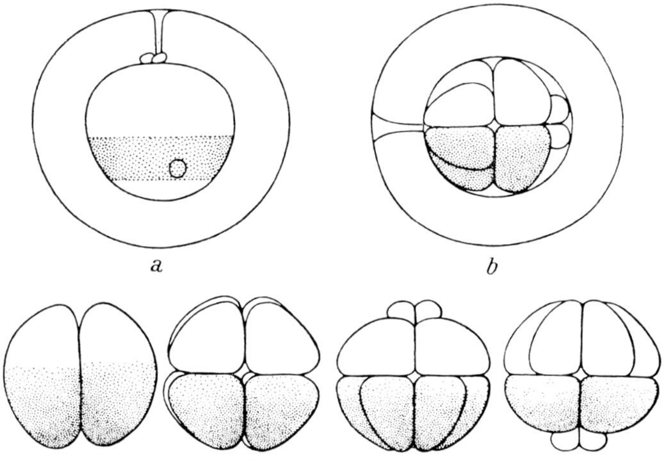

Fig. 52. Micromere formation in relation to the new stratification in centrifuged eggs: (a) normal egg with jelly-coat and the micropyle in which the two polar bodies lie; the nucleus is seen beyond the pigment ring; (b) 16-cell stage of a centrifuged egg, the centrifugal pole is below; from the position of the micropyle it can be seen that the micromeres cleaved at the opposite, the previous vegetal pole; (c) 2-cell stage of a centrifuged egg: the cleavage plane is perpendicular to the stratification; (d) 8-cell stage of a centrifuged egg; (e) 16-cell stage of a centrifuged egg in which the previous animal pole corresponds to the centrifugal pole; (f) 16-cell stage of a centrifuged egg in which the previous animal pole corresponds to the centripetal pole. (From Kühn 1965 after Boveri 1901(a); Morgan and Spooner 1909(b); and Morgan 1927 (c–f).)

By transferring eggs to diluted sea water or by shaking them, cleavage is retarded, but not the development or change of the unknown factors which influence the position of the spindle within the egg (Hörstadius 1928). It is therefore possible that eggs may reach the two-cell stage at the time when normal eggs already have four blastomeres (fig. 53, E2).

In normal cleavage the spindle of the third cleavage is oriented perpendicularly to the equator. This cleavage results in four animal and four vegetal cells. It would be expected that retarded eggs in the 2-cell stage would start the following cleavage with spindles parallel to the equator, as in normal cleavage, leading to four blastomeres. But they divide according to their 'timetable' (their age after fertilization) with spindles perpendicular to the equator, into two animal and two vegetal blastomeres (fig. 53, E2, 3). The orientation of the spindle is therefore not determined by the number of completed divisions but by the 'timetable' of development, probably a system of changing factors not influenced by dilution of the ionic environment. Similarly, after the next cleavage-step normal eggs have 4 micromeres among 12 other blastomeres, and shaken eggs also form micromeres, but only two among 6 other blastomeres. It seems that the vegetal cortex determined the spindle position of this cleavage in all cases. Links were found like those shown in row C in fig. 53, with oblique spindles, probably because the retardation of cleavage was not as long as a whole cleavage interval.

The cleavage pattern can also be altered by placing the eggs in sea water with 1–2% ether (Wilson 1902). Ether treatment impedes the development of the astral rays, but probably does not influence the formation and function of the spindle fibres, as anaphase movement is not stopped. Eggs transferred to normal sea water after the first nuclear division was completed without plasm division can develop to normal blastulae. The spindle axis of the second division is often perpendicular to the previous one. The blastomeres are arranged after a subsequent plasm division in a typical tetrahedron pattern (fig. 54c).

Nothing is known about the motive-forces causing the centrioles to move towards the future poles of the division figure. It may be (a) the stratification of the egg plasm, (b) the influence of the cortex or (c) the nuclear membrane and (d) still unknown factors as well. Costello (1961) drew attention to the fact that the axis of the daughter centriole is perpendicular to the axis of the parent centriole. The spindle orientation in cleavage with crossed spindles, as in turbellars or annelids or ether-treated sea urchin eggs, may be understood as being dependent on the genealogy of the centrioles. In Wilson's experiments ether treatment would have excluded the factors which are normally responsible for spindle orientation in sea urchin eggs, permitting the centrioles to take up their crossed position after the formation of the daughter centriole. After prolonged ether treatment the later cleavages also seem to be characterized by a crossed position of the spindle axes and absence of plasm division. Finally, many regularly distributed nuclei form a blastoderm, as in arthropode eggs (fig. 54d). Brought back to sea water the plasm finally cleaves and subsequently a nearly normal morula has often been observed. No attention has been paid to the occurrence of micromeres. Further development is very likely to be abnormal (Wilson 1902).

The cleavage of plasm can be prevented mechanically too, by pressure (Boveri) or shaking (Wilson). But in these cases the spindle orientation of the second cleavage is normal, i.e. the axes are parallel and not in the tetraeder position.

Separation of the centrioles can be effectively suppressed by shaking the

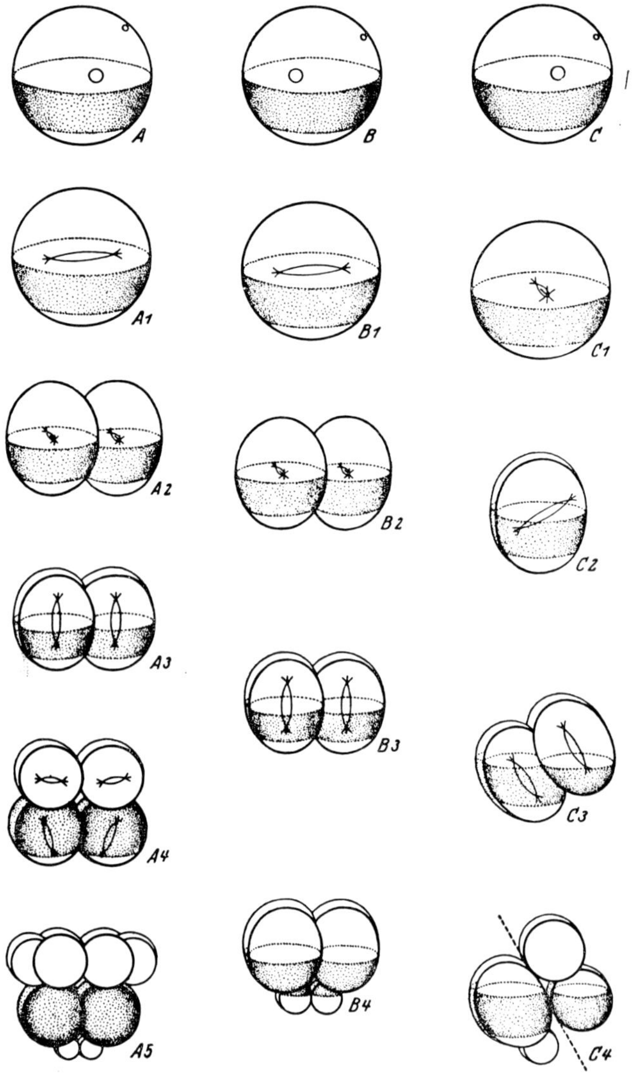

Fig. 53. Normal cleavage (A), retarded cleavage (B-E) and cleavage of meridional halves (F). B: Transferred to diluted sea water, the cleavage is more and more retarded. At the time of micromere cleavage these eggs have reached only the 4-cell stage (B_3) in which the micromeres are cleaving (B_4). The spindle orientation in B_3 is wrong! C: Another type of cleavage retardation by diluted sea water. With greater retardation the eggs reach the stage between A_2 and A_3 with two cells and their spindle orientation is a compromise between those of the A-series and is therefore oblique. Only two cells give off micromeres (C_4). The spindle orientation in the cells of the left side of C_3 is wrong! D: Similar cleavage

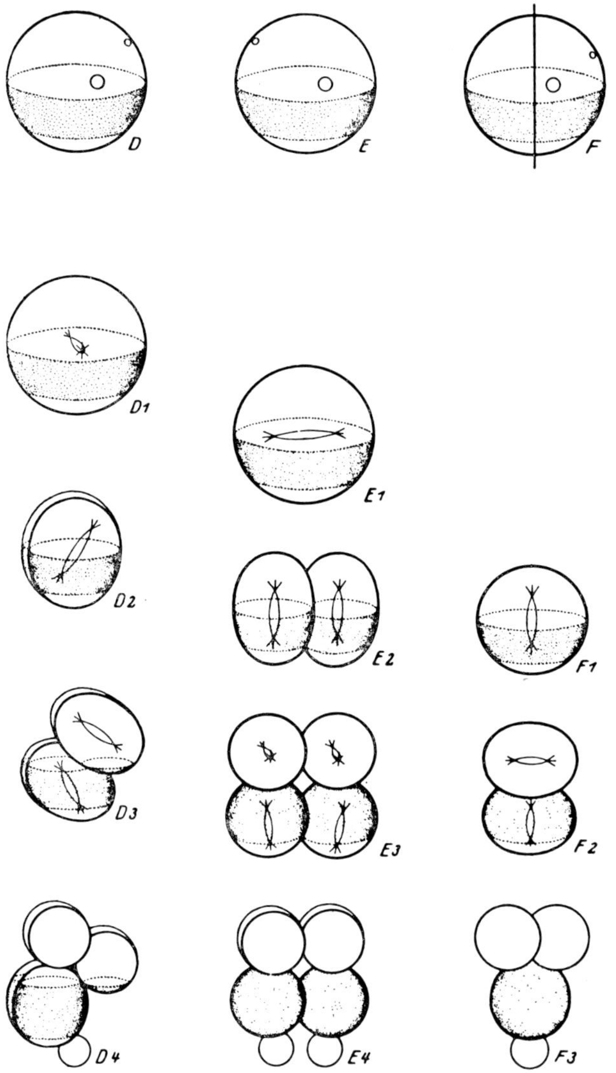

type in eggs which had been shaken vigorously. E: The same treatment can cause omission of the first cleavage and the sequence of subsequent divisions corresponds to the timetable of normal cleavage (compare B_4 and E_4). F: Cleavage in an egg which had been divided meridionally with a glass needle (vertical line in F). The cleavage type corresponds with that of a quarter, not a half of the egg. (After Hörstadius 1928.)

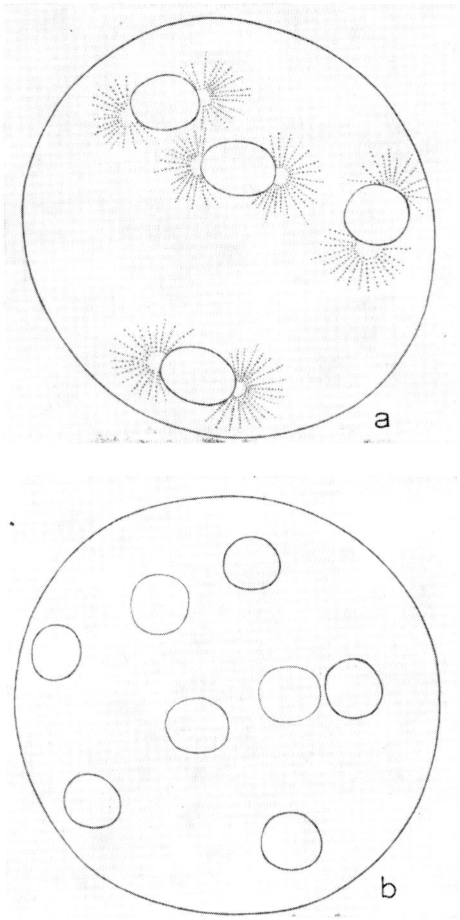

Fig. 54. (For the legend see p. 418.)

eggs vigorously 4 to 7 min after fertilization. A monaster develops, to which the spindle later becomes attached. No division occurs and the nucleus seems to be tetraploid. An amphiaster is frequently found in the next cleavage and the cleavage pattern corresponds to that of the controls (Boveri 1903). Painter continued this work in 1915 and found that the separation of centrioles or the formation of a daughter centriole can be blocked for different lengths of time, i.e. the monaster remains for different lengths of time in the shaken eggs. As soon as an amphiaster is formed, normal cleavage is resumed with spindle orientation corresponding to the 'timetable'. If for instance amphiaster formation is suppressed until the 4-cell stage in controls, in the first cleavage the spindle axis is perpendicular

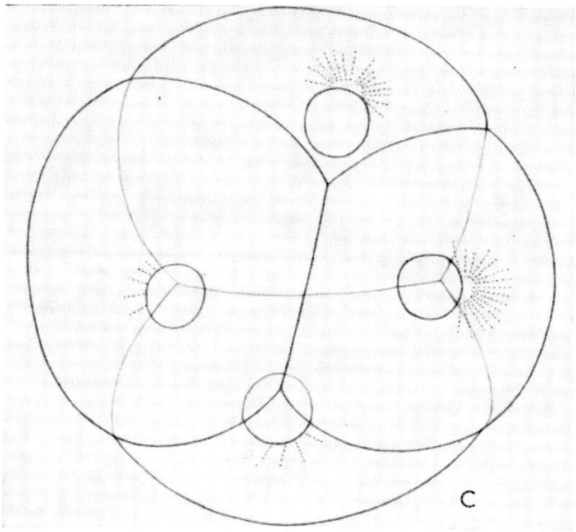

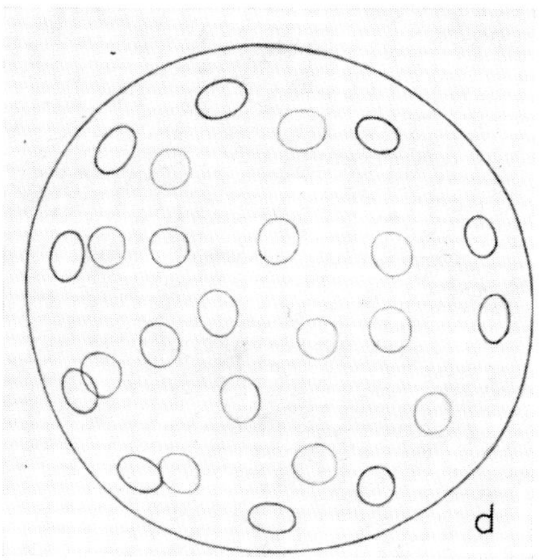

Fig. 54 *(continued)*. (For the legend see p. 418.)

to the equator and in the subsequent cleavage the micromeres cleave off. Painter obtained the same result by treatment with 0.008% phenylurethane, which impedes nuclear division. In this case the processes determining spindle orientation remain uninfluenced, as after suppression of centriole separation or division. Lowering the temperature or using KCN, which stops respiration, similarly suppresses all processes in the cell and after restoral of normal conditions, cleavage starts as in eggs without interruption of the sequences of mitosis.

The cleavage pattern of the echinid egg can also be altered by other means.

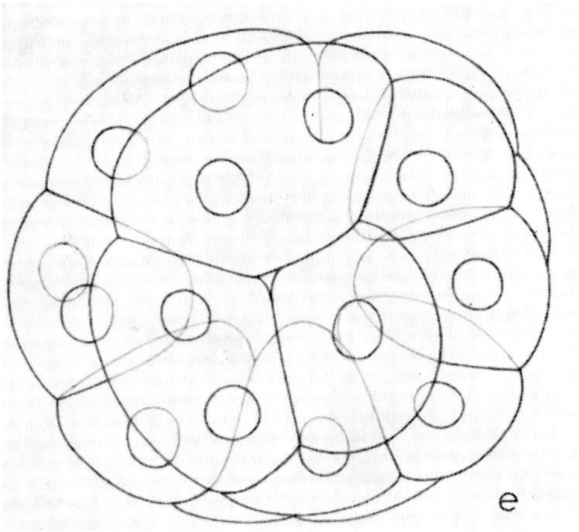

Fig. 54 *(continued)*. Cleavage of ether-treated eggs. As long as the eggs are in 2.5% ether in sea water, there is no fission of cytoplasm (a, b, d). After transfer into normal sea water the failure of plasm division is abolished (c, e); (a) and (c): Typical tetrahedron cleavage after crossed spindle axis; (e): transient binucleate blastomeres in an egg which had been placed into sea water after the 16-cell stage. (After Wilson 1902.)

The results are always the same: after treatment a spindle orientation which seems to be anticipated is found, like the separation of micromeres in the four-cell stage instead of the eight-cell stage. This was found to be the case after heat-shock and ultraviolet irradiation of eggs (Ikeda 1965) and actinomycin treatment (Czihak 1968) as well. Ikeda stated that the cleavage-rhythm is correlated with a cyclic change in the sulphydryl groups of water-soluble proteins: the greatest amount of SH-groups is found in metaphase. After ether treatment, heat-shock or UV irradiation cleavage resumes when the amount of SH-groups increases and is completed when it declines (fig. 55).

It is very probable that cytoplasmic processes are active during the first steps of development, independently of changes in the nucleus. The egg cortex also seems to be altered during the first hours of development insofar as the cortex seems to control the spindle orientation of micromere formation. Centrifugation experiments described in the previous chapter have shown that it is obviously the cortex which controls micromere formation. Centrifugation stratifies the cytoplasm, translocating all the cytoplasmic inclusions. The cortex, or parts of it, remains unchanged. And as even later centrifugation the micromeres always appear near to the vegetal pole, it must be assumed that the cortex is responsible for the location of the spindles leading to micromere cleavage. This is of course not valid for the position of the spindle in other cleavages.

'... at the time of fertilization progressive changes which go on independently of the nucleus and cleavage are initiated in the cytoplasm of the

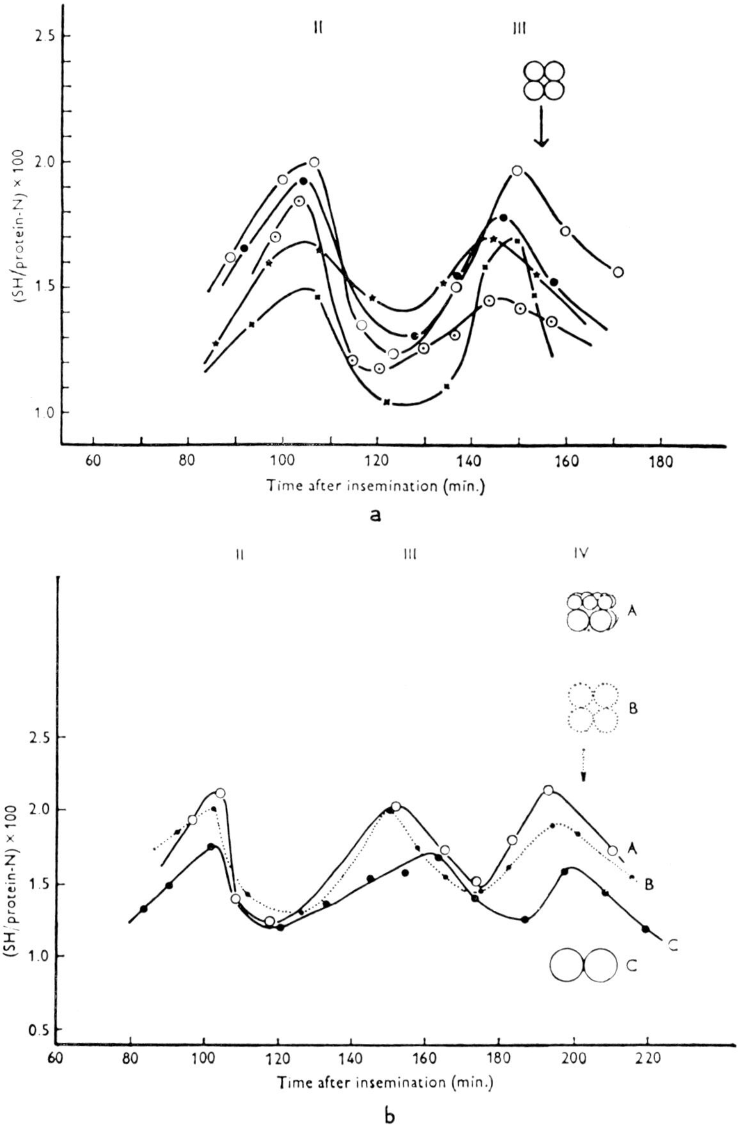

Fig. 55. Variations in the amount of SH-groups of water-soluble proteins in relation to cleavage. Cleavage is completed when the peak is transgressed (arrow) (a): open circles indicate the SH-cycles of controls, the other 5 UV-irradiated series, which had omitted one cleavage and divided into 4 (150–160 min p.f.) when controls underwent the third cleavage; (b) the same correspondence between the SH-cycles and cleavage is found if UV-irradiated eggs miss more than one cleavage, A: controls, B: omission of two cleavages, C: omission of three cleavages. (After Ikeda 1965.)

eggs and ... these changes determine the position of the spindles in the egg and consequently in its blastomeres' (Painter 1915).

12.6. Development of isolated blastomeres and blastomere groups

Chabry's method of observing the development of isolated ascidian blastomeres was used by Driesch for sea urchins and led to the famous publications of 1892 and 1893, which greatly influenced scientific thought in those days. Isolated blastomeres of the two- and four-cell stage (half- and quarter-embryos) cleave as in normal eggs. One blastomere of the four-cell stage, for instance, gives rise to 2 mesomeres, 1 macro- and 1 micromere (cf. fig. 53, F3). The further development of such parts was usually normal, i.e. normal plutei of smaller size arose. It was astonishing that halves of the 2- and 4-cell stage did not exhibit any defect on one side greater than the normal right-left differences. These results seem plausible today, as we have many reasons to assume that the above-mentioned halves and quarters con-

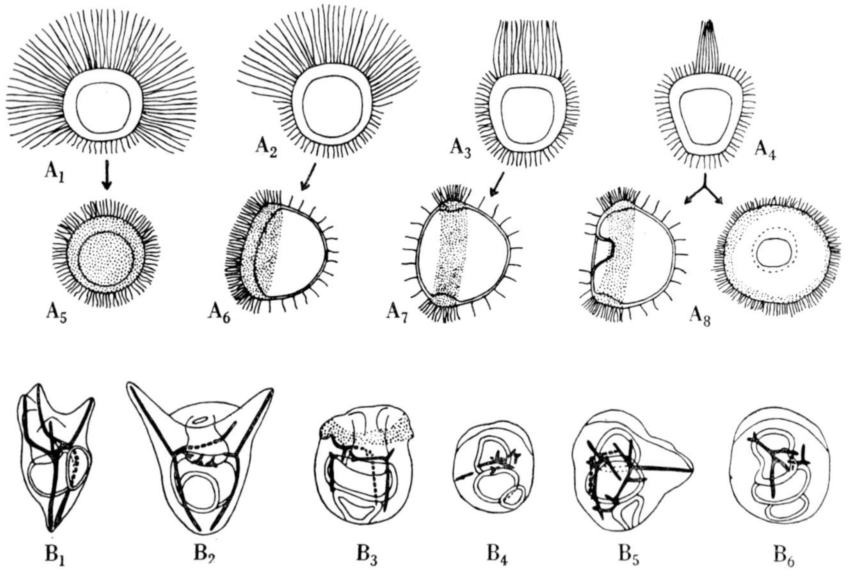

Fig. 56. A: development of isolated animal halves into strongly animalized larvae with enlarged ciliary tuft; rarely a mouth groove (A_8) is formed (cf. fig. 59e). B: development of vegetal halves into more or less normal (B_1) or vegetalized larvae (B_2–B_6). (After Hörstadius 1939.)

tain the same ratios of animal and vegetal cytoplasm and therefore also the same ratios of gene-derepressors, and therefore the qualitatively complete genetic information of the egg.

Driesch's experiments were completed by Hörstadius (1935), who isolated parts of embryos at later stages. The 4 animal blastomeres of an eight-cell stage, isolated with fine glass needles, develop to blastulae which cannot gastrulate and are therefore called 'dauerblastulae'. The ciliary tuft is much larger and a fraction is used to characterize the ratio between the surface of the embryo covered by stiff cilia and the surface covered by soft cilia, e.g. $\frac{3}{4}$ indicates that only $\frac{1}{4}$ is not covered by stiff cilia (cf. fig. 56 A1). Dauerblastulae (permanent blastulae) with moderate enlargement of the ciliary tuft will have a ciliary band around the oral field and in some cases a mouth groove too (fig. 56, A_8; fig. 57, B_3). The differentiation of the mouth is therefore independent of the presence of an intestine. In such larvae the characteristics of the normal animal half appear to be reinforced, and such larvae with a larger ciliary tuft and with or without a small intestine are therefore called animalized larvae.

Isolated vegetal halves develop to the forms shown in fig. 56, B_1–B_6. Here we find nearly normal plutei with regular skeleton and mouth opening or ovoid larvae with enlarged gut, called vegetalized larvae, often without mouth opening (fig. 56, B_4–B_6) and an irregular skeleton. Transitions between these two extreme groups are also found (fig. 56B_3).

Equatorially divided 16-cell stages develop like the equatorially divided 8-cell stages described above. In both cases the animal halves lack vegetal cytoplasm (and vice versa) and thus the gene-derepressors localized there which control the differentiation of entoderm and mesoderm as well.

Since in some cases vegetal halves develop to rather normal plutei, which animal halves never do, it can be assumed that animal gene-derepressors concentrated around the animal pole spread somewhat beyond the equator into the later macromeres.

The meridional halves of the 8- and 16-cell stages develop like the same halves of earlier cleavage stages. The missing left or right, ventral or dorsal parts of the larvae are substituted by regulation. Driesch introduced the famous designation 'harmonic equipotential system' to express that each part of the embryo contains all the potencies in the same amount and is capable of forming a harmonic larva. That this is not the case was shown by observing the development of animal halves, which cannot substitute vegetal organs and often cannot even differentiate their own anlagen such as that of the mouth opening.

Isolation experiments were resumed recently by Hagström and Lönning (1965), who reared isolated single blastomeres of the 16-cell stage. An isolated macro- or mesomere can develop into a small blastula. But an isolated micromere dies after two cleavages. Therefore there must be an exchange at least between macro- and micromeres. Perhaps the yolk-poor micromeres receive some nutrient substances from the macromeres. That there is exchange between all the blastomeres will be shown in the following chapter.

12.7. *Development of blastomere combinations*

In a series of splendidly executed experiments, which are among the most exciting in developmental physiology, Hörstadius combined different blastomere groups and observed their subsequent development (1935, reviewed 1939). The fertilization membrane was removed as described on p. 364 and the embryos transferred to Ca-free sea water 30 min before the operation, because Ca-free sea water loosens the connections between the blastomeres (this is not necessary with Psammechinus). The blastomeres were afterward separated with finest Spemann's glass needles, transferred to normal sea water and combined with others by putting two blastomere groups together in a groove scratched into cellophane and laying a small glass sphere over them to prevent dislocation.

Animal halves isolated from the 32-cell stage develop like isolated mesomeres to permanent blastulae with enlarged ciliary tuft (fig. 57A). By the 32-cell stage the macromeres have divided into two garlands of vegetal cells (fig. 49F, veg_1 and veg_2). If animal halves of the 32-cell stage are combined with the veg_1-garland, enlargement of the ciliary tuft is suppressed, the larvae are less animalized and always have a ciliary band surrounding the oral field and a mouth-anlage (fig. 57, B_1–B_3). As has been said in the previous chapter, animalization means overdevelopment of animal properties such as the ciliary tuft, and vegetalization overdevelopment of the archenteron. By the combination of animal cells with vegetal ones the tendency towards animal self-differentiation is repressed. This is even more clearly seen in the following experiment in which animal halves were combined with the veg_2-cell group (fig. 57D). This combination is able to develop into a normal pluteus with slightly reduced skeleton. Thus the tendency towards animal self-differentiation of the animal half must have been even more repressed. We must assume that there exists an exchange of substances or an effect of unknown factors between the blastomeres. These substances

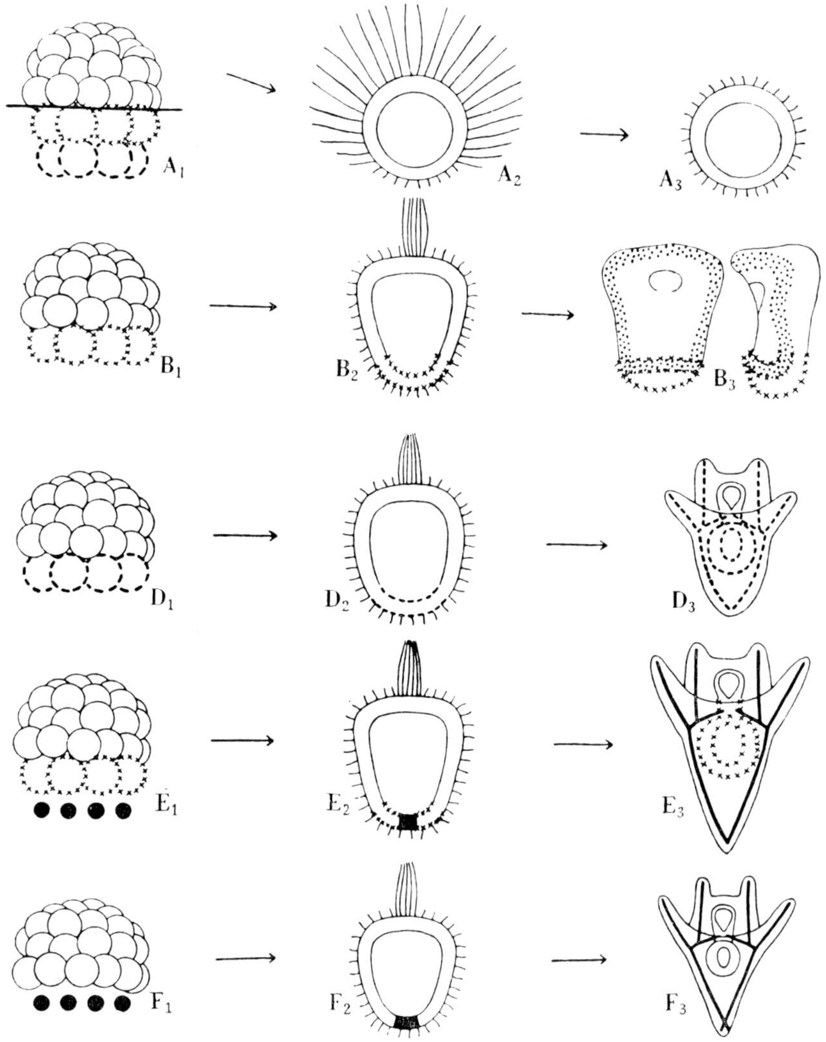

Fig. 57. Development of animal halves (A_1) combined with different blastomeres of the vegetal half (B–F); A: animal half alone; B: $an_1+an_2+veg_1$; D: $an_1+an_2+veg_2$; E: $an_1+an_2+veg_1+$micromeres; F: an_1+an_2+micromeres. (After Hörstadius 1939.)

or factors seem to be in higher concentration in the veg_2-cells than in the veg_1-cells. If animal halves are combined with veg_1-cells and micromeres (fig. 57E), normal plutei develop; animal halves combined with micromeres alone (fig. 57F), however, grow to normal plutei with a smaller archenteron.

The ability to repress animal differentiation is therefore even more pronounced in micromeres than in veg_1- or veg_2-cells: hence this ability increases in a gradient towards the vegetal pole!

Furthermore, these experiments showed that single blastomere garlands develop differently in different blastomere combinations. From the veg_2-cells in contact with animal halves not only the archenteron is formed but also the primary mesenchyme is differentiating into skeleton-forming cells. And in the combination of animal halves and micromeres a part of the animal cells which normally develop into body wall differentiate into entoderm. This differentiation is therefore considered to be induced by the micromeres, which themselves normally develop into primary mesenchyme and finally skeleton-forming cells.

The animal substances or factors determining animal differentiation are also distributed in a gradient. This is evidenced by the following experiments in which different numbers of micromeres were combined with single garlands of blastomeres from the animal or vegetal area of the 32-cell stage.

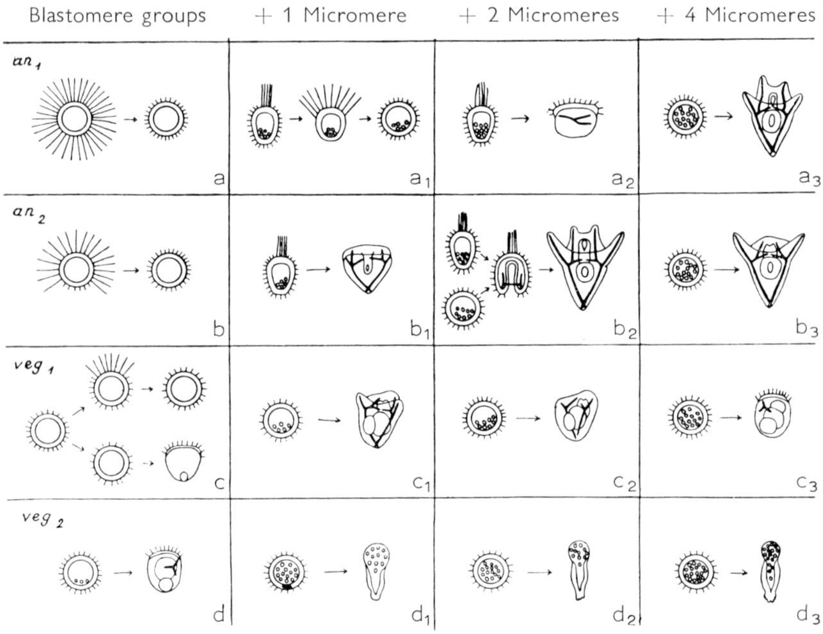

Fig. 58. Development of isolated blastomere groups of the 64-cell stage (a–d) in combination with 1, 2 or 4 micromeres (a_1–b_3). a_3 and b_2 are rather normal, a_1,a_2,b_1 are animalized, the others are vegetalized. (After Hörstadius 1935.)

The an_1-cells alone develop to a permanent blastula with a $\frac{4}{4}$ ciliary tuft (fig. 58a). One micromere combined with an_1-cells is already able to repress animal tendencies: this is seen in the smaller ciliary tuft, which attains a maximum of $\frac{1}{3}$ (fig. $58a_1$). On the other hand the influence of the an_1-cells is so strong that one micromere alone cannot develop primary mesenchyme in this combination. Animal substances or factors thus also repress vegetal tendencies! Only in the combination of an_1-cells with 4 micromeres are these tendencies (factors, substances) balanced so that a normal pluteus develops (fig. $58a_3$).

The animal tendencies of the an_2-cell garland are less pronounced. This is seen just by observing the differentiation of the isolate, which grows to a blastula with only $\frac{3}{4}$ ciliary tuft (fig. 58b). And two micromeres combined with an_2-cells are alone sufficient to balance animal and vegetal factors leading to the development of a normal pluteus (fig. $58b_2$). In the combination of an_2-cells with 4 micromeres slightly vegetalized plutei – the oral arms are reduced – develop (fig. $58b_3$).

The group of veg_1-cells develops to permanent blastulae with $\frac{1}{4}$ ciliary tuft or to gastrulae with a tiny archenteron (fig. 58c). In normal development and if reared in isolation, veg_1 behave like animal cells. The designation 'veg' is therefore misleading! Combination with one micromere is sufficient to produce slightly vegetalized larvae. Veg_1-cells combined with 2 or more micromeres will develop to strongly vegetalized larvae (fig. $58c_2$, c_3). The group of veg_2-cells alone differentiates to a strongly vegetalized larva and in combination with micromeres to highly vegetalized monsters (fig. 58d). The stretched larvae are finally composed of almost nothing but entoderm bearing a very small sphere which can perhaps be considered to be or to contain ectoderm. The veg_2-cells obviously do not contain animal factors.

By this series of brilliant experiments Runnström's ingenious idea of the double gradient system became valid: he postulated that two gradients control the first differentiations in sea urchin embryos. Runnström's hypothesis (1929) proposed that animal and animalizing and vegetal and vegetalizing factors are distributed in the manner of a gradient with maximum activity at the animal or vegetal pole (fig. 59a). This model proved to be of great value for the understanding of combination experiments. The ability of veg_1-cells to differentiate into animal ones, for instance, is so poor, that one micromere is already sufficient to restore the normal balance necessary for differentiation into a normal larva. The oral arms are reduced in larvae grown from veg_1-cells combined with one micromere: the animal factors

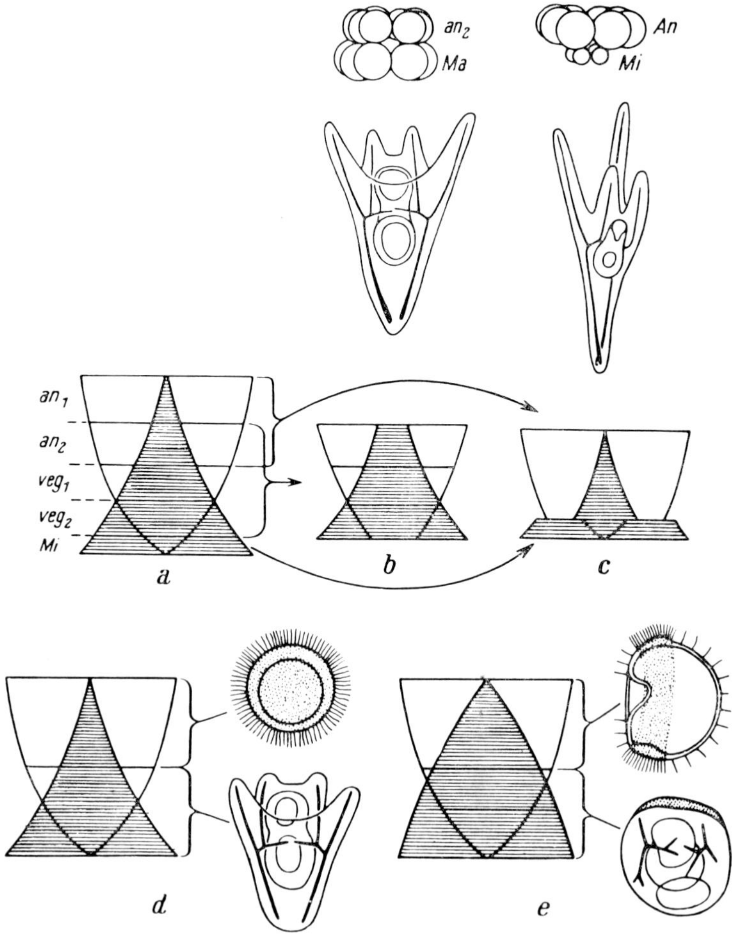

Fig. 59. Schematic representation of the animal and vegetal gradient system (a). In an_1 for instance the animal gradient (blank) is strongest and the vegetal (shaded) weakest; the different garlands of blastomeres contain different mixing ratios. If an_1 and the micromeres are removed (b) the ratio of the gradients is the same as in normal eggs and a normal pluteus can develop. A rather normal pluteus can also develop if the two maxima are combined by putting an_1, an_2 and the micromeres together. (c): It may therefore be assumed that it is not the composition of certain blastomeres which is essential for normal development, but the ratio between the two gradients represented by different blastomeres. Individual variations can be explained by casual differences in the gradient system: in (e) the vegetal gradient is stronger than in (d), so the cases in which vegetal halves develop to slightly vegetalized larvae (e, below) and animal halves to dauerblastulae with a mouth groove (e, above) can be explained in contrast to the 'normal' gradient expression with which the animal halves develop into completely ciliated dauerblastulae (d, above) and the vegetal ones into normal plutei (d, below). Ma: macromeres, Mi: micromeres. (From Kühn 1965 after Von Ubisch 1936; and Hörstadius 1936.)

of veg_1 (supporting weak animalization in a $\frac{1}{4}$ dauerblastula) are so weak that they are overwhelmed by the strong vegetal forces of a single micromere. It is clear that vegetalization will become stronger if veg_1-cells are combined with two or more micromeres. The skeleton becomes more and more irregular, because the normal arrangement is dependent on normal, undisturbed ectoderm.

The group of isolated veg_2-cells of course develop into vegetalized larvae. According to the scheme (fig. 59a) they contain much more of the vegetal factor(s) than of the animal ones. The two factors are balanced not only in normal eggs but also when the two maxima are removed, as in the combination $an_2 + veg_1 + veg_2$ (fig. 59b) and also if the two maxima (an_1 + micromeres) are combined (cf. fig. 59c). Therefore – so we conclude – these combinations develop almost normally.

Variations in the results of combination experiments are often observed. They may be understood by assuming that in single individuals one gradient is stronger than the other (fig. 59d, e).

The inductive property of the micromeres has been briefly mentioned; it can be demonstrated in combination experiments. Hörstadius (1935) implanted isolated vitally stained micromeres into different regions of complete 32-cell stages. The implanted micromeres differentiated like those of the host into primary mesenchyme (figs. 60–62, black cells). But next to the normally situated archenteron of the host an additional archenteron was found at the place where the micromeres have previously been implanted;

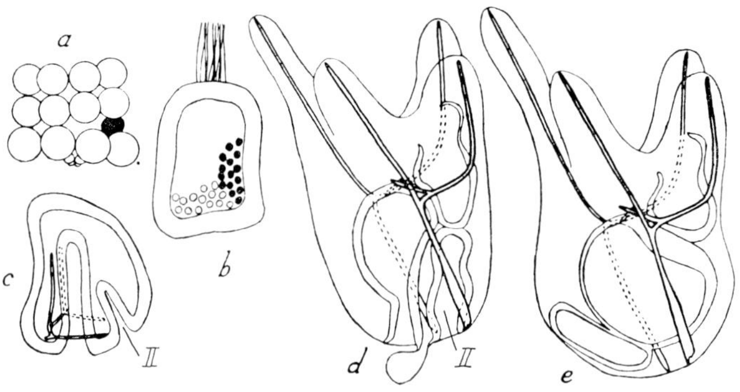

Fig. 60. Transplantation of micromeres (black) between animal and vegetal cells of a 32-cell stage (a). The implanted micromeres immigrate like those of the host into the blastocoel (b) and induce a secondary archenteron (c) which later fuses with that of the host (d,e). II: Secondary archenteron. (After Hörstadius 1935.)

the two archenterons fused to one if they were lying close together. This happened when the micromeres had been implanted at the equator (fig. 60). But the two archenterons remained separate if the micromeres were implanted between the an_1- and an_2-garlands (fig. 61). In these cases the im-

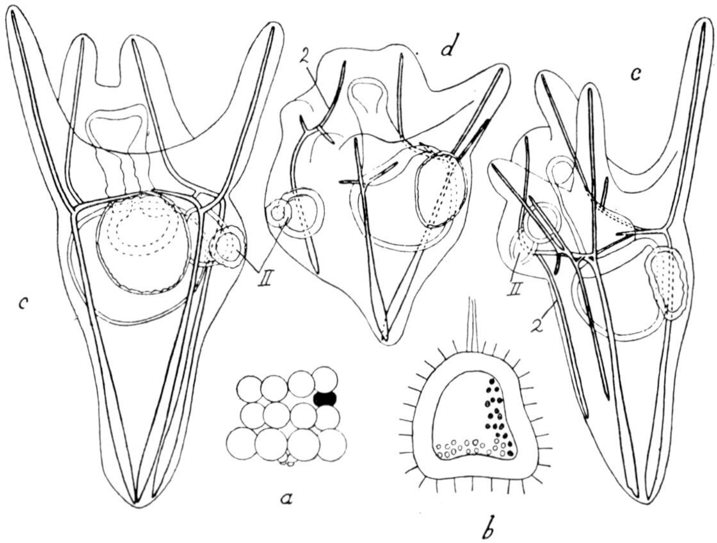

Fig. 61. Transplantation of micromeres between an_1 and an_2 (a). Parts of a secondary skeleton are formed (2), the induced secondary archenteron (II) is much smaller than that in fig. 60 and does not fuse with that of the host (c–e). (After Hörstadius 1935.)

planted micromeres were also able to give rise to an additional separate skeleton. When they were implanted into the animal pole (fig. 62) an additional skeleton was later found, but the supernumerary archenteron was very tiny. This can be understood in view of the strong counteraction of animal factors in this region. In any case, the micromeres, which themselves differentiate to skeletoblasts, brought about the differentiation of an archenteron in the neighbour-cells by induction. This is a classic example of induction, i.e. the triggering of a differentiation by adjacent inductor-cells which themselves do not differentiate in the same way. Once again it is obvious that there must be some form of exchange between the blastomeres.

It is not yet clear whether the vegetal and archenteron-inducing factors are identical.

In this connexion some modern experiments claim our greatest interest. Dissociated cells (obtained by shaking embryos with fertilization membrane

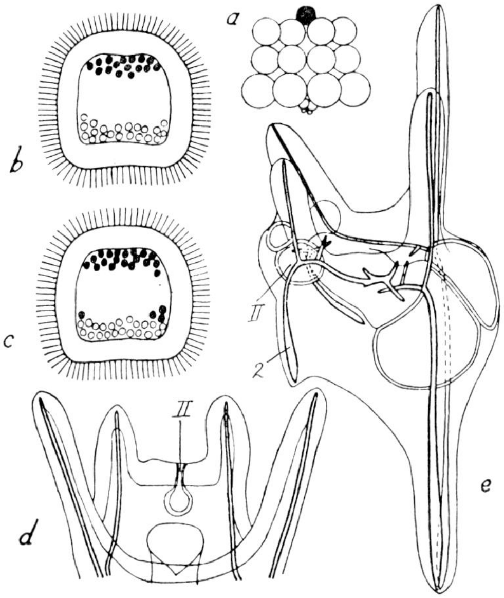

Fig. 62. Micromeres implanted into the animal pole (a) later produce a secondary skeleton (2, e) and induce a very small secondary intestine (II) (d,e). The counteraction of the animal gradient against the induction capacity of the implanted micromeres is clearly seen in figs. 60–62. (After Hörstadius 1935.)

removed in Ca-free sea water) are capable of reaggregation like sponge cells. Giudice and Just have performed such experiments (1962) with gastrulae and blastulae. The isolated cells rounded off and after transfer into normal sea water they at first formed small aggregates. When the suspension was slightly agitated, the aggregates became larger within a short time. The surface of these aggregates flattened and soon showed the normal junction between them, and developed cilia characteristic for ectoderm cells. According to studies with the electron microscope, the inner cells of the aggregates are able to form a new archenteron (Millonig and Giudice 1967). The further development of aggregation-gastrulae was observed until the pluteus stage in some cases. It is very probable that only aggregates which by chance have approximately the normal amount of ecto- and entoderm cells are capable of developing into a pluteus. It cannot be stated whether aggregates of cells dedifferentiate and differentiate de novo or whether they conserve their degree of differentiation throughout aggregation. Blastula cells isolated from animalized or vegetalized larvae develop after aggregation, if at all, into animalized or vegetalized larvae (Giudice 1963).

12.8. The different gradients

It was demonstrated in the previous chapter that combination of different blastomere groups can alter the development towards animalization or vegetalization. It was further stated that Runnström's double gradient hypothesis proved to be of great value in explaining animalization or vegetalization.

Gradients are properties of the embryo which are not uniformly distributed but gradually diminish or increase in one direction; therefore gradients have a centre of greatest distinction. The centres of the two gradients postulated to control animal and vegetal differentiation in sea urchins, coincide with the animal (sometimes marked by the polar bodies) and vegetal poles respectively.

What can these postulated gradients be? We still do not know, but the general opinion is of an unequal, gradually changing distribution of substances in the egg or a change in the physical status of these. These substances are thought to have the possibility of evoking specific gene actions leading to specific differentiations (see p. 447). That these events can be altered experimentally will be demonstrated in the chapter on animalization and vegetalization (see p. 440).

Child (1946), succeeded in detecting two metabolic gradients in sea urchin and starfish embryos along the animal-vegetal axis. He brought the hypothesis of the chemical nature of the double gradient system from the realm of pure speculation to the threshold of experimental analysis. But these metabolic gradients are rather the expression of the activity of other gradients, because they make their debut later in development. Child found that methylene blue and Janus green, used as vital dyes, were reduced by lack of oxygene in the blastula. The progress of bleaching was taken as the expression of a gradient action. In a similar study Hörstadius (1952) completed our knowledge by combining observations of the reduction of Janus green with transplantation experiments. Confirming Child's results, he found that in the mesenchyme blastula the reduction starts at the vegetal pole and proceeds to the equator; then a new gradient, starting from the animal pole, becomes visible (fig. 63A_4). The same was observed in the gastrula stage (fig. 63B). Isolated vegetal halves behave very similarly. It is not surprising that a remnant of the gradient starting from the animal pole was found, as many vegetal halves are capable of normal development (fig. 64, cf. fig. 59d). In animal halves only one gradient is active (fig. 65): we have seen in fig. 56 that animal halves can never develop beyond the

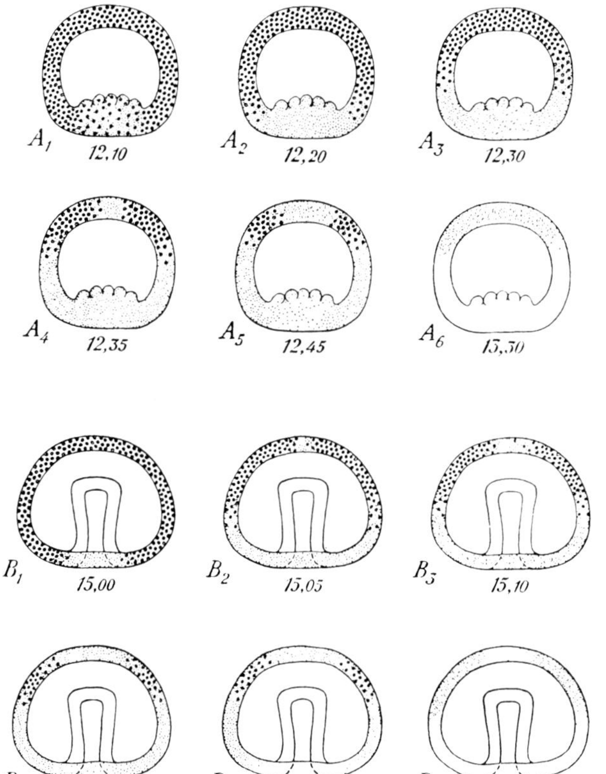

Fig. 63. Reduction gradients of the vital dye Janus green as a function of time in the blastula (A) and gastrula (B); the numerals indicate time. Reduction of the dye from blue-green to red starts at the vegetal pole (A_1, B_1). Later a second reduction gradient starts in the animal pole (A_4,B_2). Finally the dye becomes bleached (A_6,B_6). The archenteron probably remained unstained. Large dots: green; small dots: red; no dots: colourless. (After Hörstadius 1952.)

blastula stage. Most interesting is the fact that the implantation of micromeres is followed by the establishment of an additional reduction centre (fig. 66C_1, arrow). Similarly the implantation of micromeres into animal halves creates the reduction gradient starting from the vegetal pole: together with the animal gradient a normal double gradient system capable of normal development, as fig. 57F on p. 423 demonstrates, is established. Blastulae vegetalized by the action of lithium (p. 440 exhibit no animal reduction gradient, and in blastulae treated with trypsin for animalization only a very weak reduction gradient starting from the vegetal pole was seen

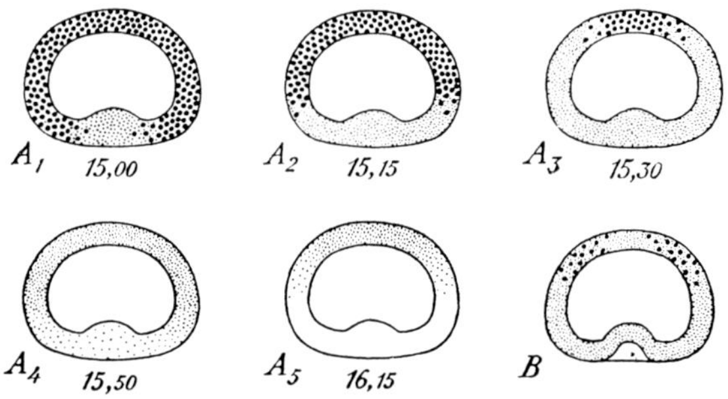

Fig. 64. Reduction gradient in isolated vegetal halves (cf. fig. 63). In most cases (A) only the vegetal gradient is found. Sometimes a weak animal gradient also occurs (cf. fig. 59d). (After Hörstadius 1952.)

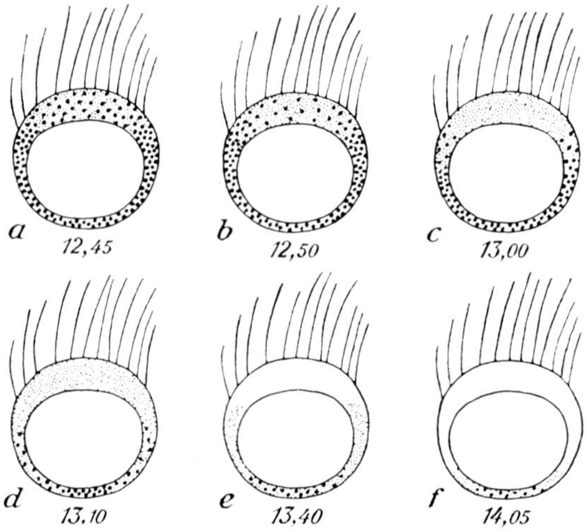

Fig. 65. Animal reduction gradient in isolated animal halves (cf. fig. 63). There is no vegetal gradient. (From Hörstadius 1952.)

by Hörstadius (1955). These experiments are beautiful supplements to his own work on the development of blastomere combinations in the light of the double gradient hypothesis of Runnström.

Another gradient was found by Child (1941), using the Nadi reaction for cytochrome oxidase activity. 'In the gastrula and probably earlier a ventro-dorsal gradation in reaction also becomes evident with ventral region in advance'. A difference in cytochrome oxidase activity is, however, recognizable even earlier (Czihak 1963). Already in the eight-cell stage a dorso-

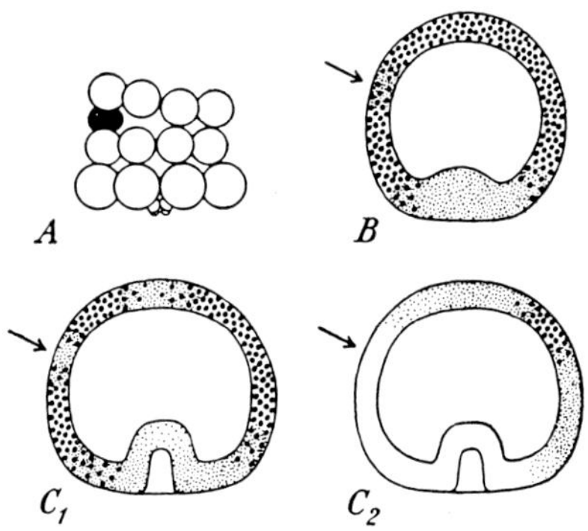

Fig. 66. The implantation of micromeres (black) between animal blastomeres (A) creates a new centre of a reduction gradient (arrows). In C_1 there are therefore three gradient centres. (After Hörstadius 1952.)

ventral gradient in cytochrome oxidase activity was found. In the mesenchyme blastula one side of the animal half (fig. 67) exhibits higher enzyme action as demonstrated by more intense dye production. Lining up the

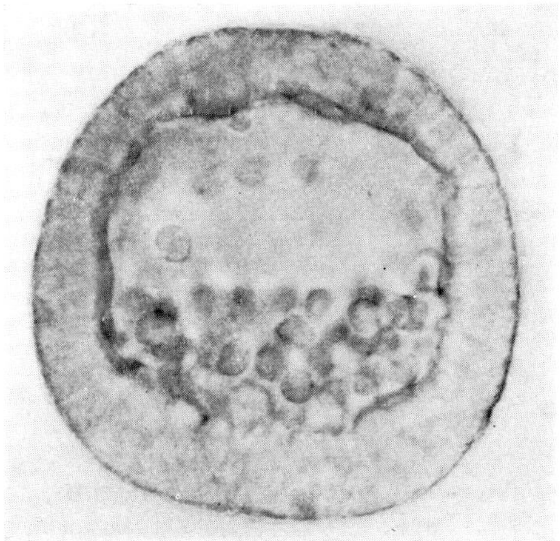

Fig. 67. Nadi reaction in the mesenchyme blastula. One side, the future oral field, is stained more strongly, demonstrating that the activity of cytochrome oxidase is highest in the oral field.

findings from the cleavage stages leading to the gastrula we may suppose that it is the future oral field (recognizable by its flattening in the gastrula), which always has a higher cytochrome oxidase activity! The oral field obviously needs a high level of respiration for the energy supply of its specific differentiation. This assumption is supported by the brilliant work of Pease (1941). Eggs of *Dendraster excentricus* in the eight-cell stage were subjected to KCN (to block cytochrome oxidase activity) for one to three hours in a diffusion gradient. Thus one part of the egg was exposed to higher KCN concentrations than the other and simultaneously vitally stained: '... In the majority of eggs the side least affected by the KCN (the unstained side) becomes the ventral region of the embryo'.

The position of the gradient of cytochrome oxidase activity within the embryo is shown schematically in fig. 68. Its location may be determined

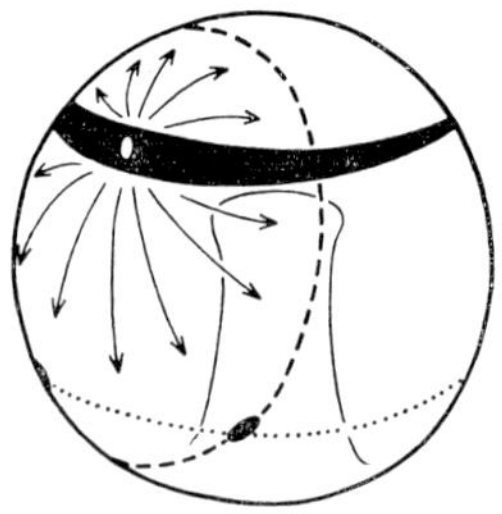

Fig. 68. Schematic representation of the oral field (cytochrome oxidase) gradient (black, centre white) which determines the position of the oral field, bilateral symmetry and position of the skeleton. Differentiation of the oral field is spreading from the centre over a quarter of the embryo's surface. The ciliary band develops on the border line (dashed) and the triradiate spicules are formed where this line crosses the circle of primary mesenchyme cells (dotted). (From Czihak 1962b.)

by a certain level in the animal-vegetal double gradient system (presented schematically in fig. 59). The centre of the oral field differentiates at the site of maximum respiration. It can be assumed that something spreads from the centre, as the behaviour of the primary mesenchyme cells (Czihak 1962b) and the structure of the cells of the oral field suggest (Czihak and Meyer 1964). A ciliary band differentiates at the border line of the oral field (fig. 68), and where the oral field comes into contact with the garland of primary mesenchyme cells, which lie at a defined level of the animal-vegetal double gradient, the two points of the attachment zones and the skeleton origin become fixed. This was concluded from the relationship between the oral field and the position of the primary mesenchyme-cell circle and their sizes in different experiments (cf. Czihak 1962b).

12.9. Symmetry

That echinoderms change such a basic property as body-symmetry during the individual life cycle was a miraculous fact for the early morphologists.

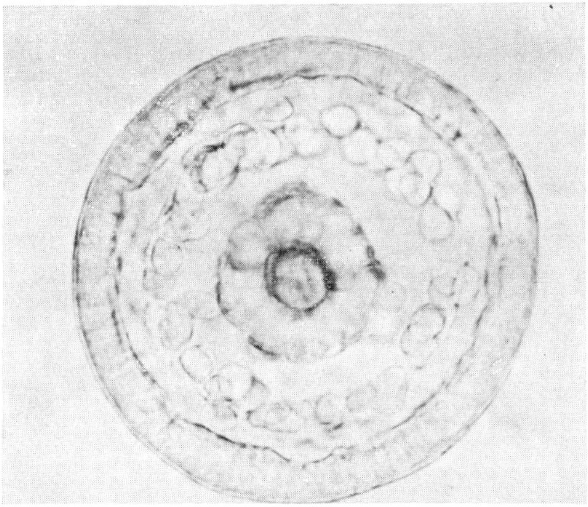

Fig. 69a. Radially symmetric larva of Psammechinus 14 hr p.f., which had been treated with 10^{-5} mol canavanine sulphate between 2 hr, 15 min and 3 hr, 15 min. The primary mesenchyme cells are rather uniformly distributed; there are no attachment zones or lateral groups of skeleton-forming cells.

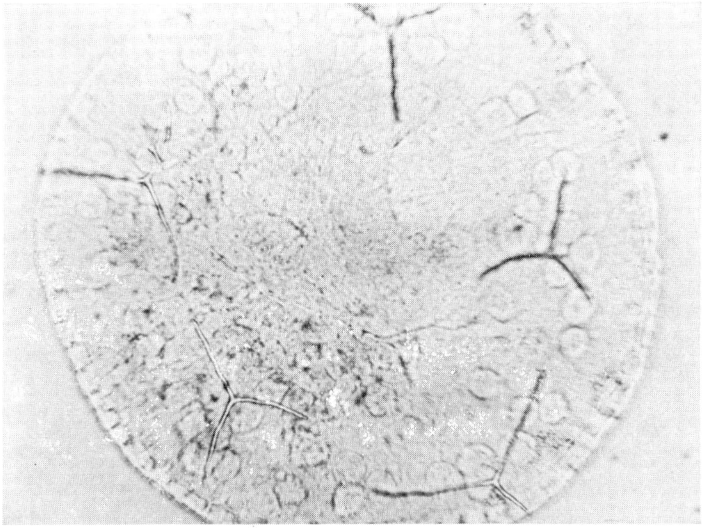

Fig. 69b. Radially symmetric larva with 5 equally developed triradiate spicules.

Now we seem to know one reason for the temporary occurrence of bilateral symmetry in the larvae: the above-mentioned oral field gradient of cytochrome oxidase activity, which could be the primer for the ground plan of the larva's body shape. If it is so, we can expect that any disturbance of the oral field gradient will influence the bilateral symmetry. In fact many cases of radial symmetry, including the suppression of bilateral symmetry, have been described, especially by Lallier in his investigations on animalization. The most completely radial larvae obtained by LiCl treatment were

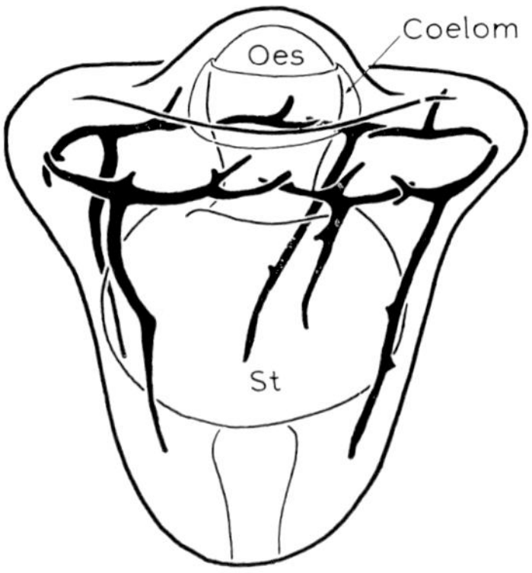

Fig. 70. Radially symmetric larva of Paracentrotus, 3.5 days p.f., which was treated 1.5–21 hr p.f. with 0.06% LiCl. Note the coelom ring and the spicules in a circular arrangement. There is still no mouth opening. (After Czihak 1962b.)

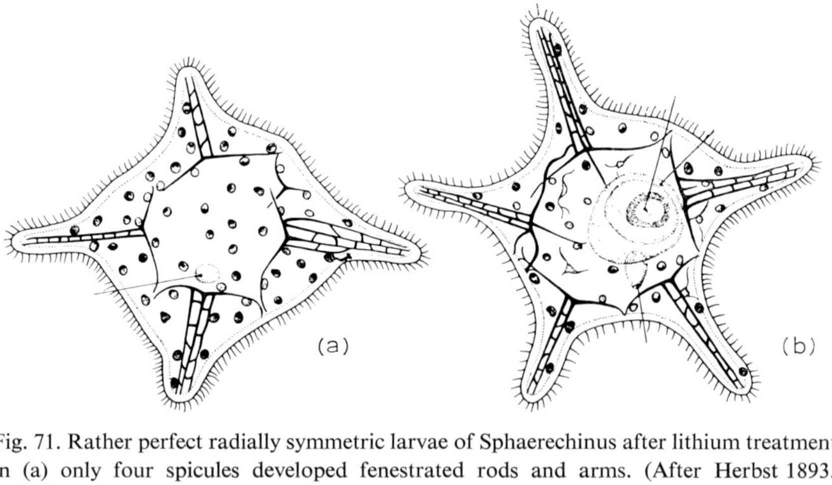

Fig. 71. Rather perfect radially symmetric larvae of Sphaerechinus after lithium treatment. In (a) only four spicules developed fenestrated rods and arms. (After Herbst 1893.)

described by Herbst (1893) and Czihak (1960b, 1962b). In these larvae the intestine is straight, the mouth opening at the animal pole or just below, the primary skeleton spicules in a larger number, often five, and equally developed (fig. 69). The coelom is found first as a terminal sac at the tip of the archenteron, then as a ring (fig. 70) around the oesophagus! Herbst's larvae had beautifully developed arms (fig. 71). With sodium lauryl sulphate it is possible to obtain intermediate forms between radial and bilateral symmetry (fig. 72).

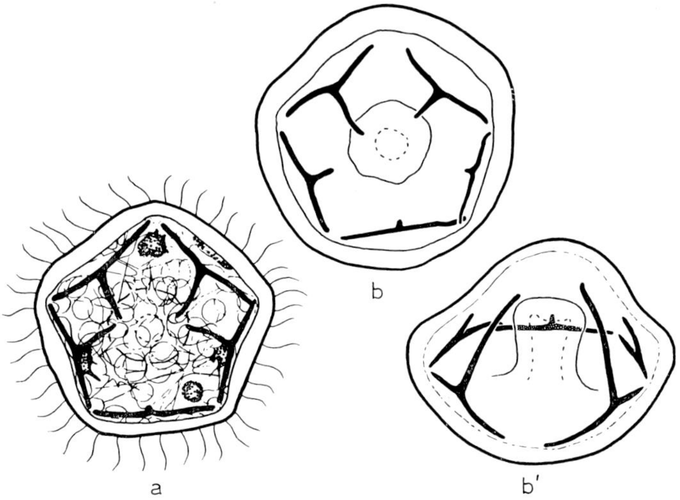

Fig. 72. Radially symmetric and intermediate forms of Paracentrotus, 3 days p.f. after treatment with sodium lauryl sulphate. In (a) a slight suggestion of bilateral symmetry is visible as one spicule is less developed. This is more pronounced in (b). (After Czihak 1962b.)

Gustafson and Sävhagen (1950) studied the moment of efficacy of this detergent during development and found full radialization after treatment of eggs aged up to 6 hr, less radialization between 6 and 9 hr and no effect at all with treatment after the age of 9 hr post fertilization (p.f.). Using 8-chloroxanthine Hörstadius and Gustafson (1954) also found 6 hr p.f. the critical time after which bilateral symmetry is determined (fig. 73).

In larvae radialized with 8-chloroxanthine the oral field gradient of cytochrome oxidase activity is abolished (fig. 74). Although in only one case the oral field gradient was examined after treatment with radializing agents (Czihak 1962b), we may assume that the presence of this gradient determines

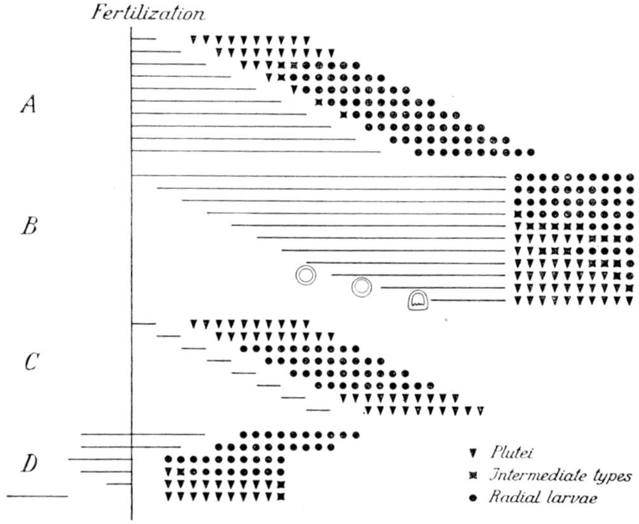

Fig. 73. Effective periods of 8-chloroxanthine for radialization. The symbols indicate percentage of normal, radialized and intermediate forms; the shortest horizontal lines correspond to one hour. Hourly treatment (C) is effective only between 3 and 6 hr p.f. Treatment before fertilization is also found to be effective (D). (After Hörstadius and Gustafson 1954.)

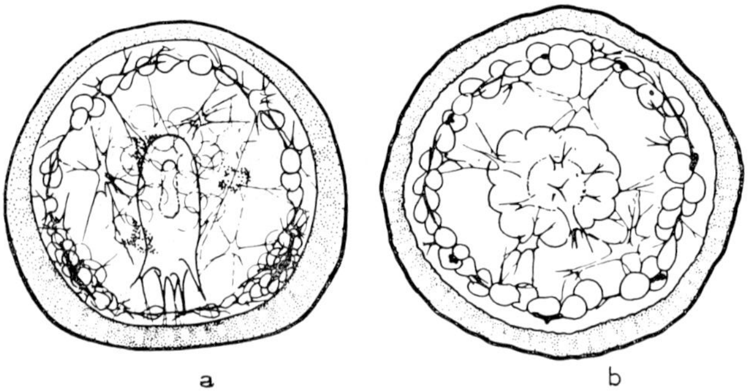

Fig. 74. Demonstration of the cytochrome oxidase gradient by vital staining with Janus green B in normal (a) and radially symmetric gastrulae (b) seen from the apical pole. There are two spicules in (a) and many in (b). (From Czihak 1962.)

bilateral symmetry. Bilateral symmetry is first distinctly marked in the ectoderm, which determines the symmetry of the other germ layers. Only one argument supporting this statement shall be mentioned here: micromeres of one species were transferred into the blastocoel of another species, of which the micromeres had been previously removed. The implanted micromeres produced a species-specific skeleton, for instance fenestrated rods, which in the host blastula otherwise never occur. But the position of the

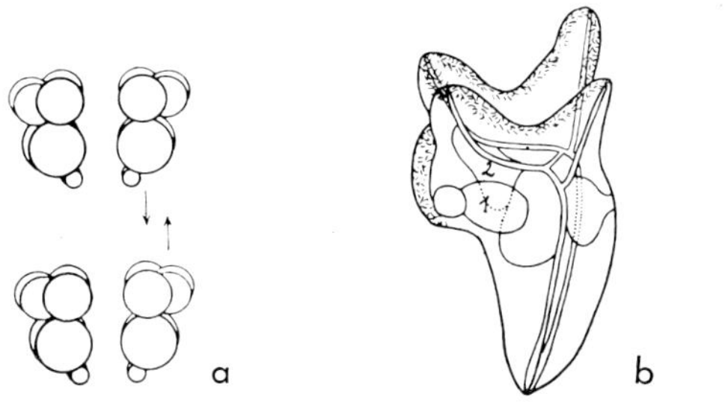

Fig. 75. (a) Meridian halves of two 16-cell stages were exchanged: 42 normal plutei developed, only 1 (b) had two oesophagi. (After Hörstadius 1957.)

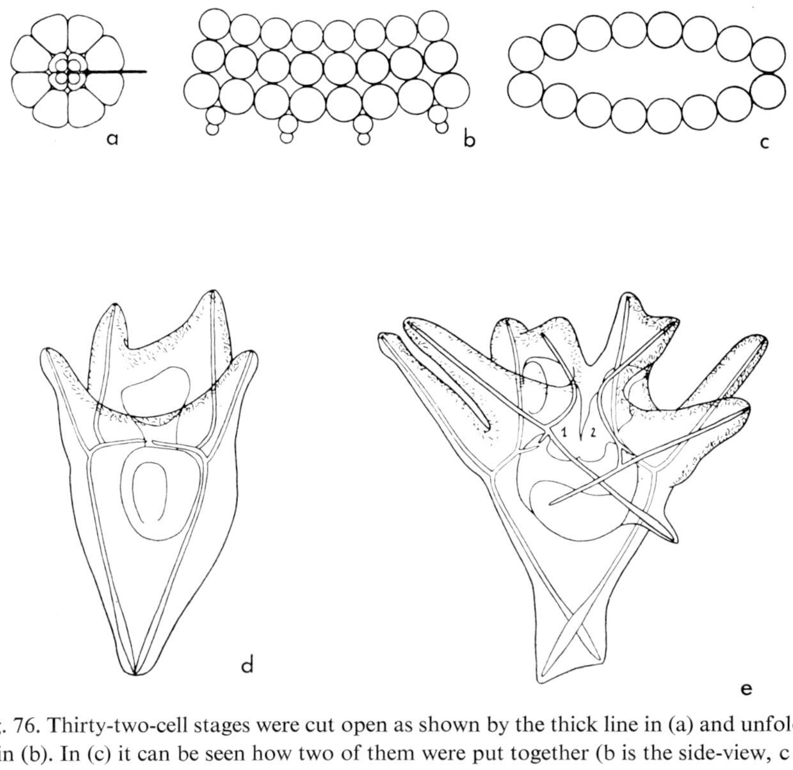

Fig. 76. Thirty-two-cell stages were cut open as shown by the thick line in (a) and unfolded as in (b). In (c) it can be seen how two of them were put together (b is the side-view, c the top-view). There developed giant plutei (d) and twin forms (e). (After Hörstadius 1957.)

arms with extraneous skeletal rods and thus the symmetry was determined by the ectoderm of the host blastula (Von Ubisch 1939).

In this connection some other combination-experiments of Hörstadius (1957) must be mentioned. Two meridional halves of the 16-cell stage were combined (fig. 75) or two complete, opened and stretched 16-cell stages (fig. 76) were united to one embryo. Thirty-six combined halves developed to normal plutei, with the exception of one pluteus which had two oesophagi (indicating the existence of two oral field centres).

Double embryos consisting of two complete eggs developed either to a normal giant pluteus or to Siamese twins (fig. 75) having two oral fields. The oral field gradient was cut at random when separating two meridian halves in the 16-cell stage. In the first case with combined halves the two combined oral field gradients united to one uniform pluteus, but in the larger embryo the two complete gradients present may remain separated, giving rise to two separate oral fields and two bilateral symmetric larvae united on the dorsal side. This is one further argument in support of the statement that a gradient in the animal halves determines the oral field centre and thereby the ground plan of bilateral symmetry of the pluteus larva, symmetry which for a short time in the life cycle suppresses the inherent radial symmetry. (For further discussion see Czihak 1962b.)

12.10. Animalization and vegetalization

In the sea urchin embryo two gradients determine the first step of differentiation: the differentiation of ectoderm and entoderm. If the two gradients are balanced, development is normal; if one is stronger, animalization or vegetalization will occur, according to whether the animal gradient or the vegetal gradient, respectively, is enhanced. Equally, it may be due to a weakening of the opposite gradient! Be that as it may, animalization is manifested by overdeveloped ectodermal organs, especially the ciliary tuft, and vegetalization by excessive development of the entoderm and enlargement oi the intestine, which often does not invaginate, thus forming an exogastrula, an embryo with the intestine hanging out. These phenomena have been already described in the chapter on blastomere combinations (pp. 422ff.), and the same effect has been found after treatment of the whole embryo with certain agents.

After Herbst's original discovery of vegetalization caused by lithium ions (1892), many substances capable of enhancing animal or vegetal differentiation were described. Lallier devoted much work to the processes of animal and vegetal differentiations, using substances of which the efficacy in biochemical reactions was or was thought to be known. But as these differentiations are the result of a long chain of different biochemical reactions involving many enzymes, this chain can be influenced at many levels, i.e. by many substances, including some very simple ones like CO (Runnström 1928b) and KCN (Czihak 1963), which elicit vegetalization and tell us that animal differentiation needs respiration, whereas vegetal differentiation obtains its energy by glycolysis, at least under experimental conditions. Lallier recently summarized his numerous experiments in some excellent reviews of which those of 1964 and 1958 are but two. In the following only some of the main topics can be described.

The number of vegetalizing agents found until now is considerably smaller than the number of substances causing animalization. Probably already in the egg the processes of animal differentiation are more advanced than the vegetal ones! It would thus be easier to disturb vegetal differentiation, as a longer and more complicated process offers a greater number of targets; further, the impairment of vegetal differentiation would favour the animal one. The strongest vegetalizing agents we know of are lithium chloride (Herbst 1892) phenazone (Lallier 1959) and chloramphenicol (Lallier 1961), an antimetabolite of protein metabolism, KCN and NaN_3 (Czihak 1963), which block cytochrome oxidase. These substances probably have no common effect, and exert their influence on different points of the processes of animal differentiation, thus favouring the vegetal one. But it must be emphasized once again that we do not know exactly whether vegetalization is brought about by suppression of animal differentiation or enhancement

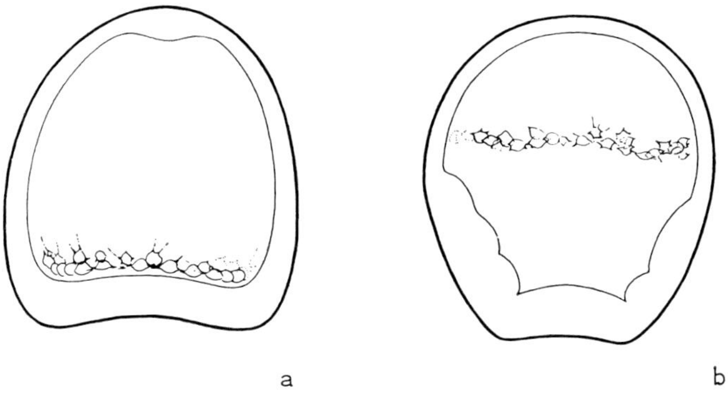

Fig. 77. Position of the chain of primary mesenchyme cells in a normal (a) and vegetalized (b) blastula. (From Czihak 1962.)

of the vegetal processes. But, especially in the case of Li, KCN and NaN_3, the first is more probable. The same is true of animalization, which can be the consequence of impairment of vegetal differentiation or stimulation of the specific animal metabolism.

Vegetalization is characterized by overdevelopment of the vegetal part of the embryo, i.e. the entoderm. This can first be seen in the blastula, in respect of the relation between the thin ectoderm and the thickened entoderm cells and the location of the primary mesenchyme cells which are shifted to a certain level in the animal-vegetal double gradient (fig. 77). Extension of the vegetal region is consequently followed by enlargement of the intestine as a weak form of alteration, and most frequently by enlargement with partial or complete exogastrulation. The latter is the inverse form of gastrulation, namely evagination instead of invagination. This phenomenon has been studied at the cellular level by time-lapse cinematography (Kinnander and Gustafson 1960). As in normal development, the cells of the vegetal pole show increasing pulsatory activity at the time of gastrulation: the pulsations in normal development are directed centripetally towards the blastocoel, in exogastrulating embryos centrifugally. The pulsations lead to enlargement of one cell-pole and subsequently to an invagination or evagination movement.

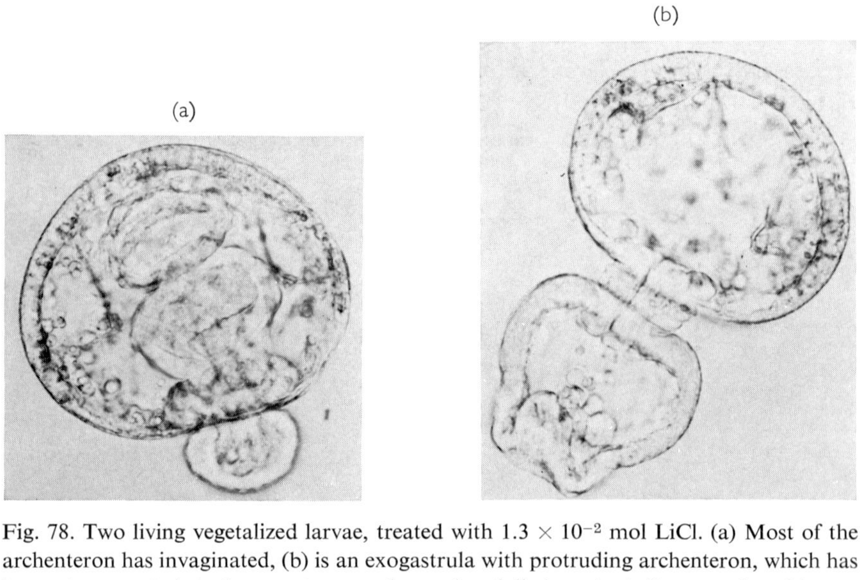

Fig. 78. Two living vegetalized larvae, treated with 1.3×10^{-2} mol LiCl. (a) Most of the archenteron has invaginated, (b) is an exogastrula with protruding archenteron, which has begun to separate into three parts: oesophagus (partially invaginated), stomach and intestine.

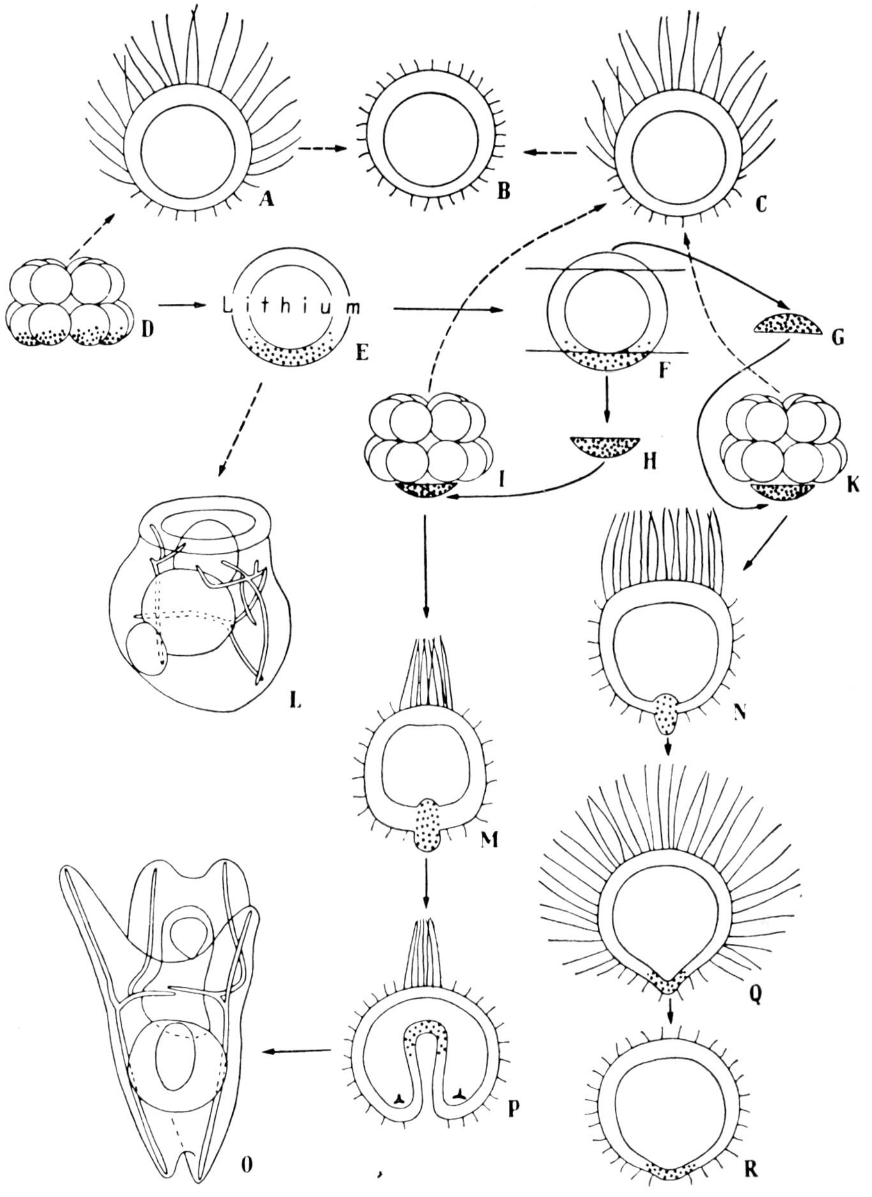

Fig. 79. Combinations of animal halves with parts of LiCl-treated blastulae and their further development. Explanation in the text. (After Hörstadius 1936.)

In vegetalized larvae the ectodermal region can differentiate into a normal pluteus shape but in stronger vegetalization the ectoderm forms a body resembling the head of a mushroom or a slightly depressed globule (fig. 78) with a ciliary ring.

Furthermore lithium also has the capacity of influencing the development of animal halves, which normally develop into 'dauerblastulae' with enlarged ciliary tuft (p. 423). Treated with lithium they often develop into normal small plutei or even to exogastrulae (Von Ubisch 1929). It must be supposed that the vegetal gradient which is also present in the animal half becomes strengthened by exposure to lithium-ions. Lallier frequently found that combination of two vegetalizing agents reinforced the vegetalization effect in a degree which was more than additive, and vegetalizing agents counteracted the animalization brought about by animalizing agents (see below).

Perhaps the most fascinating experiment concerning alteration of the gradient system was executed by Hörstadius (1936) (see fig. 79), who treated animal halves with lithium. Animal halves without lithium treatment developed as usual into 'dauerblastulae' (A, B), but under the influence of lithium they became normalized or even slightly vegetalized larvae (L). Lithium-treated animal halves were grown to the blastula stage (F). The animal pole was then removed (G) and after vital staining transplanted to the vegetal pole of an animal half (K). Such halves developed into strongly animalized larvae (N-R). Lithium was not capable of altering the animalizing capacity of the cells of the animal pole. In another experiment the vegetal pole of a lithium-treated animal half (H) was transplanted onto an animal half (I). From this combination a normal pluteus developed (M-P-O). This event is just the same as if micromeres had been transplanted onto an animal half. It means that the properties of the cells of the vegetal pole of an animal half after lithium treatment are in some respects the same as those of micromeres.

Of the many known animalizing substances the most effective are zinc ions (Lallier 1955), trypsin (Hörstadius 1949; Runnström and Immers 1966), thiouralic acid (Lallier 1952), stains with sulphonic groups such as Evans blue (Lallier 1957), iodosobenzoic acid (Runnström and Kriszat 1952) and thiocyanate (Lindahl 1936). Animalized larvae obtained by treatment with one of these substances develop like animal halves (cf. p. 424, fig. 58a) to blastulae with an enlarged ciliary tuft and more or less suppressed invagination of the small remnants of the entoderm (fig. 80). Less distinct stages of animalizaton are recognizable by the slight enlargement of the ciliary tuft. Hörstadius (1935) employed the fractions $\frac{1}{4}$–$\frac{3}{4}$ to characterize the

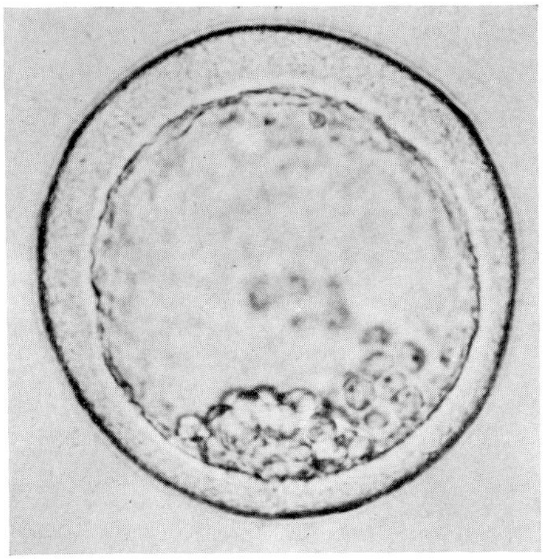

Fig. 80. Strongly animalized larva with $\frac{1}{2}$ ciliary tuft (to be seen only in phase contrast!) after treatment with 5.10^{-4} mol $ZnCl_2$. The small group of cells in the blastocoel may represent primary mesenchyme cells. Twenty hours p.f.

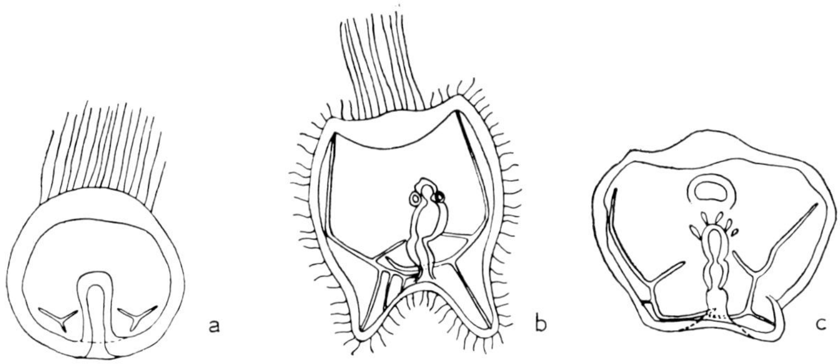

Fig. 81. Two- (a) and four- (b,c) day-old larvae of Paracentrotus, animalized by treating the unfertilized eggs with NaSCN. (a) $\frac{1}{4}$ ciliary tuft, (b) tiny archenteron, which does not touch the mouth groove (c). (After Lindahl 1936.)

enlargement. A further typical feature of animalized larvae is the reduction of the size of the intestine (fig. 81). An attempt to explain the different degrees of animalization and vegetalization by different expressions of the animal and vegetal gradients is shown in fig. 82. Only slightly animalized or vegetalized larvae are capable of living for any length of time.

Many attempts have been made to clarify the mode of action of the various agents causing animalization or vegetalization. As the animal and the vegetal halves differ in their protein composition (Ranzi 1962),

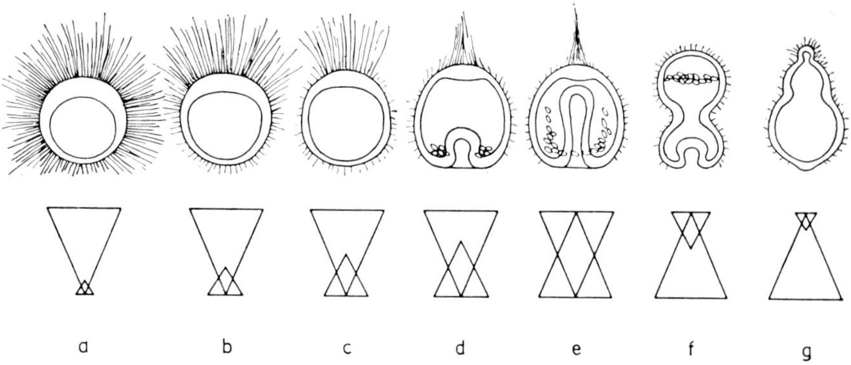

Fig. 82. Animalized (a–d) and vegetalized (f,g) larvae and the assumed expression of gradients symbolized by opposite triangles. (a) Strong animalization with $\frac{3}{4}$ ciliary tuft; (b) and (c) lesser vegetalized larvae with $\frac{1}{3}$ and $\frac{1}{4}$ ciliary tuft; (d) very weak animalization; (e) normal larva: the triangle of the animal gradient should be somewhat larger, as in fig. 59a, d, e; (f) vegetalized larva: note the position of the primary mesenchyme cells and the partially invaginated archenteron; (g) very strong vegetalization: only the terminal knob can be considered to contain ectoderm. (After Gustafson 1965.)

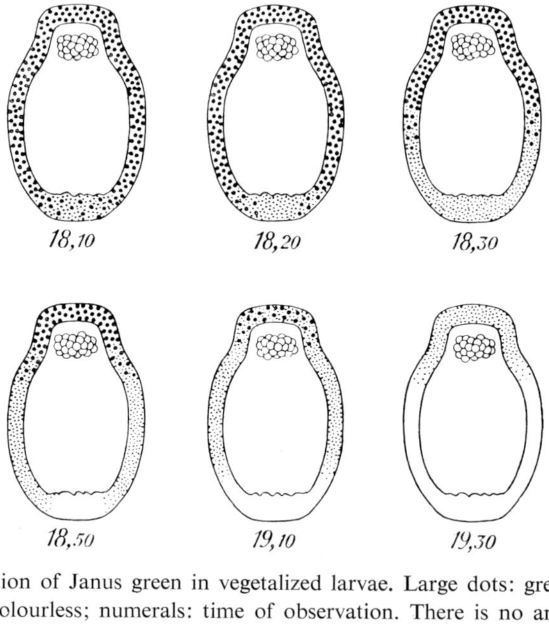

Fig. 83. Reduction of Janus green in vegetalized larvae. Large dots: green; small dots: red; no dots: colourless; numerals: time of observation. There is no animal reduction gradient. Cf. fig. 63. (After Hörstadius 1955.)

it must be presumed that each change in the gradient system finally affects the protein synthesis. However, for a given agent one can never know whether the effect on the protein synthesis is a primary one or the culmination of a disturbance elsewhere in the metabolism. Lithium, for instance, was found to have a strong influence on respiration (Lindahl 1936): it totally inhibits the uptake of oxygen in the later developmental stages; but in the first hours of development the respiration is only slightly reduced in the presence of lithium (Lindahl 1939). This is remarkable as KCN and CO are very effective vegetalizers. The effect on the carbohydrate metabolism can be seen directly as alteration of reduction gradients (cf. p. 430) in vegetalized larvae, as shown by Hörstadius (1955) (fig. 83). The vegetal reduction gradient is much stronger, and the animal one absent! In agreement with this observation is the fact that in trypsin-treated, animalized larvae (fig. 84), the animal reduction gradient is enhanced, and the vegetal very weak.

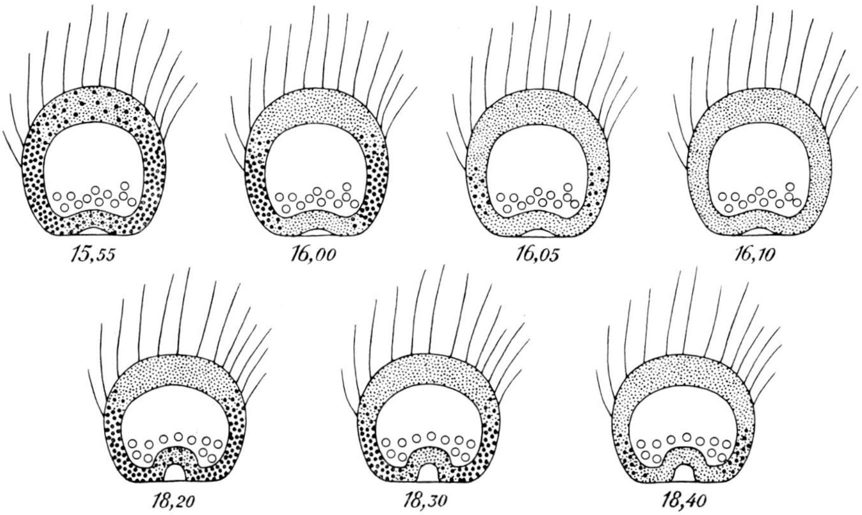

Fig. 84. Change of colour in vitally stained animalized larvae. Symbols are the same as in the previous figure. The animal gradient is much stronger than the vegetal one. No bleaching of the dye was observed. Cf. fig. 63. (After Hörstadius 1955.)

On the other hand, it must be reported that an influence on gene activation is also known, as the vegetalizing action of lithium is abolished in the presence of actinomycin (Runnström and Markman 1966). An increase of vegetalization as a result of longer treatment with both agents was demonstrated by Lallier (1963). In the first case, lithium seems to stimulate gene transcription for vegetal differentiation, which can be annulled by actinomycin, a specific (but also toxic) suppressor of gene transcription blocking mRNA formation.

Contradictory results were also published with regard to the final effect of lithium on protein synthesis. Incorporation of amino acids into homogenates of *Paracentrotus lividus* eggs has a high potassium optimum. Lithium was found to be a competitor of potassium (Runnström 1928b) and amino acid incorporation is very low in its presence (Molinaro and Hultin 1965). Mastrangelo (1965) found no effect of lithium on amino acid incorporation in whole eggs, however.

Finally, another interesting attempt to gain an insight into the mode of action of gradient-influencing substances may be noted. Lallier, as already mentioned, employed various polysulphonated dyes to study animalization. The position of $NaNO_3$-groups in the dyes proved to be of great importance for their potency as animalizing agents (fig. 85). The action of zinc ions can

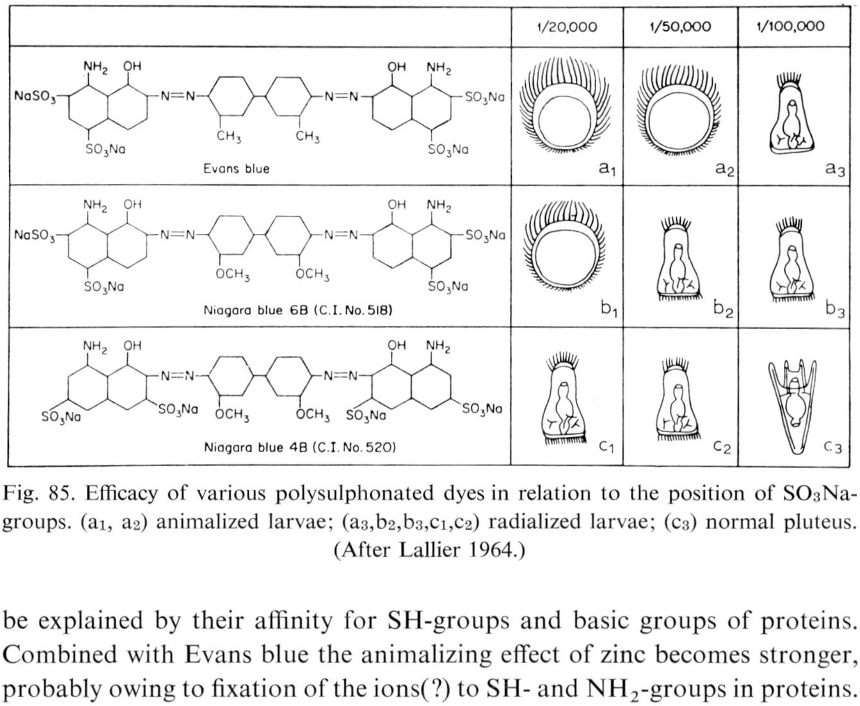

Fig. 85. Efficacy of various polysulphonated dyes in relation to the position of SO_3Na-groups. (a_1, a_2) animalized larvae; (a_3,b_2,b_3,c_1,c_2) radialized larvae; (c_3) normal pluteus. (After Lallier 1964.)

be explained by their affinity for SH-groups and basic groups of proteins. Combined with Evans blue the animalizing effect of zinc becomes stronger, probably owing to fixation of the ions(?) to SH- and NH_2-groups in proteins. To summarize therefore, some first steps towards the understanding of animal and vegetal differentiation and their experimental alteration have been made, but in the long chain of reactions leading to these differentiations, most connections are still unknown.

A remarkable step towards the elucidation of animalization and vegetalization can be seen in recent work of Hörstadius et al. (1967). Homogenates of unfertilized eggs were separated on Dowex 50-W-X2 columns. Two fractions eluted at pH 9 and added to culture dishes exhibited pronounced animalizing or vegetalizing effects on both animal and vegetal halves of cleavage stages obtained by Hörstadius' glass needle technique. A mixture of both fractions in a certain proportion did not influence the development of the halves.

12.11. Respiration and carbohydrate metabolism

Generally one is inclined to assume that there is no difference in the basic

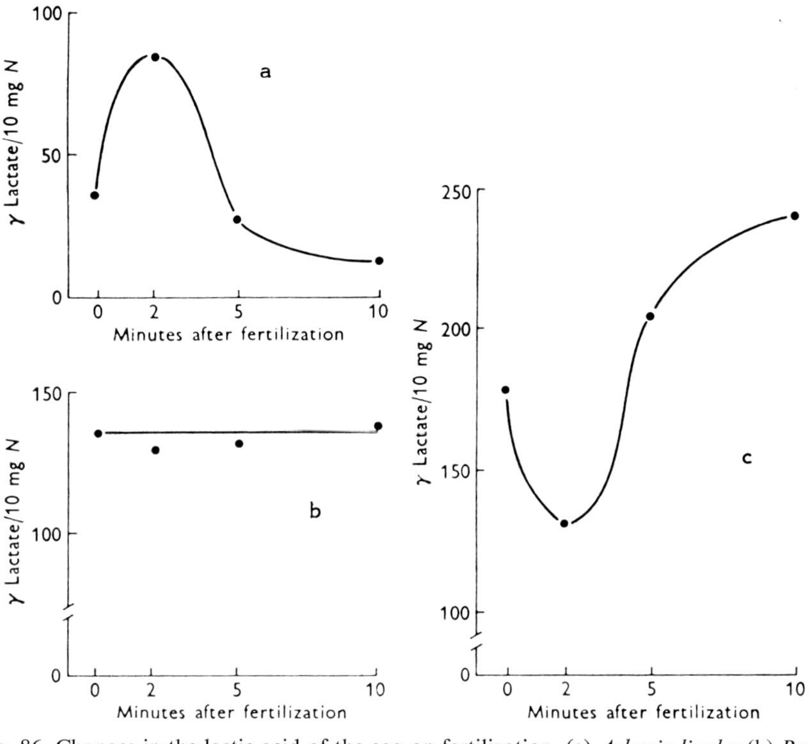

Fig. 86. Changes in the lactic acid of the egg on fertilization, (a) *Arbacia lixula*; (b) *Paracentrotus lividus*, (c) *Psammechinus miliaris*. (After Aketa 1964.)

processes of metabolism among the sea urchin species. The contrary is shown by Aketa's research (1964) on lactic acid production some minutes after fertilization. In Paracentrotus eggs the amount of lactic acid remains almost constant, but Arbacia and Psammechinus exhibit inverse fluctuations (fig. 86). A species difference was also found by Immers and Runnström (1960), between Paracentrotus and Psammechinus in respect of decoupled phosphorylation (see p. 456). One should be aware of these facts when comparing results dealing with carbohydrate metabolism of different species.

Oxygen consumption only partially reflects the total metabolism. With the rotating platinum electrode accurate measurements of oxygen uptake during the first minutes after fertilization were possible. In the unfertilized eggs of the Japanese sea urchins *Hemicentrotus pulcherrimus* and *Pseudocentrotus depressus* oxygen uptake is low and remains almost constant between 5 and 55 min after spawning and, according to Yasumasu and Nakano (1963), for more than 3 hr. After fertilization O_2 uptake is greatly increased within some seconds after a lag of less than one minute, as can be seen in fig. 87. The maximum rate is achieved at the time of membrane elevation;

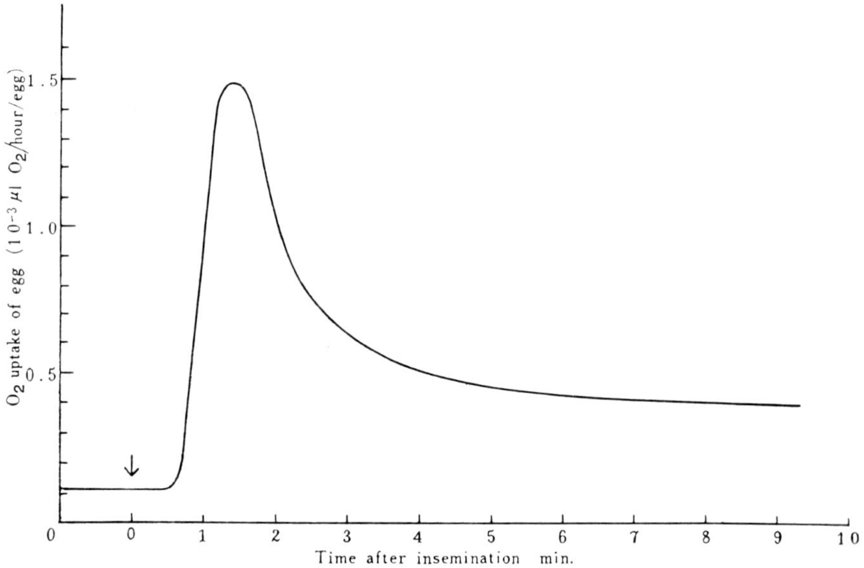

Fig. 87. Oxygen uptake of Pseudocentrotus eggs at fertilization. (After Ohnishi and Sugiyama 1963.)

it decreases afterwards and remains nearly constant between 4 and 10 min after fertilization (Ohnishi and Sugiyama 1963).

In the unfertilized egg there may be a block of the pathway between glycogen and glucose-6-phosphate, which is removed by fertilization, when

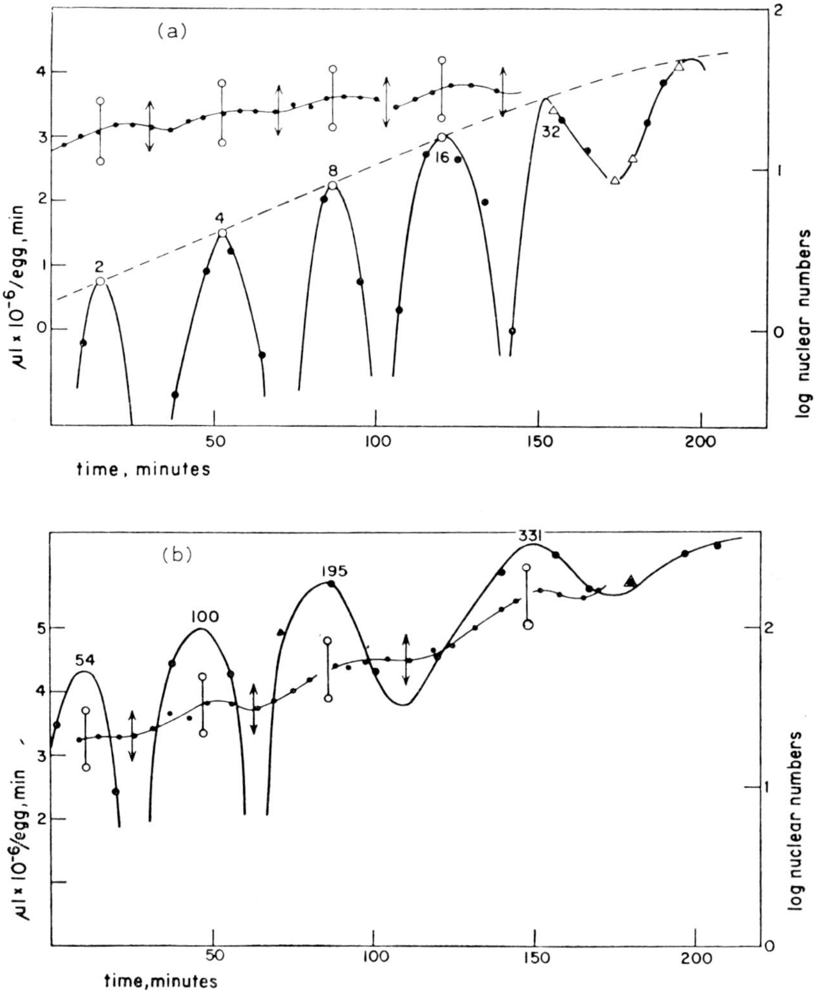

Fig. 88. Relation between respiration (upper curve) and mitotic cycle (single peaks). Numbers indicate cell stage. The open circles at the tip of the peak show when the maximum of interphase nuclei is found. This is indicated in the respiration curve by a vertical line with two circles, the middle of mitosis by a vertical double arrow. Fluctuations in the respiration curve follow mitotic events. (From Holter and Zeuthen 1957.)

the content of glucose-6-phosphate increases (Aketa et al. 1964). This group of workers stated that the 'rate of hexose phosphate synthesis appears as a primary limiting factor involved in the respiration rise upon fertilization'.

Microfluctuations in the respiratory level were found by Holter and Zeuthen (1957), during the mitotic cycle of the first cleavages of the Mediterranean *Psammechinus microtuberculatus*, using the diver method. The maximum oxygen uptake was found in the interphase of every cleavage,

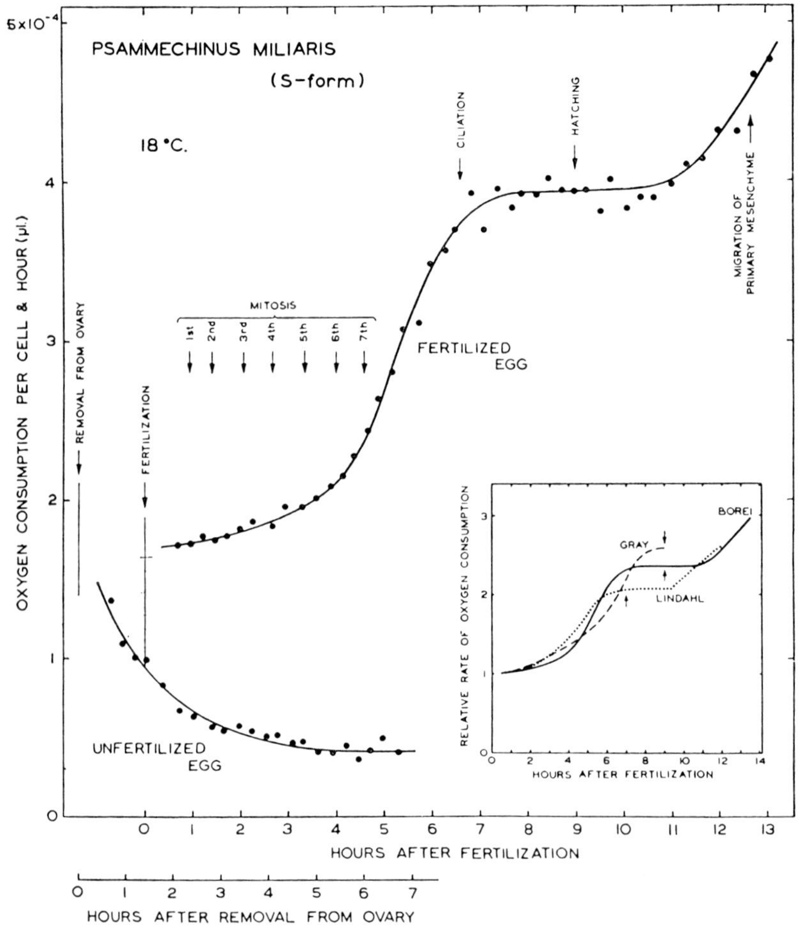

Fig. 89. Oxygen consumption of unfertilized and fertilized eggs of Psammechinus. The insert at the bottom right is for comparison with similar results by Gray and Lindahl. Note the plateau in the respiration curve with hatching. (From Borei 1948.)

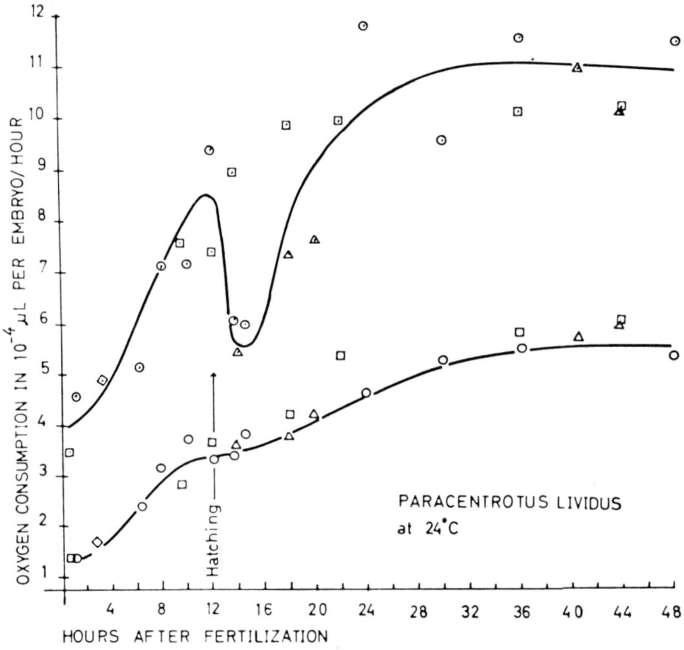

Fig. 93. Normal respiration and that released by addition of 2,4–dinitrophenol to the final concentration of 5×10^{-5} mol. Respiration in the presence of DNP is considerably greater and instead of the blastula plateau a deep slope is found. (After Immers and Runnström 1960.)

respiration released after treatment with 2,4-dinitrophenol, a decoupling agent of phosphorylation. Under the influence of this substance the respiratory activity is considerably higher and instead of the blastula plateau a steep slope indicates a sharp drop in oxygen consumption (fig. 93). During the gastrula stage the high rate of respiration is resumed.

The authors finally state that the intensity of respiration is controlled by the rate of ATP consumption, as the addition of the phosphate acceptor ADP has the same effect as 2,4-dinitrophenol. Aketa et al. (1964) have suggested that a block of the conversion of glycogen to glucose-6-phosphate is released upon fertilization. The data given by Giudice et al. (1962) and Berg (1965), in respect to protein synthesis are comparable to the respiratory activity and obviously closely related, as ATP is the energy source for such synthetic activities.

Gustafson and Hasselberg (1951) devoted intensive studies on enzyme activity to different developmental stages. They found enzymes with constant activity throughout development, namely:

pyrophosphatase (0–45 hr p.f.)
hexametaphosphatase (0–45 hr p.f.)
phosphomonoesterase (0–40 hr p.f.)
aldolase (0–45 hr p.f.)
adenosine deaminase (0–45 hr p.f.)
phenylsulphatase (0–50 hr p.f.)

Other enzymes were found to increase in activity:

apyrase (after blastula stage, lower with previous Li treatment)
succino and malic dehydrogenase
glutaminase (after blastula stage; here too Li treatment before blastula stage reduced the rise in glutaminase activity in gastrula stage)
cathepsin II (rising after blastula stage, smaller increase after Li treatment)

Possibly the mRNAs for the first group of enzymes or the enzymes themselves are present in the egg and those for the second group are synthesized

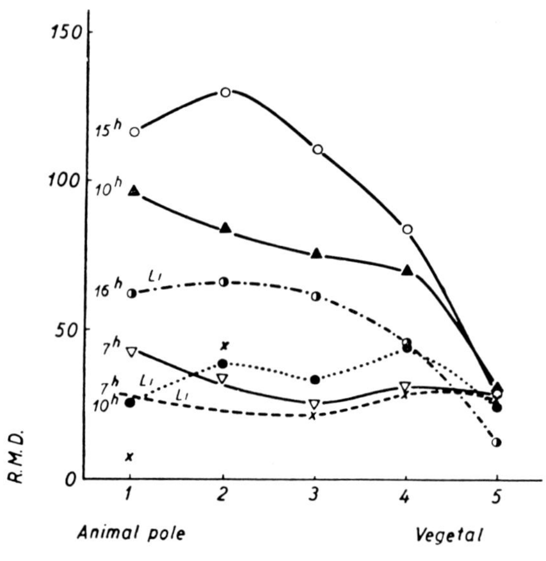

Fig. 94. Density of stainable mitochondria between the animal and vegetal poles of normal and LiCl-treated blastulae. Note the maximal density beyond the animal pole and the decreasing numbers towards the vegetal pole in the 15-hr blastula. (After Gustafson and Lenicque 1952.)

during cleavage. 'It has been suggested that the enzymes which rise in activity at the mesenchyme blastula stage are localized in mitochondria, whereas enzymes with constant activity are non-mitochondrial'. It is interesting in this connection that Berg and Long (1964) found mitochondria of equal size in the blastomeres of the 16-cell stage, but in the early gastrula larger ones in the vegetal part of the embryo, and also an enlargement of the mitochondria of the animal region after treatment with LiCl (vegetalization); they concluded 'that a differential growth of mitochondria occurs during, or more likely after, the cleavage period'. Using Nile blue sulphate as a vital stain, Gustafson and Lenicque (1952) demonstrated a mitochondrial gradient in gastrula stages (fig. 94). The maximum number of stained mitochondria was found near the animal pole. In vegetalized larvae the number of *stainable* mitochondria is far lower due to the transformation of animal (i.e. small, stainable) mitochondria into those of the vegetal type (large, less or not stainable).

But Berg et al. (1962) found no evidence of an unequal distribution of mitochondria in ultrathin sections of early gastrulae of Lytechinus. If there are no species differences, it can therefore be concluded that the number of mitochondria is not very different in animal and vegetal cells, but that those in the animal half are smaller and in some respects more active (stainable) than the larger ones of the vegetal half. However, it must be mentioned that Shaver (1956) detected no regional difference in stainability of mitochondria with Janus green B, a dye indicating cytochrome oxydase activity. This agrees nicely with the fact that no difference in respiration was found between animal and vegetal halves from 6 to 28 hr p.f. (Lindahl and Holter 1940). De Vincentiis et al. (1966) reported, however, that in later cleavage stages there is a considerable difference between the animal and vegetal halves with respect to respiration (fig. 95) but that in the mesenchyme blastula stage the contrary occurs. In both stages the respiration released in animal halves by the addition of dinitrophenol is much higher than in controls. Another fact somewhat in contradiction to these results is a gradient of the Nadi reaction observed by Child (1941). 'From late cleavage and blastula stages to the gastrula the reaction progresses from the apical (animal) region basipetally, constituting a basipetal gradient both in time and in depth of colour'. This oral field gradient was studied more completely by Czihak (1962b) (see pp. 432ff.) and related to the excellent work of Pease.

The latter author subjected cleaving eggs to graded concentrations of KCN, and demonstrated that the differentiation of the oral field needs a high respiration rate and will occur in those parts of the blastula wall which

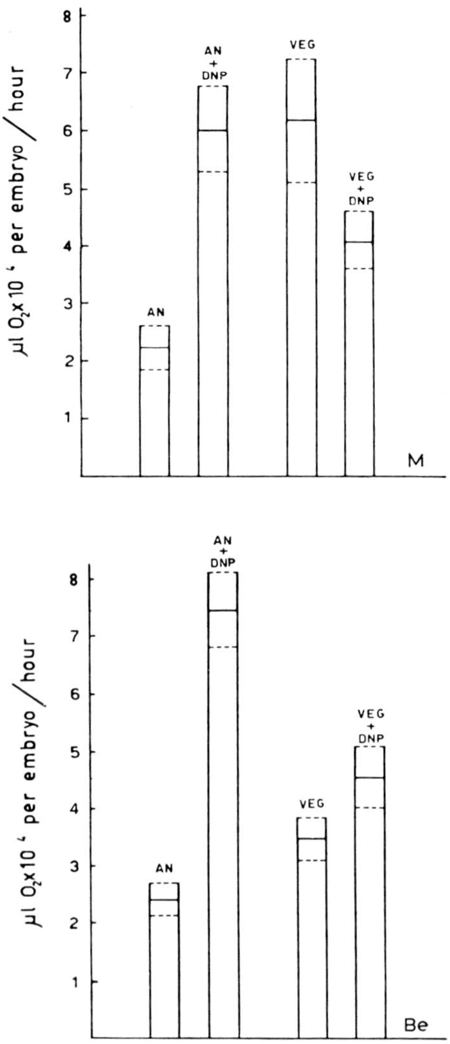

Fig. 95. Normal and, by addition of 2,4-dinitrophenol (DNP), released respiration of animal (AN) and vegetal (VEG) halves in the morula (left) and blastula (right). (After De Vincentiis et al. 1966.)

are less influenced by KCN, which blocks cytochrome oxydase. The results concerning respiration and mitochondrial activity are contradictory at first glance, but it may be that the higher mitochondrial activity in the future oral field is balanced by a lower activity of the lateral and dorsal side, so that the animal half as a whole has the same mitochondrial activity as the total ventral half.

12.12. Proteins and amino acids

Each organ is different in its protein inventary, and since in biochemical terms differentiation means synthesis of specific proteins, great efforts have been devoted to the study of amino acid and protein metabolism in development. The proteins of the yolk become split up and new ones are formed until the embryo is able to provide the protein synthesizing machinery with amino acids taken in with the food. In the unfertilized egg the incorporation of radioactive amino acids into protein is very low. It rose from 77 cpm to 198 cpm 20 min p.f. in a certain number of eggs (Hultin and Bergstrand 1960). The initiation of protein synthesis is not due to newly formed mRNA because (1) halves of the egg without a nucleus exhibit an increase in protein synthesis after artificial parthenogenetic activation (Brachet et al. 1963) and fertilization, as Denny and Tyler (1964) have shown (table 1).

TABLE 1

Effect of treatment with butyric acid (4.10^{-3} mol, 1 min) on incorporation of C^{14}-amino acids by intact non-nucleate fragments of eggs of *Lytechinus pictus* in comparison with fertilization.

Experiment	Counts per minute Untreated (U)		Treated (T)		Fertilized (F)		Ratios T/U	F/U
1	25,650	25,230	72,750	72,300			2.8	
5	9,540	6,700	20,753	17,467	19,348	18,451	2.3	2.3
7	1,005	833	8,885	7,526	8,969	8,122	8.9	9.3
8	999	879	4,627	4,307	7,645	7,404	4.8	8.0

'Values for experiment 1 are per mg protein; for the other experiments quantities of egg fragments were not determined but amounts were the same for treated as for untreated samples in each experiment. Incubations were for $\frac{1}{4}$ to 2 hr at 20 °C in 50 cm^3 of a 2.5 μc/ml sea water solution of C^{14}-valine per ml of egg-fragment suspension, except for experiment 5 in which C^{14}-phenylalanine was used.' (From Denny and Tyler 1964.)

Only in experiment 8 did the increase after fertilization differ from that after activation.

(2) Actinomycin does not prevent protein synthesis until the blastula

stage, even when RNA-synthesis is strongly suppressed, according to the results by Gross and Cousineau (1963) presented in table 2.

TABLE 2

Effect of actinomycin on uptake of C^{14}-uracil and valine in early cleavage.

Observation	cpm/10^6 eggs control	act. D	% inhibition by act. D
1st hr incorporation, uracil	2.13×10^3	1.23×10^3	42
end of 2nd hr	7.58×10^3	1.57×10^3	79
1st hr incorporation, valine	7.13×10^4	7.16×10^4	−0.4
end of 2nd hr	20.4×10^4	19.5×10^4	4.4

'Cell density: 1.18×10^4/ml L-valine-1-C^{14}; 41 mμM/ml, 0.24 μc/ml, uracil-2-C^{14}; 21 mμM/ml, 0.66 μc/ml. Actinomycin D, 65 μg/ml, added 30 min before fertilization.'

Thus there is no doubt that mRNA (see p. 486) is present in the unfertilized egg in inactive form. Intact ribosomes are present in the unfertilized egg, since the incorporation of phenylalanine into proteins was greatly stimulated after the addition of polyuridylic acid as artificial messenger to the 12,000 *g* supernatant in experiments performed by Wilt and Hultin (1962) (fig. 96).

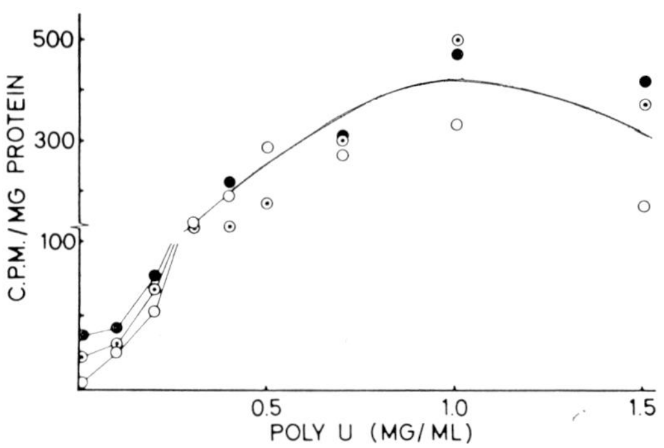

Fig. 96. Incorporation of C^{14}-phenylalanine into proteins of the 12.000 *g* supernatant of Psammechinus eggs with increasing concentrations of added polyuridylic acid. Open circles: unfertilized eggs; circles with dot: eggs 30 min p.f.; filled circles: eggs 120 min p.f. (From Wilt and Hultin 1962.)

The incapability of the unfertilized egg to synthesize larger amounts of proteins is sensitive to trypsin. Table 3 from Monroy et al. (1965a) shows this clearly.

TABLE 3

Effect of treatment with trypsin on the ability of ribosomes from unfertilized eggs to incorporate amino acids into proteins.

Expt. no.	Amino acid(s)	Addition	Cpm/mg Control	Ribosomal RNA Trypsin-treated
1	Phenylalanine	None	0	122
	Phenylalanine	Poly U	270	340
2	Phenylalanine	None	0	72
	Phenylalanine	Poly U	360	2200
	AH	None	25	260
	AH	RNA blastula	0	475
3	AA	None	18	448
	AA	RNA unfert. eggs	9	644

'Radioactive amino acids: C^{14}-phenylalanine (7.8 $\mu C/\mu M$), 1μC; C^{14} algal protein hydrolysate (AH), 1μC; 1μC total of a mixture of C^{14}-phenylalanine, lysine, valine, glycine and glutamic acid (AA). In experiment 1, 0.193 mg, in experiment 2, 0.483 mg and in experiment 3, 0.327 mg of ribosomes were used.'

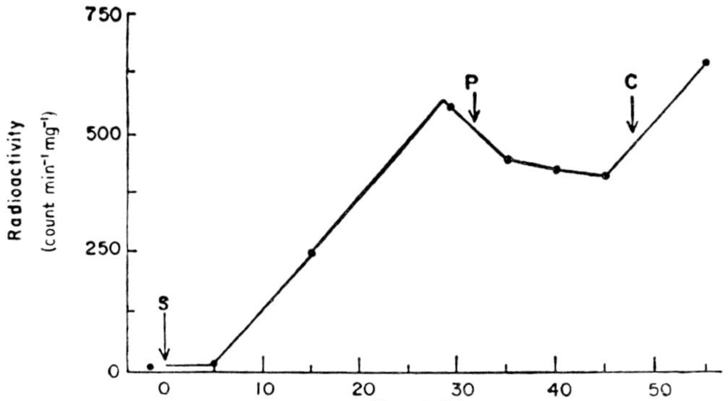

Fig. 97. Incorporation of C^{14}-leucine into protein during the first division cycle. Radioactivity in cpm/mg protein. S: addition of sperm; P: prophase; C: anaphase. (From Sofer et al. 1966.)

These authors assume that the mRNA synthesized during oogenesis becomes attached to ribosomes, which are then coated by a protein capsule, the latter being attacked by trypsin. The action of proteolytic enzymes could liberate in vitro and perhaps also in vivo the polysomes ready for protein synthesis. The presence of polyribosomes in the cytoplasm of the unfertilized egg was established by Stafford et al. (1964) and Burny et al. (1965).

As mentioned above, protein synthesis starts only a few minutes after fertilization. According to the uptake of labelled precursors, synthesis is considerably intensified during cleavage. These events were studied in detail after short pulse incorporations by several groups. Sofer et al. (1966) found

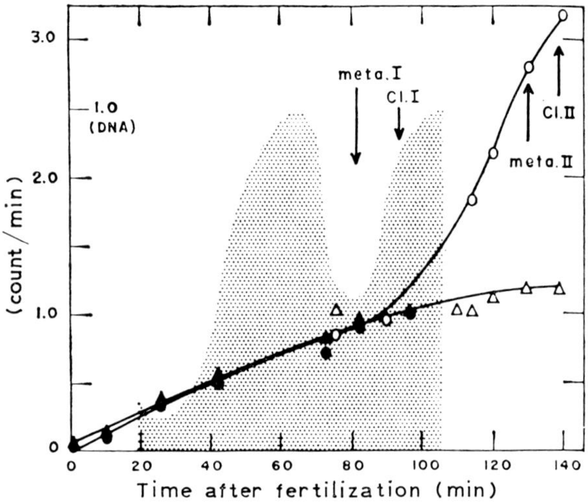

Fig. 98. Relative rates of total uptake and incorporation into polymer for L-leucine-C^{14} and of incorporation into DNA for thymidine-2-C^{14}. The data for leucine represent two different experiments (open and filled symbols), covering the first and second division cycles. All rates (count/min incorporated during a 5-min exposure) are normalized to 1.0 at the time 50% of the cells showed their first cleavage furrow (Cl. I) and are plotted against the values 0 to 3.0 on the ordinate. Triangles, total uptake; circles, counts in TCA-insoluble material. First and second metaphases and furrowing periods are indicated by 'meta.' or 'Cl.' followed by the appropriate number. DNA synthesis is represented by the shaded region: its rates are normalized to 1.0 at the observed maximum and are referred to the inner ordinate. Data points (not shown) were at the same times as those indicated for amino acid uptake, and the upper boundary of the shaded region is the curve drawn through such points. The rate change at 'meta. I' is large. The area under the shaded region may not accurately represent the amount of DNA synthesis, since the penetration rate for thymidine was not measured, and it may increase during the first cleavage cycle. (From Gross and Fry 1966.)

an increase of leucine incorporation until the prophase of the first cleavage and a decrease until anaphase, followed by a new increase (fig. 97). As the synchrony is never perfect, it may be assumed that the peak and the slope of amino acid incorporation is much more different in reality. Even if the anaphase movement of chromosomes is prevented by the addition of colchicine, the form of the curve is very similar. This oscillation is in accordance with findings of a similar fluctuation in the incorporation of uridine into RNA (Czihak 1968). The contrary case was reported by Gross and Fry (1966) who found a continuity of protein synthesis through the cleavage metaphase. It cannot yet be decided whether this is the expression of lesser synchrony or a species difference. It can also be that there is a not yet detected slope at about 60 min (fig. 98, from the paper of these authors). Furthermore the uptake of valine rises independently throughout mitosis. The great variations in fig. 99 should be noted (Mitchinson and Cummins 1966).

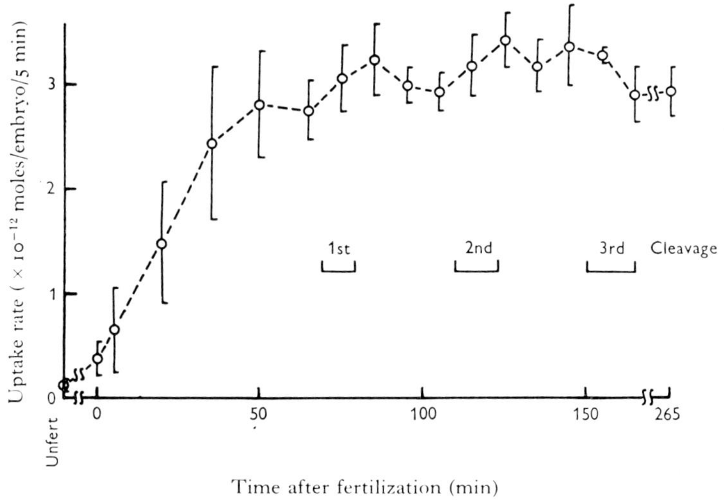

Fig. 99. *Uptake* of C^{14}-valine over 5 min. The part incorporated into proteins was not measured. A typical saturation curve. (After Mitchinson and Cummins 1966.)

A rather complete independence of protein synthesis from the nucleus was also found by other methods. The incorporation of 20 amino acids into non-nucleate fragments and whole eggs of two species was measured. But obviously there is little agreement in individual experiments with three different batches (fig. 100, Tyler 1966). Remarkable differences between

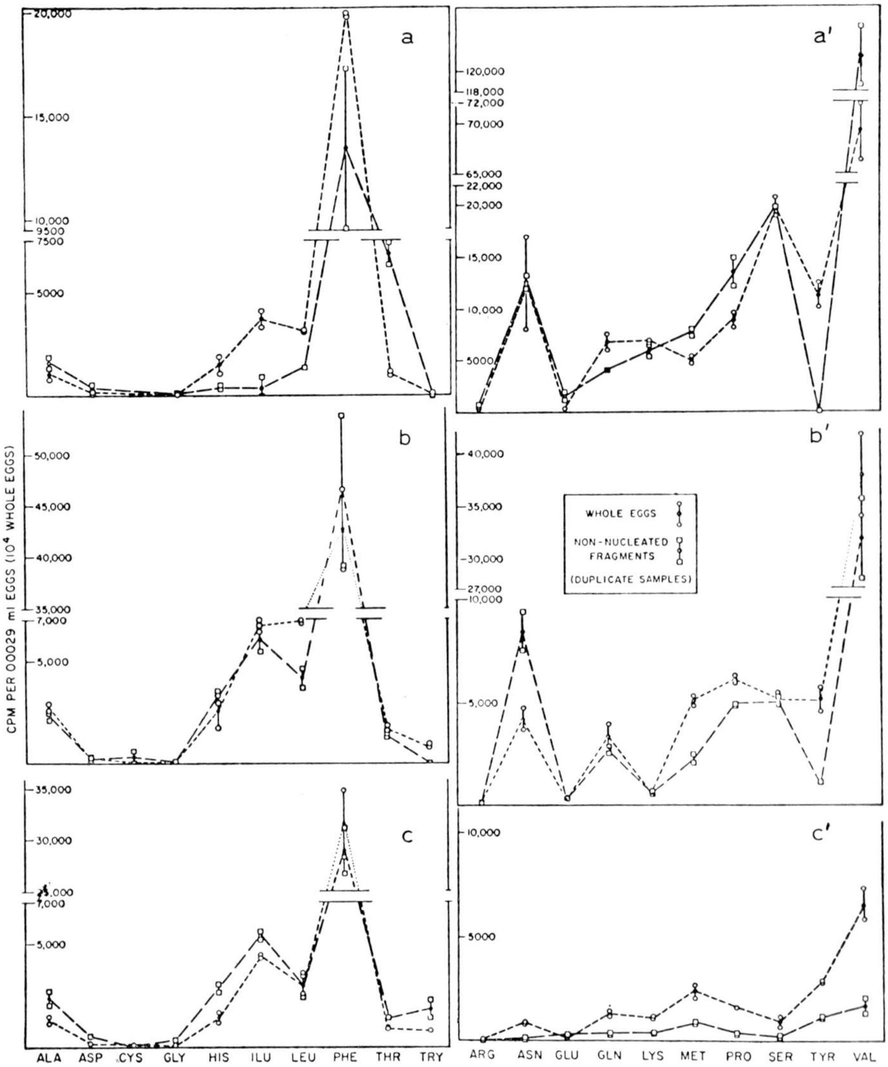

Fig. 100. Incorporation of C^{14}-labelled amino acids into protein over one hour by non-nucleate fragments and whole eggs of *Strongylocentrotus purpuratus* (a–c) at $\frac{1}{2}$ hour after activation with butyric acid. The individual amino acids were tested in groups of 10 in each experiment in duplicate samples of whole eggs and of non-nucleate fragments, and at 0.4 μc/ml. Three experiments with one group are presented in the three charts on the left (a–c) and three with the other group on the right (a′–c′). Specific activities in cpm and names for abbreviations are: 4.24 (alanine), 22.5 (aspartic acid), 7.70 (cystine), 5.25 (glycine), 22 (histidine), 6.16 (isoleucine), 6.3 (leucine), 360 (phenylalanine), 5.15 (threonine), 8.95 (tryptophan), 9.23 (arginine), 22.5 (asparagine), 205 (glutamic acid), 4.05 (glutamine), 7.1 (lysine, except 43.8 in upper right chart), 15.6 (methionine), 7.1 (proline), 1.7 (serine),

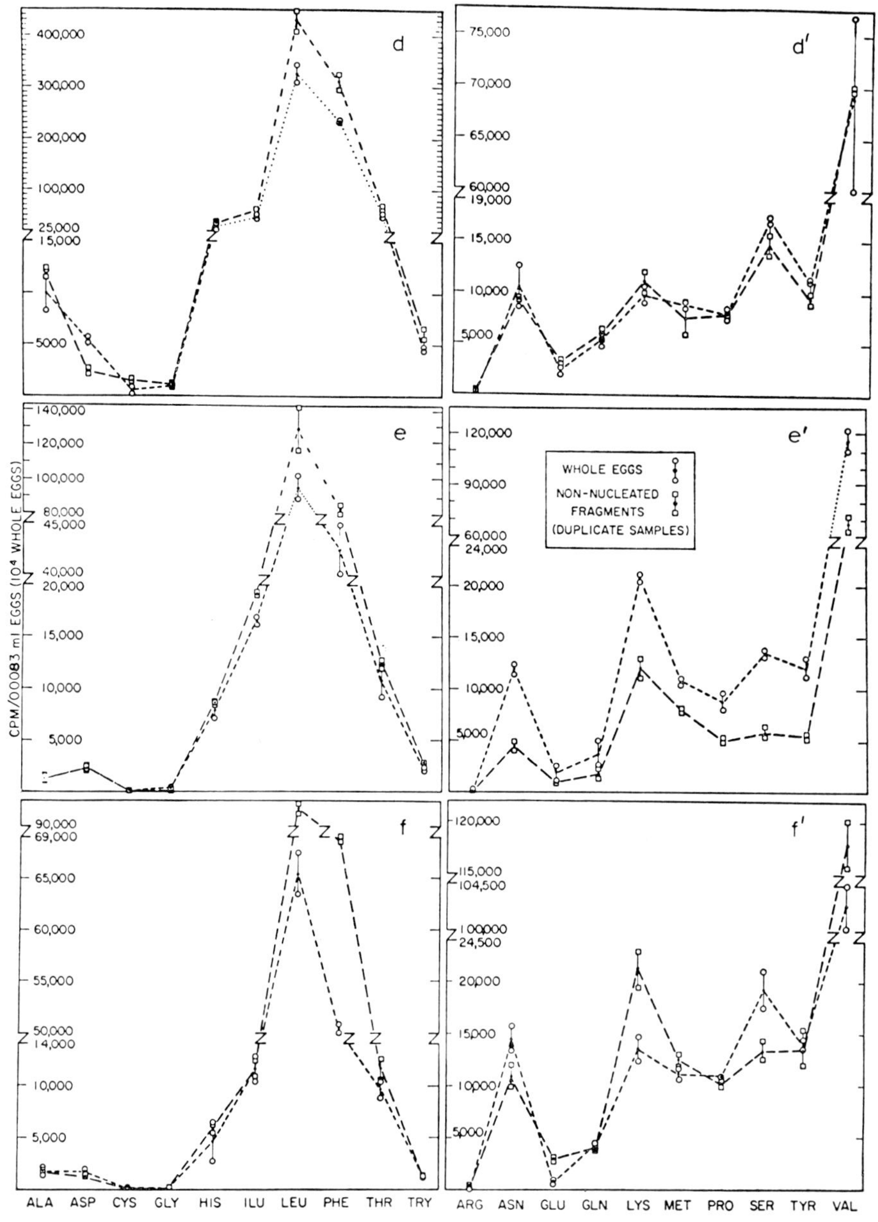

11.54 (tyrosine) and 200 (valine). For the purpose of enabling the pattern of incorporation to be more readily visualized, the averages are connected by dashed lines. The lines have no other significance. (d–f′), same for eggs of *Lytechinus pictus* and with following differences in specific activities: 160 (leucine), 25.6 (threonine) and 43.8 (lysine). (After Tyler 1966.)

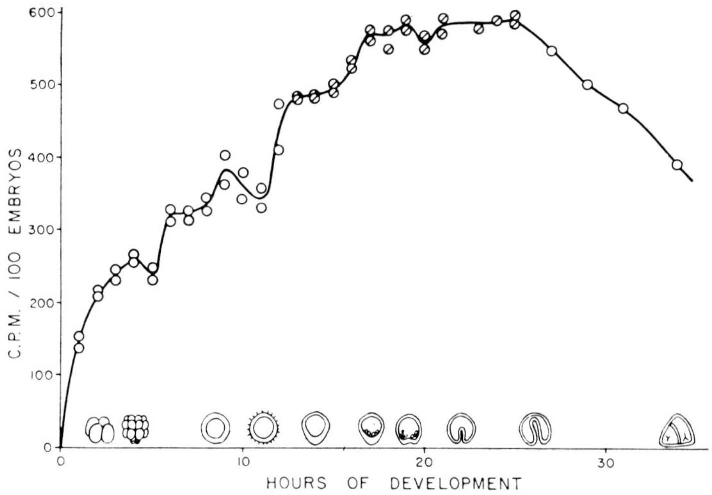

Fig. 101. *Incorporation* of C^{14}-valine in 15-min pulses with 80 embryos. There seems to be no significant difference between figs. 99 and 101. (After Berg 1965.)

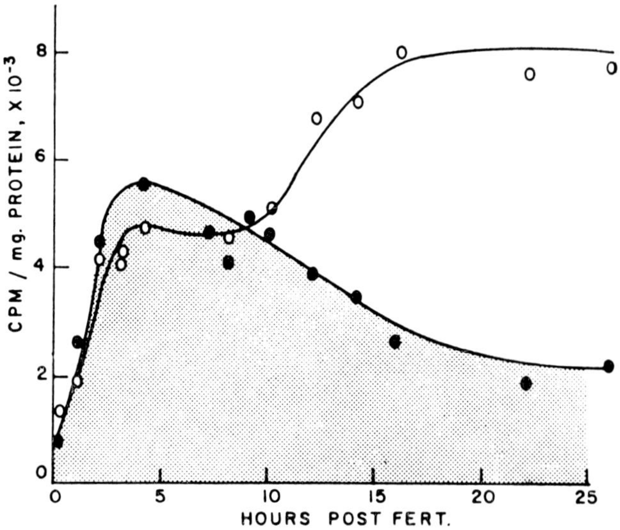

Fig. 102. Effect of actinomycin on incorporation rates of C^{14}-valine. Open circles: controls in 20-min pulse incorporations; filled circles: embryos pretreated for 3 hr before fertilization with 20 μg actinomycin-D. There is no significant difference in the incorporation rates until the early blastula stage. (After Gross et al. 1964.)

whole eggs and non-nucleated fragments in the uptake of asparagine in one batch (fig. 100b), for instance, were not seen in the others (fig. 100a′). Generally speaking one must be aware of individual differences, besides species differences, when studying protein synthesis. Differences in incorporation rates may be due to different 'nutrition' levels of the oocytes and eggs in the ovary.

It was Hultin who first studied protein synthesis by labelling over a longish period of development. But with the progress of science the points of measurement along the line of development have become denser, so that the results presented here are those of Berg (1965) (fig. 101). There is a distinct slope of lower incorporation in the 64-cell stage and a plateau or second decline in the blastula stages before hatching. The shape of the curve thus roughly corresponds to the respiration curve (cf. fig. 89). A very similar result can be seen in a paper by Gross et al. (fig. 102). It should be noted that figs. 101 and 102 refer to the uptake of valine into proteins. Fig. 100 (Tyler) shows that the uptake of valine may differ considerably

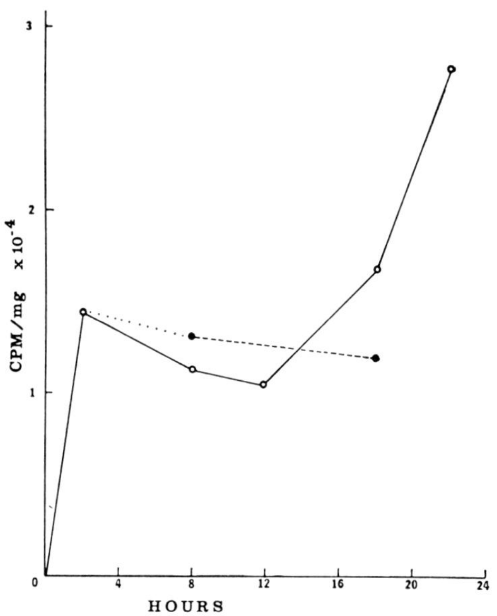

Fig. 103. Time chart of incorporation of C^{14}-yeast protein hydrolyzate into soluble protein extracts of *Arbacia punctulata* embryos. Open circles represent incorporation into untreated control embryos; closed circles represent incorporation by embryos grown in 10 μg/ml actinomycin-D since fertilization. Hatching occurs at 8 hr, gastrulation at about 16 hr. (From Ellis 1966.)

TABLE 4

Uptake of C^{14}-valine in the presence of inhibitors.

A. Inhibitors added 5 min after fertilization

Time labelling period in min after fertilization	Dinitrophenol			Sodium azide		Puromycin	
	10^{-5}M	5×10^{-5}M	2.5×10^{-4}M	10^{-3}M	5×10^{-3}M	10^{-4}M	4×10^{-3}M
20–25	87	65	22	64	0	115	65
40–45	92	51	28	20	0	102	96
60–65	79	43	16	29	0	101	87

B. Inhibitors added 60 min after fertilization

Time of labelling period in min after fertilization	Dinitrophenol		Sodium azide	
	5×10^{-5}M	2.5×10^{-4}M	10^{-3}M	5×10^{-3}M
120–125	45	41	79	39
180–185	37	33	64	24

C. Inhibitor added 30 min before fertilization

Time of labelling period in min after fertilization	Puromycin 10^{-4}M
20–25	49
40–45	64
60–65	66

'Aliquots of embryos labelled for 5 min with C^{14}-valine (16.7 μg/ml) at various times after fertilization in the presence of 2,4-dinitrophenol, sodium azide or puromycin. Results given as amount of radioactivity taken up in the presence of inhibitor as a percentage of that taken up in controls without inhibitor'. (From Mitchinson and Cummins 1966.)

in individual batches and that sometimes the incorporation of phenylalanine exceeds that of valine. Thus figs. 101 and 102 can only be understood as representing incorporation of valine into protein, and not total protein synthesis. If this is kept in mind, the differences between the above-mentioned papers and that of Ellis (1966) can be better evaluated. In fig. 103 the uptake of several amino acids into protein is higher in early cleavage than in the blastula stage. It could well be that in the early cleavage stages the relationship between incorporated valine and other amino acids is quite different from that between incorporated amino acids at later stages of development. Gross et al. and Ellis stated that the presence of actinomycin slightly increases protein synthesis in the cleavage and blastula stages but that has a negative influence after the blastula plateau of amino acid incorporation. The increase in protein synthesis thus needs new mRNA. If this new messenger is synthesized beforehand, it remains inactive until hatching!

Finally, it is also necessary to bear in mind that in all experiments with incorporation of labelled precursors, processes of simple diffusion depending on permeability or active transport are involved. The presence of an active transport mechanism for valine was demonstrated by Mitchinson and Cummins (1966). The concentration in the egg was calculated to be at least 5.7×10^{-3} mol taken up from the surrounding sea water with 1.4×10^{-4} mol. This transport mechanism coupled to the metabolism is sensitive to metabolic inhibitors, as table 4 shows.

The transport mechanism itself or the dependence of certain metabolic events seems to change, as the data of this table show.

Kavanau (1954) studied the protein metabolism by other means: quantitative determination of free amino acids and of single amino acids in peptides and proteins after hydrolization in 6n HCl at 120 °C for 10 hr. The 17 amino acids assayed showed a remarkably uniform developmental pattern of concentration changes, which were generally abrupt and either complementary or supplementary to one another. This means that a loss of a certain amino acid in proteins (fig. 105) is accompanied by an increase of this amino acid in the soluble pool of free amino acids (fig. 104), and vice versa. Kavanau found 'at least three major and three minor periods of new protein synthesis and four periods of intense yolk-protein breakdown'. These periods are shown in fig. 106, where after a period of slower protein synthesis an abrupt first increase in the eight-hour blastula (stage 10) can be seen followed by similar ones in the mesenchyme blastula and during invagination of the archenteron.

The first attempts to study specific proteins in sea urchin development

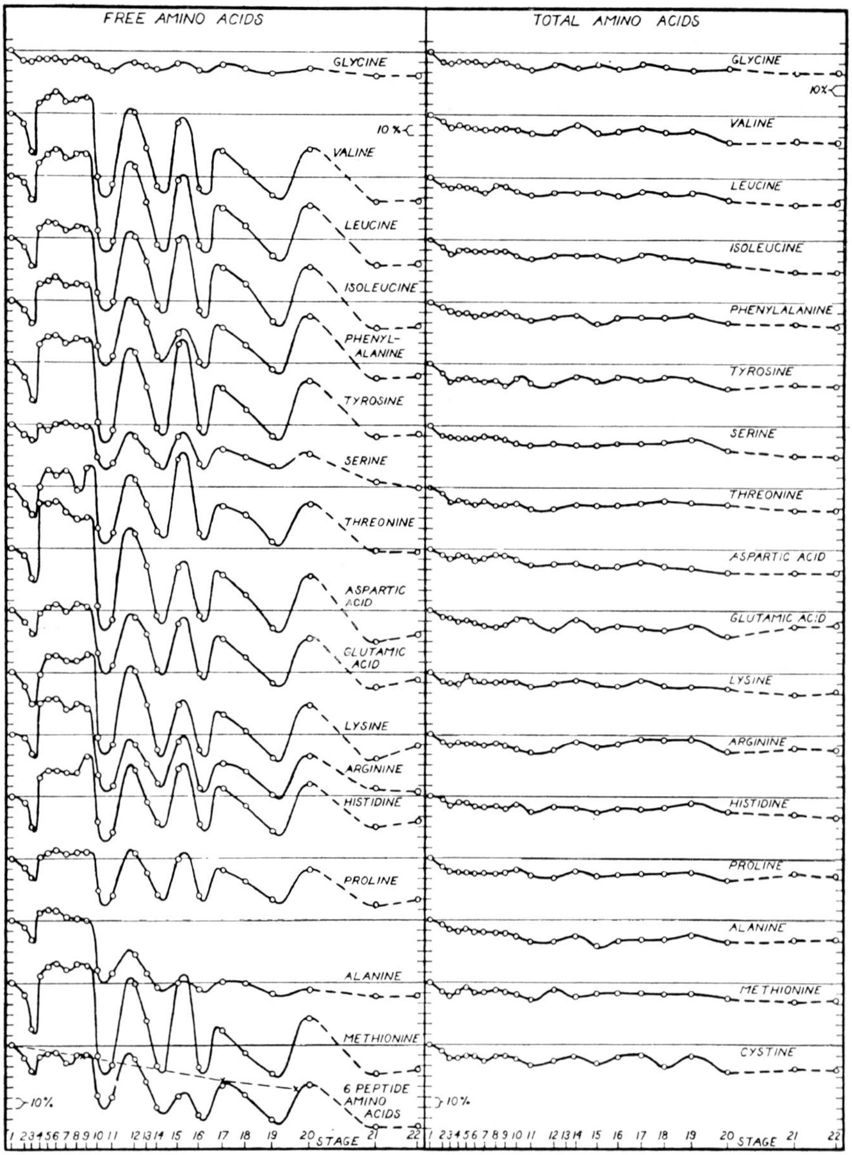

Fig. 104. Changes in the relative amounts of free and total amino acids expressed in per cent (one step: 10%) of the value in the unfertilized egg. See following figure for further explanations.

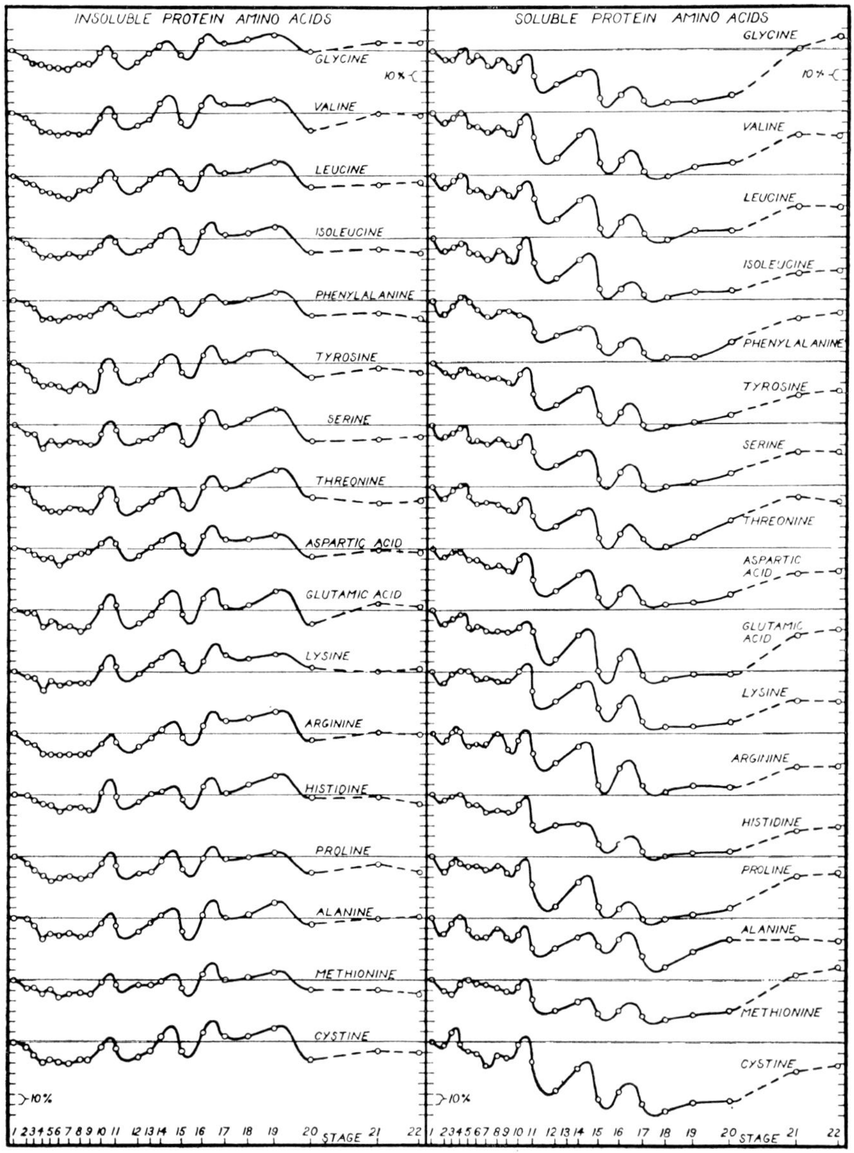

Fig. 105. Developmental changes in the relative amounts of amino acids of soluble and insoluble proteins. See also previous figure. Numbers of stages: 1: unfertilized, 2: fertilized, 3: 2 cells, 4: 4 cells, 5: 8 cells, 6: 16 cells, 7: 32 cells, 8: morula, 9–11: blastula before hatching, 12: hatched blastula, 13–14: mesenchyme blastula, 15–16: gastrulation, 17: gastrula, 18–19: prism, 20–22: plutei. (After Kavanau 1954.)

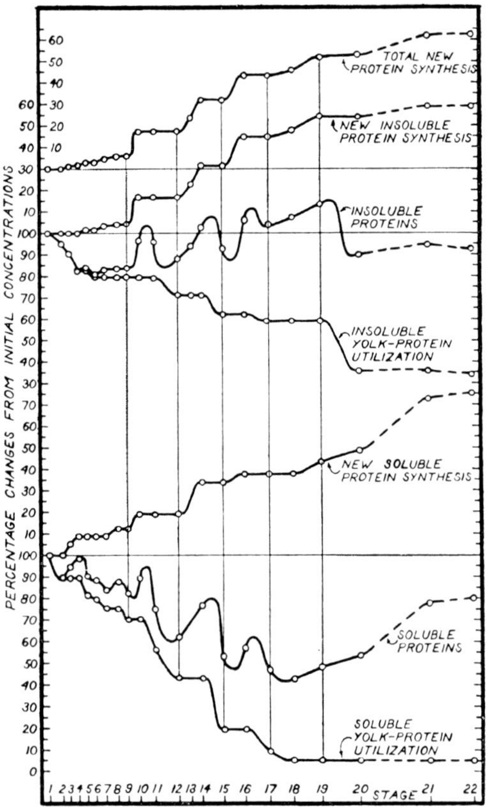

Fig. 106. Changes in proteins during development. Stages as in the previous figure. (After Kavanau 1954.)

were carried out by Ranzi and co-workers by salting out of soluble proteins with increasing concentrations of ammonium sulphate. Comparing the optical density (OD) of the protein solution without salt at λ 280 nm with the OD after addition of various amounts of ammonium sulphate, Ranzi and Citterio (1956) obtained the following diagrams (fig. 107). At first glance one can see that the proteins of the egg and cleavage stages are roughly very similar and that there is also a rough similarity between the proteins of the blastula and gastrula stages and even between the prism, pluteus and adult. The latter fact shows that the method cannot be very suitable for detection of protein differences. But nevertheless it is noteworthy that the proteins of the egg and cleavage stages are different from those of the blastula and gastrula. A similar result was presented by Ishida and Yasumasu (1957), who used immunochemical methods to study the protein status from the egg to the gastrula. Their diagram (fig. 108) shows that all the proteins present in the egg and cleavage stages disappear during gastrulation, and that new proteins, characteristic for the gastrula, are synthesized. No or only small differences were found by disc electrophoresis (Pfohl and Monroy 1962). Terman and Gross (1965) confirmed this, but demonstrated changes in the uptake of labelled amino acids into the proteins of certain bands, visible at gastrulation. These changes are not affected by actinomycin: 'this is additional evidence for the presence of a stored program for protein synthesis in the unfertilized egg'.

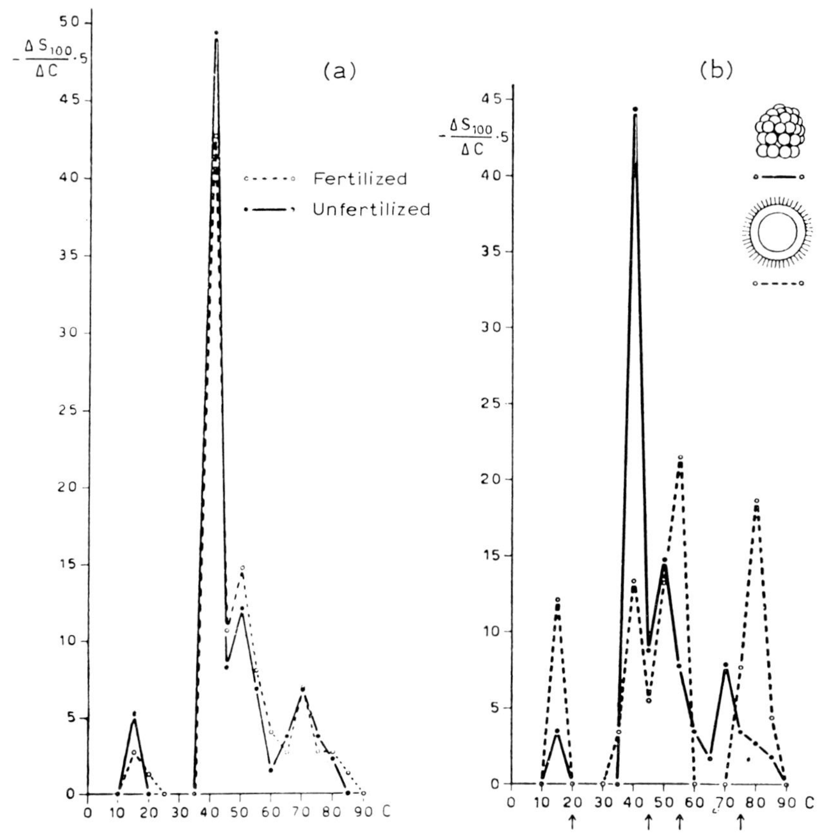

Fig. 107. (For the legend see p. 476.)

Concerning protein synthesis, the compiled results are in complete agreement with regard to the following facts: (1) Protein synthesis slowly increases during cleavage, and is unaffected by actinomycin. (2) There is intense synthetic activity in the late blastula stage, the new proteins being different from those of earlier developmental stages.

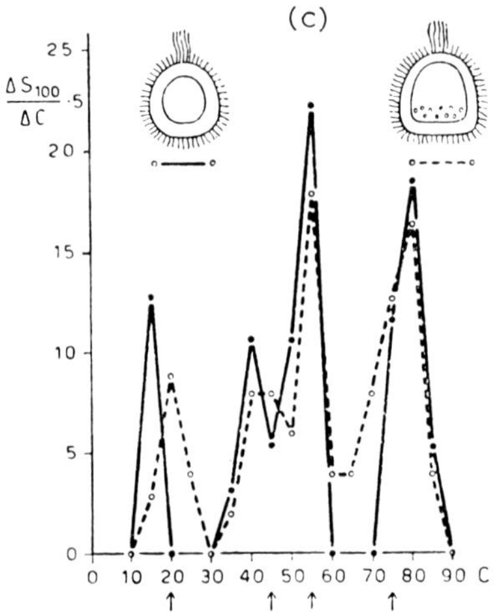

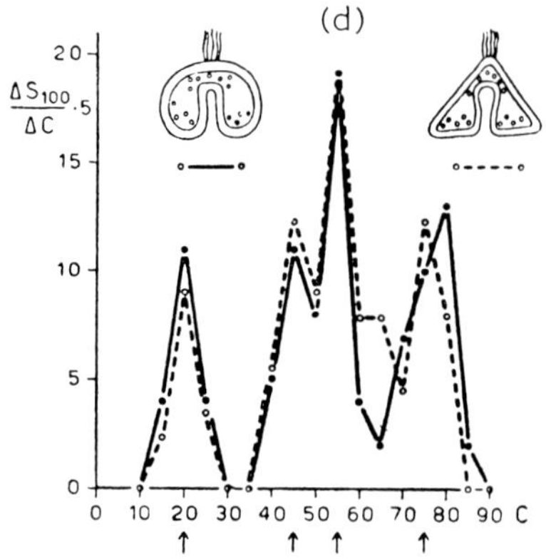

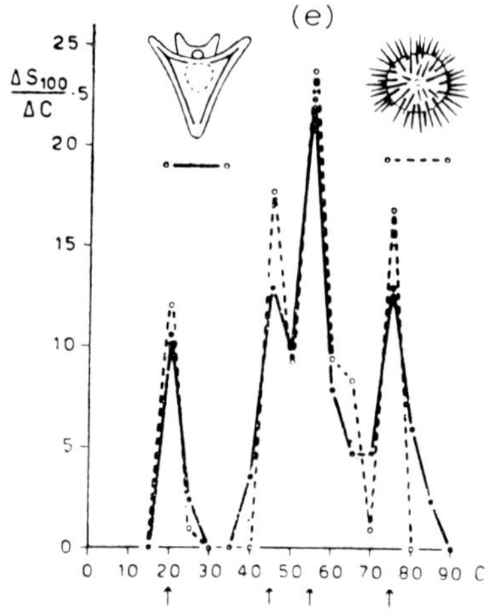

Fig. 107 *(continued)*. Precipitation of proteins by various concentrations of ammonium sulphate. S_{100} expresses complete solubility at the salt concentration zero (C = 0). Each diagram represents two different developmental stages. The arrows below the abscissa indicate the four precipitable fractions of the adult. (After Ranzi and Citterio 1956.)

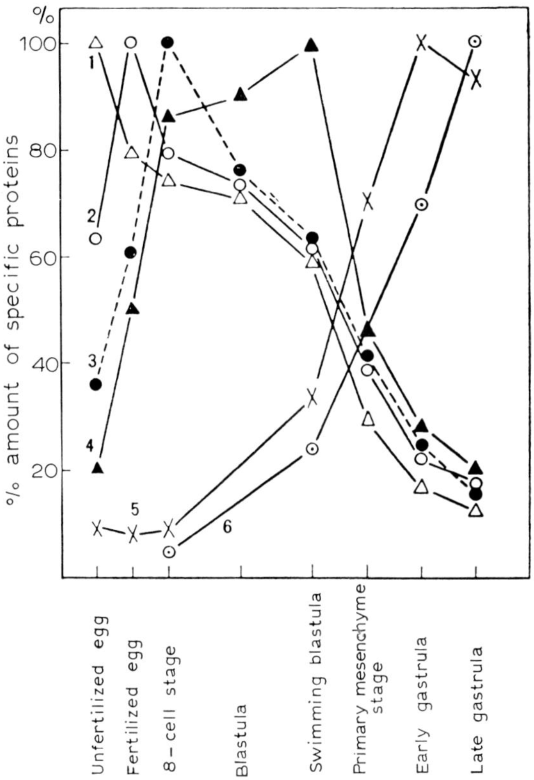

Fig. 108. Changes in the amount of stage-specific proteins revealed by immunological techniques. The proteins of the cleavage stages (1–3) decrease before gastrulation and new ones appear. (After Ishida and Yasumasu 1957.)

12.13. Nucleic acids

Development can be characterized in biochemical terms as the sequence of transcription of the coded information contained in DNA. This transcription, performed by the synthesis of messenger-RNA on one strand of DNA, is probably the primary event of any step in development.

We have all witnessed the rise of molecular biology in the last decade. It might therefore astonish the reader to hear that as early as in 1910, Masing studied nucleic acids of the sea urchin egg and found: ‘that the unfertilized sea urchin egg (*Arbacia pustulosa*) contains abundant nucleic acid and that during cleavage the nucleic acid content does not increase, despite considerable enlargement of the visible nuclear masses’.

12.14. Deoxyribonucleic acids and related enzymes

As early as in 1931(a) Brachet published the following table for the content of DNA at different developmental stages:

	mg DNA per g dry wt
Unfertilized eggs	<0.1
Fertilized eggs	<0.1
Blastula 14.5 hr p.f.	2.6
Blastula 17 hr p.f.	3.3
Gastrula 20 hr p.f.	6.2

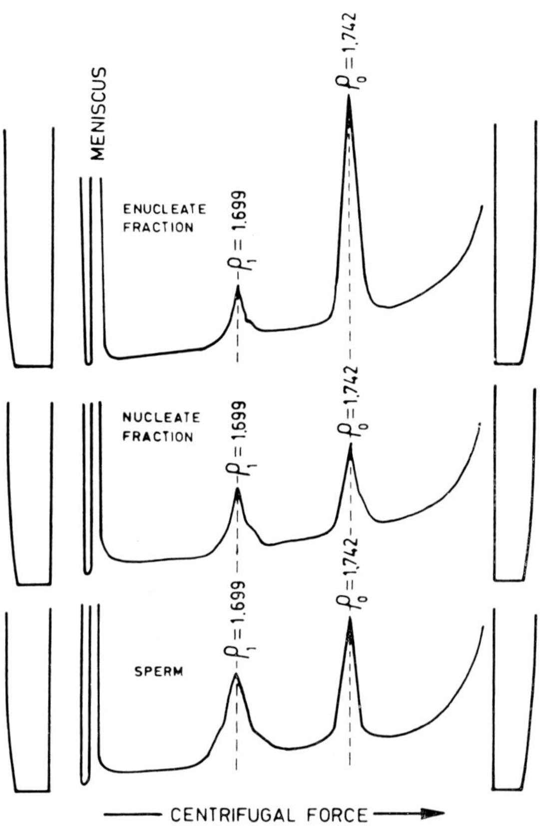

Fig. 109. Equal distribution of nucleic acids from egg halves and sperm on the CsCl density gradient. (From Bibring et al. 1965.)

The egg and sperm of an organism capable of complete development must contain at least one double strand of DNA, bearing all the genetic information in which everything that is to be constructed throughout the life of the individual is coded. It was astonishing to see that the egg of the sea urchin, like the eggs of other organisms, contains much more DNA than does the sperm head (Elson and Chargaff 1952).

It was speculated that the egg could use the 'reserve-DNa' for the multiplication of its own nuclei, all the more so since DNAse is found predomi-

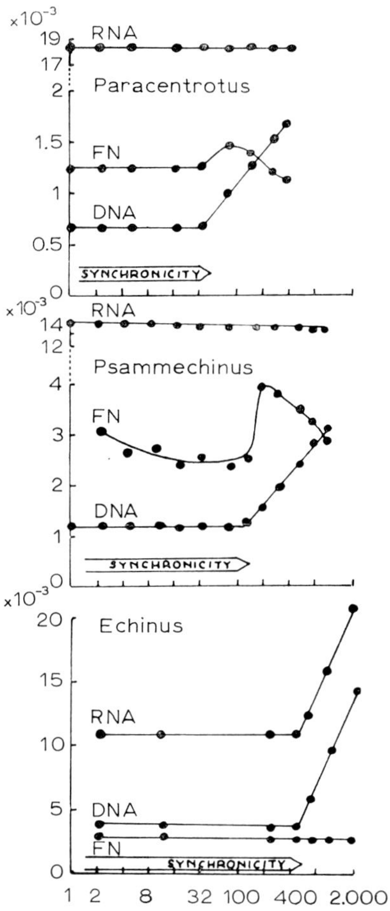

Fig. 110. Changes in the amount of DNA, RNA and free nucleotides in embryos of three species. Below the abscissa, numbers of cells. The synchrony of cleavage coincides with the period of constant DNA. (After Agrell and Persson 1956.)

nantly, if not exclusively, in the cytoplasm and not in the nucleus (Mazia 1949). Egg halves (one with nucleus, one without) obtained by centrifugation (see pp. 406f.) contain DNA of high molecular weight and the same density of 1.699 as the sperm (Bibring et al. 1965) shown in fig. 109. The content of cytosine plus guanine was calculated by the same authors to be 38% of the *Arbacia lixula* DNA. Nothing is known of the function of the cytoplasmic DNA. It should be borne in mind that the cytoplasm of the egg is apportioned to many blastomeres and that the cytoplasmic DNA thus is destined for many cells, the mitochondria of which thus perhaps finally contain no more DNA than any somatic cell. Dawid (1966) has discussed this for the frog embryo. The fact that the egg cytoplasm contains more DNA than the nucleus was endorsed by studying the increase in the DNA content in the total egg over many cleavages. If the DNA content of the cytoplasm of the egg is much higher, then the duplication of the nuclear DNA in the first cleavages cannot bring about a noticeable increase in the total DNA content. This was in fact found to be the case by Agrell and Persson (1956) fig. 110, Baltzer and Chen (1960) and Olsson (1965).

Radioactive thymidine is stored in the unfertilized egg (Esper 1962) and becomes incorporated into acid-insoluble substance, probably DNA. As fig. 111, from Hinegardner et al. (1964), demonstrates, in the two-cell stage (between 100 and 120 min p.f.) about 55,000 cpm thymidine are incorporated into a large number of embryos. Before the first cleavage, however, 11,000 cpm are incorporated. We should subtract the 11,000 cpm from the 55,000 cpm of the two-cell stage (leaving 44,000 cpm), since 11,000 cpm were always

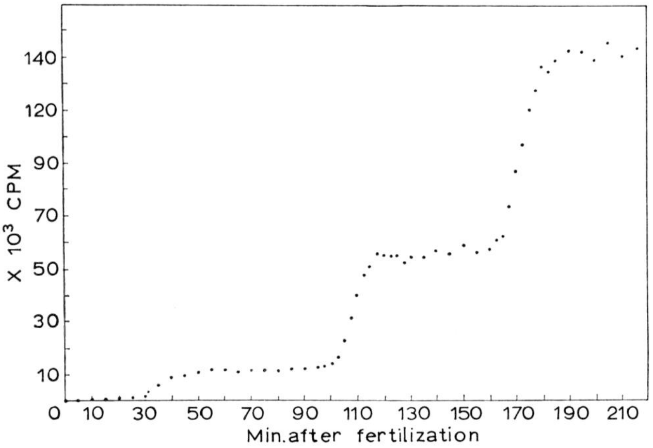

Fig. 111. Thymidine incorporation during early cleavage. (From Hinegardner et al. 1964.)

present at the beginning of the two-cell stage. Taking the number of eggs used in the experiment as x, each egg incorporated $44{,}000/x$ and each diploid nucleus $44{,}000/2x$, or $22{,}000/x$. In the mononuclear egg, however, we obtain the value of $11{,}000/x$ for the single nucleus, being exactly half the amount incorporated into advanced nuclei. Thus it seems possible that only one of the 'haploid' pronuclei, the male or the female, has synthesized DNA. Whether one of the gamete-nuclei already has the double amount of DNA before fertilization or one of the nuclei takes up exclusively pooled DNA or its fission product out of the cytoplasm whereas the other uses externally administered thymidine cannot be decided.

DNA synthesis starts in early telophase and proceeds until the individual 'Kernbläschen' are fused to one nucleus, at least after the first and second cleavage (fig. 112). At least in early cleavage the G_1-period, the period

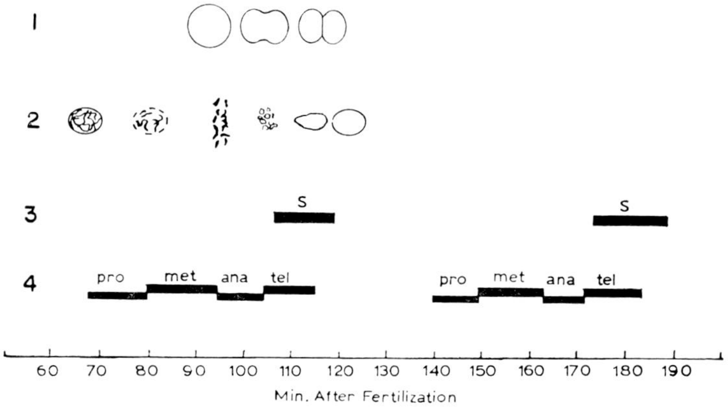

Fig. 112. Time and duration of DNA synthesis (3) in relation to mitotic phases (2). (From Hinegardner et al. 1964.)

between completion of the division and DNA synthesis (S-period), is lacking. These findings have recently been confirmed by Nagano and Mano (1968).

The activity of thymidine kinase corresponds to the progress of DNA synthesis in that the phosphorylation of thymidine is increased just before and during the S-period (Hansen-Delkeskamp and Duspiva 1966) (fig. 113). This was confirmed and similar changes in activity were observed for thymidilate kinase by Nagano and Mano (1968). Once again it can be noted that the individual variations among different batches are high.

Methylation of DNA by labelled methylmethionine was found by Comb (1965) as late as in the gastrula of 16 hr and on a larger scale even later in

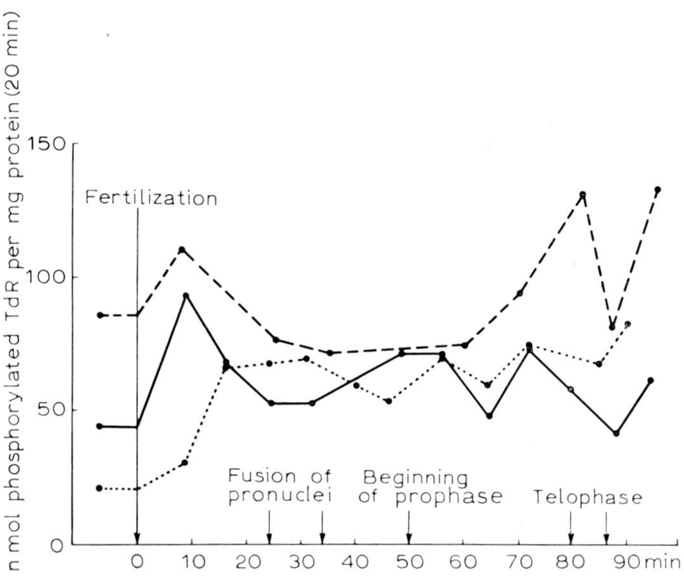

Fig. 113. Activity of thymidine kinase during early cleavage in three different batches of eggs. (From Hansen-Delkeskamp and Duspiva 1966.)

the pluteus of 23 hr. But according to Scarano et al. (1965) the methyl group of 5-methylcytosine in DNA is already taken over from methyl-methionine in the cleavage stages.

It was surprising to discover that the chromosomes of closely related species of animals can contain very different amounts of DNA and exhibit a different incorporation rate of labelled thymidine. In sea urchins Baltzer and Chen (1960) found the rate of DNA-synthesis of *Paracentrotus lividus* to be about twice as high as that of *Arbacia lixula*.

12.15. Ribonucleic acid

As early as 1931 Brachet (1931b) published data on the pentose content (thought to be equivalent to the RNA content), demonstrating a sharp decrease:

	mg pentose per g dry wt
Unfertilized eggs	6.35
Gastrulae (20 hr p.f.)	3.25
Plutei (40 hr p.f.)	2.60

and concluded that 'the values obtained clearly suggest the idea that the considerable reduction in pentose ... is bound up with synthesis of thymonucleic acid, which occurs at the same time'.

In 1948 Schmidt et al. confirmed Brachet's finding of an increasing DNA-content, but found a nearly constant and high level of RNA in different developmental stages.

	10^{-3} μgP per embryo		ratio
	DNA	RNA	DNA/RNA
Unfertilized egg	0.6	22.0	0.05
Morula 8 hr p.f.	3.4	22.8	0.17
Pluteus 24 hr p.f.	10.7	21.5	0.46

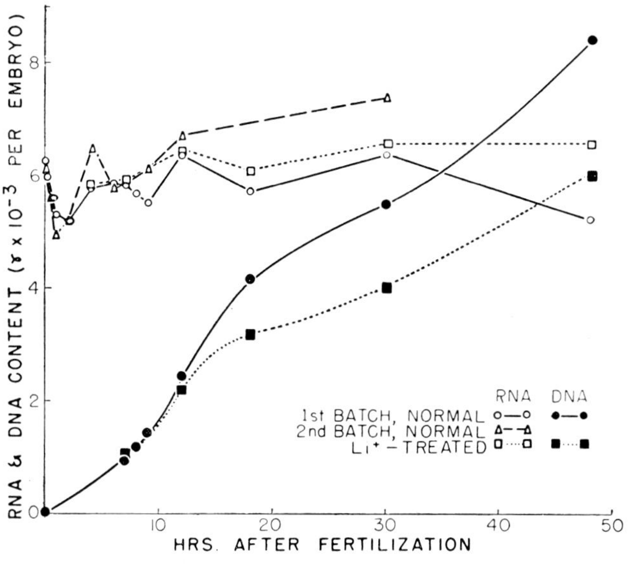

Fig. 114. DNA and RNA content up to pluteus stage. (After Elson et al. 1954.)

Elson et al. (1954) found fluctuations in the RNA content within certain limits (fig. 114). The RNA composition of nucleotides was practically constant (table 5).

TABLE 5

Composition of RNA of *Paracentrotus lividus* eggs and embryos.

Moles nucleotide per 10 moles adenylic acid*

	1st batch; normal			2nd batch; normal		
	GA	CA	UA	GA	CA	UA
Unfertilized eggs	13.3	12.3	9.3	12.2	12.4	8.9
All stages, † mean	13.75	12.17	9.23	11.99	12.19	9.09
Standard error of mean	± 0.14	± 0.16	±0.08	± 0.16	± 0.17	±0.08
Coefficient of variability	3.6	4.8	3.0	3.7	3.8	2.4
No. of stages analyzed		14			9	

	3rd batch; normal			Li⁺-treated		
	GA	CA	UA	GA	CA	UA
Unfertilized eggs	12.7	12.2	9.4			
All stages, † mean	12.25	12.00	9.20	13.35	11.68	9.48
Standard error of mean				± 0.26	± 0.24	±0.17
Coefficient of variability				4.3	4.6	4.1
No. of stages analyzed		2			6	

'* GA, guanylic acid; CA, cytidylic acid; UA, uridylic acid.
† This includes the unfertilized eggs. The following developmental stages were investigated: first batch, unfertilized eggs; embryos, 10 and 30 min (1 cell), 1-hr (2 cells), 2-hr (4 cells), 4-hr (32 to 64 cells), 6-, 7- and 8-hr (blastula), 9-hr (mesenchyme blastula), 12-hr (onset of gastrulation), 18-hr (gastrula), 30-hr (oral opening formed), and 48-hr (free swimming pluteus). Second batch, unfertilized eggs, 30 min, 1-, 2-, 4-, 6-, 9-, 12- and 30-hr embryos. Third batch, unfertilized eggs and 30-hr embryos. Li^+-treated 4-, 7-, 12-, 18-, 30- and 48-hr embryos.' (From Elson et al. 1954.)

Changes in RNA content of the fertilized egg were confirmed for the first hour of development by Olsson (1960): a decrease in RNA was found to occur as little as five minutes after fertilization, accompanied by an increase in acid-soluble nucleotides. As nuclease activity (not specified by Olsson) rises simultaneously, the RNA present in the egg seems to be split immediately after fertilization and the nucleotides set free are probably used for new RNA synthesis starting some 10 or 15 min after fertilization (see p. 487).

TABLE 6

Stimulation of L-leucine-C^{14} incorporation with ribosomes from unfertilized and fertilized eggs by poly UC.

Ribosomal source	Additions	L-leucine incorporation (cpm/mg protein)	Ratio of incorporations: stimulated / control
Unfertilized egg	None	15	
	Poly UC	367	24
	Poly U	39	2.6*
2-cell stage (90 min)	None	103	
	Poly UC	297	3

* As compared to 16 for phenylalanine at the same concentration of poly U.

TABLE 7

Stimulation of L-phenylalanine-C^{14} incorporation with ribosomes from embryos of various stages of development by poly U.

Ribosomal source	Additions	L-phenylalanine incorporation (cpm/mg protein)	Ratio of incorporations: stimulated / control
Unfertilized egg	None	21	
	Poly U	3615	173
2-cell stage (90 min)	None	113	
	Poly U	3010	27
Blastula (12 hr)	None	117	
	Poly U	1600	14

From Nemer (1962).

A new approach to the study of RNA in the egg was made possible by Wilt and Hultin (1962) and Nemer (1962) by using artificial messenger-RNAs, consisting only of uridine (poly U) or uridine plus cytidine (poly UC).

It can be seen that addition of both synthetic messengers (tables 6, 7) elicits an increase in amino acid incorporation into proteins. This increase is very small in fertilized eggs or developmental stages but very high (173-fold) in homogenates of unfertilized eggs. These data permit the conclusion that in unfertilized eggs many ribosomes are accessible for mRNA; this means that the protein synthesizing machinery is ready, but not functioning, because either no maternal messenger is present or it is present in inactivated form (cf. p. 462). Stimulation of protein synthesis by the addition of artificial messenger was confirmed by Wilt and Hultin (1962), although a difference between fertilized and unfertilized eggs was found only with low poly U concentrations (0.1 mg/ml) and not with 0.5–0.7 mg/ml as in Nemer's experiment. Wilt and Hultin draw attention to the fact that the amounts of poly U required for ... 'phenylalanine incorporation were considerably greater than that required in the *E. coli* system' where it was detected first, and that 'poly U had a relatively moderate affinity to the microsomes'.

These experiments lend support to the previous work of Hultin (Hultin and Bergstrand 1960; Hultin 1961) in which a great increase in amino acid incorporation into proteins was described after fertilization (see p. 461). The question was thus whether mRNA is present in the unfertilized egg or is newly synthesized.

Gross and Cousineau (1963) used actinomycin to suppress new mRNA synthesis. They found that relatively high concentrations are necessary to obtain 96% inhibition.

TABLE 8

Inhibition of C^{14}-uracil incorporation in actinomycin-treated embryos.

Conc. act. D μg/ml	cpm/10^6 eggs	% inhibition	Development at $5\frac{1}{2}$ hr
0	840	0	Normal early blastulae
1.4	304	64	Normal early blastulae
5.7	197	77	Very irregular blastulae, slight cleavage delay
24	54	94	"Morulae" and multicellular masses; no normal forms
96	36	96	Many cleavages, but no differentiation

From Gross and Cousineau (1963).

However, one never knows how much of the exogenous actinomycin actually penetrates into the cell. A second experiment shows that the permeability for actinomycin may change, because in the first hour of development 65 μg/ml actinomycin inhibit only 42 % uracil incorporation, whereas after two hours 79 % of the incorporation into RNA is inhibited.

In connection with this difference in the penetration or action of actinomycin the different susceptibility of early RNA synthesis to actinomycin inhibition should also be mentioned.

TABLE 9

Uptake of C^{14}-valine in normal and actinomycin-treated embryos.

Time post-fertilization (hr)	dpm/10^6 embryos $\times$ 10^{-5} Control	Act. D 115 μg/ml	Developmental stage Control	Act. D
1	1.19	0.96	2-cell	2-cell
3	3.00	2.98	15-, 32-cell	Irregular 8-, 16-cell
5	3.99	4.13	Early blastula	32-cell and above, irregular
12	5.06	6.06	Gastrula	Undifferentiated multicellular mass

From Gross and Cousineau (1963).

In these first two hours of development as well as in later developmental stages, no significant inhibition of valine uptake into proteins was detected with actinomycin (table 9). It must be concluded that the mRNA guiding this protein synthesis is already present in the unfertilized egg, or it is synthesized in the first hours in the resistant or unaffected 48 % of the RNA synthesis occurring in the presence of actinomycin. Later, however, Gross et al. (1964) found that actinomycin D inhibits only uridine incorporation into RNA-types heavier than 4S (in the ultracentrifuge), whereas in controls other RNA-types also took up the marker. All RNA synthesized in the presence of actinomycin seems to be tRNA (fig. 115b). In the region of 18S the RNA of the control eggs exhibits a peak of radioactivity and a continuous incorporation level in all the fractions is found (fig. 115a). Such a continuous level of radioactivity may result from the presence of many types of RNA of different molecular size, probably containing different types of mRNA.

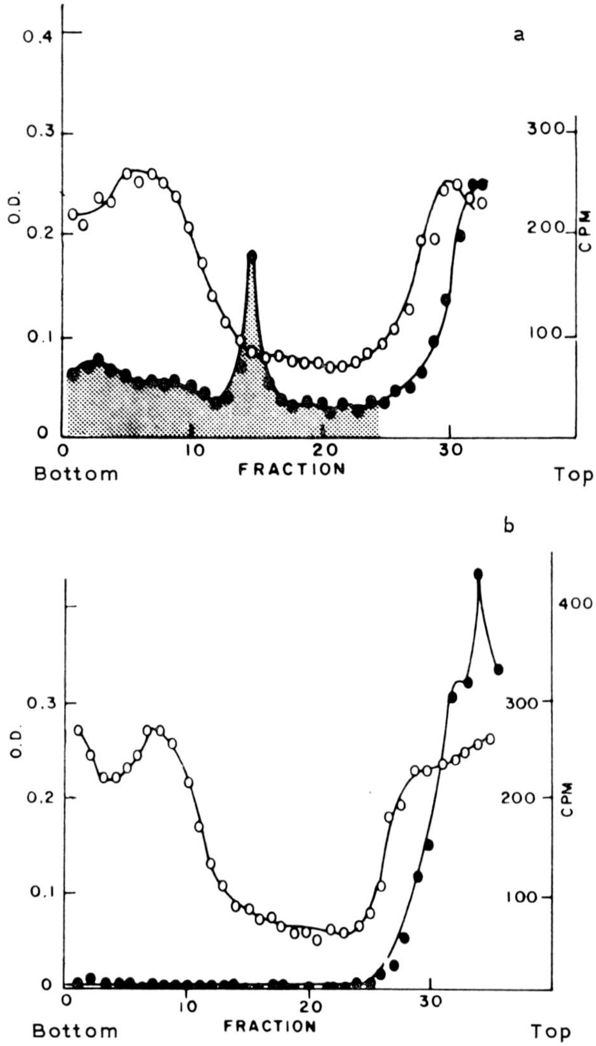

Fig. 115. Distribution of radioactivity in RNA after uridine incorporation 20 min p.f. in the sucrose density gradient. (a) The shaded area represents different size classes of RNA. (b) Pretreatment with actinomycin-D suppresses RNA-synthesis with the exception of 4S tRNA which is formed in fractions above 25. Open circles: optical density; filled circles: radioactivity. (After Gross et al. 1964.)

What can the task of this new messenger be? Probably the same as that of the mRNA already present in the unfertilized egg (see pp. 462, 486).

Gross et al. (1964) described development until late cleavage stages in the presence of actinomycin. Denny and Tyler (1964) divided eggs by centrifugation into nucleate and non-nucleate halves and found the same rate of protein synthesis after artificial parthenogenesis and normal fertilization. We are therefore convinced today that the messenger needed for the first hours of development is present at least in part in the cytoplasm of the unfertilized egg.

The ribosome apparatus for protein synthesis is ready and the mRNA which controls development until the end of cleavage phase is present. The new fascinating problem which therefore arose was why in fact no protein synthesis occurs in the unfertilized egg, or what prevents the protein synthesizing machinery from functioning when the principal components necessary for it are present?

Monroy and his coworkers, guided by the idea that the activity of mRNA in the egg could be inhibited by a protein capsule surrounding the molecule, tried to attack the latter with proteolytic enzymes. Using trypsin, Monroy et al. (1965b) were able to increase the amino acid incorporation of ribosomes to a high level.

The following facts are the most interesting: although untreated ribosomes do not take up phenylalanine, they do so to a great extent after addition of polyuridylic acid as artificial messenger: if the ribosomes are coated with protein, polyuridylic acid is able to come through the envelope. The RNA of the blastula, for instance, lacks this ability. It is further note-

TABLE 10

Change of distribution of template RNA activity in subcellular fractions resulting from fertilization and tryptic digestion.

cpm incorporated	With RNA (0.5 mg) derived from H	S	M	P	Without RNA
Unfertilized	256	158	80	334	38
Fertilized	307	252	328	96	202
Unfertilized, trypsinized	321	218	317	136	73

H, unfertilized whole homogenate; S, 105,000 *g* supernatant; M, microsomes; P, 12,000 *g* precipitate in the following fractionation. *Hemicentrotus pulcherrimus* eggs developed for 30 min after fertilization at 20 °C. Trypsinization was stopped by soybean trypsin inhibitor. The table includes in the last line the values of ribosomal activity without addition of RNA for comparison. (From Mano and Nagano 1966.)

worthy that the presence of several amino acids also increases their uptake into proteins and finally that all this synthetic capacity of the ribosomes of the unfertilized egg is considerably increased after trypsin treatment of the ribosomal fraction. The interpretation given was 'that the mRNA that is synthesized during oogenesis becomes attached to ribosomes. The mRNA-ribosome complex, however, is made inactive by a protein coat, the activity being released by the removal of the coat'. And it can be further suggested that ribosomes without mRNA are coated as well, as the RNA of the blastula is able to attach to free ribosomes after trypsin treatment. It is worthy of note that in 1949 Lundblad described a transient activation of proteases immediately following fertilization. Mano and Nagano (1966) recently amplified these studies, as shown in table 10.

The highest template activity of RNA is found in the 12,000 *g* pellet, i.e. in relatively large particles of the unfertilized egg. With fertilization or trypsin treatment this maternal RNA becomes liberated from the 12,000 *g* particles and is found mainly in the ribosome fraction. 'These facts suggest that the maternal RNA resides in some particles other than ribosomes... which sediment faster... the activation by the trypsin treatment may be due not to the removal of some inhibitory proteins from the ribosomes but to the release of the maternal RNA from other particles besides ribosomes'. And the RNA set free by trypsinization 'may combine with ribosomes to form polyribosomes spontaneously'. Neither the 12,000 *g* pellet nor the microsome fraction had the same protein-synthesizing capacity as the two fractions combined (see table 11).

TABLE 11

Effect of combination of the particulate fractions on the stimulation of protein synthesis by trypsinization.

	P	M	P + M cpm incorporated
Nontreated	10	22	38
Trypsinized	74	51	113

Abbreviations are the same as in table 10. (From Mano and Nagano 1966.)

These heavy particles are thought to be a kind of 'informosomes'. They seem to have maternal RNA attached to them and 'this may be favourable for the protection of the RNA... for a long survival awaiting fertilization'.

It is very probable that a trypsin-like protease also triggers the start of protein synthesis in vivo. In a similar manner to Lundblad, Mano (1966) noted that a 'pH 8 protease was activated temporarily after fertilization' and was 'strongly inhibited by soy bean trypsin inhibitor'. '... it is believed to occur in the same particles as those which contain the maternal RNA'. And these particles are perhaps one type of 'informosomes', a new class of cellular components assumed to exist since Spirin and coworkers began to speak of them in 1964.

According to the investigations of Spirin and Nemer (1965), informosomes are ribonucleoprotein (RNP) particles which are lighter than polyribosomes, the latter being engaged in protein synthesis, and heavier than single ribosomes. They have mRNA attached to them but are not active in protein synthesis. They are thought to become active later and therefore can be regarded as a conservation unit for newly synthesized mRNA. This definition is derived from the following facts: (1) Sedimentation profile (fig. 116): there are several distinct and reproducible peaks of particles between 75S ribosomes and 18S ribosome subunits having incorporated uridine.

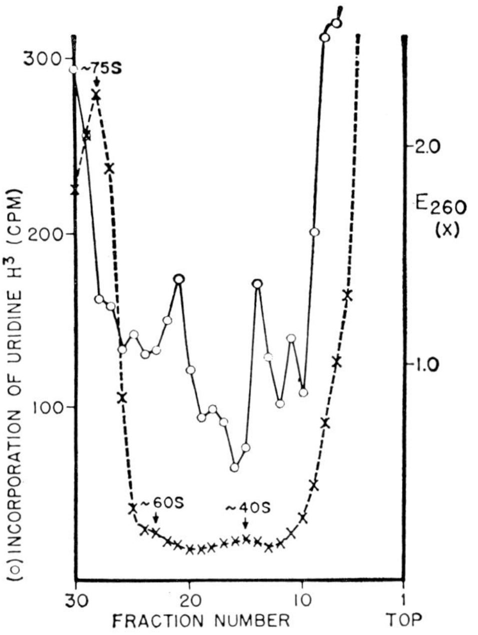

Fig. 116. Centrifugation of the 12,000 *g* supernatant of embryos which had incorporated uridine for 2 hr beginning with the four-cell stage. (From Spirin and Nemer 1965.)

(2) Phenol extraction of these RNP particles reveals a mRNA annealing in a high percentage with DNA. (3) The particles active in protein synthesis contain little newly synthesized mRNA, while those with newly synthesized mRNA are slow in protein synthesis (fig. 117). (4) After RNase treatment no

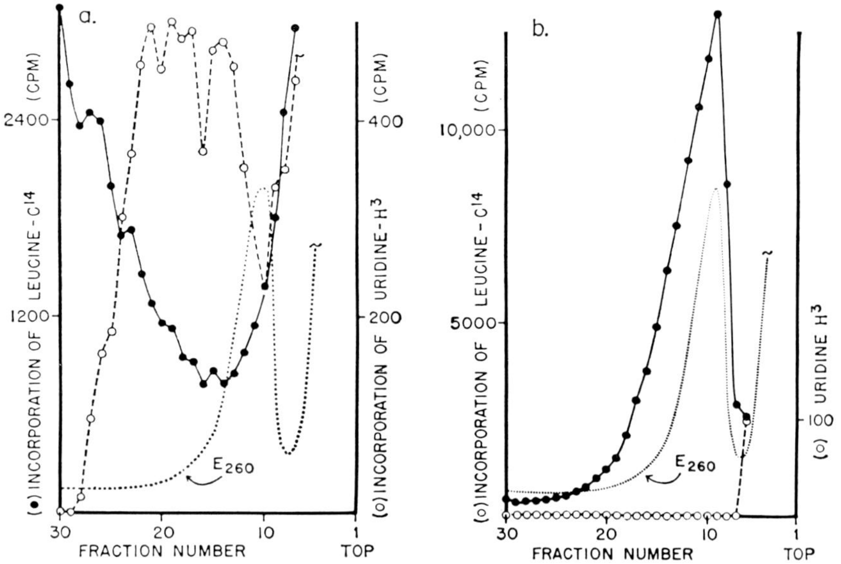

Fig. 117. Incorporation of uridine and leucine in the polyribosomes of cleavage stages. In (b) after treatment with ribonuclease the peak of radioactivity corresponds to the ribosome peak in optical density. (From Spirin and Nemer 1965.)

radioactivity is sedimentable (the RNA is split up into very light oligonucleotides by the enzyme action) and the radioactivity of proteins (after incorporation of leucine) is found to coincide with the ribosomal peak: this demonstrates that by the fission of RNA the ribosomal units of polyribosomes (and informosomes?) are set free.

Now we can assume that in the sea urchin egg there are two types of particles with attached mRNA that differ from both ribosomes and polyribosomes, at least in size. A common feature of Mano and Nagano's and Spirin and Nemer's particles (informosomes) is that they bear mRNA and regulate little or no protein synthesis. In perfectly executed experiments Slater and Spiegelmann (1966b) estimated the amount of mRNA present in the unfertilized egg by comparing the priming activity of the egg RNA with the RNA of the MS-2 phage. 'Thus, since MS-2 RNA may be regarded as a 100 percent translatable message, the bulk RNA extracted from the unfertilized eggs possesses a relative template activity of approximately 4–5 per cent'. The rest is of course rRNA and tRNA.

Nemer and Infante (1965) distinguished three distinct size classes of the newly synthesized mRNA of cleavage stages after density gradient centrifugation (fig. 118). According to recent work by this team (Infante and

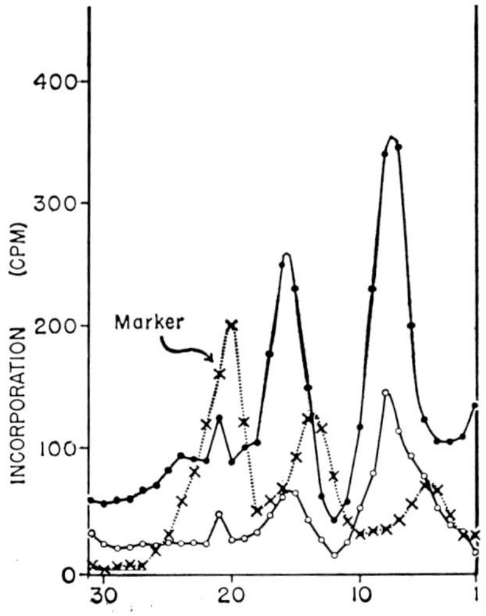

Fig. 118. Incorporation of uridine in RNA of polyribosomes (filled circles). The RNA of the three peaks forms hybrids with DNA (open circles). The curve connecting crosses represents two peaks of ribosomal RNA and one peak of 4S RNA. (After Nemer and Infante 1965.)

Nemer 1967) large polyribosomes of about 300S appear soon after fertilization with accumulation 2 hr after fertilization (fig. 119) and regress to a constant level, which is maintained until the blastula stage of 12 hr p.f. Later a new accumulation of this size class of polysomes is noticeable. The first accumulation is insensitive to actinomycin, indicating that maternal RNA present in the unfertilized egg may form these 300 S polysome. About 4 hr p.f. a new size class of polysomes appears, sedimenting at 200S, which are lacking in the presence of actinomycin. Therefore it is probably the above-mentioned mRNA of the cleavage stages, that forms the new 200S polyribosomes. If the relationship between the 200S and 300S polysomes is calculated and compared with the data of valine incorporation into proteins (Gross et al. 1964) in the absence and presence of actinomycin,

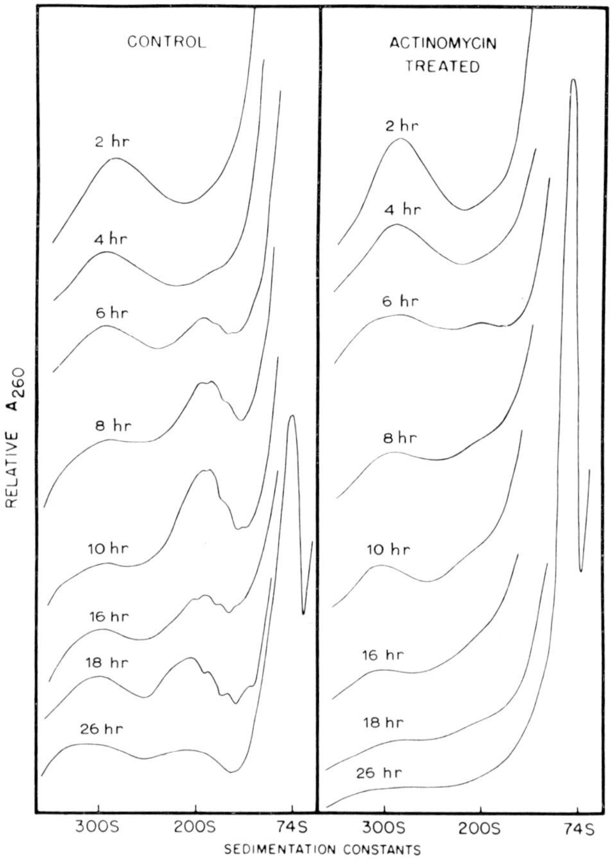

Fig. 119. Sedimentation profiles of polyribosomes of different developmental stages. At about 6 hr a new 200S-class, lacking in actinomycin-treated embryos, becomes visible (After Infante and Nemer 1967.)

a very good agreement is found. Although newly formed mRNA may already be synthesized in the early cleavage stages, it obviously contributes little if at all to protein synthesis and embryonic development until the early blastula. This assumption is further supported by the fact that amino acid incorporation per ribosome is considerably greater for 300S than for 200S polysomes. Infante and Nemer assumed 'that the RNA templates synthesized at this time are accumulated in the form of inactive polysomes, and that the extensive cellular changes, associated with the formation of the mesenchyme, blastula and early gastrula, are brought about by the activation and use of these accumulated polysomal templates'.

Other experiments support a connection between the RNA synthesized during early cleavage and the control of development beyond the blastula stage. The observation of Agrell (1958) that the micromeres become extremely rich in RNA after the fourth cleavage served as starting point for a series of investigations. It was established by autoradiography that in the period of the 16-cell stage only the micromeres have a prolonged interphase (cf. p. 377) and incorporate uridine into the RNA of the nuclei. This RNA is later found in the cytoplasm of the micromeres (Czihak 1965). On pp. 427f., figs. 60–62, it is shown that transplanted micromeres have the capacity of inducing a supernumerary archenteron in the host blastula (Hörstadius 1935). If RNA synthesis in the micromeres is influenced by antimetabolites of RNA synthesis such as 8-azaguanine, which according to Bamberger et al. (1963) is exclusively incorporated into RNA, the blastula is not able to gastrulate! This antimetabolite does not give rise to general inhibition of differentiation, as other organs are indeed capable of further differentiation. The primary mesenchyme, for instance, differentiates into skeletoblasts and forms a normal skeleton. And as the primary mesenchyme develops from the micromeres these findings demonstrate that the RNA synthesized normally or with 8-azaguanine is not needed for differentiation of primary mesenchyme and skeleton-forming cells but for differentiation of the adjacent macromeres into entoderm cells (Czihak 1965, 1968). In the light of our present knowledge it is very probable that the actual RNA of the micromeres is the entoderm-inducting and gastrulation-triggering substance. It is not yet clear which type of RNA is produced by the micromeres of the 16-cell stage.

It must be noted here that some authors call into question that RNA synthesis occurs at all in the cleavage stages. Brachet et al. (1963) for instance, using P^{32}, concluded that 'no incorporation could be detected in whatsoever RNA before the blastula stage'. But this stands in contradiction with their own findings stating that: 'From the 4-cell stage on, the nuclear take up uridine H^3 at a rate of $\frac{1}{3}$ in DNA and $\frac{2}{3}$ in nuclear RNA' (Ficq et al. 1963). Slater and Spiegelmann (1966a) stated 'that the early events of embryogenesis do not alter the major physical or chemical characteristics of the bulk RNA native to the unfertilized egg'. But as the ribosomal RNA content is very high and thus hides from view minor components with the same sedimentation characteristics, one cannot expect to detect alterations in mRNA only in optical density.

The above-mentioned incorporation of uridine into DNA has been already established by many authors: Nigon and Gillot (1963), Ficq et al. (1963)

and Czihak et al. (1967). In this case uridine must be transformed into thymidine by methylation in the 5-position, as the radioactivity is found in thymidine after hydrolysis of the DNA peak after separation of the nucleic acids on hydroxylapatite columns (Czihak et al. 1967).

With regard to transfer RNA, characterized by the CCA-terminal group, Glišin and Glišin (1964) detected the first synthesis in sea urchin development even before the 8-cell stage (fig. 120, cf. fig. 115). Later, before the

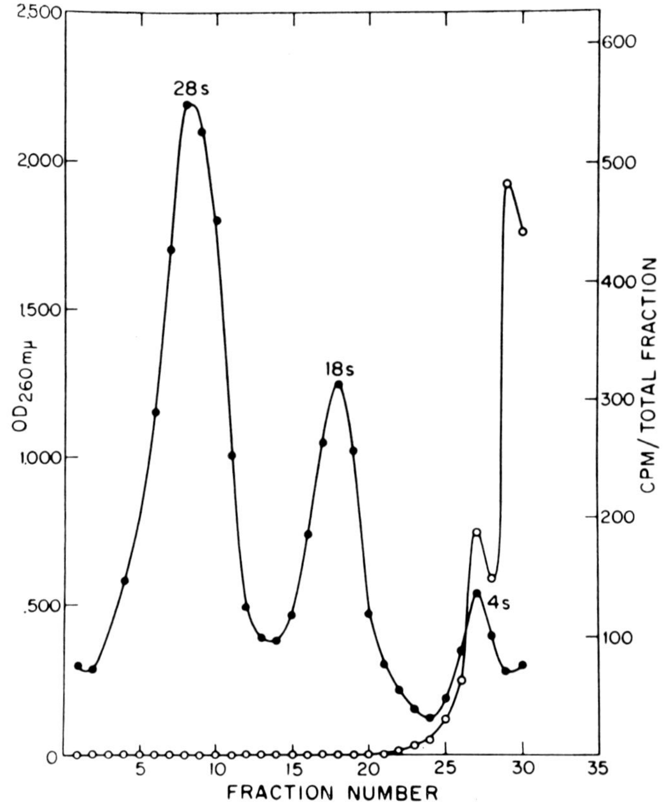

Fig. 120. Radioactivity of RNA (open circles) in sucrose gradient. Eggs had incorporated P^{32} up to the 4-cell stage only into 4S RNA. Filled circles: optical density. Two clear ribosomal peaks (28S and 18S). (From Glišin and Glišin 1964.)

64-cell stage, synthesis of tRNA was found to be increased (fig. 121) and moreover a slow synthesis of different-sized types of RNA spread over the gradient was seen. The same picture has been obtained with preparations

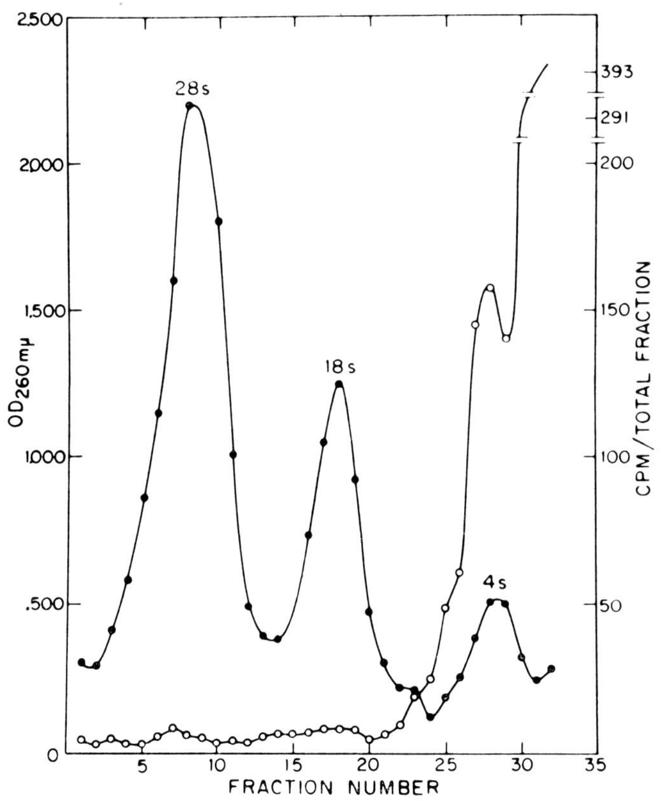

Fig. 121. The same as in fig. 120 for the 32-cell stage. Here radioactivity is found not only in 4S RNA but also in RNA of different size classes in the fractions below 20. (After Glišin and Glišin 1964.)

of blastula RNA, not only by Glišin and Glišin but also by Gross et al. 1965. 'The newly formed high-molecular weight RNA is most likely to be messenger RNA' (Glišin and Glišin 1964), '... most of the RNA made during the period of cleavage is DNA like, and probably messenger RNA' (Gross et al. 1965).

Thus we already have considerable evidence from different experiments that mRNA is already synthesized in the cleavage stages and that this mRNA is responsible for the next step of differentiation which consists in the separation of ectoderm and entoderm in the process of gastrulation (Czihak 1965; Barros et al. 1966; Giudice et al 1968).

As for ribosomal RNA, evidence has been presented by Giudice and

Mutolo (1967) that it is not synthesized in great quantities before the midgastrula stage. All the rRNA necessary for development until the early gastrula with its increased protein synthesis is therefore provided by the egg.

Nemer and Infante (1967) suppose that after the blastula stage half of the rRNA is replaced within 30 hr, as in *Strongylocentrotus purpuratus* the 18S rRNA of the egg can be split into 13S RNA by heating and this type of heat-convertible RNA decreases after the blastula stage and is probably replaced by newly-formed rRNA.

This manuscript was finished in 1968. Therefore only a part of the literature published in 1968 was taken into account.

References

AFZELIUS B. A., 1956. The ultrastructure of the cortical granules and their products in the sea urchin egg as studied with the electron microscope. Exptl. Cell Res. *10*, 257–285.

AGRELL I., 1956. A mitotic gradient as the cause of the early differentiation in the sea urchin embryo. Bertil Hanström Zool. Pap., Lund, 27–34.

AGRELL I., 1958. A cytoplasmic production of ribonucleic acid during the cell cycle of the micromeres in the sea urchin embryo. Arkiv Zool. *11*, 435–440.

AGRELL I. and H. PERSSON, 1956. Changes in the amount of nucleic acids and free nucleotides during early embryonic development of sea urchins. Nature *178*, 1398–1399.

AKETA K., 1964. Some comparative remarks on the transient change in lactic acid content in sea urchin eggs following fertilization. Exptl. Cell Res. *34*, 192–194.

AKETA K., R. BIANCHETTI, E. MARRÉ and A. MONROY, 1964. Hexose monophosphate level as a limiting factor for respiration in unfertilized sea urchin eggs. Biochim. Biophys. Acta *86*, 211–215.

BÄCKSTRÖM S., 1959. Activity of glucose-6-phosphate dehydrogenase in sea urchin embryos of different developmental trends. Exptl. Cell Res. *18*, 347–356.

BÄCKSTRÖM S., 1963. 6-phosphogluconate dehydrogenase in sea urchin embryos. Exptl. Cell Res. *32*, 566–569.

BALTZER F. and P. S. CHEN, 1960. Über das zytologische Verhalten und die Synthese der Nukleinsäuren bei den Seeigelbastarden *Paracentrotus* ♀ × *Arbacia* ♂ und *Paracentrotus* ♀ × *Sphaerechinus* ♂. Rev. Suisse Zool. *67*, 183–194.

BALTZER F., P. TARDENT and P. S. CHEN, 1967. About the DNA-synthesis during the early development of *Paracentrotus lividus*, *Arbacia lixula* and their hybrids. Experientia *23*, 777–779.

BAMBERGER J. W., W. E. MARTIN, L. W. STEARNS and W. B. JOLLEY, 1963. Effect of 8-azaguanine on cleavage and nucleic acid metabolism in sea urchin, *Strongylocentrotus purpuratus*, embryos. Exptl. Cell Res. *31*, 266–274.

BARROS C., G. S. HAND JR. and A. MONROY, 1966. Control of gastrulation in the starfish *Asterias forbesii*. Exptl. Cell Res. *43*, 167–183.

BERG W. E., 1965. Rates of protein synthesis in whole and half embryos of the sea urchin. Exptl. Cell Res. *40*, 469–489.

BERG W. E. and N. D. LONG, 1964. Regional differences of mitochondrial size in the sea urchin embryo. Exptl. Cell Res. *33*, 422–437.

BERG W. E., A. TAYLOR and W. J. HUMPHREYS, 1962. Distribution of mitochondria in echinoderm embryos as determined by electron microscopy. Develop. Biol. *4*, 165–176.

BIBRING T., J. BRACHET, F. S. GAETA and F. GRAZIOSI, 1965. Some physical properties of cytoplasmic deoxyribonucleic acid in unfertilized eggs of *Arbacia lixula*. Biochim. Biophys. Acta *108*, 641–651.

BOREI H., 1948. Respiration of oocytes, unfertilized eggs and fertilized eggs from *Psammechinus* and *Asterias*. Biol. Bull. *95*, 124–150.

BOVERI T., 1895. Über das Verhalten der Centrosomen bei der Befruchtung des Seeigel-Eies. Nebst allgemeinen Bemerkungen über Centrosomen und Verwandtes. Würzburg, 75 p.

BOVERI T., 1901. Die Polarität von Ovozyte, Ei und Larve des *Strongylocentrotus lividus*. Zool. Jahrb. Abt. Anat. Ontog. Tiere *14*, 630–653.

BOVERI T., 1903: Über das Verhalten des Protoplasma bei monocentrischen Mitosen. Sitz. Ber. Phys. Med. Ges. Würzburg.

BRACHET J., 1931a. La synthèse de l'acide thymonucléique pendant le développement de l'oeuf d'oursin. Compt. Rend. Soc. Biol. *108*, 813–815.

BRACHET J., 1931b. L'évolution des pentoses pendant le développement de l'oeuf d'oursin. Compt. Rend. Soc. Biol. *108*, 1167–1169.

BRACHET J., M. DECROLY, A. FICQ and J. QUERTIER, 1963. Ribonucleic acid metabolism in unfertilized and fertilized sea urchin eggs. Biochim. Biophys. Acta *72*, 660–662.

BRACHET J., A. FICQ and R. TENCER, 1963. Amino acid incorporation into proteins of nucleate and anucleate fragments of sea urchin eggs: effect of parthenogenetic activation. Exptl. Cell Res. *32*, 168–170.

BURNY A., G. MARBAIX, J. QUERTIER and J. BRACHET, 1965. Demonstration of functional polyribosomes in nucleate and anucleate fragments of sea urchin eggs following parthenogenetic activation. Biochim. Biophys. Acta *103*, 526–528.

CALLAN H. G., 1949. Cleavage rate, oxygen consumption and ribose nucleic acid content of sea urchin eggs. Biochim. Biophys. Acta *3*, 92–102.

CHILD C. M., 1936. Differential reduction of vital dyes in the early development of echinoderms. Arch. Entwicklungsmech. Organ. *185*, 426–456.

CHILD C. M., 1941. Formation and reduction of indophenol blue in development of an echinoderm. Proc. Natl. Acad. Sci. US *27*, 523–528.

COM D. G., 1965. Methylation of nucleic acids during sea urchin embryo development. J. Mol. Biol. *11*, 851–855.

COSTELLO D. P., 1961. On the orientation of centrioles in dividing cells, and its significance: a new contribution to spindle mechanics. Biol. Bull. *120*, 285–312.

CZIHAK G., 1960a. Untersuchungen über die Coelomanlagen und die Metamorphose des Pluteus von *Psammechinus miliaris* (Gmelin). Zool. Jahrb. Abt. Anat. Ontog. Tiere *78*, 235–256.

CZIHAK G., 1960b. Pseudoradiärsymmetrische Seeigelplutei. Arch. Entwicklungsmech. Organ. *152*, 593–601.

CZIHAK G., 1961. Ein neuer Gradient in der Pluteusentwicklung. Arch. Entwicklungsmech. Organ. *153*, 353–356.

CZIHAK G., 1962a. Entwicklungsphysiologische Untersuchungen an Echiniden (Topochemie der Blastula und Gastrula, Entwicklung der Bilateral- und Radiärsymmetrie und der Coelomdivertikel). Arch. Entwicklungsmech. Organ. *154*, 29–55.

CZIHAK G., 1962b. Entwicklungsphysiologie der Echinodermen. Fortschr. Zool. *14*, 238–267.

CZIHAK G., 1963. Entwicklungsphysiologische Untersuchungen an Echiniden (Verteilung und Bedeutung der Cytochromoxydase). Arch. Entwicklungsmech. Organ. *154*, 272–292.

CZIHAK G., 1965a. Evidences for inductive properties of the micromere-RNA in sea urchin embryos. Naturwissenschaften *52*, 141–142.

CZIHAK G., 1965b. Entwicklungsphysiologische Untersuchungen an Echiniden (Ribonucleinsäure-Synthese in den Mikromeren und Entodermdifferenzierung. Ein Beitrag zum Problem der Induktion). Arch. Entwicklungsmech. Organ. *156*, 504–524.

CZIHAK G., 1968. Molekularbiologische Aspekte der frühen Embryonalentwicklung von Seeigeln. Eine Übersicht mit neuen Ergebnissen und einer Hypothese zur Erklärung der Regulationsfähigkeit. Pubbl. Staz. Zool. Napoli *36*, 321–345.

CZIHAK G. and G. F. MEYER, 1964. Differentiation of the oral field of the sea urchin embryo. Nature *201*, 315.

CZIHAK G., H. G. WITTMANN and I. HINDENNACH, 1967. Uridineinbau in die Nucleinsäuren von Furchungsstadien der Eier des Seeigels *Paracentrotus lividus*. Z. Naturforsch. *22b*, 1176–1182.

DAWID I., 1966. Evidence for the mitochondrial origin of frog egg cytoplasmic DNA. Proc. Natl. Acad. Sci. US *56*, 269–276.

DENNY P. and A. TYLER, 1964. Activation of protein biosynthesis in non-nucleate fragments of sea urchin eggs. Biochem. Biophys. Res. Commun. *14*, 245–249.

DE VINCENTIIS M., S. HÖRSTADIUS and J. RUNNSTRÖM, 1966. Studies on controlled and released respiration in animal and vegetal halves of the embryo of the sea urchin *Paracentrotus lividus*. Exptl. Cell Res. *41*, 535–544.

DRIESCH H., 1892. Entwicklungsmechanische Studien. I. Der Wert der beiden ersten Furchungszellen in der Echinodermenentwicklung. Experimentelle Erzeugung von Theil- und Doppelbildungen. II. Über die Beziehungen des Lichtes zur ersten Etappe der thierischen Formbildung. Z. Wiss. Zool. *53*, 160–178, 178–184.

DRIESCH H., 1893. Entwicklungsmechanische Studien III-VI. III. Die Verminderung des Furchungsmaterials und ihre Folgen (weiteres über Theilbildungen). IV. Experimentelle Veränderungen des Types der Furchung und ihre Folgen (Wirkungen von Wärmezufuhr und von Druck). V. Von der Furchung doppelt befruchteter Eier. VI. Über einige allgemeine Fragen der theoretischen Morphologie. Z. Wiss. Zool. *55*, 1–62.

ELLIS C. H., 1966. The genetic control of sea urchin development: a chromatographic study of protein synthesis in the *Arbacia punctulata* embryo. J. Exptl. Zool. *163*, 1–22.

ELSON D. and E. CHARGAFF, 1952. On the deoxyribonucleic acid content of sea urchin gametes. Experientia *8*, 143–145.

ELSON D., T. GUSTAFSON and E. CHARGAFF, 1954. Ribonucleic acid of the sea urchin during embryonic development. J. Biol. Chem. *209*, 285–294.

ESPER H., 1962. Uptake of H^3-thymidine by eggs of *Arbacia punctulata*. Biol. Bull. *123*, 475–476.

FICQ A., F. AIELLO and E. SCARANO, 1963. Métabolisme des acides nucléiques dans l'oeuf d'oursin en développement. Exptl. Cell Res. *29*, 128–136.

FRY H. J., 1936. Studies of the mitotic figure. V. The time schedule of mitotic changes in developing *Arbacia* eggs. Biol. Bull. *70*, 28–35.

GIUDICE G., 1963. Aggregation of cells isolated from vegetalized and animalized sea urchin embryos. Experientia *19*, 83–84.

GIUDICE G. and R. JUST, 1962. Restitution of whole larvae from disaggregated cells of sea urchin embryos. Develop. Biol. *5*, 402–411.

GIUDICE G. and V. MUTOLO, 1967. Synthesis of ribosomal RNA during sea-urchin development. Biochim. Biophys. Acta *138*, 276–285.

GIUDICE G., M. L. VITTORELLI and A. MONROY, 1962. Investigations on the protein metabolism during the early development of the sea urchin. Acta Embryol. Morphol. Exptl. *5*, 113–122.

GLIŠIN V. R. and M. V. GLIŠIN, 1964. Ribonucleic acid metabolism following fertilization in sea urchin eggs. Proc. Natl. Acad. Sci. US *52*, 1548–1553.

GROSS P. R. and G. H. COUSINEAU, 1963. Effects of actinomycin D on macromolecule synthesis and early development in sea urchin eggs. Biochem. Biophys. Res. Commun. *10*, 321–326.

GROSS P. R. and B. J. FRY, 1966. Continuity of protein synthesis through cleavage metaphase. Science *153*, 749–751.

GROSS P. R., K. KRÆMER and L. I. MALKIN, 1965. Base composition of RNA synthesized during cleavage of the sea urchin embryo. Biochem. Biophys. Res. Commun. *18*, 569–575.

GROSS P. R., L. MALKIN and W. MOYER, 1964. Templates for the first proteins of embryonic development. Proc. Natl. Acad. Sci. US *51*, 407–414.

GROSS P. R., W. SPINDEL and G. H. COUSINEAU, 1964. Mitosis and macromolecule synthesis in cells exposed to D_2O. Pure Appl. Chem. *8*, 483–497.

GUSTAFSON T., 1965. Morphogenetic significance of biochemical patterns in sea urchin embryos, in 'The biochemistry of animal development,' ed. by R. Weber. Academic Press, *1*, 139–202.

GUSTAFSON T. and I. HASSELBERG, 1951. Studies on enzymes in the developing sea urchin egg. Exptl. Cell Res. *2*, 642–672.

GUSTAFSON T. and P. LENICQUE, 1952. Studies on mitochondria in the developing sea urchin egg. Exptl. Cell Res. *3*, 251–274.

GUSTAFSON T. and R. SÄVHAGEN, 1950. Studies on the determination of the oral side of the sea urchin egg. I. The effect of some detergents on the development. Arkiv Zool. *42A*, 1–6.

GUSTAFSON T. and L. WOLPERT, 1967. Cellular movement and contact in sea urchin morphogenesis. Biol. Rev. *42*, 442–498.

HAGSTRÖM B. E. and S. LÖNNING, 1965. Studies on cleavage and development of isolated sea urchin blastomeres. Sarsia *18*, 1–9.

HAGSTRÖM B. E. and S. LÖNNING, 1968. Personal communication.

HANSEN-DELKESKAMP E. and F. DUSPIVA, 1966. Aktivitätsverlauf der enzymatischen Phosphorylierung von Thymidin während der Entwicklung des Seeigels *Psammechinus miliaris* von der Befruchtung bis zum Zweizeller. Experientia *22*, 381–382.

HARVEY E. B., 1949. The growth and metamorphosis of the *Arbacia punctulata* pluteus, and late development of the white halves of centrifuged eggs. Biol. Bull. *97*, 287–299.

HARVEY E. B., 1956. The American *Arbacia* and other sea urchins. 298 pp, Princeton University Press.

HERBST C., 1892. Experimentelle Untersuchungen über den Einfluß der veränderten chemischen Zusammensetzung des umgebenden Mediums auf die Entwicklung der Thiere. I. Versuche an Seeigeleiern. Z. Wiss. Zool. *55*, 446–518.

HERBST C., 1893. Experimentelle Untersuchungen über den Einfluß der veränderten chemischen Zusammensetzung des umgebenden Mediums auf die Entwicklung der Thiere. II. Weiteres über die morphologische Wirkung der Lithiumsalze und ihre theoretische Bedeutung. Pubbl. Staz. Zool. Napoli *11*, 137–220.

HINEGARDNER R. T., B. RAO and D. E. FELDMANN, 1964. The DNA synthetic period during early development of the sea urchin egg. Exptl. Cell Res. *36*, 53–61.

HOLTER H. and E. ZEUTHEN, 1957. Dynamics of early echinoderm development as observed by phase contrast microscopy and correlated with respiration measurements. Pubbl. Staz. Zool. Napoli *29*, 285–306.

HÖRSTADIUS S., 1928. Über die Determination des Keimes bei Echinodermen. Acta Zool. (Stockholm) *9*, 1–191.

HÖRSTADIUS S., 1935. Über die Determination im Verlaufe der Eiachse bei Seeigeln. Pubbl. Staz. Zool. Napoli *14*, 251–479.

HÖRSTADIUS S., 1936. Weitere Studien über die Determination im Verlaufe der Eiachse bei Seeigeln. Arch. Entwicklungsmech. Organ. *135*, 40–68.

HÖRSTADIUS S., 1939. The mechanics of sea urchin development, studied by operative methods. Biol. Rev. *14*, 132–179.

HÖRSTADIUS S., 1949. Experimental researches on the developmental physiology of the sea urchin. Pubbl. Staz. Zool. Napoli Suppl. *21*, 131–172.

HÖRSTADIUS S., 1952. Induction and inhibition of reduction gradients by the micromeres in the sea urchin egg. J. Exptl. Zool. *120*, 421–436.

HÖRSTADIUS S., 1955. Reduction gradients in animalized and vegetalized sea urchin eggs. J. Exptl. Zool. *129*, 249–256.

HÖRSTADIUS S., 1957. On the regulation of bilateral symmetry in plutei with exchanged meridional halves and in giant plutei. J. Embryol. Exptl. Morphol. *5*, 60–73.

HÖRSTADIUS S. and T. GUSTAFSON, 1954. The effect of three antimetabolites on sea-urchin development. J. Embryol. Exptl. Morphol. *2*, 216–226.

HÖRSTADIUS S., L. JOSEFSSON and J. RUNNSTRÖM, 1967. Morphogenetic agents from unfertilized eggs of the sea urchin *Paracentrotus lividus*. Develop. Biol. *16*, 189–202.

HULTIN T., 1953. Incorporation of C^{14}-labelled carbonate and acetate into sea urchin embryos. Arkiv Kemi *6*, 195–200.

HULTIN T., 1961. Activation of ribosomes in sea urchin eggs in response to fertilization. Exptl. Cell Res. *25*, 405–417.

HULTIN T. and A. BERGSTRAND, 1960. Incorporation of C^{14}-L-leucine into protein by cell-free systems from sea urchin embryos at different stages of development. Develop. Biol. *2*, 61–75.

IKEDA M., 1965. Behaviour of sulfhydryl groups of sea urchin eggs under the blockage of cell division by UV and heat shocks. Exptl. Cell Res. *40*, 282–291.

IMMERS J. and J. RUNNSTRÖM, 1960. Release of respiratory control by 2,4-dinitrophenol in different stages of sea urchin development. Develop. Biol. *2*, 90–104.

INFANTE A. and M. NEMER, 1967. Accumulation of newly synthesized RNA templates in a unique class of polyribosomes during embryogenesis. Proc. Natl. Acad. Sci. US *58*, 681–688.

ISHIDA J. and I. YASUMASU, 1957. Changes in protein specificity determined by protective enzyme test during embryonic development of the sea urchin and the fresh-water fish. J. Fac. Sci. Univ. Tokyo Sect. IV, *8*, 95–107.

ISHIKAWA M., 1962. The relation between the incorporation of phosphorus and the induction of cleavage by hypertonic treatment in the sea urchin egg. Embryologia (Nagoya) *7*, 109–126.

JANDORF B. J. and M. E. KRAHL, 1942. Studies on cell metabolism and cell division. VIII. The diphosphopyridine nucleotide (cozymase) content of eggs of *Arbacia punctulata*. J. Gen. Physiol. *25*, 749–754.

KAVANAU L., 1954. Amino acid metabolism in the early development of the sea urchin *Paracentrotus lividus*. Exptl. Cell Res. *7*, 530–557.

KINNANDER H. and T. GUSTAFSON, 1960. Further studies on the cellular basis of gastrulation in the sea urchin larva. Exptl. Cell Res. *19*, 278–296.

KRAHL M. E., 1950. Metabolic activities and cleavage of eggs of the sea urchin, *Arbacia punctulata*. A review 1932–49. Biol. Bull. *98*, 175–217.

KRAHL M. E., 1956. Oxydative pathways for glucose in eggs of the sea urchin. Biochim. Biophys. Acta *20*, 27–32.

KRAHL M. E., A. K. KELTCH, C. E. NEUBECK and G. CLOWES, 1941. Studies on cell metabolism and cell division. V. Cytochrome oxidase activity in the eggs of *Arbacia punctulata*. J. Gen. Physiol. *24*, 597–617.

KRAHL M. E., A. K. KELTCH, C. P. WALTERS and G. CLOWES, 1954. Activity of glucose-6-phosphate and 6-phosphogluconate dehydrogenases in relation to glycolytic enzymes of *Arbacia* eggs. Biol. Bull. *107*, 315–316.

KRAHL M. E., A. K. KELTCH, C. P. WALTERS and G. CLOWES, 1955. Glucose-6-phosphate and 6-phosphogluconate dehydrogenases from eggs of the sea urchin, *Arbacia punctulata*. J. Gen. Physiol. *38*, 431–439.

KUHL W. and G. KUHL, 1950. Neue Ergebnisse zur Cytodynamik der Befruchtung und Furchung des Eies von *Psammechinus miliaris*. Zool. Jahrb. Abt. Anat. Ontog. Tiere *70*.

KÜHN A., 1965. Vorlesungen über Entwicklungsphysiologie, 591 pp. Springer, Berlin-Heidelberg.

LALLIER R., 1952. Recherches sur le problème de la détermination chez les echinodermes. Experientia *8*, 271.

LALLIER R., 1955. Effets des ions zinc et cadmium sur le développement de l'oeuf d'oursin *Paracentrotus lividus*. Arch. Biol. (Liège) *66*, 75–102.

LALLIER R., 1957. Recherches sur l'animalisation de l'oeuf de l'oursin *Paracentrotus lividus* par les dérivés polysulfoniques. Pubbl. Staz. Zool. Napoli *30*, 185–209.

LALLIER R., 1958. Analyse expérimentale de la différentiation embryonnaire chez les echinodermes. Experientia *14*, 309–315.

LALLIER R., 1959. Végétativisation de l'oeuf de l'oursin par la phénazone. Experientia *15*, 228–229.

LALLIER R., 1961. Les effets du chloramphénicol sur la détermination embryonnaire de l'oeuf de l'oursin *Paracentrotus lividus*. Compt. Rend. Acad. Sci. Paris *253*, 3060–3062.

LALLIER R., 1963. Effets de l'actinomycine D sur le développement normal et sur les modifications expérimentales de la morphogénèse de l'oeuf de l'oursin *Paracentrotus lividus*. Experientia *19*, 572–573.

LALLIER R., 1964. Biochemical aspects of animalization and vegetalization in the sea urchin. Advanc. Morphogenesis *3*, 147–196.

LINDAHL P. E., 1936. Zur Kenntnis der physiologischen Grundlagen der Determination im Seeigelkeim. Acta Zool. (Stockholm) *17*, 179–365.

LINDAHL P. E., 1939. Zur Kenntnis der Entwicklungsphysiologie des Seeigeleies. Z. Vergleich. Physiol. *27*, 233–250.

LINDAHL P. E. and H. HOLTER, 1940. Die Atmung animaler und vegetativer Keimhälften von *Paracentrotus lividus*. Compt. Rend. Trav. Lab. Carlsberg *25*, 257–287.

LOEB J., 1916. The organism as a whole. Putnam's Sons, New York.

LØVTRUP S., 1966. The chemical basis of sea urchin embryogenesis (Symposion). Bull. Schweiz. Akad. Med. Wiss. *22*, 201–276.

LUNDBLAD G., 1949. Proteolytic activity in eggs and sperms from sea-urchins. Nature *163*, 643.

MANO Y., 1966. Role of a trypsin-like protease in 'informosomes' in a trigger mechanism of activation of protein synthesis by fertilization in sea urchin eggs. Biochem. Biophys. Res. Commun. *25*, 216–221.

MANO Y. and H. NAGANO, 1966. Release of maternal RNA from some particles as a mechanism of activation of protein synthesis by fertilization in sea urchin eggs. Biochem. Biophys. Res. Commun. *25*, 210–215.

MASING E., 1910. Über das Verhalten der Nucleinsäure bei der Furchung des Seeigeleies. Z. physiol. Chem. *67*, 161.

MASTRANGELO M., 1965. A study of the vegetalizing action of tyroxine on the sea urchin embryo. Dissertation, Yale University.

MAZIA D., 1949. The distribution of deoxyribonuclease in the developing embryo *(Arbacia punctulata)*. J. Cellular Comp. Physiol. *34*, 17–32.

MILLONIG G. A. and G. GIUDICE, 1967. Electron microscopic study of the reaggregation of cells dissociated from sea urchin embryos. Develop. Biol. *15*, 91–101.

MITCHINSON J. M. and J. E. CUMMINS, 1966. The uptake of valine and cystidine by sea urchin embryos and its relation to the cell surface. J. Cell Sci. *1*, 35–47.

MOLINARO M. and T. HULTIN, 1965. The anomalous potassium requirement of amino acid incorporation system from sea urchin eggs. Exptl. Cell Res. *38*, 398–411.

MONROY A. and R. MAGGIO, 1964. Biochemical studies on the early development of the sea urchin. Advanc. Morphogenesis *3*, 95–145.

MONROY A., R. MAGGIO and A. M. RINALDI, 1965a. Experimentally induced activation of the ribosomes of the unfertilized sea urchin egg. Proc. Natl. Acad. Sci. US *54*, 107–111.

MONROY A., R. MAGGIO and A. M. RINALDI, 1965b. Attivazione in vitro dei ribosomi delle uove vergini di riccio di mare. Rend. Sci. Fis. Mat. Nat. *38*, 962–967.

MOORE A. R., 1933. Is cleavage rate a function of the cytoplasm or of the nucleus? J. Exptl. Biol. *10*, 230–236.

MORGAN T. H. and G. B. SPOONER, 1909. The polarity of the centrifuged egg. Arch. Entwicklungsmech. Organ. *28*, 104–117.

NAGANO H. and Y. MANO, 1968. Thymidine kinase, thymidylate kinase and ^{32}Pi and (^{14}C) thymidine incorporation into DNA during early embryogenesis of the sea urchin. Biochim. Biophys. Acta *157*, 546–557.

NEMER M., 1962. Interrelation of messenger polyribonucleotides and ribosomes in the sea urchin egg during embryonic development. Biochem. Biophys. Res. Commun. *8*, 511–515.

NEMER M. and A. A. INFANTE, 1965. Messenger RNA in early sea-urchin embryos: size classes. Science *150*, 217–221.

NEMER M. and A. A. INFANTE, 1967. Ribosomal ribonucleic acid of the sea urchin egg and its fate during embryogenesis. J. Mol. Biol. *27*, 73–86.

NIGON V. and S. GILLOT, 1963. Étude autoradiographique du métabolisme des acides nucléiques et des protéines au cours du développement de l'oeuf chez *Arbacia lixula*. Cahiers Biol. Marine *4*, 277–298.

OHNISHI T. and M. SUGIYAMA, 1963. Polarographic studies of oxygen uptake of sea urchin eggs. Embryologia (Nagoya) *8*, 79–88.

OKAZAKI K., 1960. Skeleton formation of sea urchin larvae. II. Organic matrix of the spicule. Embryologia (Nagoya) *5*, 283–320.

OLSSON T., 1960. Estimations of ribonucleic acid, acid-soluble nucleotides and nuclease activity during the caryogamic period of the sea urchin *Paracentrotus lividus*. Arkiv Zool. *13*, 89–93.

OLSSON T., 1965. Changes in metabolism of RNA during the early embryonic development of the sea urchin. Nature *206*, 843–844.

PAINTER T. S., 1915. An experimental study in cleavage. J. Exptl. Zool. *18*, 299–323.

PEASE D., 1941. Echinoderm bilateral determination in chemical concentration gradients. I. The effects of cyanide, ferricyanide, iodoacetate, picrate, dinitrophenol, urethane, iodine, malonate, etc. J. Exptl. Zool. *86*, 381–404.

PFOHL R. J. and A. MONROY, 1962. Changes in some properties in the course of the development of *Arbacia punctulata*. Biol. Bull. *123*, 477.

RANZI S., 1962. The proteins in embryonic and larval development. Advanc. Morphogenesis *2*, 211–257.

RANZI S. and P. CITTERIO, 1956. Sulle proteine dell'*Arbacia lixula* nello sviluppo embrionale e postembrionale. Rend. Ist. Lombardo Sci. Lettere *90*, 515–521.

RUNNSTRÖM J., 1928a. Plasmabau und Determination bei dem Ei von *Paracentrotus lividus*. Roux' Arch. Entwicklungsmech. Organ. *113*, 556–581.

RUNNSTRÖM J., 1928b. Zur experimentellen Analyse der Wirkung des Lithiums auf den Seeigelkeim. Acta Zool. (Stockholm) *9*, 365–424.

RUNNSTRÖM J., 1929. Über Selbstdifferenzierung und Induktion bei dem Seeigelkeim. Roux' Arch. Entwicklungsmech. Organ. *117*, 123–145.

RUNNSTRÖM J. and J. IMMERS, 1966. The animalizing action of trypsin on embryos of the sea urchin *(Psammechinus miliaris, Paracentrotus lividus)*. Arch. Biol. (Liège) *77*, 365–410.

RUNNSTRÖM J. and G. KRISZAT, 1952. Animalizing action of iodosobenzoic acid in the sea urchin development. Exptl. Cell Res. *3*, 497–499.

RUNNSTRÖM J. and B. MARKMAN, 1966. Gene dependency of vegetalisation in sea urchin embryos treated with lithium. Biol. Bull. *130*, 402–414.

SCARANO E., M. IACCARINO and P. GRIPPO, 1965. On methylation of DNA during development of the sea urchin embryo. J. Mol. Biol. *14*, 603–607.

SCHMIDT G., L. HECHT and S. J. THANNHAUSER, 1948. The behavior of the nucleic acids during the early development of the sea urchin egg *(Arbacia)*. J. Gen. Physiol. *31*, 203–207.

SELENKA E., 1883. Studien zur Entwicklungsgeschichte der Thiere. II. Heft. Die Keimblätter der Echinodermen. Wiesbaden.

SHAVER J. R., 1956. Mitochondrial populations during development of the sea urchin. Exptl. Cell Res. *11*, 548–559.

SLATER D. W. and S. SPIEGELMAN, 1966a. A chemical and physical characterization of echinoid RNA during early embryogenesis. Biophys. J. *6*, 385–404.

SLATER D. W. and S. SPIEGELMAN, 1966b. An estimation of genetic messages in the unfertilized echinoid egg. Proc. Natl. Acad. Sci. US *56*, 164–170.

SOFER W. H., J. F. GEORGE and R. M. IVERSON, 1966. Rate of protein synthesis: regulation during first division cycle of sea urchin eggs. Science *153*, 1644–1645.

SPIRIN A. S. and M. NEMER, 1965. Messenger RNA in early sea-urchin embryos: cytoplasmic particles. Science *150*, 214–217.

STAFFORD D., W. H. SOFER and R. M. IVERSON, 1964. Demonstration of polyribosomes after fertilization of the sea urchin egg. Proc. Natl. Acad. Sci. US *52*, 313–316.

TERMAN S. A. and P. R. GROSS, 1965. Translation-level control of protein synthesis during early development. Biochem. Biophys. Res. Commun. *21*, 595–600.

TYLER A., 1966. Incorporation of amino acids into protein by artificially activated nonnucleate fragments of sea urchin eggs. Biol. Bull. *130*, 450–461.

VON UBISCH L., 1929. Über die Determination der larvalen Organe und der Imaginalanlage bei Seeigeln. Roux' Arch. Entwicklungsmech. Organ. *117*, 80–122.

VON UBISCH L., 1933. Untersuchungen über Formbildung. IV. Über Plutei ohne Darm, Fortsätze ohne Skelet und die Grenzen der Virtuellen Keimbezirke. Roux' Arch. Entwicklungsmech. Organ. *129*, 45–67.

VON UBISCH L., 1937. Die normale Skelettbildung bei *Echinocyamus pusillus* und *Psammechinus miliaris* und die Bedeutung dieser Vorgänge für die Analyse der Skelette von Keimblattchimären. Z. Wiss. Zool. *149*, 402–476.

VON UBISCH L., 1939. Keimblattchimärenforschung an Seeigellarven. Biol. Rev. *14*, 88–103.

VON UBISCH L., 1959. Die Entwicklung von *Genocidaris maculata* und *Sphaerechinus granularis*, sowie Bastarde und Merogone von *Genocidaris*. Pubbl. Staz. Zool. Napoli *31*, 159–208.

WHEELER G. E., 1963. Radially oriented cleavage in blastula and in gastrula of the starfish *Asterias forbesi*. Biol. Bull. *125*, 582–590.

WILSON E. B., 1902. Experimental studies in cytology. II. Some phenomena of fertilization and cell-division in etherized eggs. III. The effect on cleavage of artificial obliteration of the first cleavage-furrow. Roux' Arch. Entwicklungsmech. Organ. *13*, 353–395.

WILT F. H. and T. HULTIN, 1962. Stimulation of phenylalanine incorporation by polyuridylic acid in homogenates of sea urchin eggs. Biochem. Biophys. Res. Commun. *9*, 313–317.

WOODLAND W., 1906. Studies in spicule formation. III. On the mode of formation of the spicular skeleton in the pluteus of *Echinus esculentus*. Quart. J. Microscop. Sci. *49*, 305–325.

YASUMASU I. and E. NAKANO, 1963. Respiratory level of sea urchin eggs before and after fertilization. Biol. Bull. *125*, 182–187.

YČAS M., 1954. The respiratory and glycolytic enzymes of sea-urchin eggs. J. Exptl. Biol. *31*, 208–217.

G. Reverberi (ed.), Experimental embryology of marine and fresh-water invertebrates

CHAPTER 13

Ascidians

G. REVERBERI

Zoological Institute, University of Palermo, Italy

13.1. Bibliographic introduction

The researcher who intends to devote his life, or at least part of it, to embryology (whether chemical or molecular) and plans to use the ascidian egg as the material for his investigations, will profit greatly by first reading Conklin's very splendid paper '*Organization and cell-lineage of the Ascidian egg*'. This paper, besides being a model of scientific literature, is a most appropriate introduction to the mysteries of embryonic development: it is rightly considered a milestone in embryology.

More recent information on the ascidian development will be found in the very valuable book by Schleip, '*Determination der Primitiventwicklung*' (1929), Kuhn's '*Entwicklungsphysiologie*' (1965), the paper '*Die Entwicklung der Monascidien*' by Von Ubisch (1952) and in '*The embryology of Ascidians*' (1961a), '*L'uovo delle Ascidie e le sue potenze organo-formative*' (1961b) and '*La mécanique du développement de l'œuf des Ascidies*' (1968) by Reverberi. With these bibliographic indications at hand the researcher can confidently enter this special embryological field. However, he should always remember that the best book is always Nature!

13.2. Historical introduction

The pioneer in research on ascidian development was the famous Russian biologist Kowalewsky. He was well prepared for this research as he had first studied the embryology of Amphioxus, which has many points of similarity with that of Ascidians. Kowalewsky's publications on the Ascidians are dated 1866–1871. Conklin expressed his opinion on them in these terms:

'Considering the time when they were written . . . they are models of accuracy'.

The papers by Van Beneden and Julin on Clavellina (1884, 1886) followed; of these Conklin writes: 'The beautiful studies of Van Beneden and Julin . . . surpass in excellence anything which had been done up to that time, and in some respect they have not been equalled by any more recent work on the development of these animals'.

Some years later a paper by Castle (1894) appeared describing the cell-lineage of Ciona: this paper must also be considered a classic of embryology. After some other publications by Lacaze-Duthiers (1874), Davidoff (1889) and Samassa (1894), Conklin published in 1905 his very valuable paper on Styela. With exemplary modesty Conklin wrote: 'it may seem superfluous to devote yet another paper to this subject, and nothing was further from my purpose when I began . . .'. Time, however, has shown that no publication was more opportune.

In the above paragraphs we have introduced some of the fundamental papers dealing with the normal development: let us now review the papers related to experimental research. First we should mention the research of Chabry (1887) on *Ascidia scabra*. This work was done at Concarneau, the marine laboratory of the Collège de France; Conklin considered it to be 'one of the first and best illustrations of the application of the experimental method to the study of embryology'. With a very rudimentary micro-manipulator invented and built by himself, Chabry was able to perforate the envelopes of the egg and to kill one of the two blastomeres of the segmented egg. The surviving blastomere, not damaged by the operation, continued to segment as though it were a part of the whole and develop: as a result Chabry obtained a *half* tadpole. This result was exactly the opposite of that obtained by Driesch with the sea urchin egg. The experiments of Chabry were extended by Conklin (1905c) who killed one or two cells at the 4-cell stage: in all cases he obtained a partial larva which lacked the potential organs represented in the killed cells. In consequence of the results obtained by Chabry and Conklin the ascidian egg was catalogued as a 'mosaic egg'.

In 1930, however, Schmidt and in 1936 Cohen and Berrill showed that more than half a larva can be obtained from a blastomere isolated from the egg at the 2-cell stage; for instance, each of the two larvae derived from the two isolated blastomeres had three palps as does a normal larva. At the same time Reverberi (1931) and Dalcq (1932) showed that each fragment of the still unsegmented egg was able to give rise to a perfectly normal larva. Finally in 1946 Reverberi and Minganti showed that the formation of the nervous system is not independent, as it should be in a 'mosaic egg', but is conditioned by the territories with which the presumptive neural territory comes in contact during development.

13.3. Technical introduction

Spermatozoa and eggs are easily obtained from animals kept in the laboratory; at times they lay them spontaneously. Deposition occurs at different hours of the day, according to the species and seems to be facilitated by light. Ripe gametes can also be obtained directly by excision of the gonoducts. These are readily visible in an animal from which the tunic has been removed. The animals are hermaphrodites having both spermio and egg gonoducts, which run parallel along the body to the atrial aperture. With the exception of Ciona autofertilization is common. In many species the eggs develop very rapidly: for instance in *A. malaca* the tadpole is obtained in 7–8 hr.

For many experimental interventions the eggs must be liberated from their envelopes; this can be done mechanically with needles (free hand) or chemically with trypsine. As however the 'naked' egg is fragile and cytolizes very easily, micrurgical interventions are performed on a thin film of agar spread on the bottom of a Syracuse dish: needles and pipettes must also be dipped in liquid agar before use. Tungsten needles are very practical, but we prefer glass needles. The use of antibiotics is not necessary: all operations should, however, be carried out under semi-sterile conditions; Syracuse dishes must be flamed before use. Filtered sea water must also be used. Development, on the other hand, is so rapid that the danger of bacterial infection is very limited.

13.4. The ovarian egg

Oogenesis has been extensively investigated in the past by such authors as Julin (1893) (Styelopsis), Crampton (1899) (Molgula), Korschelt and Heider (1902), Bluntschli (1904) (Cynthia), Schaxel (1910) (Ciona), Hirschler (1917), Wernicke (1919), Parat and Bhattacharya (1926), Harvey (1927) (Ciona), Jägersten (1935), Knaben (1936) (Corella), Tucker (1942) (Styela), Pérèz (1954). In the present review only investigations using the electron microscope and histochemistry will be reported.

13.4.1. The oocyte has been investigated under the electron microscope by Kessel and Kemp (1962), Kessel (1962), Kessel and Beams (1965),

Hsu (1962a, 1963) and Mancuso (1964, 1965, 1967). According to these authors the oocyte has a large germinal vesicle, a cytoplasm crowded with

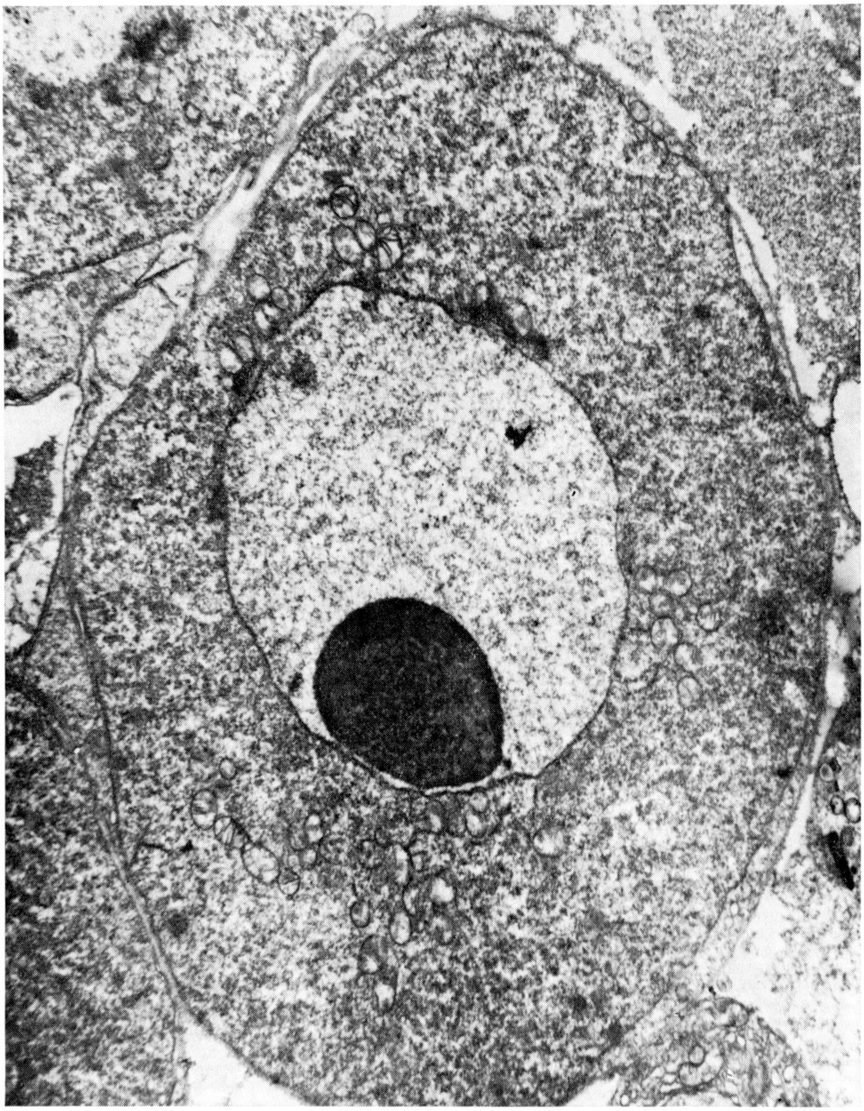

Fig. 1. Young oocyte of Ciona. Note the large germinal vesicle, the eccentrically placed nucleolus, the cytoplasm crowded with vesicles of the endoplasmic reticulum and mitochondria (original electron micrograph).

vesicular elements of endoplasmic reticulum, mitochondria, yolk and lipidic granules (Fig. 1). In very young oocytes, a peculiar formation can be seen around the nucleus: this is the yolk nucleus. In vitellogenetic oocytes the mitochondria are mostly situated at the periphery. The 'cortex' does not have cortical granules or microvilli; it is only raised in short processes or spines. Immersed in the cortex are the 'test-cells'. Peripherally there is the chorial membrane with the 'follicular cells'. This is only the general aspect of the oocyte; certain structures will be described in detail below.

13.4.2. The oocytes have not been thoroughly investigated cytochemically, and only a few substances have been detected. According to Pérèz (1954) the polysaccharide content is high in young oocytes, less in actively growing oocytes and still less in ripe oocytes; it is contributed in part by the perivitelline cells. De Vincentiis (1960a, b), studying the distribution of RNA, noticed a basophilic perinuclear area, which is supposedly responsible for the genesis of the yolk. The nucleolar RNA differs from that of the cytoplasm. Some metachromatic granules are observed near the nucleus and, probably correspond to the Golgi body. Other cytochemical investigations (for nucleic acids, mucopolysaccharides and basic proteins) were performed by Cowden and Markert (1961) and Cowden (1961, 1962). According to Mansueto (1964), besides the perinuclear area of RNA another area of RNA is present at the periphery: this RNA seems to derive from the 'test-cells'.

The study of oogenesis raises many questions.

(1) The first is that of the origin of the cellular envelopes. In this regard three hypotheses have been formulated. According to Julin (1893) the hermaphroditic gonad derives from a mesodermic mass; the exterior cells give rise to the wall of the gonad, the more interior ones to the primordial germ cells. These at first constitute a multinuclear syncytium, but later individualize and divide into two layers, the upper one of which gives rise to the primordial egg cells and the lower to the primordial sperm cells. The primordial egg cells divide several times; finally each cell gives rise to an oogonium and three follicular cells. Two layers of cells derive from the follicular cells: the exterior one, which will form the definitive 'follicular cells' and the interior one, which will form the 'test-cells'. The 'follicular cells', after a new division, give rise to an exterior and an interior follicular sheet.

Korschelt and Heider (1902) and more recently Tucker (1942) supported the observations of Julin. According to these authors the egg, as well the auxiliary cells, derives from the germinal epithelium. The cell destined to give rise to an egg grows, bulges and becomes surrounded by the other cells. These give rise to the follicular and test cells by transversal division (fig. 2). Conklin agrees with Van Beneden and Julin.

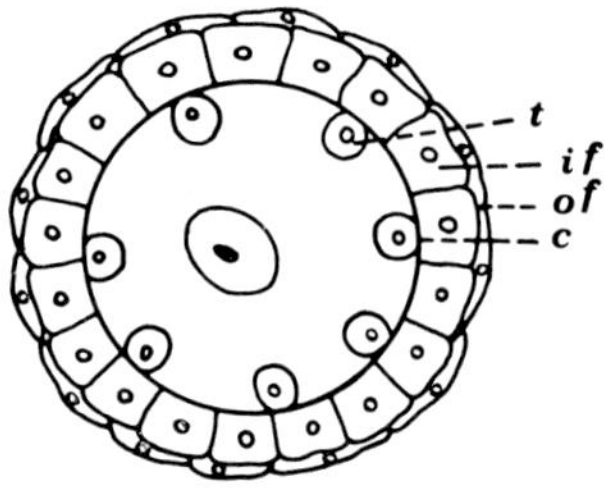

Fig. 2. Oocyte of Styela inside its membranes: t = test cells; if = inner follicle cells; of = outer follicle cells; c = vitelline membrane. (Redrawn from Kessel 1962.)

A different view is held by Knaben (1936). He maintains that the egg and the auxiliary cells do not arise from the same source; the eggs would originate in the germinal epithelium while the follicular and test cells would derive from some mesodermic migrating cells (fig. 3). These multiply amitotically and form the 'primary follicle cells' from which, by division, the 'test cells' form. This view is also held by Spek (1927).

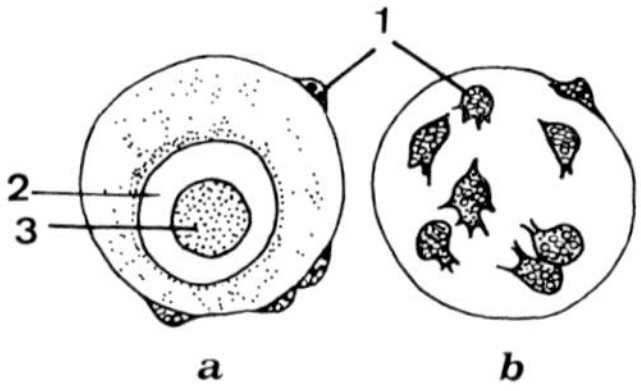

Fig. 3. Oocyte of Corella on which the cells from which the membranes will be formed are migrating (1); (2) germinal vesicle; (3) nucleolus. (Redrawn from Knaben 1936.)

A third view has been recently presented by Pérèz (1954) who claims that the eggs and follicular cells derive from the germinal epithelium, the test cells, however, from some blood cells. This view is supported also by De Vincentiis (1962) and Kalk (1963). De Vincentiis observed that under ultraviolet light the 'test cells' of Ciona show a yellow fluorescence which changes to blue in the same way as blood cells. From the above references it can be seen that the problem of the origin of the auxiliary cells has not been solved by morphological or histochemical techniques. The question has been considered with the electron microscope by Mancuso (1965) who also affirms that the 'test cells' derive from mesodermic migrant cells.

This problem was not directly considered by Kessel and Kemp (1962), Kessel (1962) and Kessel and Beams (1965). However, they made some important contributions concerning the function of the 'test cells'. They observed that ripe oocytes are yellow-orange because of a pigment localized in the 'test cells'. Examined by electron microscope these show a cytoplasm with oval masses of parallel rods: at high magnification the single rods are seen to be composed of numerous small spheres (30–40 Å) connected to form a chain. The masses are elaborated by the test cells themselves. They originate in the Golgi body, which in these cells is exceptionally conspicuous. These oval masses are probably poured into the egg, which thus becomes pigmented (fig. 4). It is important to recall that after fertilization, the yellow pigment collects to form the yellow crescent. According to De Vincentiis (1960c) the 'test cells' also possess granules which stain metachromatically. It seems that these granules are mucopolysaccharide in nature. The test cells could thus be considered to correspond to the cortical granules observed in many types of eggs.

(2) Another question which arises from the investigations on the oocytes is the origin of the yolk, for which many ooplasmic constituents have been held responsible in the past. Crampton (1899), Hirschler (1917) and Harvey (1927) have studied this question in Ascidians. Crampton (1899) laid great emphasis on the 'yolk nucleus', a granular mass situated around the nucleus which stains intensely with plasm stains. Schaxel (1910), on the other hand, attributed much significance to some chromatin bodies emitted from the nucleus. Loyez (1909), Bluntschli (1904), Wernicke (1919), Hirschler (1917), Parat and Bhattacharya (1926) for their part, leaned towards the mitochondria. In their view these took some materials formed outside the oocyte and transformed them into the yolk. The origin of these materials was seldom considered in the past. According to Harvey part of them would derive from some osmiophilic granules that are present in the test cells. When the oocytes begin to synthesize the yolk, these granules liquefy and their substance is poured into the cytoplasm, while other materials would derive from the Golgi apparatus. The yolk formation (in Ciona) would be as follows: 'into the cytoplasm of the growing oocyte are poured two materials in solution, a lipoid constituent from the granules of the test cells, and an albuminous material secreted by the Golgi apparatus. The molecules of these two compounds become aggregated around and finally inside the

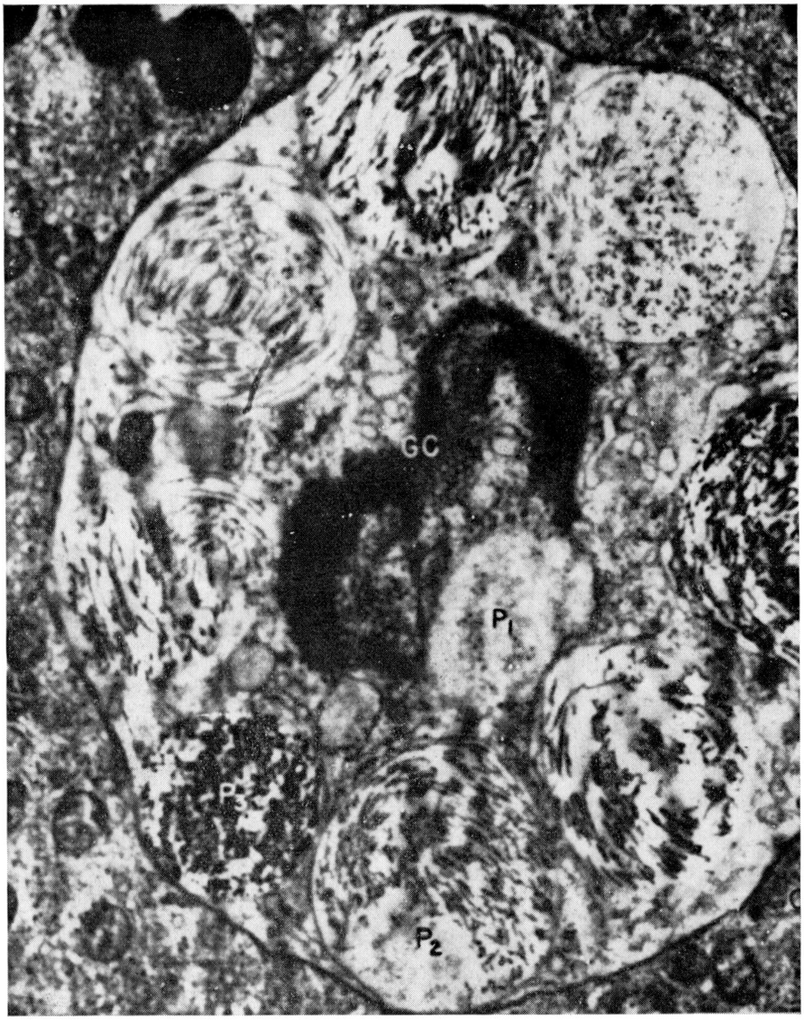

Fig. 4. Electron micrograph of a test cell of Styela. GC = Golgi complex; P_1, P_2, P_3, pigment granules at different stages of maturation (Kessel and Beams 1965).

mitochondria which are scattered in the cells. Here they are compounded or mixed into yolk, the material of mitochondria taking part in the formation of the former, but possibly retaining its identity as a material. The primary origin of the constituents derived from the Golgi apparatus, may quite conceivably be the nucleolus.'

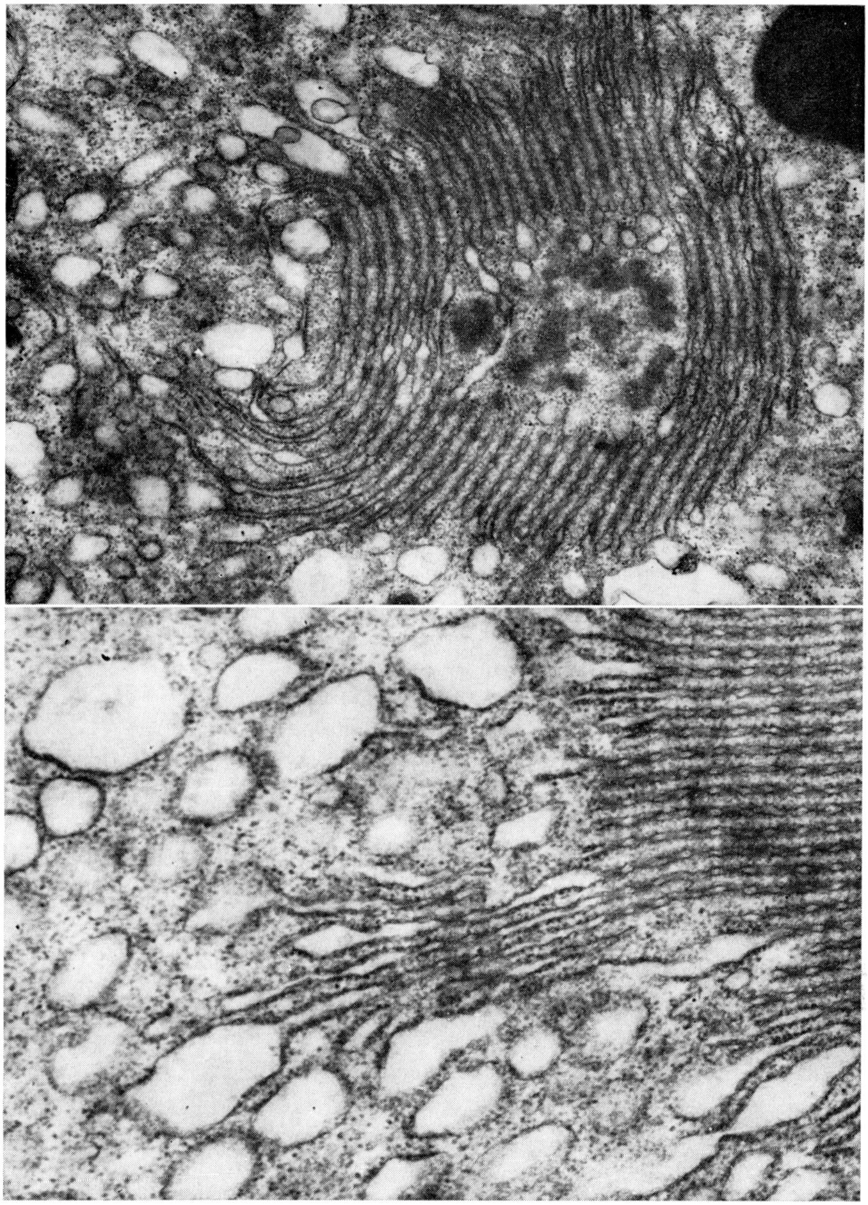

Fig. 5. Vitellogenetic oocytes of Ciona examined by electron microscope: notice the numerous 'lamellae annulatae' in the cytoplasm (Mancuso 1967).

The question has also been investigated in the electron microscopic studies of Hsu (1962a) on Bolthenia and of Mancuso (1967) on Ciona. Mancuso observed that the first yolk granules appear in the proximity of the nucleus, where many 'multivesicular bodies' (which disappear at the same time) are also situated. He maintains that the yolk granules derive from these. But this is not only the origin; many other structures such as the 'lamellae annulatae', the Golgi complex, the mitochondria and finally the pinocytotic vesicles are implicated in the genesis of the yolk granules. The 'lamellae annulatae' are very abundant in oocytes in full vitellogenesis (fig. 5). They are found throughout the cytoplasm, where they form 'complexes' with the endoplasmic reticulum. Vitelline material is observed within these complexes. This material is, however, elaborated in other cells, from which it is passed to the oocytes by means of pinocytosis. The mechanism of formation of the pinosomes is as follows: the material accumulates on the plasma membrane of the oocyte; the membrane invaginates and pockets are formed in which yolk granules are afterwards organized (fig. 6).

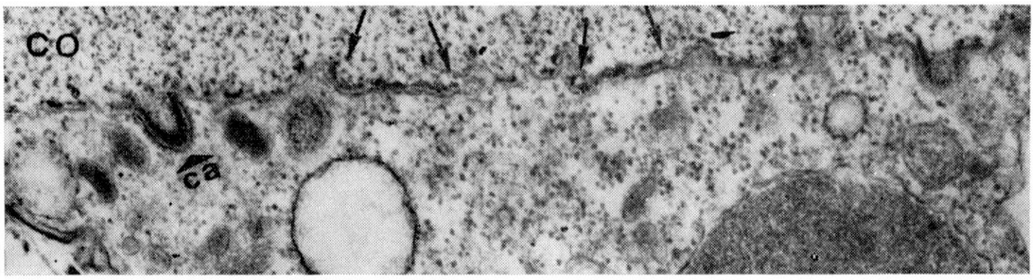

Fig. 6. Pynocinotic formations in a Ciona oocyte: CO = chorion; ca = a caveola (Mancuso 1967).

(3) Another problem, which involves many aspects of present-day cytology (and some of molecular biology), concerns the origin of the ribosomes and the endoplasmic reticulum. The problem has been studied by Hsu (1962b) in the oocytes of Bolthenia with the electron microscope. According to this author the ribosomes are formed earlier than the endoplasmic reticulum; they derive from the nuclear membrane. As is known, this is composed of an exterior and an interior layer. During the growth of the oocyte the exterior one desquamates continuously, the scales passing into the ooplasm. Here they disintegrate liberating the attached ribosomes: micrographs very often show ribosomes aligned in fine threads.

The endoplasmic reticulum is present only in oocytes which have already entered into oogenesis. It also derives from the nuclear membrane, which forms the 'annulatae lamellae'; these open at their lateral ends, detach, and constitute the reticulum. According to the author the nuclear membrane should not be imagined as a static structure. On the contrary, it is continuously in activity throughout the oocyte's development and undergoes the following changes: (a) loss of the conventional two-membrane condition by the nuclear envelope; (b) formation of intranuclear vesicles; (c) intensive formation of ribosomes involving the outer membrane of the nuclear envelope; (d) formation of the endoplasmic reticulum by the outer nuclear membrane; (e) formation of the 'annulatae lamellae' on both sides of the nuclear envelope. These morphological changes should be regarded as expressions of a differential gene activity. The author hypothesizes that the function of the intranuclear vesicles and annulatae lamellae is to 'activate other genes or, as finished information, to find temporary shelter'. As a consequence of the continuous production of the above-mentioned intranuclear formations, the germinal vesicle not only increases in volume, but probably becomes different from the original nucleus.

(4) Biochemical investigations of the oocytes are rare. Such an investigation on Ciona should not be difficult as the ovary is large and easily removable. However, up to now, only a few investigations have been done along these lines. One of them concerns the content of 'free' amino acids (Ferrini et al. 1963). A peculiar and still unexplained finding is the large quantity of taurine and sarcosine present; these two amino acids constitute about two-thirds of the total pool.

Another biochemical investigation deals with the content of activated amino acids. According to Molinaro (1964) this content is very high in oocytes (a demonstration of an intense protein synthesis); activated histidine is particularly abundant. Other biochemical studies have been undertaken on the content of inorganic phosphate, ATP, and 6-glucose phosphate (D'Anna and Metafora 1965). The data obtained are the following: 4.45 μg/mg protein for inorganic phosphate; 0.44 μg/mg protein for ATP and 0.58 μg/mg protein for 6-glucose phosphate.

13.5. The unfertilized egg

At the end of its growth the oocyte leaves the ovary and enters the oviduct; here it remains for several hours, and possibly days. In the oviduct the germinal vesicle collapses and a meiotic spindle forms. The spindle is very small, and is situated immediately under the cortex: it remains in this (metaphasic) state until fertilization. The polocytes are emitted after fertilization in Ciona, Phallusia, Ascidia, Ascidiella and probably many other species.

TABLE 1

Yolk spheres alone pigmented	Approx. diam. of ovum (mm)
(a) Eggs clear and colourless	
Ascidiella aspersa	0.17
Ascidiella scabra	0.17
Diazona violacea	0.16
(b) Eggs clear with a faint greenish tinge	
Phallusia mammillata	0.16
Ascidia mentula	0.15
Eggs clear with a bright red pigment	
Ascidia conchilega	0.13
Eggs translucent with greenish or reddish pigment	
Ciona intestinalis	0.16
(c) Eggs semi-opaque but without pigment	
Boltenia echinata	0.17
Ascidia prunum	0.18
Ascidia obliqua	0.15
(d) Eggs opaque and pigmented. Yellowish pigment occasionally present	
Polycarpa fibrosa	0.16
Molgula manhattensis	0.11
Molgula ampulloides	0.11
Molgula arenata	0.11
Molgula occulta	0.11
Molgula bleizi	0.15
Clavelina lepadiformis	0.26
Yellow or reddish pigment	
Tethyum pyriforme americanum	0.26
Molgula citrina	0.20
Stolonica socialis	0.72
Reddish purple pigment	
Molgula robusta	0.12
Molgula oculata	0.11
Molgula retortiformis	0.18
Polycarpa rustica	0.18
Styelopsis grossularia	0.48
Boltenia ovifera	0.16
Distomus variolosus	0.59
Amaroucium nordmanni	0.38

TABLE 1 (*continued*)

Mitochondria pigmented	
	Approx. diam. of ovum (mm)
Two cases only are known with certainty:	
Styela partita	0.15
Yolk spheres opaque but colourless. Mitochondria yellow	
Boltenia hirsuta	0.18
Yolk spheres opaque but colourless. Mitochondria occasionally colourless but usually orange	

Mature animals have an oviduct so full of eggs that the slightest laceration lets them out. The eggs float in sea water: this is due to the vacuolation of the follicular and test cells. In some Ascidians (Clavellina) the eggs develop in the atrial cavity up to the stage of swimming larva. Eggs of different species have particular characteristics (size, colour, envelopes, etc.). Some data on these aspects are shown in table 1, taken from Berrill. As shown in the table some species have colourless eggs while in others the eggs are green, red or orange, depending on the presence or absence of pigments. As mentioned, the egg is protected by several envelopes: the outermost is the chorion, which contains numerous vacuolated cells, the 'follicular' cells; in Ciona these have the appearance of long digitate cones with a drop of oil at the tip.

Another envelope is made up of the 'test cells' which are also vacuolated and pigmented. They float more or less freely in the liquid of the perivitelline space. Sometimes, however, (in Phallusia) they are pressed against the surface of the egg by another structureless membrane. The follicular cells stain red (metachromasia) and the 'test cells' green with toluidine-blue. The nucleus of the unfertilized egg is in meiotic metaphase, the spindle is barrel-shaped and without centrioles: the chromosomes are small and not numerous. Data on their form, number and structure will be found in a paper by Minganti (1956). *A. aspersa*, *A. scabra*, and Phallusia possess 16 chromosomes, and *A. mentula* and Ciona 19. Styelopsis, according to Julin (1893), has only 4 chromosomes. If treated with Janus green or with the Nadi reaction the egg stains evenly with the exception of a small area of the animal pole. The coloration is indicative of some mitochondrial enzymes; it can therefore be assumed that the mitochondria are distributed all over the surface of the egg. This has been confirmed by electron microscopy (Mancuso 1963). The other components of the egg such as the yolk and pigment granules, the lipids, the endoplasmic reticulum and the ribosomes are not differentially distributed: this means that the unfertilized egg has no architecture. According to Ursprung and Schabtach (1964) the plasma membrane has no microvilli: no cortical granules are observed under it. The endoplasmic reticulum is represented by spherical or oval vesicles which have ribosomes on their surface. Granules larger than ribosomes (probably glycogen) are also observed: they are not precisely localized. Lipidic granules are quite scanty, while the yolk granules, of which there seem to be two different types, are very numerous.

Metabolically the unfertilized egg is more or less inert, but it exerts a strong action on the spermatozoa. This is due to the production of a substance which diffuses from the egg and is released to the membranes and water. Under the effect of this substance the spermatozoa become extremely active. This can be demonstrated by depriving the eggs of Ciona of their membranes and placing them in a fresh spermatic suspension. Under these circumstances the eggs can be seen moving in all directions, rotating and rising in a sort of dance caused by the frenetic movements of the spermatozoa which have been activated by the eggs. This influence of the eggs on the

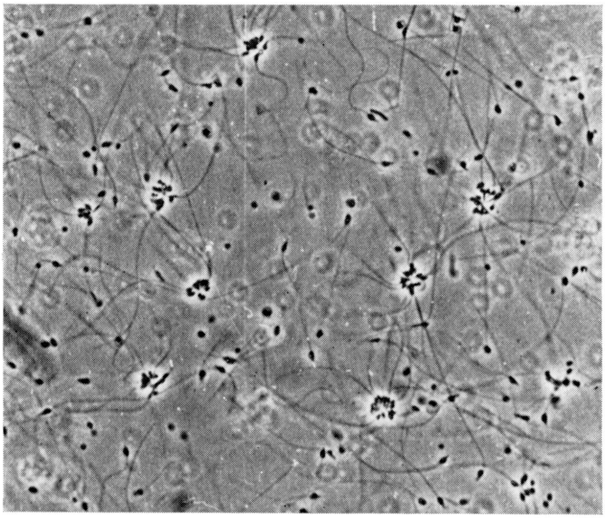

Fig. 7. Ciona spermatozoa agglutinated by homologous fertilizines (Metafora and Restivo 1964).

spermatozoa can be demonstrated by another experiment. If a 'naked' egg of Ciona is pricked with a fine needle it disintegrates into a multitude of granules and any resting spermatozoa in its vicinity are immediately reactivated. From these examples it is evident that the eggs release fertilizine into the water. The presence of fertilizines in the egg has been investigated by Minganti (1951), Metafora and Restivo (1963, 1964) and Metafora (1966) (fig. 7).

Lillie had noticed that the 'egg-water' of some animals causes agglutination, activation and probably attraction of the spermatozoa. The activating and agglutinating substance in Ciona is not bound to particles but is recovered from the supernatant after homogenization of the eggs. This substance has a low molecular weight, and is resistant to heat, trypsine and ultraviolet radiations. While specific, is also active on the spermatozoa of foreign species, causing agglutination. In time the unfertilized egg becomes less fertilizable, and after 72 hr it can no longer be fertilized. Old eggs do not develop normally. There is no explanation for this fact; possibly the inactivation of some enzymes, the accumulation of metabolites or asphyxia are the responsible factors. At present however, the problem must be considered unresolved. In aged eggs the cortex becomes sticky and as a consequence eggs which come in contact often fuse.

13.6. The spermatozoa

Here we shall report only some recent investigations using the electron microscope and cytochemistry. An interesting study with the electron microscope was carried out by Ezell (1963) and Ursprung and Schabtach (1964) in spermatozoa of Ciona and *Ascidia nigra*. A very unexpected finding is the occurrence of a single gigantic mitochondrion which is not situated in the intermediate piece, but in the head, around the nucleus (fig. 8).

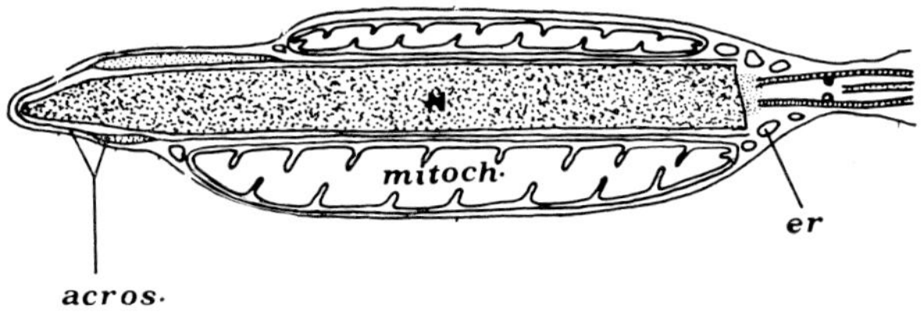

Fig. 8. Longitudinal section of a sperm of *Ascidia nigra:* a single gigantic mitochondria is situated in the head of the sperm surrounding the nucleus. acros. = acrosome; mitoch. = mitochondrion; er = endoplasmic reticulum; N = nucleus. (Redrawn from Ursprung and Schabtach 1964.)

The significance of this observation is not clear; nor is it known whether this situation occurs in other Ascidians or is peculiar to the spermatozoa of Ciona.

The mitochondrion is expelled at the moment of entry of the spermatozoon at fertilization; this fact recalls an observation by earlier cytologists regarding the so-called 'lateral body'.

In the Ciona spermatozoon the acrosome is evident. It is not known whether it produces a filament at fertilization. Nor is anything known about the mechanism of the entrance of the spermatozoon into the egg. With regard to the position of the mitochondrion in the head a cytochemical investigation of Reverberi and Restivo (1957) should be mentioned. The spermatozoa were broken into two parts, head and tail, and the suspension was then centrifuged in order to separate the heads from the tails. The Nadi reaction was done on each separated suspension and yielded a positive result only in the suspension containing the heads. The same research showed that the spermatozoa possess much glycogen, which is utilized to obtain energy for movement and is localized in the tail.

13.7. Fertilization

13.7.1. In the Ascidian monospermy is the rule: polyspermy occurs only in aged or otherwise damaged eggs. That this block to polyspermy does not reside in the egg membranes is evidenced by the fact that demembranated eggs are also penetrated by only one spermatozoon: the block is evidently localized in the cortex. The spermatozoon enters at the vegetal pole or very near it: Conklin attributes a structural peculiarity to that region ('the peripheral layer of the protoplasm is a little thicker at this pole than elsewhere'), but it is also observed that egg fragments obtained from the animal pole are easily penetrated by the spermatozoon.

After fertilization the egg is no longer inert. The first indication that the egg has been fertilized is a rapid modification of its shape, which occurs a few minutes after the entrance of the spermatozoon. This modification can easily be followed in demembranated ('naked') eggs and resembles a muscular contraction. The egg abruptly loses its spherical shape, elongates, becomes pyriform and amoeboid; a lobe forms at the vegetal pole. These changes last 15–20 sec, after which the egg reassumes its original spherical shape (fig. 9). The factors responsible have not been investigated; the change in permeability of the egg is a possible cause.

The cortex contracts, as can be clearly demonstrated if some chalk

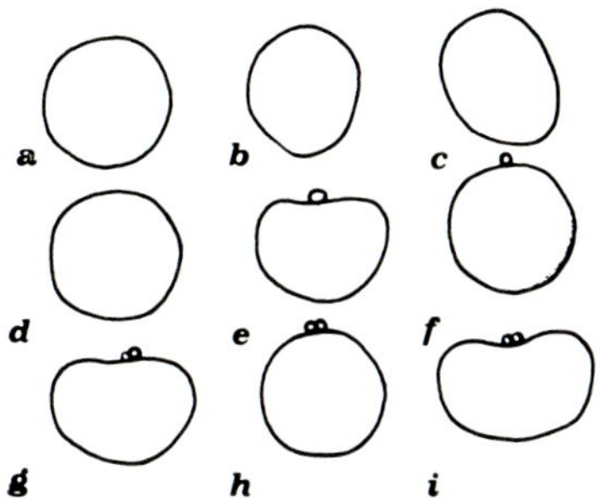

Fig. 9. Morphological modifications of an Ascidian egg at fertilization and maturation. (From Reverberi 1961.)

granules are applied to the egg. At the moment the egg contracts the granules shift from their original position, sometimes collecting together, sometimes scattering: occasionally a composite granule is broken in two or more smaller granules. Conklin, who saw some of these modifications in Cynthia, observed that the test cells, which form a sheet around the egg, gather at the vegetal hemisphere exactly in correspondence with the yellow crescent as a result of these violent modifications. Conklin explained this movement of the test cells by the ooplasmic movements, which will be reported soon.

13.7.2. Besides the above-mentioned morphological modifications, more relevant changes at fertilization are those connected with the redistribution of the pigmented plasms. They have been accurately described by Conklin in Cynthia. In this egg there are five differently coloured plasms, unevenly mixed in the unfertilized egg. At fertilization, however, they are separated and finally segregated in definite regions of the egg (fig. 10). The yellow plasm is segregated in the form of a crescent, immediately below the equator of the egg: it marks the posterior part of the future embryo. The grey plasm is segregated, again in the form of a crescent, at the very opposite region: this marks the anterior region of the future larva. Just below it, immediately under the equator, a third plasmatic constituent is segregated, also in the form of a crescent. The clear protoplasm is segregated in the animal region and the grey yolky one in the vegetal region.

In consequence of this segregation, five different regions are recognizable in the fertilized egg. Since the musculature, nervous system, notochord, ectoderm and intestine differentiate from these regions, the plasms have been designated meso-, neuro-, chordo-, ecto- and endoplasm respectively. The question whether these plasms are responsible for the structures men-

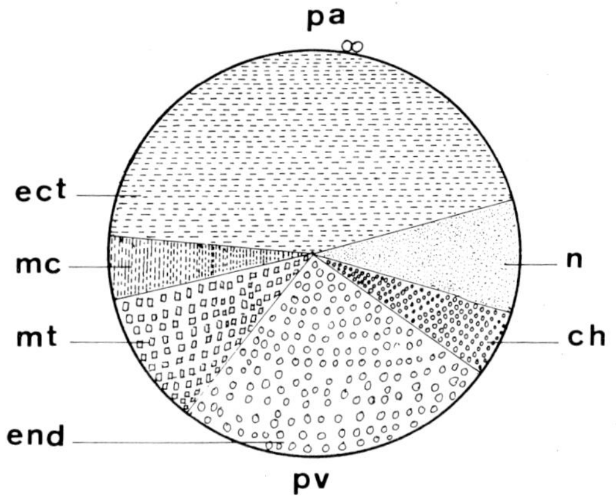

Fig. 10. Plasm segregation at fertilization. pa = animal pole; pv = vegetal pole; n = neuroplasm; ch = chordoplasm; end = endoplasm; mt = myoplasm; mc = mesenchyme; ect = ectoplasm. (Redrawn from Conklin.)

tioned cannot be answered by the methods of descriptive embryology. If one displaces the plasms by centrifugation the organs to which they give rise (La Spina 1958) are not displaced. Conklin did not consider these five plasms to be organ-forming, but interpreted them as indicators of some chemical, physical or metabolic peculiarity of the surrounding cytoplasm, much as the variegation of a lawn reveals local differences in the composition of the soil (Weiss 1939, p. 231).

This peculiar plasmatic segregation, which seems to be so significant for differentiation, raises certain questions. By what mechanism are the plasms segregated? What is the morphological and chemical constitution of the individual plasms? Costello (1948) tried to answer the first question by adopting Téorell's views on the diffusion effect, but his explanation meets with unsurmountable difficulties. As for the morphological and biochemical constitution of the differently coloured plasms, only a few indications are available. According to Conklin, the yellow crescent is made up of mitochondria and some pigment granules. The presence of mitochondria is confirmed by the results obtained on treating the eggs with Janus green. With this vital stain the unfertilized egg (Phallusia) shows a diffuse green coloration; at fertilization the green coloration becomes concentrated in the region of the yellow crescent (fig. 11). We have no information on the other plasms.

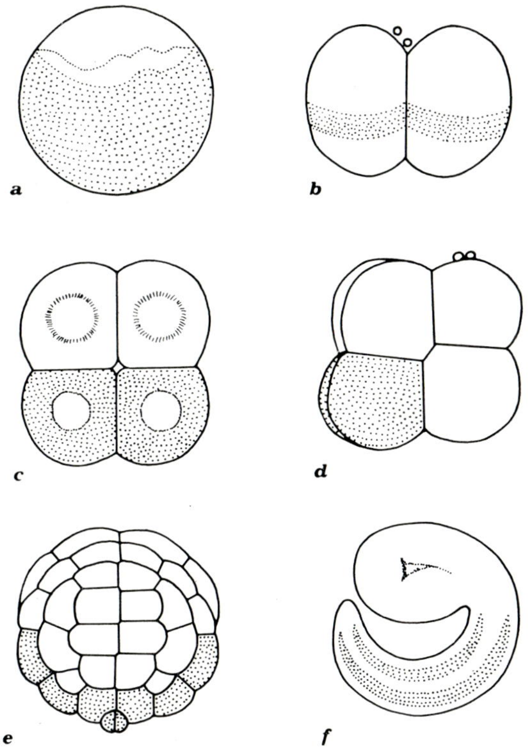

Fig. 11. An egg of Phallusia was vitally stained with Janus green: the stain was gradually segregated in the cells responsible for the formation of the larval musculature. (Redrawn from Reverberi 1956.)

13.7.3. Other modifications of the egg at fertilization have also been reported: Ries (1939) stated that the transparency of the Phallusia egg is modified but we have not been able to confirm this observation.

According to Von Ubisch the test-cells are also modified at fertilization, in accordance with Knaben's affirmation that they elaborate and pour the enzymes which are needed for the digestion of the chorion into the follicular space: these observations have not been confirmed, however.

13.7.4. The respiratory activity of the fertilized egg has been determined by Tyler and Humason (1937), Minganti (1957), Runnström (1930) and Lentini (1961). As shown in fig. 12 this activity is much greater than that of the unfertilized egg.

Connected with fertilization is the problem of autosterility in Ciona. Since

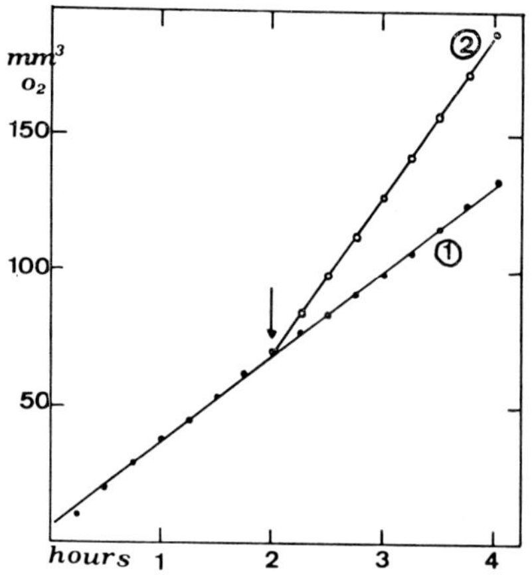

Fig. 12. Respiration curve of the ascidian egg before (1) and after fertilization (2). (Redrawn from Minganti 1957.)

the Ascidians are hermaphrodites, with sexual products ripe at the same time, autofertilization commonly occurs; however, in Ciona it is blocked. The factors underlying autosterility have been accurately investigated by Morgan, but no satisfactory solution has been presented. According to Morgan the spermatozoon is able to enter the vitelline space, but is not capable of penetrating the egg. On the other hand, 'naked' eggs are 100% autofertilized. These results call attention to the perivitelline content. Autofertilization has also been obtained after treating the eggs with Versene as shown in table 2 (Ortolani 1957b).

TABLE 2

Lots	Versene M	Autofertil. after Versene treatment	Controls autofertil.	Controls heterofertil.
10	0.001	67%	3.8%	81%
12	0.0001	36%	1.5%	79%
22	0.0005	63%	8.0%	85%

The action of Versene has not been explained; it possibly removes the Ca ions from the perivitelline liquids.

However the problem is further complicated by the fact that the efficacy of the block to autofertilization varies in the different geographical strains:

The Sicilian strains are 12% autofertile (Esposito Seu 1949), but strains from other countries show other values. Animals from the same geographical area also show different degrees of autofertilization.

The egg of *Ascidiella aspersa* is very interesting. The unfertilized eggs float on the water and are fertilized only under these conditions. If they are demembranated they sink to the bottom, but in this new condition they only are slightly fertilizable. Probably the spermatozoa need some chemical stimulation from the membrane (or vitelline fluids) to penetrate the egg.

The ascidian egg does not develop parthenogenetically. All attempts to obtain such a result by means of violent shaking, various degrees of concentration or dilution of sea water or solutions of sodium or magnesium chloride of varying strength have been unsuccessful. This negative result does not rule out the possibility that the egg can develop gynogenetically.

Ortolani has obtained good preliminary results (unpubl.) with the eggs

TABLE 3

Hybridizations in the Ascidians (Minganti 1959b).

Crosses	Developed eggs (%)
Vital hybrids	
Ascidia malaca ♀ × *Ciona intestinalis* ♂	8
Ascidia malaca ♀ × *Phallusia mamillata* ♂	72
Ascidia malaca ♀ × *Ascidia mentula* ♂	87
Ascidia mentula ♀ × *Phallusia mamillata* ♂	11
Lethal hybrids	
Ciona intestinalis ♀ × *Phallusia mamillata* ♂	5
Ciona intestinalis ♀ × *Ascidia mentula* ♂	4
Phallusia mamillata ♀ × *Ciona intestinalis* ♂	5
Ascidia mentula ♀ × *Ascidia malaca* ♂	1
Ascidiella aspersa ♀ × *Ciona intestinalis* ♂	6
Ascidiella aspersa ♀ × *Ascidia malaca* ♂	72
Ascidiella aspersa ♀ × *Ascidia mentula* ♂	41
Partially vital hybrids	
Ciona intestinalis ♀ × *Ascidia malaca* ♂	45
Phallusia mamillata ♀ × *Ascidia mentula* ♂	27
Phallusia mamillata ♀ × *Ascidia malaca* ♂	89
Ascidiella aspersa ♀ × *Phallusia mamillata* ♂	81

of *Ascidia malaca*. The eggs of *A. malaca* are sufficiently clear and the position of the nuclei can be individuated after fertilization. If the egg is properly fragmented it is possible to obtain fragments possessing only the maternal nucleus which are capable of development. The fragment emits the polocytes, segments as a miniature egg, and if sufficiently large, gives rise to a complete, small larva. Interspecific hybridizations are possible in Ascidians. Reverberi (1933) first noted that 'naked' eggs of Ciona can be penetrated by spermatozoa of Phallusia and vice versa. The hybrids develop up to the gastrula stage but afterwards die: death is accompanied by elimination of the paternal chromatin.

This research has been continued by Minganti (1959a, b) who produced 15 hybrid combinations with 5 different species (table 3). Some of them were more or less vital while others were lethal. Lethal hybrids die at gastrulation. The vital ones show abnormal development up to the swimming larva and have matrocline characteristics. The heterospecific chromosomes which persist in the cells seem to be completely afunctional.

Transplantation of the lethal blastomeres onto a maternal embryo revitalizes them and they become able to differentiate. Androgenetic hybrids have also been obtained: the combination *A. malaca (♀)* × *Phallusia (♂)* was vital, giving rise to larvae which unexpectedly presented matrocline features (palps) (fig. 13).

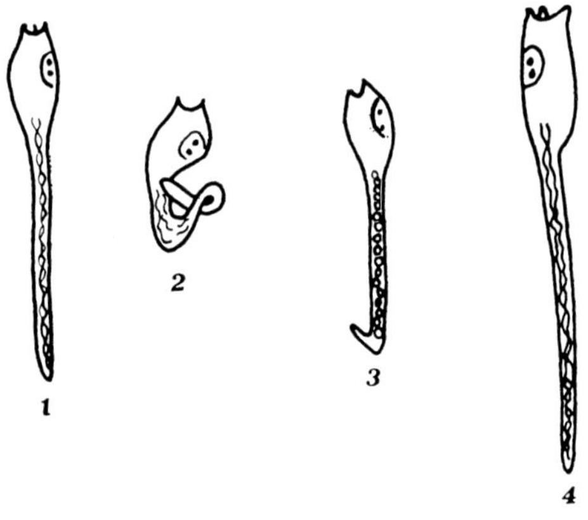

Fig. 13. Androgenetic hybrids of *A. malaca* (♀) × *Phallusia* (♂). (1) a normal larva of *A. malaca;* (4) a normal larva of *Phallusia;* (2–3) their androgenetic hybrids. (Redrawn from Minganti 1959a.)

13.8. The potentials of the unsegmented egg

Once liberated from its envelopes the egg can be manipulated at will: it can be broken, or partially aspirated and its blastomeres disarticulated, transplanted, rotated and recombined. These operations have been very useful for the analysis of the potentials of the egg.

As mentioned, the ascidian egg has long been considered the 'prototype' of the 'mosaic' egg: however, this view can no longer be rigidly held. We shall report here some arguments on which this assertion is based. We may

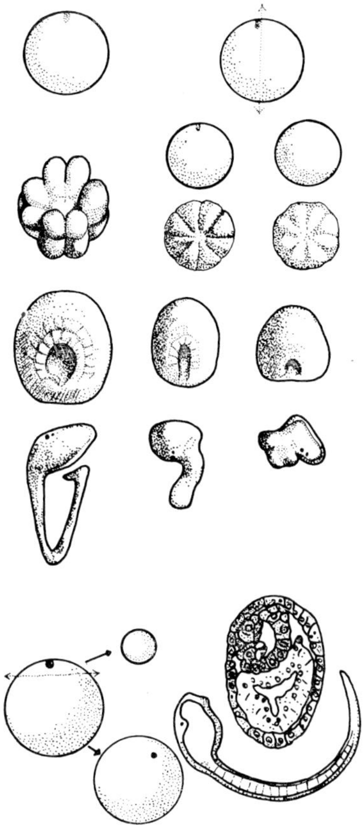

Fig. 14. An unfertilized Ascidian egg was fragmented by a vertical or latitudinal plane. The results are indicated in comparison with the development of an entire egg. (Redrawn from Dalcq 1932.)

begin by reporting the results obtained by Dalcq (1932) with the unfertilized egg. He broke the egg free hand with oriented cuts into two fragments, which were fertilized and their development followed (fig. 14). The fragments cleaved in the typical pattern, gastrulated normally, and gave rise to larvae. The fragments obtained by meridional cuts gave rise to normal twin larvae; with the fragments obtained by equatorial section, however, the results were less clear. This did not permit Dalcq to draw a precise conclusion. He reported that the unfertilized egg already has a structure and must consequently be considered a mosaic egg; however it is also capable of some regulation, which, however, is not 'essential' but 'topographic'. On the other hand, Reverberi (1931) obtained normal tadpoles from fragments of freshly fertilized Ciona eggs and concluded that the egg is totipotent.

The egg acquires a structure only a few minutes after fertilization: the fragments no longer behave equally. The supra-equatorial fragments do not cleave typically but radially and give rise terminally to permanent blastulae;

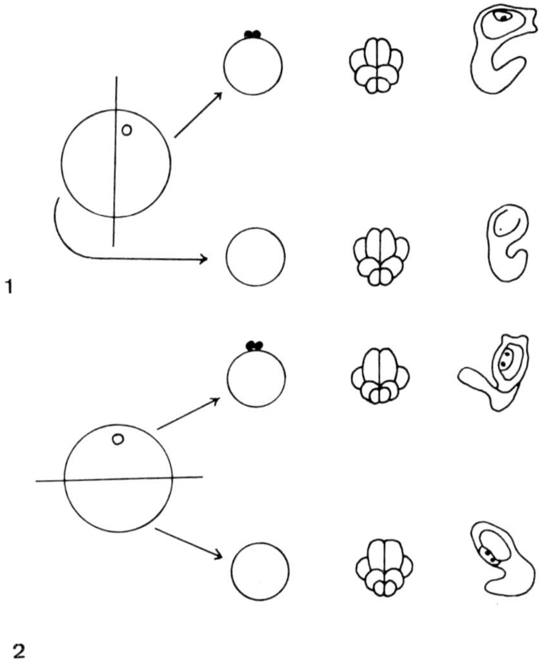

Fig. 15a. Twins obtained from both fragments of an unfertilized egg. In (1) the egg was fragmented meridionally; in (2) the egg was fragmented equatorially. (Redrawn from Reverberi and Ortolani 1962.)

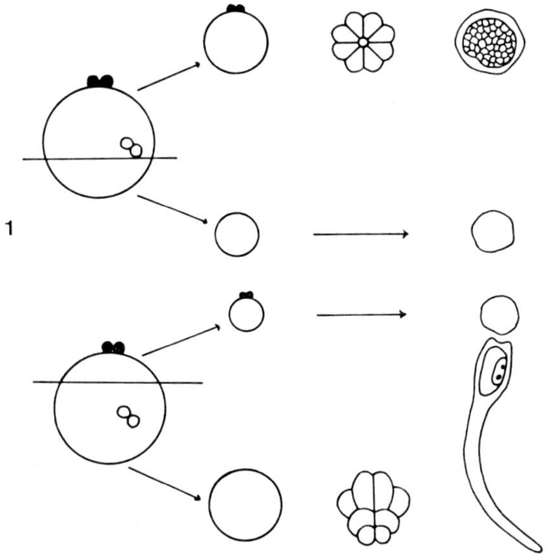

2

Fig. 15b. After the extrusion of the 2nd polar lobe the egg was fragmented by a sub-equatorial (1) or supra-equatorial section. The results are shown on the right. (Redrawn from Reverberi and Ortolani 1962.)

the sub-equatorial fragments, on the other hand, cleave typically and give rise to normal larvae (Reverberi 1937). The experiments repeated both on unfertilized and fertilized eggs of *A. malaca* (Reverberi and Ortolani 1962) yielded the same result; whatever the plane of section, two fragments of equal size of the same unfertilized egg gave rise to two complete larvae (fig. 15a); on the other hand, only the vegetal fragments of the fertilized egg segmented typically and gave rise to normal larvae (fig. 15b).

The result is more impressive with fragments obtained from centrifuged eggs. These fragments are made up of different plasms according to their specific weight. Some fragments can also be lacking in certain plasms. 'Heavy' and 'dark' fragments generally have pigments, yolk and mitochondria: the 'light' and 'hyaline' fragments, on the other hand, clear plasm and oils. The totipotence of the unfertilized egg is demonstrated by another result. Farinella-Ferruzza (1966) has been able to obtain one normal larva from two fused eggs. This result was obtained from demembranated eggs which had lain in sea water for some time (24–48 hr). Under these conditions their

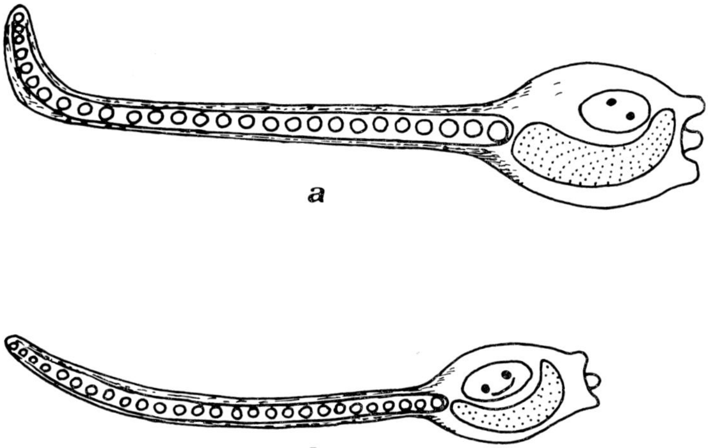

Fig. 16. A giant tadpole of *A. malaca* obtained from two fused unfertilized eggs: (a) giant larva; (b) the normal larva. (Redrawn from Farinella-Ferruzza 1966.)

cortex becomes sticky and if two eggs touch, they form a single gigantic egg which can then be fertilized (fig. 16). The egg buds off 4 polar globules, cleaves typically and finally gives rise to a gigantic but otherwise normal larva.

An additional proof of the totipotence of the unfertilized egg is as follows. With a micropipette inserted into the egg up to one-third of the cytoplasm can be removed by aspiration. If fertilized, the egg develops normally and becomes a normal larva.

13.9. Normal development

The normal development of the ascidian egg (as already mentioned) has been described by many authors. We shall report the fundamental steps of this development (table 4).

The first polocyte is extruded 10 min after fertilization, the second after 20 min and the egg cleaves after 40 min. The first cleavage is meridional and divides equally the three crescents already formed. It coincides with the bilateral plane of symmetry of the larva. The two formed blastomeres are quantitatively and qualitatively equal: according to Conklin's terminology they are indicated by AB2, *AB2*. The second cleavage is also meridional and perpendicular to the first: the four resulting blastomeres have the same size, but contain different plasms. The posterior ones are designated B.3, *B.3* and

TABLE 4

The cell-lineage up to the 32-cell stage.

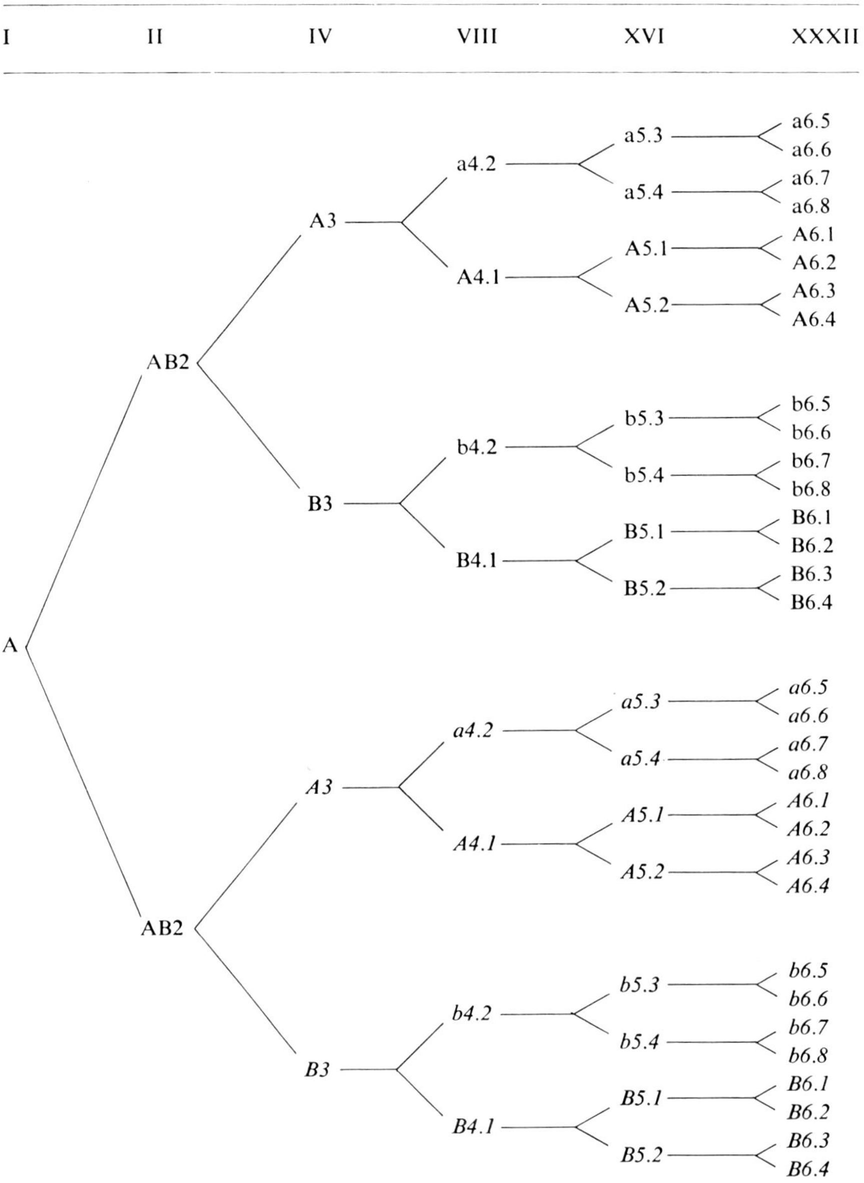

contain all the material of the yellow crescent, the anterior ones are designated A3, *A3* and contain all the material of the two anterior crescents. The third cleavage is equatorial: the eight cells so formed are more or less equal in size, but are constitutionally different. The anterior animal blastomeres, indicated by a4.2, *a4.2*, contain nearly all the neuroplasm, and part of the ectoplasm; the animal posterior blastomeres, indicated by b4.2, *b4.2*, contain only ectoplasm; the vegetal posterior blastomeres, indicated by B4.1, *B4.1*, contain all the mesoplasm (yellow crescent) and part of the entoplasm; and finally, the vegetal anterior blastomeres, A4.1, *A4.1*, have the chordoplasm, part of the endoplasm and part of the neuroplasm.

Fig. 17 shows an egg at this stage, with an indication of the different

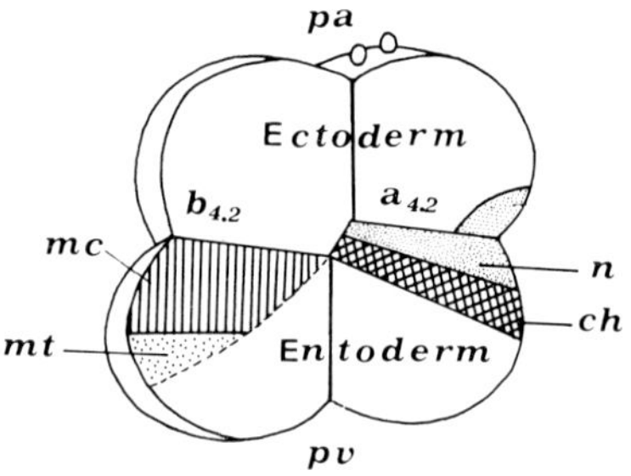

Fig. 17. Topographic distribution of the organ-forming territories in an egg at the 8-cell stage: n = neural system; ch = chorda; mt = musculature; mc = mesenchyme; pa = animal pole; pv = vegetal pole. (Redrawn from Ortolani 1954.)

plasms. The map, constructed from the results obtained using coloured chalk granules (Ortolani 1954), agrees in its essential lines with that of Conklin. At this stage the egg can be easily orientated taking into consideration the localization of the polar globules and the position of the posterior vegetal blastomeres (which protrude under the posterior animal ones).

The fourth segmentation occurs earlier in the blastomeres of the vegetal hemisphere; it is also vertical: the egg now has 16 cells. Those of the supra-equatorial hemisphere are more or less equal in size; those of the sub-equatorial hemisphere, however, are different; B5.2, *B5.2* are very small, are hyaline and protrude under b5.4, *b5.4*. Sometimes called 'micromeres', they are a landmark for orienting the egg. The denominations of the single blastomeres are indicated in fig. 18.

By the fifth and sixth segmentation, 32 and 64 cells respectively are formed: this stage is equivalent to a blastula, although this blastula is not spherical

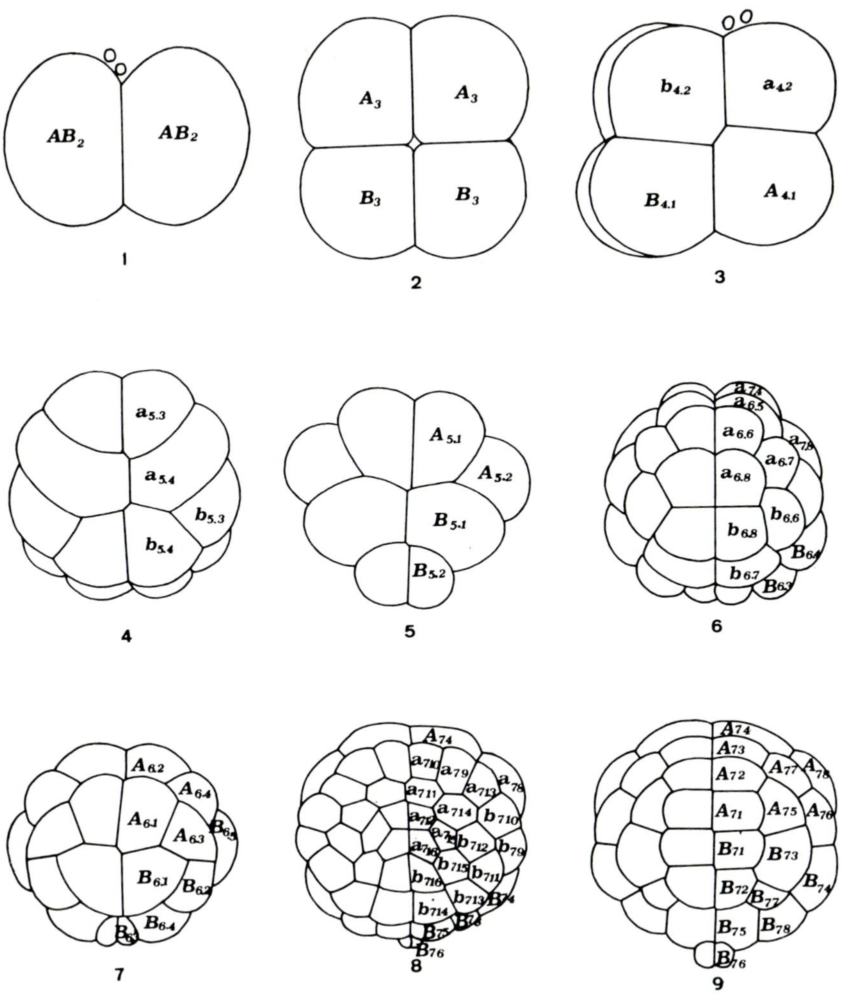

Fig. 18. Development of an ascidian egg up to 64 cells. In 4, 5, 6, 7, 8, 9, the blastomeres are seen from the animal and vegetal poles respectively. (Redrawn from Conklin 1905a.)

and does not have a cavity. According to Conklin the fate of the blastomeres at this stage is the following:

animal hemisphere = 26 epidermic cells + 6 neural cells;

vegetal hemisphere = 4 neural cells + 10 entodermic cells + 10 mesenchymatic cells + 4 muscular cells + 4 notochordal cells.

But Ortolani has shown that A7.6 and *A7.6* are notochordal cells and A7.5, *A7.5.* and B7.6, *B7.6* muscular cells. The presumptive fate of the 32 cells of the vegetal hemisphere, according to Ortolani, is the following: 8 muscular + 4 mesenchymal + 6 chordal + 4 neural + 10 entodermic. At the 64- or 128-cell stage the egg enters into gastrulation. The first cells involved in the invagination are the entodermic A7.5, *A7.5*, A7.2, *A7.2*, A7.1, *A7.1*. As the result of their invagination the notochordal cells constitute the dorsal blastoporic lip: the lateral blastoporic lips are formed by the muscular and mesenchymal cells. The chorda cells form the blastoporic dorsal lip only for a while; soon they invaginate, contributing to the formation of the archenteric roof. The neural plate forms on it. The notochordal cells however do not remain in the archenteric roof for long: they soon shift caudally between the lateral masses of mesenchymal and muscular cells. At the end of neurulation a neural tube is formed: the tube is large anteriorly, forming a vesicle on the walls of which the two pigmented sensorial organs differentiate (the eye made up of many cells, and the otholite). A swimming larva is shown (with organs indicated) in fig. 19.

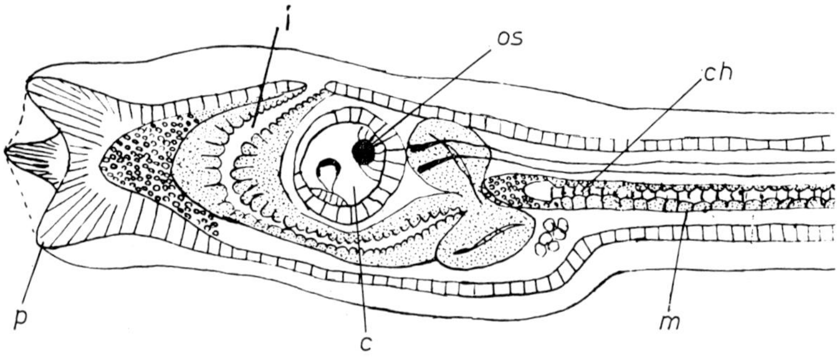

Fig. 19. A swimming ascidian larva: p = palps; c = brain; m = musculature; ch = chorda; os = eye; i = intestine.

13.10. The potentials of the segmented egg

The results of the development of isolated blastomeres from an egg at the 2- or 4-cell stage have already been reported; each blastomere yields exactly what would have originated if it had been left *in situ*. This result indicates that at the 2-4-cell stage the egg is a complex of self-differentiating parts, a 'mosaic' system. One could suppose that as the development of the egg proceeds this mosaicism becomes more and more rigid. This conclusion, however, is not correct. In fact, experimental analysis of the destiny of the blastomeres at the 8-cell stage yields unexpected results. Let us dissociate the homologous pairs of blastomeres at this stage and follow their development. The results are shown in fig. 20.

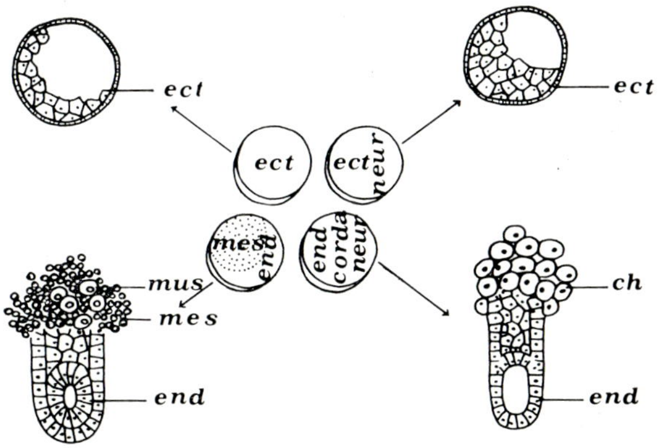

Fig. 20. The four disarticulated couples of homologous blastomeres of an egg at the 8-cell stage; the development is indicated: ch = chorda; end = entoderm; ect = ectoderm; mus = musculature; mes = mesenchyme (Reverberi and Minganti 1946).

With the exception of the pair formed by the anterior animal blastomeres, the other pairs differentiate according to their presumptive destiny. The anterior animal blastomeres, if left *in situ,* give rise to the larval brain, the sensorial organs (eye and otholite) and the palps. If isolated, however, these structures no longer differentiate. Evidently their differentiation is conditioned. In order to discover how this occurs we performed the following experiments: combination of the two anterior animal blastomeres with (a) the posterior animal blastomeres, (b) the posterior vegetal blastomeres and (c) the anterior vegetal blastomeres (fig. 21). Only combination (c) gave rise to a larval form having brain, sensorial spots and palps; (a) (which is equivalent to an animal half) gave only an ectodermic blastula; while (b) developed as a form possessing only musculature, entoderm, and mesenchyme. The same result was observed in an experiment in which the egg was more complete. At the 8-cell stage the couples were removed from their connection with the anterior animal blastomeres (fig. 22).

Brain, sensorial spots, and palps formed *only* when the anterior vegetal blastomeres were present. The anterior vegetal blastomeres must thus be considered necessary for the differentiation of the brain, sensorial organs and palps from the anterior animal ones. The situation is similar to that of Amphibians, where the differentiation of the neural system is dependent on another territory (blastoporal lip).

This fact raises a question. Have the anterior vegetal blastomeres of the

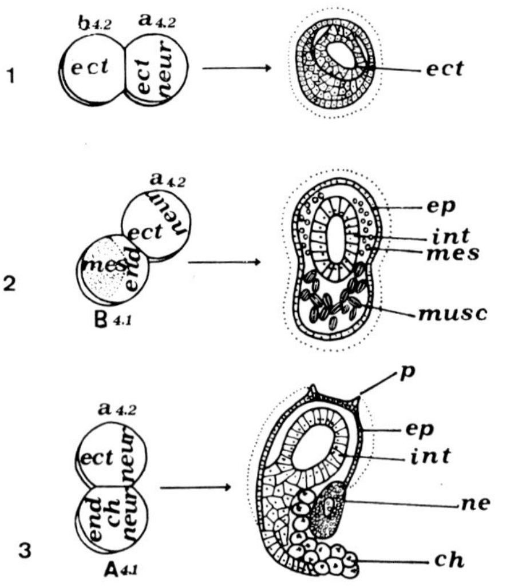

Fig. 21.

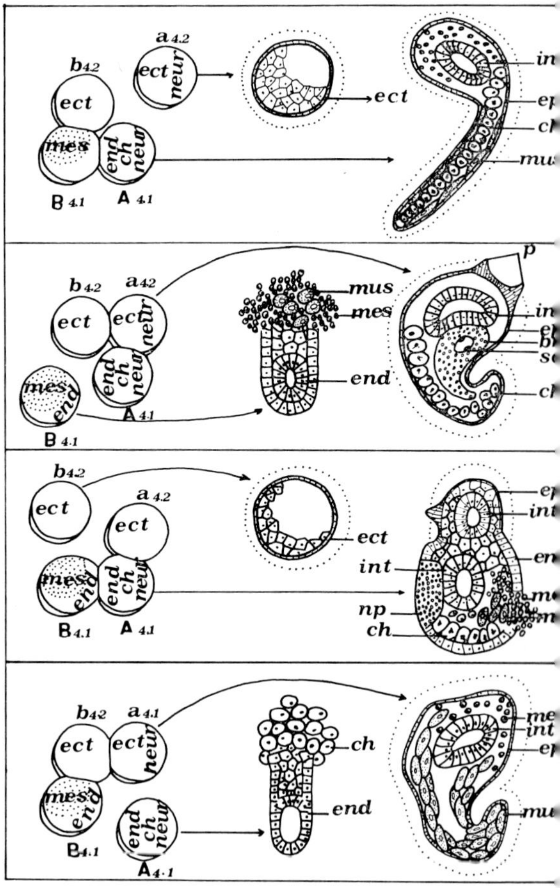

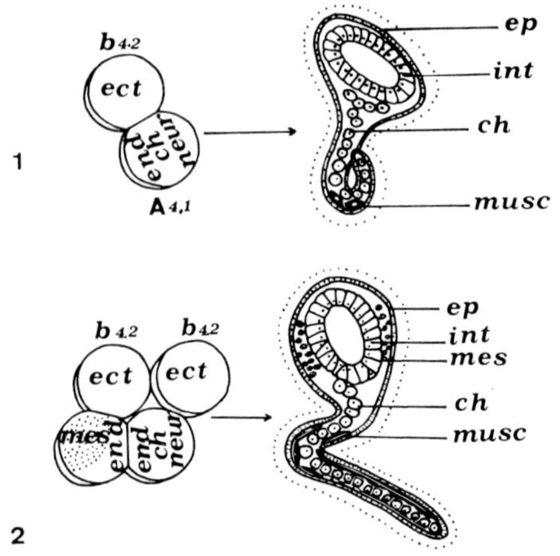

Fig. 23.

Fig. 22.

Figs. 21, 22, 23. Various blastomeric combinations: the results of their development are indicated. (Redrawn from Reverberi and Minganti 1946.) br = brain; ch = chorda; ect = ectoderm; end = entoderm; ep = epidermis; int = intestine; mes = mesenchyme; musc = musculature; neu = neural system, np = neural plate; p = palps; so = sensorial organs.

ascidian egg the capacity of an 'organizer' as is the case of the blastoporal lip of the Amphibians? The problem was again studied experimentally: at the 8-cell stage the pair of the anterior animal blastomeres was removed and a pair of posterior animal blastomeres taken from another egg transplanted in its place (fig. 23). The egg thus has a double dose of epidermic material: it might be hoped that part of it, under the stimulus of the anterior vegetal blastomeres, will be induced to form a neural system with its sensorial spots. The result, however, is negative. The larvae that form possess neither brain nor pigmented spots. Evidently the 'non-competent' ectoderm does not respond to the 'stimulation' of the anterior vegetal blastomeres. The anterior vegetal blastomeres do not possess the capacity of inducing but only of evocating. The question then arises as to the time when the evocative stimulus is released.

The problem was systematically studied by Ortolani (1959), who destroyed the derivatives of the two anterior vegetal blastomeres in eggs at the 16-32-64-cell stages respectively. From these experiments she concluded that the destiny of the 'competent' neural ectoderm is not fixed up to the 64-cell stage.

At the 128-cell stage the egg is in gastrulation, and the anterior entodermic cells, as well the chorda cells, are interiorly under the ectoderm, which is competent to give rise to the neural plate. It seems that the evocation is exerted at this stage when the anterior entodermic cells, as well the chorda cells, form the roof of the archenteron. The condition is similar to that observed in Amphibians, with the difference that in Amphibians the archenteric roof is formed from the mesoderm as well as the notochord. Thus a new question arises; is the 'evocative' power exerted by the entodermic or the chordal cells? This question has been considered by Von Ubisch who performed the following operations on the egg at the 64-cell stage: (a) destruction of the entoderm; (b) destruction of the mesoderm; (c) destruction of the notochordal cells. In all these operations he obtained larvae with brain and palps; he consequently concluded that the formation of these organs is autonomous and that their differentiation is not dependent on induction or evocation from other territories. It must, however, be noted that the type of operation indicated by (b) is out of the question because, as mentioned, the mesoderm is not responsible for the evocation. Concerning operation (c) it should be noted that Von Ubisch did not destroy all the notochordal cells; in fact following the map of Conklin he considered cells A7.6 and *A7.6* to be mesenchymal; according to Ortolani, however, these are notochordal cells. Finally, the result obtained from (c) is not contrary to our deductions. In fact, we consider both entodermic and notochordal cells to be equally responsible for evocation: the presence of either would be sufficient for the evocation.

In conclusion, the process by which the brain forms in the Chordates is much the same. In Ascidians it is more evocative than inductive: in Amphioxus it is inductive (it 'induces' the 'non-competent' ectoderm to become the neural system). In vertebrates it is inductive and organizing. In both Amphioxus (Tung et al. 1962) and Amphibians the inductor is represented by the notochord and mesoderm; in Ascidians by the notochord and the (anterior) entoderm. In Amphioxus and vertebrates the notochord and mesoderm make up the roof of the archenteron: in Ascidians the notochord participates only for a short time in the roof of the archenteron. It is under the stimulus of the archenteric roof that the superposed ectoderm is caused to differentiate into the neural system. In Amphibians, the inducing power seems due to a chemical substance: is the evocative power also exerted by a chemical substance in the Ascidians? Unfortunately we cannot answer this question.

13.11. Cellular differentiation

Doubtless the problem most extensively investigated today by biologists is cell differentiation. The problem arises from the fact that the egg starts as an undifferentiated cell and after a while becomes many differently differentiated cells. The problem has not been extensively studied in Ascidians. These nevertheless seem to be good material because differentiation of the cells of the dividing egg occurs very early.

We shall report here on some relevant investigations. Let us return to the egg at the 8-cell stage: as we have seen we can tell at this stage what will be the inevitable destiny of each cell. Thus for instance, we may say that the entire musculature of the larva will differentiate from the posterior vegetal blastomeres.

At this stage, of course, there is no trace of musculature in these blastomeres: it must, however, be accepted that there are the 'precursors'. It is possible to imagine that these 'precursors' are 'changed' into musculature by a more or less long chain of chemical reactions. The interest of the research is to follow, step by step, all the transformations which make a chemical substance become an organic structure. Unfortunately we still know nothing about these processes.

We have some information on the role of the posterior vegetal blastomeres. Ries (1937) found that they give an intense positive Nadi reaction while the others remain more or less colourless. This result has a meaning and in fact Ries tried to explain it. As the Nadi reaction is an enzymatic reaction ('indophenol-oxidases') he thought that these enzymes were localized differentially in the posterior vegetal blastomeres. Ries held these enzymes responsible for differentiation of the musculature because after displacing them by centrifugation he obtained larvae with displaced musculature. Later Reverberi showed that the enzymes noticed by Ries are 'cytochrome-oxidases', which are indubitably localized in the mitochondria. However he attributed an energetic rather than a plastic role to the mitochondria in the differentiation of musculature. Inhibiting the cytochrome oxidase by sodium azide (a specific inhibitor) produces larvae which, although they move poorly, possess a differentiated musculature. The presence of a high content of cytochrome-oxidase in the posterior vegetal blastomeres has been confirmed by Berg (1956), using microchemical methods (table 5): on the other

TABLE 5

Ciona, 4-cell stage.

	Ratio post./ant. blastomeres	Reference
Cytochrome oxidase	2.45	Berg 1956
Succinic dehydrogenase	1.90	Berg 1957
Apyrase	1.43	Berg 1957
Acid phosphatase	0.90	Berg 1957
Ribonucleic acid	1.09	Berg 1957
Proteins	1.01	Berg 1957
Number of mitochondria	2.32	Berg and Humphreys 1960

hand Berg and Humphreys (1960) and Mancuso (1964) have shown in electron microscopic studies that these blastomeres are particularly rich in mitochondria. Moreover De Vincentiis and Ortolani (1962) have observed greater respiratory activity in the posterior vegetal blastomeres than in the others (fig. 24).

In any case the observations reported here excluded the 'direct' participation of cytochrome-oxidase in the differentiation of the musculature: the function of the enzyme is only to provide the energy for differentiation. This conclusion is also valid for other enzymes, for instance those associated with differentiation of the larval eye. The eye on an ascidian larva is a rather

Sketch of operation.

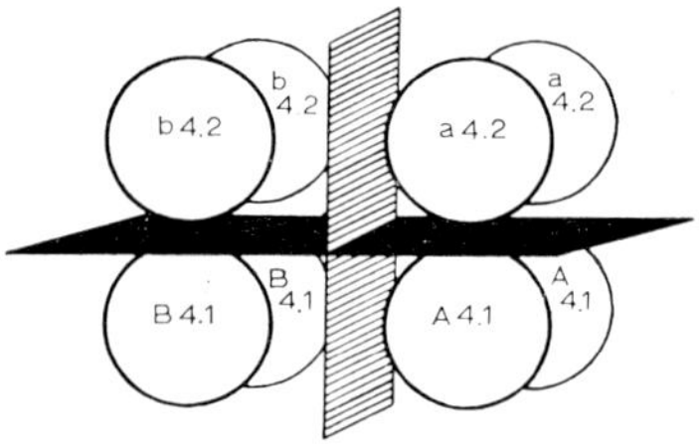

	b4.2	B4.1	a4.2	A4.1
Number of determinations	35	47	37	28
Means ± standard error	0.347±0.016	0.430±0.017	0.249±0.012	0.241±0.009
Standard deviation	±0.009	±0.121	±0.073	±0.049

Fig. 24. At the 8-cell stage the four homologous couples of blastomeres were separated and their respiration was determined (De Vincentiis and Ortolani 1964).

complex structure, being made up of numerous sensitive pigmented cells and lens. The pigment is a melanin. The territory destined to give rise to the eye can be indicated long before its differentiation. In fact the specific cytochemical reaction of dopa-oxidase yields a strong positive reaction in the still open neural folds (Minganti 1951). If the dopa-oxidase activity is blocked specifically the eye is still formed although it does not have any pigment. The same result is obtained with acetyl-cholinesterase. According to Durante (1956), in the developing Ciona embryo the territory of the musculature can be localized by the cytochemical reaction specific for this enzyme. Blocking the enzyme with eserine or neostigmine completely paralyzes the larva; the musculature, however, is present!

The cases reported above show clearly that these enzymes are not directly implicated in the processes of cellular differentiation. Differentiation is certainly due to an enzymatic process, but which enzymes are responsible? Can they be blocked by specific inhibitors? The problem was investigated in another way.

Since, in the final analysis, differentiation is the synthesis of specific proteins, we tried interfering with this synthesis. Our idea was to give a 'wrong' amino acid instead of the normal one to a synthesizing protein. However, this is not easy to do. It is necessary to persuade an amino-acyl synthetase to combine not with its specific amino acid but with another, so as to offer the mRNA a 'wrong' amino-acyl tRNA. We hoped to achieve this result by altering the usual pool of amino acids in the egg, letting it develop in sea water containing an unusual solution of a particular amino acid. The results obtained from these attempts cannot be considered satisfactory: only the treatment with cysteine produced abnormal larvae. Although its action was revealed to be stronger at gastrulation, which is the period of intense protein synthesis, it is doubtful if this was really due to an unusual incorporation of cysteine in the forming proteins. Other attempts to solve the problem were made by trying to influence the process of transcription or translation with actinomycin or puromycin. Puromycin blocks segmentation; in chloramphenicol however the eggs develop although the larvae are distorted and smaller than controls (Fiasconaro 1963, 1964).

Actinomycin D interferes with the process of transcription. Eggs treated with large doses of actinomycin (10 μg) cleave normally but stop at the gastrula stage. With smaller doses (1–3) they develop to abnormal larvae; the organs of the larvae, however, are well differentiated (Librera 1964). Some interesting observations were made by Smith (1967) on the eggs of *Ascidia nigra*. According to this author there are two intense periods of protein synthesis: one immediately after fertilization, the other at gastrulation. The first is insensitive, the second sensitive to actinomycin. This latter period corresponds to intense RNA synthesis.

The author maintains that protein synthesis in the first period is directed by the instructions released during oogenesis (stable mRNA); that of the second period, on the other hand, is based on new instructions (new mRNA). Evidently one could interfere with differentiation by modifying the composition of the mRNA. This result could be achieved by modifying the composition of the pool of bases or nucleotides from which the egg obtains the materials for the synthesis of RNA.

We allowed the eggs to develop their nucleosides, their nucleotides and their analogues in solutions of purines and pyrimidines. Adenine blocks segmentation (0.05–0.02%): lower concentrations block the development at gastrulation (Waddington and Mancuso 1955), guanine acts similarly: thymine does not alter the development (5 mM); uracil improves it (Mancuso 1962). Thymidine, even in strong concentrations (2 mM), does not modify development; adenosine, guanosine, cytidine and uridine, however, produce abnormalities (Cusimano-Carollo 1961). Purinic as well pyrimidinic nucleotides (0.5 mM) generally affect development: the nucleotide of the uracil, however, improves it (Cusimano-Carollo 1964).

Of some interest are the results with synthetic homopolyribonucleosides. Treatment of the eggs with solutions of poly-U produces abnormal larvae. If, however, phenylalanine (which is the corresponding amino acid in the genetic code) is added to poly-U, normal larvae are obtained. Poly-U + valine or lysine (or glycine which does not correspond to U U U in the code), produces abnormal larvae. The same happens with poly-C: alone it is lethal, poly-C + proline, however, produces normal larvae (fig. 25). This result would suggest that, as expected, the 'code' is also valid for the Ascidians (Pisanò 1966).

Poly-U mg/ml	0.5 mM phe	Development	No. exper.	Hatching
0.5	−		26	+
0.5	+		19	+
1	−		22	−
1	+		20	+
1.5	−		21	−
1.5	+	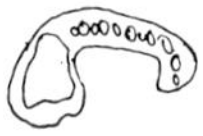	21	+
Control		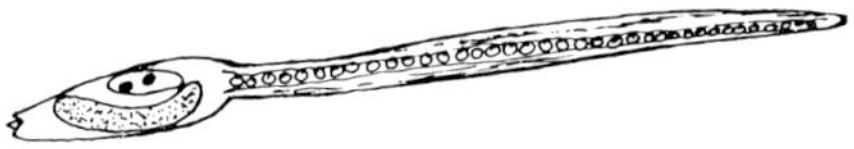		

Fig. 25. Effects of poly-U with or without phenylalanine on development. The addition of phenylalanine improves development (Pisanò 1966).

References

BERG W. E., 1956. Cytochrome oxidase in anterior and posterior blastomeres of *Ciona intestinalis*. Biol. Bull. *110*, 1–7.

BERG W. E., 1957. Chemical analysis of anterior and posterior blastomeres of *Ciona intestinalis*. Biol. Bull. *113*, 365–375.

BERG W. E. and W. J. HUMPHREYS, 1960. Electron microscopy of four-cell stages of the Ascidians Ciona and Styela. Develop. Biol. *2*, 42–60.

BERRILL N. J., 1950. The Tunicata. Ray Society, London (1950).

BLUNTSCHLI H., 1904. Beobachtungen am Ovarialei der Monascidie *Cynthia microcosmus*. Morphol. Jahrb. *32*, 391–450.

CASTLE W. E., 1894. On the cell lineage of the Ascidian egg. Proc. Amer. Acad. Arts Sci. *30*, 200–216.

CHABRY L., 1887. Contribution à l'embriologie normale et tératologique des Ascidies simples. J. Anat. Physiol. Paris *23*, 167–319.

COHEN A. and N. J. BERRILL, 1936. The development of isolated blastomeres of the Ascidian egg. J. Exptl. Zool. *74*, 91–117.

CONKLIN E. G., 1905a. The organization and cell-lineage of the Ascidian egg. J. Acad. Natl. Sci. Phila. *13*, 1–119.

CONKLIN E. G., 1905b. Organ-forming substances in the eggs of Ascidians. Biol. Bull. *8*, 205–230.

CONKLIN E. G., 1905c. Mosaic development in Ascidian eggs. J. Exptl. Zool. *2*, 145–223.

COSTELLO P. D., 1948. Ooplasmic segregation in relation to differentiation. Ann. N.Y. Acad. Sci. 49, *663–683*.

COWDEN R. R., 1961. A comparative cytochemical study of oocyte growth and development in two species of Ascidians. Acta Embryol. Morphol. Exptl. *4*, 123–141.

COWDEN R. R., 1962. A comparative cytochemical study of oocyte growth and development in five ascidian species. Trans. Microscop. Soc. *81*, 149–158.

COWDEN R. R. and C. L. MARKERT, 1961. A cytochemical study of the development of *Ascidia nigra*. Acta Embryol. Morphol. Exptl. *4*, 142–160.

CRAMPTON H. E., 1899. Studies on the early history of the ascidian egg. J. Morphol. (Suppl.) *15*, 29–56.

CUSIMANO-CAROLLO T., 1961. Results of further experiments with purinic and pyrimidinic nucleosides in the development of Ascidian eggs. Acta Embryol. Morphol. Exptl. *4*, 305–312.

CUSIMANO-CAROLLO T., 1964. Sull'azione di nucleotidi purinici e pirimidinici sullo sviluppo dell'uovo di Ascidie. Ric. Sci. *34*, (II-A) 335–348.

DALCQ A. M., 1932. Étude des localisations germinales dans l'œuf vierge d'Ascidie par des expériences de mérogonie. Arch. Anat. Microscop. Morphol. Exptl. *28*, 223–333.

D'ANNA T. and S. METAFORA, 1965. Fosfati inorganici, esteri fosforici labili, ATP e glucoso-6-fosfato nell'ovario di *Ciona intestinalis*. Rend. Accad. Naz. Lincei *38*, s. 8, 227–230.

DAVIDOFF M., 1889. Untersuchungen zur Entwicklungsgeschichte der Distaplia magnilarva Della Valle einer zusammengesetzen Ascidie. Pubbl. Staz. Zool. Napoli *9*, 113–178.

DE LACAZE-DUTHIERS, H. 1874. Les Ascidies simples des côtes de France. Arch. Zool. Exptl. Gen. *3*, 119–174.

DE VINCENTIIS M., 1960a. Observations morphologiques et cytochimiques sur quelques aspects de l'ovogénèse des Ascidies. Symp. Germ Cells Develop. 291–298.

DE VINCENTIIS M., 1960b. Osservazioni citologiche e citochimiche su alcuni momenti della oogenesi di *Ciona intestinalis*. Rend. Ist. Sci. Univ. Camerino *1*, 145–150.

DE VINCENTIIS M., 1960c. Indagini spettrografiche e citochimiche per la caratterizzazione di un materiale fluorescente presente nelle cellule testacee di uova di Ascidia (*Phallusia mamillata* e *Ciona intestinalis*). Rend. Ist. Sci. Univ. Camerino *1*, 151–161.

DE VINCENTIIS M., 1962. Ulteriori indagini istospettrografiche e citochimiche su alcuni aspetti dell'ovogenesi di *Ciona intestinalis*. Atti Soc. Peloritana Sci. Fis. Mat. Nat. *8*, 190–198.

DE VINCENTIIS M. and G. ORTOLANI, 1962. Sul consumo di O_2 delle metà dell'uovo di *Phallusia mamillata*, allo stadio 8 (ricerche microrespirometriche). Rend. Accad. Naz. Lincei *32*, s. 8, 529–533.

DE VINCENTIIS M. and G. ORTOLANI, 1964. Sul consumo di O_2 delle coppie di blastomeri di *Phallusia mamillata* allo stadio di 8 blastomeri (ricerche microrespirometriche). Rend. Accad. Naz. Lincei *35*, s. 8, 604–608.

DURANTE M., 1956. Cholinesterase in development of *Ciona intestinalis* (Ascidia). Experientia *12*, 307–310.

DURANTE M., 1959. Sulla localizzazione istochimica della acetilcolinesterasi lungo lo sviluppo di alcune Ascidie e in Appendicularie. Acta Embryol. Morphol. Exptl. *2*, 234–243.

ESPOSITO SEU M., 1949. Autofecondazione e fecondazione incrociata nelle Ascidie. Rend. Accad. Naz. Lincei *8*, s. 8, 502–505.

EZELL S. D., 1963. The lateral body of *Ciona intestinalis* spermatozoa. Exptl. Cell Res. *30*, 615–617.

FARINELLA-FERRUZZA N., 1965. Sullo sviluppo di uova 'invecchiate' nelle Ascidie. Rend. Accad. Naz. Lincei *39*, s. 8, 338–343.

FARINELLA-FERRUZZA N., 1966. Fusioni e sviluppo di uova vergini di *Ascidia malaca*. Acta Med. Romana IV, 54–57.

FERRINI U., M. L. MARCANTE, A. CAPUTO, S. MINAFRA and G. REVERBERI, 1963. Amino acids in the ovary of *Ciona intestinalis*. Acta Embryol. Morphol. Exptl. *6*, 283–288.

FIASCONARO A. M., 1963. Effects of chloramphenicol on the embryonic development. Acta Embryol. Morphol. Exptl. *6*, 149–157.

FIASCONARO A. M., 1964. L'azione della puromicina sullo sviluppo embrionale delle Ascidie. Acta Embryol. Morphol. Exptl. *7*, 201–216.

GERZELI G., 1962. Istochimica delle cellule follicolari e delle cellule testali in diverse Ascidiacee. Rend. Ist. Lombardo Sci. Lettere *96*, 165–178.

HARVEY L. A., 1927. The history of cytoplasmic inclusions of the egg of *Ciona intestinalis* during oogenesis and fertilization. Proc. Roy. Soc. (London), Ser. B *101*, 137–161.

HIRSCHLER J., 1917. Über die Plasmakomponenten der weiblichen Geschlechtzellen. Arch. Mikroskop. Anat. *89*, 1–58.

HSU W. S., 1962a. An electron microscopic study on the origin of yolk in the oocytes of the ascidian *Boltenia villosa*. Cellule *62*, 140–165.

HSU W. S., 1962b. The site of ribosome formation in the oocytes of the ascidian, Boltenia villosa. Z. Zellforsch. *58*, 17–26.

HSU W. S., 1963. The nuclear envelope in the developing oocytes of the tunicata *Boltenia villosa*. Z. Zellforsch. *58*, 660–678.

JÄGERSTEN G., 1935. Untersuchungen über den strukturellen Aufbau der Eizelle. Zool. Bidrag, Uppsala *16*, 1–282.

JULIN C., 1893. Les Ascidiens des côtes du Boulonnais. 1. Recherches sur l'anatomie et l'embryogénie de *Styelopsis grossularia*. Bull. Soc. Franc. Belg. *24*, 208–259.

KALK M., 1963. Cytoplasmic transmission of a vanadium compound in a tunicate oocyte, visible with electronmicroscopy. Acta Embryol. Morphol. Exptl. *6*, 289–303.

KESSEL R. G., 1962. Fine structure of pigment inclusions in the test cells of the ovary of Styela. J. Cell Biol. *12*, 637–640.

KESSEL R. G. and H. W. BEAMS, 1965. An unusual configuration of the Golgi complex in pigment-producing 'test' cells of the ovary of the tunicate, Styela. J. Cell Biol. *25*, 55–67.

KESSEL R. G. and N. KEMP, 1962. An electron microscope study on the oocyte, test cells and follicular envelope of the tunicata *Molgula manhattensis*. J. Ultrastruct. Res. *6*, 57–76.

KNABEN N., 1936. Über Entwicklung und Funktion der Testazellen bei *Corella parallelogramma* Müll. Bergens Mus. Arbok No 1, 1–33.

KORSCHELT E. and K. HEIDER, 1902. Lehrbuch der vergleichenden Entwicklungsgeschichte der wirbellosen Tiere. Jena, p. 539.

KOWALEWSKY A., 1866. Entwicklungsgeschichte der einfachen Ascidien. Mém. Acad. Sci. St. Pétersbourg Sér. 7, *10*, 1–19.

KOWALESWKY A., 1871. Weitere Studien über die Entwicklungsgeschichte der einfachen Ascidien. Arch. Mikroskop. Anat. *7*, 101–130.

KÜHN A., 1965. Vorlesungen über Entwicklungsphysiologie. Springer Verlag, Berlin.

LA SPINA R., 1958. Lo spostamento dei plasmi mediante centrifugazione nell'uovo vergine di Ascidie e il conseguente sviluppo. Acta Embryol. Morphol. Exptl. *3*, 66–78.

LENTINI R., 1961. The oxygen uptake of *Ciona intestinalis* eggs during development in normal and experimental conditions. Acta Embryol. Morphol. Exptl. *4*, 209–218.

LIBRERA E., 1964. Effects of actinomycin D on the embryonic development of *Ascidia malaca*. Acta Embryol. Morphol. Exptl. *7*, 242–248.

LOYEZ M., 1909. Les premiers stades de la vitellogénèse chez quelques Tuniciers. Compt. Rend. Ass. Anat. *11*, 189–195.

MANCUSO V., 1962. L'azione di basi puriniche e pirimidiniche nello sviluppo dell'uovo di *Ciona intestinalis*. Acta Embryol. Morphol. Exptl. *5*, 71–81.

MANCUSO V., 1963. Distribution of the components of normal unfertilized eggs of *Ciona intestinalis* examined at the electron microscope. Acta Embryol. Morphol. Exptl. *6*, 260–274.

MANCUSO V., 1964. The distribution of the ooplasmic components in the unfertilized, fertilized and 16-cell stage egg of *Ciona intestinalis*. Acta Embryol. Morphol. Exptl. *7*, 71–82.

MANCUSO V., 1965. An electron microscope study of the test cells and follicle cells of *Ciona intestinalis* during oogenesis. Acta Embryol. Morphol. Exptl. *8*, 239–266.

MANCUSO V., Ulteriori dati sulla vitellogenesi in *Ciona intestinalis*. Acta Embryol. Morphol. Exptl. *9*, 255–269.

MANSUETO C., 1964. Ricerche citologiche e citochimiche concernenti la oogenesi di *Ciona intestinalis*. Ric. Sci. (II-B), 529–542.

METAFORA S., 1966. Effetti dell'acqua di uova sul metabolismo degli spermi di *Ciona intestinalis* (Ascidie). Rend. Accad. Naz. Lincei 40, s. 8, 290–295.

METAFORA S. and F. RESTIVO, 1963. Le fertilizine dell'uovo di Ascidie. Rend. Accad. Naz. Lincei *34*, s. 8, 439–442.

METAFORA S. and F. RESTIVO, 1964. Le fertilizine dell'uovo di Ciona. Ric. Sci. *34* (II-B), 5–24.

MINGANTI A., 1950. Esperimenti di ibridazione interspecifica nelle Ascidie. Pubbl. Staz. Zool. Napoli *22*, 293–307.

MINGANTI A., 1951. Esperienze sulle fertilizine nelle Ascidie. Pubbl. Staz. Zool. Napoli *23*, 58–65.

MINGANTI A., 1956. I cromosomi nei Tunicati. Boll. Zool. *23*, 299–315.

MINGANTI A., 1957. Experiments on the respiration of Phallusia eggs and embryos (Ascidians). Acta Embryol. Morphol. Exptl. *1*, 150–163.

MINGANTI A., 1959a. Androgenetic hybrids in Ascidians. 1. *Ascidia malaca* (♀) × *Phallusia mamillata* ♂. Acta Embryol. Morphol. Exptl. *2*, 244–256.

MINGANTI A., 1959b. Lo sviluppo embrionale e il comportamento dei cromosomi in ibridi tra 5 specie di Ascidie. Acta Embryol. Morphol. Exptl. 2, 269–301.

MINGANTI A., 1960. Recent investigations on the development of Ascidians. Symp. Germ Cells Develop., 255–276.

MOLINARO M., 1964. Amino acid activation in Ciona ovary and developing egg. Experientia *20*, 216.

MORGAN T. H., 1939. The genetic and the physiological problems of self-sterility in Ciona. III. Induced self-fertilization. J. Exptl. Zool. *80*, 19–55.

ORTOLANI G., 1954. Risultati definitivi sulla distribuzione dei territori presuntivi degli organi, nei germi di Ascidie allo stadio VIII, determinati con le marche al carbone. Pubbl. Staz. Zool. Napoli *25*, 161–187.

ORTOLANI G., 1957a. Il territorio precoce della corda nelle Ascidie. Acta Embryol. Morphol. Exptl. *1*, 33–36.

ORTOLANI G., 1957b. Rimozione dell'autosterilità in *Ciona intestinalis* mediante trattamento con versene. Ric. Sci. *27*, 2150–2153.

ORTOLANI G., 1959. Ricerche sulla induzione del sistema nervoso nelle larve delle Ascidie. Boll. Zool. *26*, 341–348.

ORTOLANI G., 1962. Territorio presuntivo del sistema nervoso nelle larve di Ascidie. Acta Embryol. Morphol. Exptl. 4, 189–198.

ORTOLANI G., 1964. Il destino delle cellule A7.6 nell'uovo delle Ascidie. Ric. Sci. *34* (II-B), 525–528.

ORTOLANI G. La ginogenesi nelle uova dell Ascidie. In press.

PARAT M. A. and D. R. BHATTACHARYA, 1926. Les constituants cytoplasmiques de la cellule génitale fémelle. Compt. Rend. Soc. Biol. *94*, 444–446.

PÉRÈZ J. M., 1954. Considérations sur le fonctionnement ovarien chez *Ciona intestinalis*. Arch. Anat. Microscop. Morphol. Exptl. 43, 58–78.

PISANÒ A., 1966. The effects of homopolyribonucleotides on the development of ascidian eggs. Acta Embryol. Morphol. Exptl. *9*, 127–131.

RESTIVO F. and G. REVERBERI, 1957. Ricerche sugli spermi di Ciona e di Patella. Localizzazione della citocromo-ossidasi, della succino-deidrogenasi e del glicogeno. Acta Embryol. Morphol. Exptl. *1*, 164–170.

REVERBERI G., 1931. Studi sperimentali sull'uovo di Ascidie. Pubbl. Staz. Zool. Napoli *11*, 168–193.

REVERBERI G., 1933. Esperimenti di incrocio fra uova di *Ciona intestinalis* e spermi di *Phallusia mamillata*. Rend. Accad. Naz. Lincei *17*, s. 6, 737–739.

REVERBERI G., 1937. Ricerche sperimentali sulla struttura dell' uovo fecondato delle Ascidie. Commentationes Pontif. Ac. Sci. *1*, 135–172.

REVERBERI G., 1956. The mitochondrial pattern in the development of the Ascidian egg. Experientia *12*, 55–60.

REVERBERI G., 1957. The role of some enzymes in the development of Ascidians, in 'Beginning of embryonic development'. A.A.A. Washington D.C., p. 319–340.

REVERBERI G., 1961a. The embryology of Ascidians. Advanc. Morphogenesis *1*, 55–101.

REVERBERI G., 1961b. L'uovo delle Ascidie e le sue potenze organo-formative. Rend. Ist. Sci. Univ. Camerino *2*, 167–219.

REVERBERI G., 1968. La mécanique du développement de l'œuf des Ascidies. Année Biol. *7*, 557–576.

REVERBERI G. and A. MINGANTI, 1946. Fenomeni di evocazione nello sviluppo dell'uovo di Ascidie. Risultati dell'indagine sperimentale sull'uovo di *Ascidiella aspersa* e di *Ascidia malaca* allo stadio di 8 blastomeri. Pubbl. Staz. Zool. Napoli *20*, 199–252.

REVERBERI G. and G. ORTOLANI, 1962. Twin larvae from halves of the same egg in Ascidians. Develop. Biol. *5*, 84–100.

RIES E., 1937. Die Verteilung von Vitamin C, Gluthation, Benzidin Peroxydase, Phenolase and Leukomethylenblau-Oxydoreductase während der frühen Embryonalentwicklung verschiedener wirbelloser Tiere. Pubbl. Staz. Zool. Napoli *16*, 363–401.

RIES E., 1939. Versuche über die Bedeutung des Substanzmosaik für die embryonalen Gewebedifferenzierung bei Ascidien. Arch. Exptl. Zellforsch. *23*, 95–121.

RUNNSTRÖM J., 1930. Atmungsmechanismus und Entwicklungserregung bei dem Seeigeleie. Protoplasma *10*, 106–173.

SAMASSA P., 1894. Zur Kenntnis der Furchung bei den Ascidien. Arch. Mikroskop. Anat. *44*, 1–15.

SCHAXEL J., 1910. Die Morphologie des Eiwachstums und der Follikelbildung bei den Ascidien. Arch. Exptl. Zellforsch. *4*, 265–308.

SCHLEIP W., 1929. Die Determination der Primitiventwicklung. Akademie Verlag. Gesell., Leipzig.

SCHMIDT G., 1930. Untersuchungen über Entwicklungsmechanik bei Ascidien. Rev. Zool. Moscou *10*, 5–16.

SMITH K. D., 1967. Genetic control of macromolecular synthesis during development of an Ascidian: *Ascidia nigra*. J. Exptl. Zool. 164, 393–405.

SPEK J., 1927. Über die Winterknospenentwicklung, Regeneration und Reduktion bei *Clavelina lepadiformis* und die Bedeutung besonderer 'omnipotenter' Zellelemente für diese Vorgänge. Arch. Entwicklungsmech. Organ. *111*, 119–172.

TUCKER G. H., 1942. The histology of the gonads and the development of the egg envelopes of an Ascidian *(Styela plicata)*. J. Morphol. *70*, 81–108.

TUNG T. C., S. C. WU and F. Y. TUNG, 1962. Experimental studies in the neural induction in Amphioxus. Scientia Sinica *11*, 805–820.

TYLER A. and W. D. HUMASON, 1937. On the energetics of differentiation. VI. Comparison of the temperature coefficients of the respiratory rates of unfertilized and of fertilized eggs. Biol. Bull. *73*, 261–279.

URSPRUNG H. and E. SCHABTACH, 1964. The fine structure of the egg of a Tunicate, *Ascidia nigra*. J. Exptl. Zool. *156*, 253–268.

VAN BENEDEN E., 1886. Recherches sur la morphologie des Tuniciers. Arch. Biol. *6*, 237–476.

VAN BENEDEN E. and C. JULIN, 1884. La segmentation chez les Ascidiens et ses rapports avec l'organisation de la larve. Bull. Acad. Belg. (Liège) *7*, 431–447.

VON UBISCH L., 1952. Die Entwicklung der Monascidien. Verhandel. Koninkl. Ned. Akad. Wetenschap. *49*, 1–56.

WADDINGTON C. H. and V. MANCUSO, 1955. The effects of some purines and purine-analogues on Ascidian embryos. Exptl. Cell Res. *9*, 369–371.

WEISS P., 1939. Principles of development. Holt & Co., N.Y.

WERNICKE W., 1919. Über die Eibildung bei den Ascidien. Zool. Jahrb. Abt. Anat. Ontog. Tiere *41*, 113–174.

G. Reverberi (ed.), Experimental embryology of marine and fresh-water invertebrates

CHAPTER 14

Amphioxus

G. REVERBERI

Zoological Institute, University of Palermo, Italy

14.1. A short historical introduction

14.1.1. The chief papers on the embryology of Amphioxus are those by Kowalewsky; they were published in 1867 and 1876, following a very beautiful piece of research undertaken at the Zoological Station of Naples. This research, when presented for doctoral qualification, met much opposition because of the unexpected conclusion that Amphioxus is a Chordate. Time has shown that Kowalewsky was right.

Kowalewsky's research was followed by that of Hatschek (1881) and carried out at Messina. According to Conklin Hatschek's paper is an 'embryological classic'. With this paper it seemed that all the embryological research on Amphioxus could be considered definitely concluded; however, time has proved the contrary. A score of publications still appeared: those of Lwoff (1892), Mac Bride (1897, 1900), Wilson (1893) and Cerfontaine (1906): these last are of exceptional importance.

In 1932 a new paper appeared under the name of Conklin: it contained much new information and in particular showed that the embryonic development of Amphioxus is comparable to that of Ascidians. Once more the research was done at the Zoological Station in Naples. Words are inadequate to describe how accurate and exhaustive the research of Conklin is.

14.1.2. The experimental embryology of Amphioxus began with Wilson. Every student of embryology knows the research of Wilson: it led the author to conclude that at the 2- and 4-cell stage the egg is equipotential.

As, however, the development of Amphioxus does not differ from that of the Ascidians a critical re-examination of Wilson's results was necessary:

in this regard it must be recalled that Chabry (1887) and Conklin (1905c, 1906, 1911, 1932, 1933) had shown that the Ascidian egg is a 'mosaic egg'. The re-examination was made by the most qualified person, Conklin himself, who published his new paper in 1933. Experimental analysis led him to assert that the 'prospective value of a blastomere is primarily a function of its organization (i.e., differentiation and localization of substances)', in other words that the 'development of Amphioxus is essentially a 'mosaic work'. The new situation was again reversed by Tung et al. (1960a, b) who, in a series of very brilliant experiments, showed the development of Amphioxus to be largely 'regulative'.

14.2. Methods of fertilizing and manipulating the eggs

It is not easy to obtain eggs of Amphioxus regularly in the laboratory. A few researchers succeeded sporadically; only Tung et al. have been able to get eggs daily (from the middle of June to the middle of July). The animals lay their eggs spontaneously only in the evening (at 6–8 p.m.); however, they also seem to be fertilizable when removed artificially from the gonads. The eggs have envelopes, but these can be easily removed so that any sort of operation can be performed on them. This was not realized by Wilson and Conklin, who limited themselves to shaking the eggs violently at the 2- or 4-cell stage in order to separate the blastomeres: the shaken eggs, however, were not followed individually in their development. As mentioned above, Wilson and Conklin did their research in Naples: at the present time neither Naples nor Messina is so rich in Amphioxus as formerly.

14.3. The ovarian egg

Very few investigations have been made on the ovarian egg. Previous observations with the light microscope were made by Cerfontaine (1906), Cowden (1963) and Guraya (1966, 1968). A preliminary investigation with the electron microscope has been made by Reverberi (1966).

Under the ordinary microscope (fig. 1), the egg shows: (a) a very large and eccentric germinal vesicle situated at the animal pole (where the egg attaches to the ovary); (b) a cytoplasm which is very rich in yolk granules (according to Cerfontaine the egg should be classified as 'telolecytic'); (c) vacuoles at the periphery, which give a reaction for mucopolysaccharides; (d) a nucleolus whose appearance is different in different periods of the oocyte growth. These few data can be complemented by those deriving

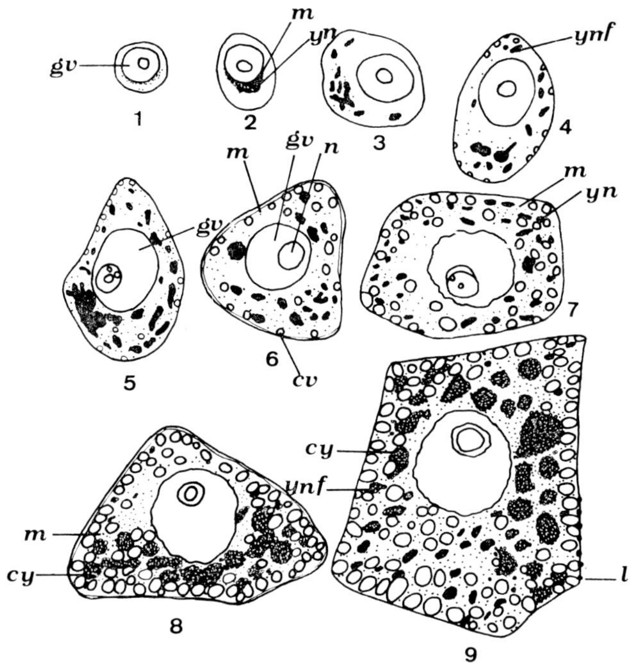

Fig. 1. Different stages of development of the oocyte of Amphioxus: the diagram illustrates the growth and distribution of the organelles. (Redrawn from Guraya, 1968.) gv = germinal vesicle; m = mitochondria; yn = yolk nucleus; ynf = yolk nucleus fragment; n = nucleolus; cv = cortical vacuoles; cy = compound yolk; l = lipids.

from observations with the electron microscope (fig. 2). The oocytes in previtellogenetic conditions show a poorly organized endoplasmic reticulum; the mitochondria are present and situated mostly around the germinal vesicle which is large and the pores of its membrane are crowded with ribosomes; the nucleolus is compact.

Oocytes in the vitellogenesis phase show a well-formed endoplasmic reticulum composed of spherical or oval vesicles with the exterior surface covered with ribosomes. Yolk granules and mitochondria are scattered all over the cytoplasm. There are a few cortical granules at the periphery. In more advanced oocytes the endoplasmic reticulum has a definite and typical appearance; the yolk granules are very large; the multivesicular bodies and the annulate lamellae are frequent; the periphery is crowded with large cortical granules in close contact which are filled with a structured substance; the plasma membrane does not have microvilli, but there are 'niches'

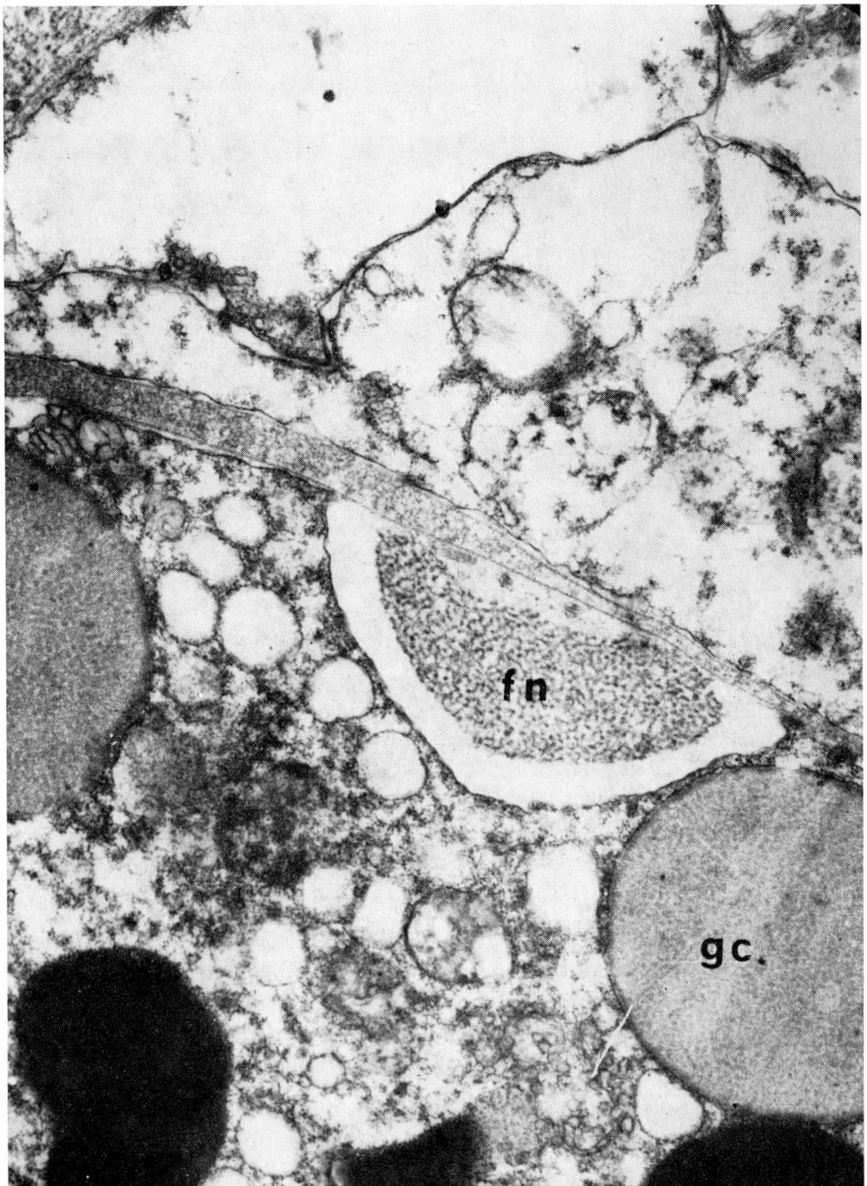

Fig. 2. Electron micrograph of the cortical region of a vitellogenetic oocyte. The layers of the external and internal follicular cells, a large cortical granule (gc), a 'niche' with its content (fn) and endoplasmic vesicles can be seen (Reverberi 1966).

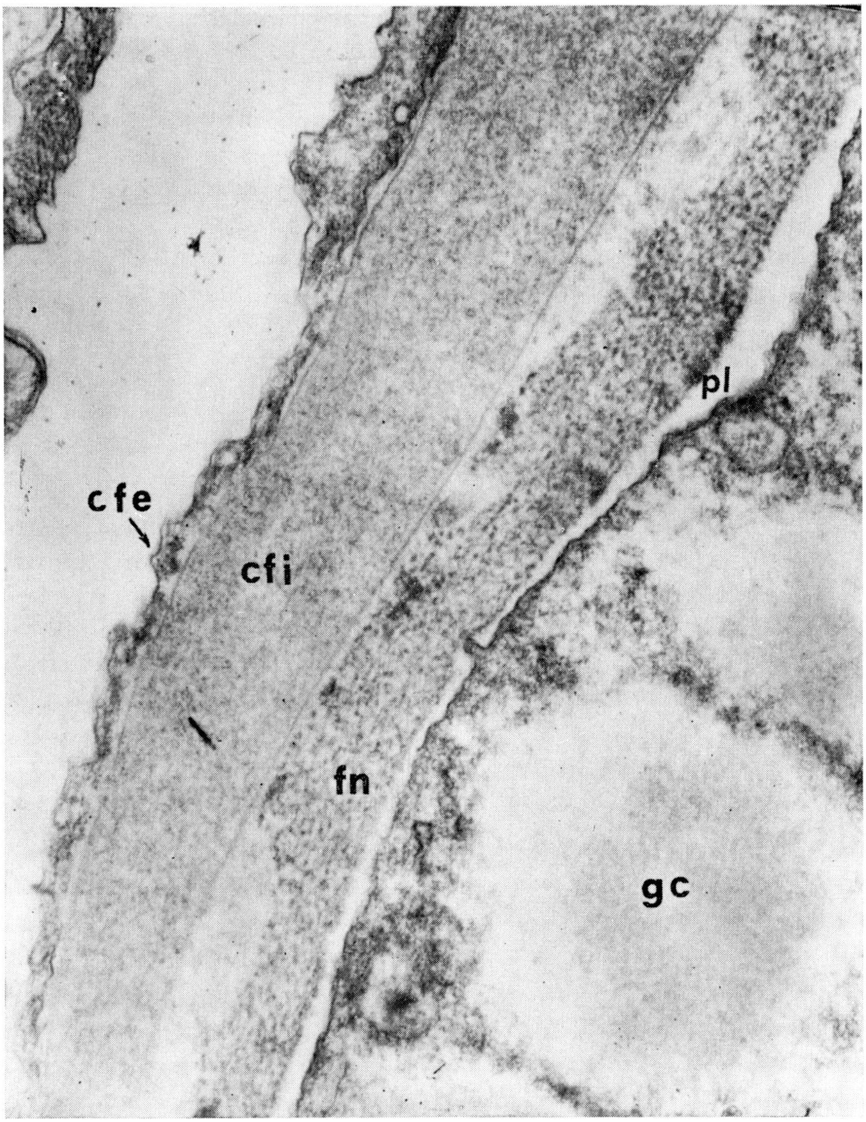

Fig. 3. Cortical region and membranes of an oocyte at a later stage of development: cfe = the external follicular layer; cfi = the internal follicular layer; fn = the layer derived from the content of the niches; gc = cortical granule; pl = plasma membrane (Reverberi 1966).

occupied by certain 'formations' which greatly recall the situation of the 'test cells' of the Ascidian egg. In more developed oocytes the niches are no longer present, the plasma membrane is smooth, and the substance which was in the niches forms a continuous layer interposed between the egg and its membranes (fig. 3). We have interpreted these formations as homologues of the 'test cells' of the Ascidians (Reverberi 1966).

Cowden (1963) and Guraya (1966) have investigated the oocyte of Amphioxus using cytochemical techniques. The latter author took particular interest in the cortical vacuoles which were thought to originate by pinocytosis. Each vacuole is limited by a membrane which gives a lipoprotein reaction. The chemical nature of the substance filling the vacuole, according to Cowden, is that of a mucopolysaccharide (it stains with alcian blue).

14.4. The mature egg

The mature eggs from the 'cavité ovarienne secondaire' (Cerfontaine), where they are situated in groups of twelve, are released in sea water. They are surrounded by a thin layer of follicular cells and by jelly which has been produced by the 'cells' of the 'cortical niches'; the germinal vesicle has faded and a small maturation spindle (without centrosomes or asters as in Ascidians) is formed. The emission of the first polocyte follows. The spermatozoon enters at the vegetal pole and after its entrance the second polocyte is emitted. This is located within the fertilization membrane and, as it remains attached to the surface of the egg for a long while, it is a good

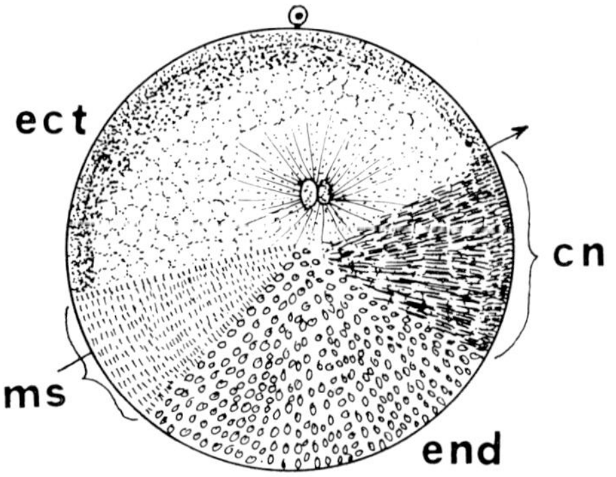

Fig. 4. Distribution of the organ-forming substances in the egg after fertilization: cn = neural and notochordal crescent; end = entoderm; ms = mesodermal crescent; ect = ectoderm. (Conklin 1933, modified.)

landmark for orientation. Another relevant event which follows fertilization is the enormous expansion of the perivitelline space: this is filled with jelly, which probably derives from the explosion of the cortical granules.

According to Conklin fertilization is followed by a migration of plasms, as in the Ascidians. The substance of the peripheral layer, which in its character, movement, and final location is so remarkably like that of the peripheral layer in the Ascidians, very probably has a similar destination (fig. 4). According to Conklin 'this substance goes into the mesoderm; therefore, the crescent around the posterior side of the egg, parallel to the first-cleavage amphiaster in Amphioxus, may be called the 'mesodermal crescent' as in the case of ascidians'. It is not necessary to emphasize these remarks of Conklin: we only wish to note that the final location of this peripheral substance in the musculature of the animal has been checked by means of coloured markers by the above-mentioned Chinese authors. It would be of extraordinary importance to see whether this material, which forms the yellow crescent, is rich in mitochondria, as is the case in Ascidians. Besides the yellow crescent formed at fertilization, two other crescents are formed at the opposite site: although ill-defined, they indicate the area from which the notochord and the neural system will develop.

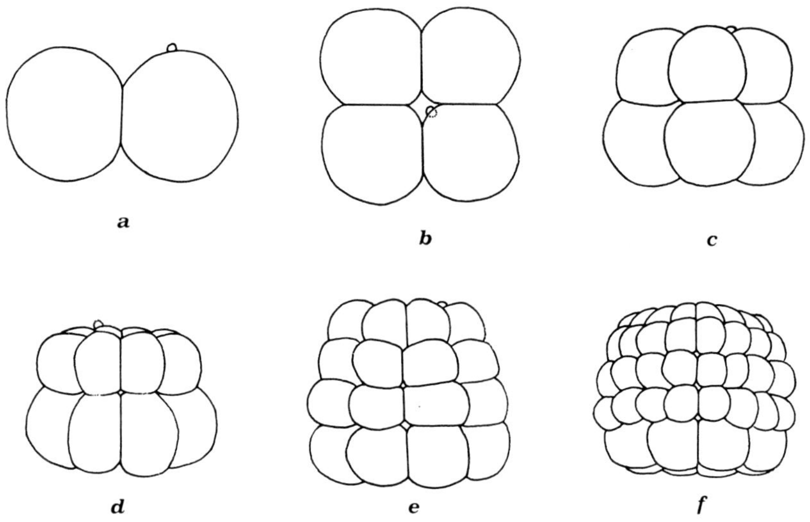

Fig. 5. Development of the egg up to 64 cells. (Redrawn from Hatschek 1881.)

14.5. Segmentation and development

According to Conklin, the first cleavage in the egg of Amphioxus occurs about $1\frac{1}{4}$ hr after fertilization: the Chinese authors, on the other hand, have observed that at 24.5 °C the first segmentation occurs about 50 min after fertilization. The first plane divides the mesodermic crescent into two symmetrical halves (fig. 5). The second cleavage, which according to Conklin occurs 45 min later (20 min according to Tung), is at right angles to the first, and divides the egg into four equal (Tung) or slightly unequal (Conklin)

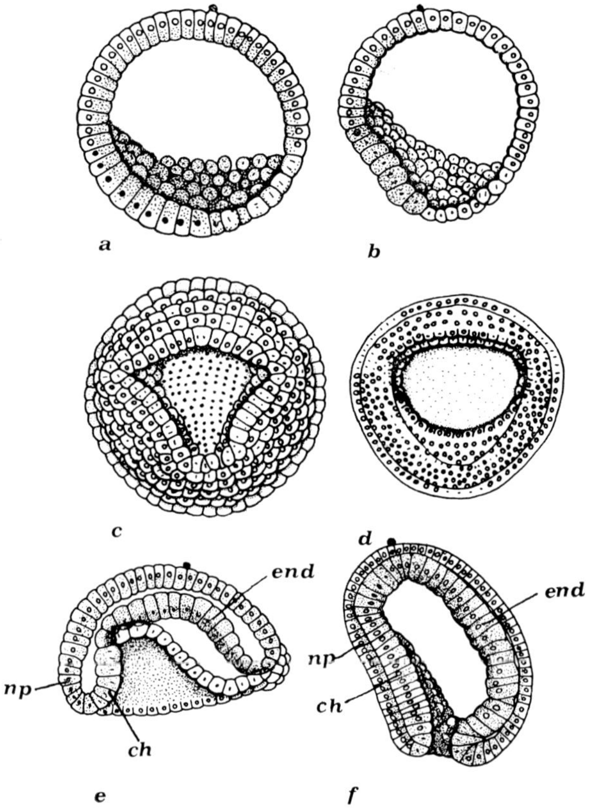

Fig. 6. Development from the blastula stage up to the end of gastrulation a, b = early and late blastula; c = early gastrula with triangular blastopore; d = later gastrula with transversely elongated blastopore; e = cup-shaped gastrula showing the neural plate (np), notochord (ch), and endoderm (end); f = elongated gastrula. (Redrawn from Conklin 1933.)

blastomeres. The mesodermal crescent is located in the posterior blastomeres, but part of it also extends into the anterior ones. The third cleavage is horizontal: according to Conklin the four animal blastomeres are smaller than the four vegetal: again, he reports that the two vegetal posterior blastomeres are smaller than the anterior ones (Tung et al. 1958, however, have not noted this difference). The fourth cleavage is meridional: 16 blastomeres are formed, arranged in two superposed crowns.

The fifth cleavage is latitudinal: as a consequence four rows of eight cells are formed, arranged one upon the other: as proposed by Tung et al., they can be designated, according to their position, as an_1, an_2: veg_1, veg_2. The sixth cleavage is again meridional: 64 cells are formed, situated in four crowns of 16 cells each: at this stage identification of the individual cells is very difficult.

With the eighth cleavage a blastula is formed: it has a large cavity which is filled with a jelly-like substance (fig. 6). The blastula is pear-shaped, the neck of the pear being constituted by some small and rapidly dividing cells, the mesodermic cells, which contain the substance of the posterior crescent. In their proximity (anteriorly) are the large entodermic cells. Anteriorly to the entodermic cells are the chordal cells.

Gastrulation begins with flattening of the blastula. The endodermal plate, anterior to the mesodermal cells, sinks into the blastocoel and a blastopore is formed; its dorsal lip is made up of the chorda cells, the lateral lips of the mesodermic cells. The mesodermic and chorda cells invaginate: on each side of the chorda plate the mesodermal grooves are formed. Finally the blastopore closes and development proceeds as follows: 'The mesodermal grooves, on

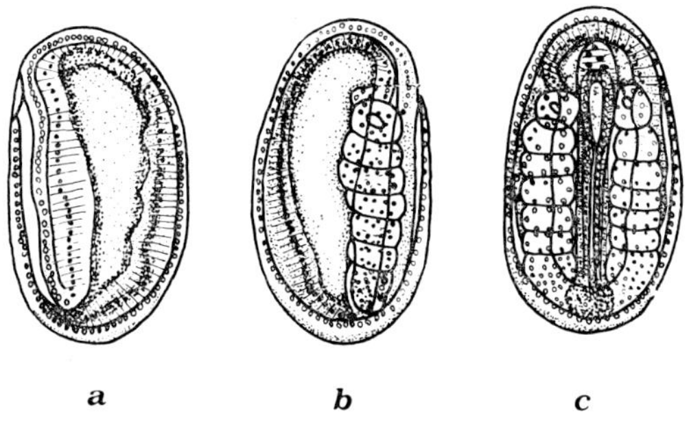

Fig. 7. (a) Optical sections in the medial plane of an embryo of 16 hr: note the notochord, the neural tube and the large archenteric cavity; (b) lateral view of an embryo of 18 hr showing somites overlying the notochord and the neural plate; (c) dorsal view of an embryo of 18 hr. (Redrawn from Conklin 1932.)

each side of the chorda plate, become constricted into a series of moniliform pouches . . . These pouches then separate from the gut and become the mesodermal somites (fig. 7). A forward growing transverse fold across the roof of the enteron in front of the first pair of somites separates the right and the left dorsal diverticula from the enteron; the left diverticulum unites with an ectodermal invagination on the left side to form the preoral pit, the right diverticulum expands to form the head or rostral cavity. The endostyle and club-shaped gland form from the right wall of the mouth cavity and the mouth opens on the left wall. The branchial anlage appears in the ventral and right wall and the first primary gill slit opens on the right-ventral side posterior to the club-shaped gland.' A larva is shown in fig. 8.

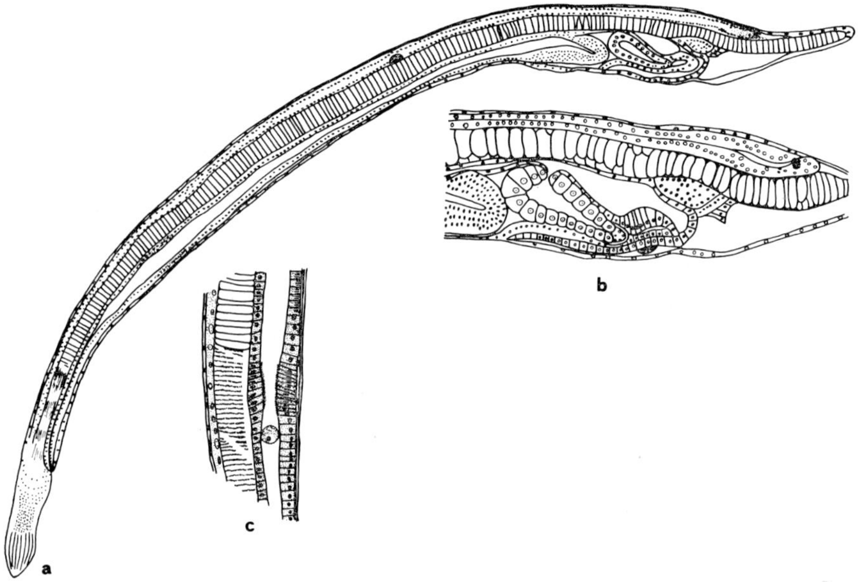

Fig. 8. (a) Median longitudinal section of a larva of 120 hr; (b) section of the anterior portion of the same larva nearer the median plane; (c) next section to the one shown in (a), note the strongly ciliated zone of the gut. (Redrawn from Conklin 1932.)

14.6. *The developmental potentials of the egg at the early cleavage stages*

Experimental analysis of the potentials of the isolated blastomeres at the 2-, 4- and 8-cell stages has been carried out by Wilson (1892, 1893). The technique used for separating the blastomeres was very rudimentary as the eggs were shaken in a test-tube half filled with water. The results were very interesting: isolated half blastomeres segmented as normal eggs, producing larvae of half-size but otherwise normal. In some cases the two blastomeres did not separate completely, and gave rise to twin embryos more or less united. An isolated quarter blastomere (one case) still yielded a small, more or less normal larva. Isolated 2/4 blastomeres (from a 4-cell stage) gave half-size complete larvae. In consequence of these results Wilson came to the conclusion that the egg of Amphioxus is 'regulative', and that its cells, at least up to the 4-cell stage, are equipotent.

Some years later Conklin, studying the normal development of the Amphioxus egg, remarked that this development is very similar to that of the Ascidians. Having obtained partial larvae from isolated blastomeres of the Ascidian egg he supposed that a similar result could be obtained with

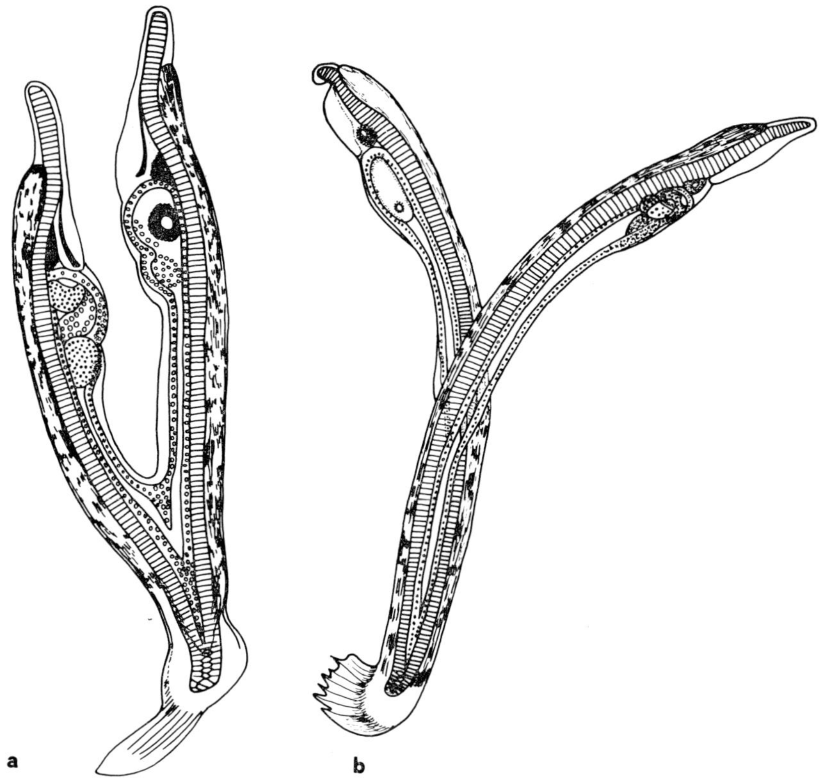

Fig. 9. (a) and (b): twins derived from the blastomeres partially separated at the 2-cell stage. (Redrawn from Conklin 1933.)

isolated blastomeres of Amphioxus. This conviction induced him to repeat the experiments of Wilson.

In fig. 9 two larvae from half-blastomeres of the 2-cell stage are shown. One can see how normal they are, 'even the asymmetrical organs of the larva, such as the right (rd) and left (ld) dorsal diverticula of the gut, which become the head cavity and preoral pit, respectively; the endostyle (es) club-shaped gland (cg) and branchial anlage (ba) which are largely on the right side; and the mouth and preoral pit which are on the left side, are all normal in position and development'. Without any hesitation he concluded that the two blastomeres of the 2-cell stage are equipotential and undifferentiated.

Isolation of 2/4 at the 4-cell stage was performed (a) along the plane of the 1st cleavage and (b) along the plane of the 2nd cleavage. Other operations consisted in isolating 1/4 anterior and 1/4 posterior. Normal larvae always developed from (a) and never from (b); from 1/4 anterior and 1/4 posterior complete larvae never formed.

'In brief, complete regulation is bilateral, i.e. it is limited to the bilateral distribution of formative substances already isolated in a half blastomere and does not lead to the formation anew of substances that have been completely isolated in later blastomeres . . . Except for this bilateral regulation, the development is a mosaic work' (Conklin 1933, p. 347).

The contrast between the conclusions of Wilson and Conklin is evident and for this reason Tung and his collaborators decided to take up the

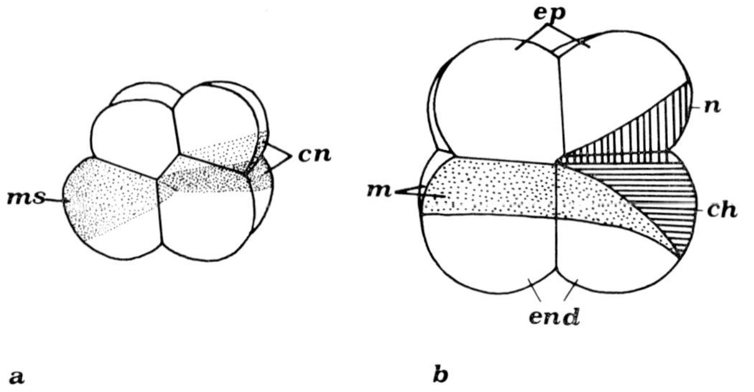

Fig. 10. Maps of the organ-forming territories of the egg at the 8-cell stage according to Conklin (a) and Tung et al.; ms = mesodermic crescent; ch = notochord (b); n = neural crescent; ch = chordal crescent; end = entoderm; ep = epidermis; m = mesodermic crescent. (Redrawn from Tung et al. 1962a.)

experiments again. However, before reporting on these experiments let us examine how the presumptive territories are distributed at the 8–32 cell stages. The map made by Conklin is shown in fig. 10: one will immediately realize how similar it is to that of Ascidians, which the same author prepared. Fig. 10 shows the map of Tung et al. (1962a), for the construction of which they used the method of vital staining. At the 8-cell stage they stained (fig. 11):

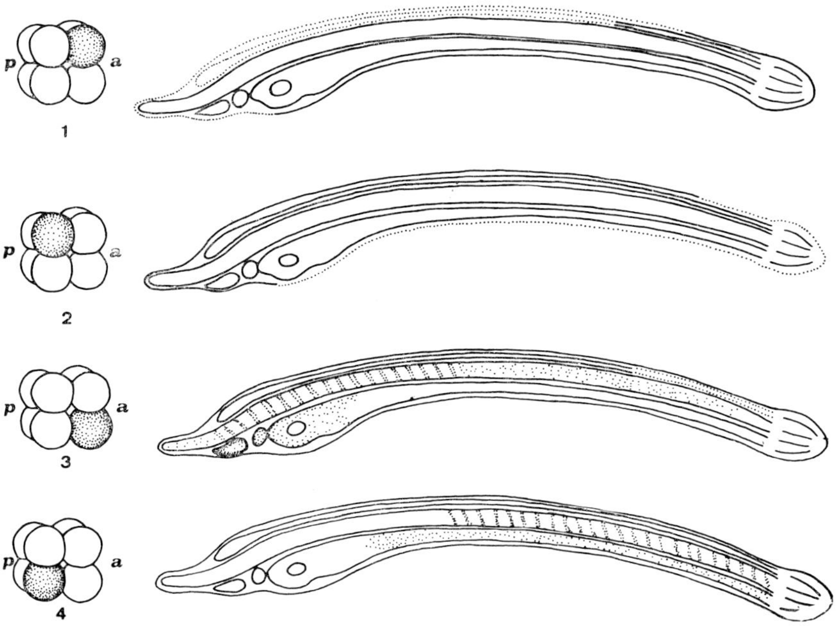

Fig. 11. The various blastomeres of an egg at the 8-cell stage were vitally stained; the territories which remained stained in the larva are indicated by fine stippling. (Redrawn from Tung et al. 1962a.)

one animal anterior blastomere: the larva had part of the epidermis and nearly all the neural system coloured (the caudal portion was unstained); *one animal posterior blastomere:* only the caudoventral epidermis stained: the neural system did not take any coloration;
one vegetal anterior blastomere: the stain was present in the caudal portion of the neural tube, in the notochord, in the somites of the anterior half of the body, in the endodermal organs of the head, and in the anterior part of the gut;

one vegetal posterior blastomere: the staining involved the somites of the posterior half of the body, and the posterior part of the gut.

At the 32-cell stage they stained (fig. 12):

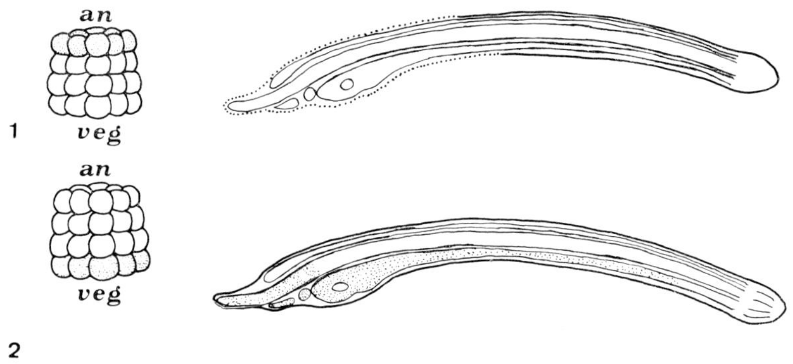

Fig. 12. Staining of an_1 or veg_2 at the 32-cell stage, and the regions which remained stained in the larvae. (Redrawn from Tung et al. 1962a.)

the blastomeres forming an_1*:* only the epidermis of the anterior half of the body became stained;
the blastomeres forming veg_2*:* the staining affected the endodermal organs (the preoral pit and the club-shaped gland; the gut with the exception of the end portion) the anterior end of the notochord and the three anterior somites. With the results obtained by vital staining they constructed the map shown in fig. 13. Comparing the map of Tung et al. with that of Conklin

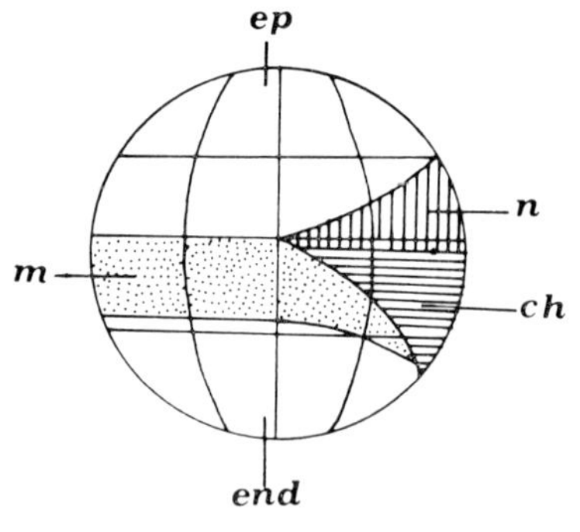

Fig. 13. Map of the presumptive organ-forming territories at the 32-cell stage; ep = epidermis; n = neural territory; ch = notochord; m = mesoderm; end = endoderm. (Redrawn from Tung et al. 1962a.)

one can see that they differ only in some minor details. An important feature, which occurs in both maps, is the extension of the mesodermic area which looks like 'a narrow belt situated just below the third cleavage furrow. It encircles the two posterior vegetal blastomeres with two long horns extending on each side along the lower margin of the area of the notochord to a point near the anterior end of the anterior vegetal blastomeres.' This area marks a fundamental difference between Amphioxus and Ascidians.

Isolation of the blastomeres at the 2-cell stage. The completely isolated blastomeres cleave as an entire egg and finally give rise to twin-larvae entirely normal except in size (fig. 14). Even the asymmetrical organs are

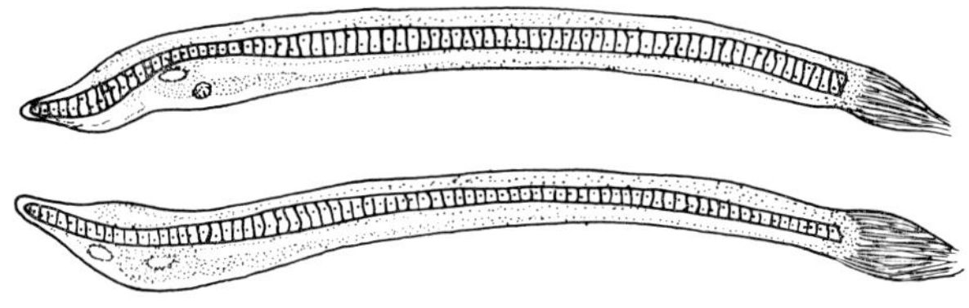

Fig. 14. Twin larvae from the blastomeres isolated at the 2-cell stage: the larvae are completely normal except in size. (Redrawn from Tung et al. 1958.)

in a normal position: unilateral reduction of the somites (on the left side in one embryo, on the right side in the other) was, however, observed and an asymmetrical development was also noted in the neural tube. These facts show that the larvae are not as normal as they appear externally.

Isolation of the blastomeres at 4-cell stage. Cleavage is like that of the normal egg: the blastula and gastrula are also normal. Fig. 15 shows the results obtained with the four isolated blastomeres; three embryos show a neural system, notochord, entoderm and somites. Tung et al. (1958) explained this result by supposing that the plane of the first cleavage formed an angle to the plane of bilateral symmetry: so three of the four blastomeres at the 4-cell stage received the substance of the chorda-neural crescent.

In another series of experiments the authors obtained only two 1/4 normal embryos: the other two 1/4 did not possess notochord and neural tube at all: probably in such cases the first cleavage plane coincided with the bilateral plane of symmetry of the egg. Other results were also obtained: in any case 1/4 blastomere can give rise to a normal embryo, as Wilson stated.

Isolation of the animal quartet (fig. 16). The four animal blastomeres were isolated at the 8-cell stage. They developed into an ectodermic ciliated ball,

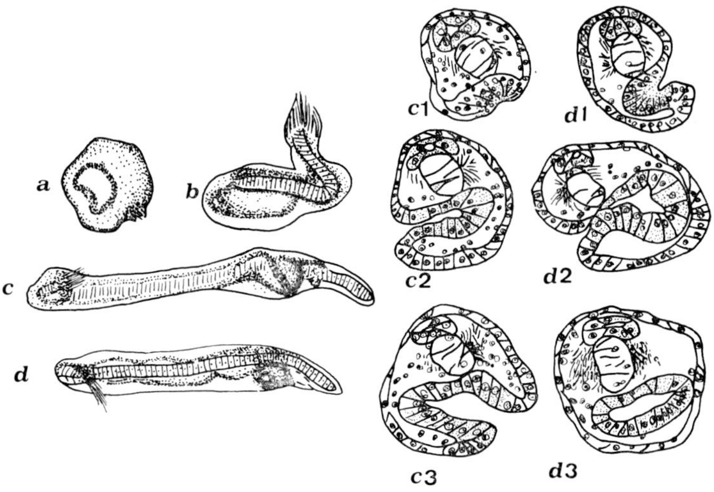

Fig. 15. Four embryos (a, b, c, d,) from the four blastomeres isolated from an egg at the 4-cell stage; in c_1–c_3 and d_1–d_3 are sections of (c) and (d) at different levels. (Redrawn from Tung et al. 1958.)

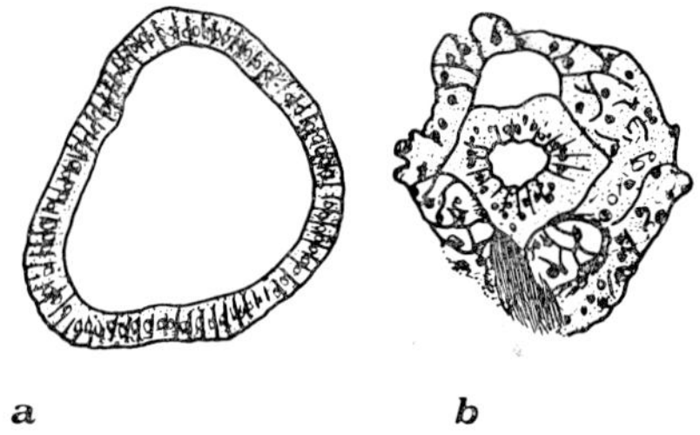

Fig. 16. Development of an animal (a) and vegetal (b) half of an egg at the 8-cell stage. In (b) the gut and the notochord are present while the ectoderm and neural system are lacking. (Redrawn from Tung et al. 1958.)

which did not show any differentiated structure: the fact that the neural system never differentiates in these embryos is of particular importance.

Isolation of the vegetal quartet. Generally the entoderm invaginates: sometimes it gives rise to an external intestinal tube. Other differentiations are the notochord and the muscles. In other words the vegetal quartet differentiates according to its presumptive destiny.

It will be noticed that the results obtained from the two last types of operation do not differ from those obtained in Ascidians by Reverberi and Minganti (1951) with regard to the differentiation of the animal quartet, and in particular its inability to give rise to a neural system should be noted.

This fact suggests that the neural system needs a stimulus from the vegetal quartet for its formation. We shall return to this question later: for the moment let us examine the other results obtained by Tung et al. after removing or combining the different blastomeric layers at the 32-cell stage (fig. 17).

As mentioned earlier, at the 32-cell stage there are four superposed rings of eight blastomeres each ($an_1 + an_2 + veg_1 + veg_2$). These rings were isolated singly and their development followed: not one of them was capable of giving rise to a complete larva. They have qualitatively different potentials: an_1 yields only epidermis; an_2 also yields epidermal structures (with some mesoderm); veg_1 yields chiefly notochord and muscle fibers; and veg_2 yields chiefly entodermal structures (no entodermal organs). One or another of the four rings (respectively an_1, an_2, veg_1, veg_2) was eliminated and the development of the remaining germ observed; the results were the following.

Removal of an_1. The remaining part develops into a typical larva with all internal structures completely normal. Thus the removal of half of the ectodermal material has no effect on normal development.

Removal of an_2. Again the larvae are quite normal: this result is important because the brain of these larvae is formed from an_1 which normally does not contribute to the formation of the brain.

Removal of veg_1. Again the larvae are perfectly normal having all organs well differentiated and in a normal position.

Removal of veg_2. The larvae are almost completely normal; only the gut and entodermal organs are underdeveloped or lacking.

From these results one can conclude that the removal of a blastomere layer has no effect on development. The processes of regulation at the 32-cell stage are thus very strong.

The same conclusion resulted from the experiments in which the four blastomere layers were separated and afterwards recombined in pairs: $an_1 + veg_2$; $an_2 + veg_1$; $an_1 + veg_1$; $an_2 + veg_2$. In all these recombinations the larvae which develop were normal (although with some reductions and disproportions). From these various experiments Tung et al. concluded: 'that each of the blastomere layers in the 32-cell stage, when isolated, developed more or less in accordance with its normal fate within the intact larva. But a combination of two or three layers may produce a complete, well-proportioned larva. The facts show that the egg of Amphioxus is not a mosaic of structures but has considerable power of regulation.'

Earlier it was remarked that the larva resulting from $an_1 + veg_1 + veg_2$ has a normal neural system and that the same result was obtained from

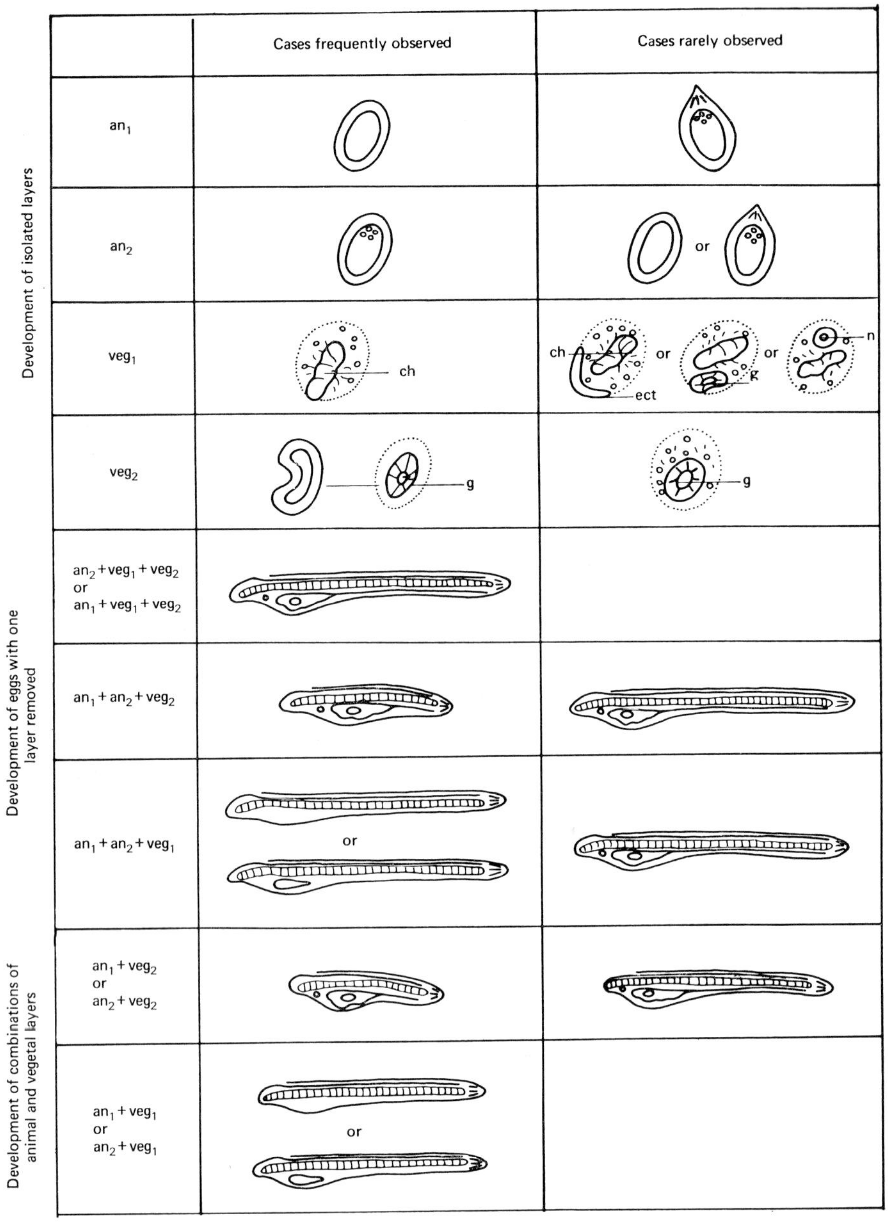

Fig. 17. Diagram illustrating the development of the different blastomeric layers when isolated or variously combined. ch = notochord; g = gut; or = alternative result; ect = ectoderm; n = neural system. (Redrawn from Tung et al. 1960b.)

$an_1 + veg_2$. On the other hand it was shown that a neural system never differentiates from an isolated an_1 because an_1 does not possess the 'formative substance' for the formation of a neural system. To explain these results it must be supposed that veg_2 acts as an inductor of an_1. To check this hypothesis Tung et al. undertook the following experiments.

Transplantation of a dorsal blastoporal lip. This induced in the host the formation of a secondary, although incomplete, embryo (fig. 18) possessing notochord, neural tube and somites.

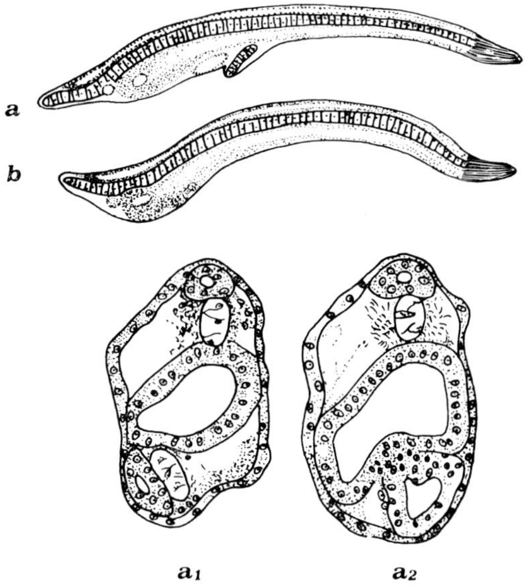

Fig. 18. Result of the transplantation of a blastoporal lip of a young gastrula (b) into the blastocoel of another gastrula (a); a secondary embryo is formed (a); (b) the normal donor; a_1 and a_2 are sections of larva (a). (Redrawn from Tung et al. 1962b.)

Transplantation of entodermic material (taken from the *central* part of the invaginated plate of a gastrula). This did not induce a secondary neural tube; the graft differentiated into an entodermal vesicle and the ectoderm in contact with this vesicle did not form a neural tube.

Transplantation of chordal or mesodermal materials. The transplants of chordal material induced a secondary embryo which, besides notochord, had a neural tube and two rows of somites. The transplants of mesodermal material did not induce any secondary neural system: the transplants remained more or less undifferentiated and the ectoderm in close contact with them did not show any modification.

From these results it must be concluded that the formation of the neural system is not independent: a stimulus from the materials contained in the vegetative anterior blastomeres of the 8-cell stage is needed. The chordal material in particular is responsible for such induction. This conclusion is confirmed by another series of experiments – rotating the animal quartet around the vegetal one (Tung et al. 1960a). From the map already referred to, one can see that at the 8-cell stage the presumptive neural material is located only in the animal anterior blastomeres and not in the posterior ones. The larva which formed was, however, quite normal. 'The fact that the nervous system is completely normal illustrates clearly that the nervous tissue is not derived from the presumptive neural material but rather, from the lateral posterior ectoderm instead. The cells of this area must have received the inductive stimulus from the vegetal half and undergone differentiation to form the brain and the neural tube, whereas the presumptive material of the brain and the neural tube in the absence of this morphogenetic influence becomes the epidermis. The result demonstrates that the formation of the nervous system in the embryo of Amphioxus requires a certain external stimulus resetted in the vegetal half of the egg.'

Another point to be considered concerns the differentiation of the notochord. The combination $an_1 + veg_2$ yields a larva possessing a well-formed notochord, yet veg_2 contains no formative material for the notochord. In fact an isolated veg_2 never forms a notochord. It must then be assumed that the differentiation of a notochord is dependent on a stimulus from the ectoderm. In this regard Tung et al. have written: 'There is no doubt that mutual interactions exist between cells of the an_1 and the veg_2 layers, and that some stimulus from the ectoderm is necessary for the differentiation of the notochord from cells of the veg_2 layer, and that the latter, in their turn, exert an effect on the ectoderm resulting in the development of the brain and the neural tube'.

Finally let us consider the differentiation of the entodermic organs. As shown earlier these derive from veg_2, but they form only when veg_2 is combined with an_1 or an_2: isolated veg_2 do not form these organs. One must therefore conclude that 'their normal differentiation requires some stimulus, which may be either a simple mechanical one . . . or a chemical phenomenon. Further investigation is necessary'.

From the remarks made above it results that in the development of Amphioxus there are many facts which immediately recall the observations of Hörstadius on the development of the sea urchin.

It will be recalled that Runnström, to explain the development of the sea urchin egg, proposed the theory of double-gradients. Could this theory be applied to the development of the Amphioxus? Tung and coworkers do not think so.

In conclusion, the statements of Wilson that the blastomeres of the Amphioxus eggs are totipotent up to the 4-cell stage and that 'in the earlier stages the morphological value of a cell may be determined by its location' not only are true, in contrast to the conclusion of Conklin 'that the prospective value of a blastomere is primarily a function of its own organization (i.e. differentiation and localization of substances)', but are extended by the results of Tung et al. 'From facts presented above', they write, 'it is clear that the interaction of the cells is not of minor importance in the development of Amphioxus'. In other words we may say that the egg of Amphioxus is neither 'totipotent' nor 'mosaic', as suggested by Wilson and Conklin: 'both factors, the internal organization and the location of a cell, are the parameters of its differentiation: mutual relations between the cells are essential for determining their destiny'.

Of course nobody intends to resurrect the assumption of Driesch of the 'entelechia'; on the other hand one cannot escape the conviction that the construction of an organism, probably all organisms, is the consequence of a causal chain of relationships, stimulations, and inductions between the forming organs. We must be grateful to Tung et al. who have contributed so much to giving us the picture of how an organism becomes constructed.

References

CERFONTAINE P., 1906. Recherches sur le développement de l'Amphioxus. Arch. Biol. (Liège) *22*, 229.

CHABRY D., 1887. Contribution à l'embryologie normale et tératologique des Ascidies simples. J. Anat. Phys. *23*, 167–319.

CONKLIN E. G., 1905a. The organization and cell-lineage of the Ascidian egg. J. Acad. Natl. Sci. Phila. *13*, 1–119.

CONKLIN E. G., 1905b. Organ-forming substances in the eggs of Ascidians. Biol. Bull. *8*, 205–230.

CONKLIN E. G., 1905c. Mosaic development in Ascidian eggs. J. Exptl. Zool. *2*, 145–223.

CONKLIN E. G., 1906. Does half of an Ascidian egg give rise to a whole larva? Arch. Entwicklungsmech. Organ. *21*, 727–753.

CONKLIN E. G., 1911. The organization of the egg and the development of single blastomeres of *Phallusia mamillata*. J. Exptl. Zool. *10*, 493–507.

CONKLIN E. G., 1932. The embryology of Amphioxus. J. Morphol. *54*, 69.

CONKLIN E. G., 1933. The development of isolated and partially separated blastomeres of Amphioxus. J. Exptl. Zool. *64*, 303.

COWDEN R. R., 1963. Cytochemical studies of oocyte growth in the Lancelet *Branchiostoma caribaeum*. Z. Zellforsch. Mikroskop. Anat. Abt. Histochem. *60*, 399.

DALCQ A., 1935. L'organisation de l'œuf chez les chordés. Étude d'embryologie causale. Gauthier-Villars, Paris.

GURAYA S., 1966. The origin and nature of cortical vacuoles in the Amphioxus egg. Z. Zellforsch. *79*, 326.

GURAYA S. S., 1968. Cytochemistry of yolk elements in the Amphioxus egg. Z. Zellforsch. *86*, 499–504.

HATSCHEK B., 1881. Studien über die Entwicklung des Amphioxus. Arb. Zool. Inst. Wien *4*, 1–88.

KOWALEWSKY A., 1867. Entwicklungsgeschichte des *Amphioxus lanceolatus*. Mem. Acad. Sci. St. Pétersbourg *11*, ser. 7, n. 4.

KOWALEWSKY A., Weitere Studien über die Entwicklungsgeschichte des *Amphioxus lanceolatus*. Arch. Mikroskop. Anat. *13*, 181.

LWOFF B., 1892. Ueber einige wichtige Punkte in der Entwicklung des Amphioxus. Biol. Centralbl. *12*, 769–774.

MAC BRIDE E. W., 1900. Further remarks on the development of Amphioxus. Quart. J. Microscop. Sci. *43*, 351.

MAC BRIDE E. W., 1897. The relationship of Amphioxus and Balanoglossus. Proc. Cambridge Phil. Soc. *9*, 309–313.

MAC BRIDE E. W., 1898. The early development of Amphioxus. Quart. J. Microscop. Sci. *40*, 589.

MORGAN T. H., 1896. The number of cells in larvae from isolated blastomeres of Amphioxus. Arch. Entwicklungsmech. Organ. *3*, 269.

REVERBERI G., 1966. L'uovo ovarico di Amfiosso al microscopio elettronico. Arch. Zool. Ital. *51*, 903.

REVERBERI G. and A. MINGANTI, 1951. Concerning the interpretation of the experimental analysis of the Ascidian development. Acta Biotheor. *9*, 197–204.

SCHLEIP W., 1929. Die Determination der Primitiventwicklung. Akad. Verl. Gesell., Leipzig.

SOBOTTA J., 1895. Die Befruchtung des Eies von Amphioxus. Anat. Anz. *11*, 129.

SOBOTTA J., 1896. Die Reifung und Befruchtung der Eier von Amphioxus. Arch. Mikroskop. Anat. *50*, 15.

TUNG T. C., S. C. WU and Y. Y. F. TUNG, 1958. The development of isolated blastomeres of Amphioxus. Scientia Sinica *7*, 1280.

TUNG T. C., S. C. WU and Y. Y. F. TUNG, 1960a. Rotation of the animal blastomeres in Amphioxus egg at the 8-cell stage. Scientia Rec. *4*, 389.

TUNG T. C., S. C. WU and Y. Y. F. TUNG 1960b. The developmental potencies of the blastomere layers in Amphioxus egg at the 32-cell stage. Scientia Sinica *9*, 119.

TUNG T. C., S. C. WU and Y. Y. F. TUNG, 1962a. The presumptive areas of the egg of Amphioxus. Scientia Sinica *11*, 629.

TUNG T. C., S. C. WU and Y. Y. F. TUNG, 1962b. Experimental studies on the neural induction in Amphioxus. Scientia Sinica *11*, 805.

VAN DER STRICHT O., 1896. La maturation et fécondation de l'œuf de l'*Amphioxus lanceolatus*. Arch. Biol. (Liège) *14*, 469.

WILSON E. B., 1892. On multiple and partial development in Amphioxus. Anat. Anz. *7*, 732.

WILSON E. B., 1893. Amphioxus and the mosaic theory of development. J. Morphol. *8*, 579.

Subject index

Author index